Fundamentals of Engineering Materials

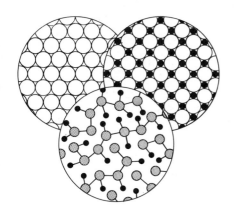

Fundamentals
of Engineering
Materials

Peter A. Thornton

Vito J. Colangelo

Benét Laboratories
U.S. Army Research and Development Center
Watervliet, N.Y.

Prentice-Hall, Inc., Englewood Cliffs, N.J. 07632

Library of Congress Cataloging in Publication Data

Thornton, Peter A.
    Fundamentals of engineering materials.

    Includes index.
    1. Materials. I. Colangelo, Vito J. II. Title.
TA403.T46 1985   620.1′1   84-17679
ISBN 0-13-338401-2

Editorial/production supervision and
    interior design: Shari Ingerman/Eileen O'Sullivan
Cover design: Lundgren Graphics, Ltd.
Manufacturing buyer: Tony Caruso

Printed in the United States of America

10 9 8 7 6 5 4 3

ISBN 0-13-338401-2

Prentice-Hall International, Inc., *London*
Prentice-Hall of Australia Pty. Limited, *Sydney*
Editora Prentice-Hall do Brasil, Ltda., *Rio de Janeiro*
Prentice-Hall Canada Inc., *Toronto*
Prentice-Hall of India Private Limited, *New Delhi*
Prentice-Hall of Japan, Inc., *Tokyo*
Prentice-Hall of Southeast Asia Pte. Ltd., *Singapore*
Whitehall Books Limited, *Wellington, New Zealand*

"The heights by great men reached and kept
Were not attained by sudden flight
But they, while their companions slept,
Were toiling upwards in the night."

—*The Ladder of St. Augustine, Longfellow*

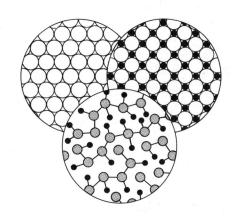

Contents

CHAPTER THIRTEEN   NONFERROUS ENGINEERING ALLOYS AND THEIR APPLICATIONS   450

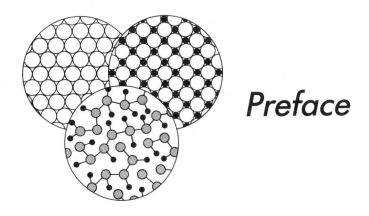

Preface

The role of materials science and engineering in our society has continued to steadily expand over the last decade. This expansion has been due in part to an increasing realization that the development of new materials and the improvement of existing materials play a crucial role in the creation of new systems. Recent technological advances, such as the artificial heart, the space shuttle, and personal computers, would not be possible without the availability of new engineering materials. Correspondingly, the scope of materials is enlarging dramatically. Never before have there been as many complex problems or as many challenges. Emerging technologies on every front are requiring and producing new materials and methods to fabricate them. The successful implementation of these materials and systems will necessarily require engineers and scientists who have been educated to design and work with these advanced materials.

This book is intended as a text for the initial course in engineering materials, recognizing that the utilization and applications of materials is interdisciplinary in nature. Technical knowledge of modern materials is required throughout a broad range of engineering and scientific activities. For many students this course may be their primary exposure to the properties and behavior of materials. In this case, the text serves to bring them to a basic level of understanding in an interdisciplinary manner. For others, it may simply be the first course in a series designed to create a fuller awareness of the characteristics of engineering materials.

A primary objective of this book is to present the fundamental aspects of engineering materials in an understandable manner, without sacrificing the underlying basic scientific principles. Essentially, we have attempted to strike a balance between theory and applications of materials in order to explain a maximum amount of scientific information, while keeping the students' level of interest aroused with pertinent examples. This is accomplished by the judicious use of practical examples, illustrative example problems, and emphasis on engineering applications in the numerous end-of-chapter study problems.

In addition to the traditional engineering materials topics found in most introductory materials books, this text covers a number of important areas that have been largely untreated by existing books of this level. These areas include specific chapters or sections on:

Fracture toughness and fatigue
Composite materials
Mixed potential theory of corrosion
Nonequilibrium solidification (rapid solidification technology)
Directionally solidified eutectics
Superconductivity
Biomedical materials
Superalloys
Construction materials

The reader may find the sections on rapidly solidified materials and directionally solidified eutectics of particular interest since they demonstrate the benefits that can occur when solidification is conducted in a controlled manner.

This book has been designed and written, first and foremost, as a teaching device. Accordingly, the material included in this text has been utilized in the classroom over a three-year period to sample student response and incorporate their pertinent suggestions. Traditional textbook format is utilized, with example problems appropriately dispersed throughout to illustrate and augment the text material. End-of-chapter problems have been selected to provide students with means for exercising their analytical ability, and to demonstrate the important relationships described in the text. Many of the problems have been designed based on actual engineering situations and thereby serve to stimulate and challenge students.

Both international units (SI) and English units have been used interchangeably in this text, in recognition of the fact that we are in a transitional period and it will undoubtedly be necessary for engineers and scientists to be familiar with both systems for many years to come. In addition to the conversions and selected property data listed inside the front and rear covers, we have also included numerous tables of materials reference data throughout the book which will be useful to the students when they become practicing engineers and scientists.

Finally, the authors gratefully acknowledge the help and assistance generously given by E. Nippes, H. Frenkel, and M. Tomozowa of Rensselaer Polytechnic Institute; J. Passmore of Jefferson College; R. Chait of the U.S. Military Academy, West Point; J. Underwood, Benét Laboratory; J. V. Lindyberg, General Electric Co.; D. L. Batey, Babcock and Wilcox; M. T. Gallivan, D. D. S.; R. L. Kennard, Air Force Wright Aeronautical Laboratories; and D. G. Baldrey, Hudson Valley College. In addition, thanks are in order to our many friends and colleagues who provided valuable insight, helpful suggestions, and illustrative materials. We are also indebted to Ellen Thornton and Betty Ann Melius for their diligence and persistence in typing the manuscript. Last, but certainly not least, we sincerely appreciate the support and understanding of our families during this project.

Peter A. Thornton
Vito J. Colangelo

Fundamentals of Engineering Materials

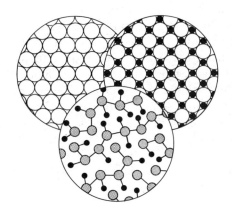

CHAPTER ONE:
The Role of Materials
in Engineering

RAISON D'ETRE—THE REASON ENGINEERING MATERIALS EXISTS AS A MODERN SCIENCE

Archeological evidence indicates that "engineered materials" have been available and utilized for the benefit of humankind since the Neolithic period, beginning about 10,000 B.C. Initially, these materials consisted of wood, stone, ceramic clays, meteoric metals, and ores, which were simply fashioned into useful objects. Later, copper metallurgy was developed in Asia Minor and resulted in such items as beads, pins, wire, awls, needles, knives, and spearheads. Indeed, many of these early tools have been [1]C-dated[1] from the period 7000–4000 B.C. The Iron Age started in the same geographical area about 2000 B.C., but was developed rather slowly until the Romans promoted the use of iron in the first and second centuries A.D. for their military and civil needs.

The point of this abridged introduction to the historical aspects of engineering materials is to show that metals, ceramics, and certain natural polymers are by no means recent discoveries. Some of these materials have been developed and utilized for thousands of years. Perhaps this is best expressed by the following passage from the first book of the Old Testament:

And they said one to another, go to, let us make brick, and burn them thoroughly. And they had brick for stone and slime had they for mortar.

Genesis XI, 3

---

[1] Carbon-14 dating is a technique that utilizes the radioactive isotope of carbon with an atomic mass equal to 14 (^{14}C) and a half-life of 5730 years. In this method, radioactivity measurements are used to establish the age of objects containing ^{14}C, especially archeological materials.

Engineering materials have, of course, undergone continuing evolution and have improved considerably since these early events. In fact, the development of materials used for engineering purposes has experienced unprecedented growth over the last few decades, and engineering materials have truly become an essential part of modern science and technology. This integral relationship is evident in virtually every product and every industry that one can think of. For example, engineering materials in the form of semiconductors play a vital role in such areas as communication systems, computers, aerospace, and *consumer* electronics. In the biomedical field, certain polymers and metals are crucial to the success of surgical implants. This is evidenced by the successful implantation of an artificial heart in a human patient.[2] The transportation industry (land, sea, air) is rapidly becoming more reliant on materials that exhibit high strength and light weight. These needs are satisfied by certain combinations of engineering materials called composites, such as fiberglass and other reinforced materials. Elevated-temperature applications such as heat engines employ metals, called "superalloys," and ceramics for increased operating efficiency. Energy exploration, such as deep-well and offshore drilling, plus energy production, such as nuclear reactors, is closely linked to the performance of modern engineering materials. In the construction industry, skyscrapers and vast networks of interstate highways are visible testimony to the proper utilization of modern engineering materials.

Undoubtedly, materials considerations have permeated virtually all facets of modern engineering and technology. In many instances, continued progress in a scientific or technical area hinges on advancements in engineering materials. Conceivably, the best example of this premise is the space shuttle orbiter, which depends crucially on modern materials for light weight, high strength, high-temperature resistance, thermal shock resistance, radiation protection, oxygen and water production, electrical power, and many other necessities.

These achievements and countless other examples are due in large part to the advancement of engineering materials as a modern science. In simple terms, this is the science dealing with the composition and the structure of materials, as well as how these factors are related to their properties or behavior. In this context, the composition of an engineering material refers to its chemical makeup, while structure includes atomic and electronic configurations; molecular and crystalline structures; microstructure, the appearance of internal structure under a microscope; and macrostructure, how the internal structure appears to the unaided eye. Understanding the relationships between these factors and the resultant mechanical or physical properties forms the basis for considering engineering materials as a modern science.

Very often, students in ostensibly non-materials-oriented disciplines question why they should spend time studying materials. We believe that students should study materials for the same reasons they study chemistry or physics: to develop an understanding of the basic scientific principles and relationships that govern the properties and behavior of engineering materials. Hopefully, such fundamental knowledge will (1) provide a sufficient foundation to study more diverse or advanced

---

[2] An artificial heart consisting essentially of a polyurethane bladder over an aluminum frame was implanted in Mr. Barney Clark on December 1, 1982, at the University of Utah Medical Center. Mr. Clark lived 112 days with this biomedical system in operation.

topics dealing with materials, (2) assist in more fully appreciating the effects of manufacturing processes on the structure and concomitant properties of materials, and (3) enable any engineer or scientist to utilize engineering materials intelligently and efficiently.

TYPES OF ENGINEERING MATERIALS

Engineering materials are commonly divided into categories based on their physical and chemical characteristics. Although there may be some similarities among different types, it is appropriate and convenient to separate engineering materials into the following categories: (1) metals, (2) ceramics, (3) polymers, and (4) composites. The reasons for these distinctions are obvious in many instances. For example, these types of materials often look different from each other, and they can certainly behave differently. Let us briefly examine each category.

Metals

Most students are familiar with metals in a general way because in ordinary day-to-day situations we are frequently exposed to, and utilize, objects made from various metals. Typical metal objects are depicted in Figure 1-1. Metals as a group are employed heavily in such diverse applications as aerospace, construction, mining, drilling, shipbuilding, transportation, heavy equipment, pressure vessels, appliances, cookware, tableware, and a host of other items too numerous to mention.

This type of engineering material can usually be distinguished from other categories by some of its more obvious traits, such as reflectivity of light, transmission of heat, conduction of electrical current, and very often, the ability to be deformed

(a)   (b)

Figure 1-1 Examples of typical applications for metals: (a) steel padlock; (b) brass gate valve; (c) alloy steel wrench.

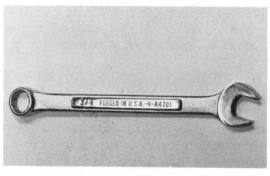

(c)

Figure 1-1 (Cont.)

without breaking (who hasn't bent a paper clip?). The elements that make up metals are located in the left portion of the periodic table (see inside front cover), and tend to exhibit electronic configurations which lend themselves to losing outer (valence) electrons. Many of these elements also have the ability to mix or dissolve with each other in the solid state, thus producing metallic alloys such as steel (a solid solution of iron and carbon). Alloys can exhibit properties vastly different from the individual elements that constitute them, and fortunately, such properties can usually be controlled by regulating the proportions of the constituent elements—not unlike the ingredients in a cookbook recipe. Admittedly, this may be a simplistic approach, but the more complex factors that influence the properties and behavior of metals and metallic materials will be postponed until later chapters, where it is appropriate to study them.

Ceramics

The term "ceramics" frequently brings to mind visions of hard, brittle objects of porcelain, china, and glass, just to mention a few familiar items. Also included in this category are some of the ordinary materials of construction: bricks, tile, and bathroom fixtures. In this case we recognize ceramics as hard, brittle materials that exhibit "glass-like" properties, and break abruptly without noticeable deformation. Such materials are produced largely by combinations of elements from both the metals and nonmetals portions of the periodic table. As stated in the preceding section, the metallic elements tend to lose outer electrons, whereas the nonmetals correspondingly tend to gain them. This results in a strong mutual attraction (coulombic) between the reacting elements, forming the basis for stable compounds which display "ceramic-like" properties. Typical ceramic materials are shown in Figure 1-2.

Yet in today's context of engineering materials, ceramics means much more than the common applications we have mentioned. In addition to these *traditional* ceramics, a group of materials called the *new* ceramics are utilized in considerably more sophisticated applications. The new ceramics we are referring to are used in magnets, semiconductors, integrated circuits, high-temperature engines and rockets, nuclear fuel elements, fuel cells, high-strength materials, transducers, and so on.

Although the demand for such ceramics on a tonnage basis is certainly not as great as that of traditional ceramics and structural metals, their unique thermal,

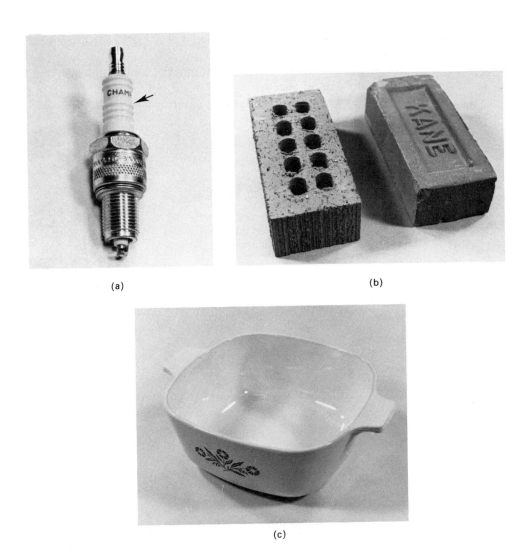

(a)

(b)

(c)

Figure 1-2 Examples of ceramics: (a) Al_2O_3 spark plug insulator (arrow); construction brick; ovenware.

electrical, and magnetic properties make them extremely attractive for certain engineering purposes. Accordingly, ceramic engineering materials, traditional and new, will continue to increase in importance rapidly as this technology emerges and the attendant requirements imposed on materials and design become more demanding.

Polymers

The term "polymer" is derived from the Greek *poly*, meaning "many," and *meros*, meaning "units." Polymers are substances composed of large numbers of molecules, joined together in chain-like fashion. A molecule is the basic (smallest) repetitive chemical or structural unit of the polymer system. The majority of engineering polymers are based on hydrocarbons—molecules that consist of hydrogen and carbon atoms in various structural arrangements.

Polymeric engineering materials consist of a large number of synthetic "plastics in addition to many natural polymers such as wood and rubber. Examples of polymers are illustrated in Figure 1-3. For instance, you are probably familiar with many of

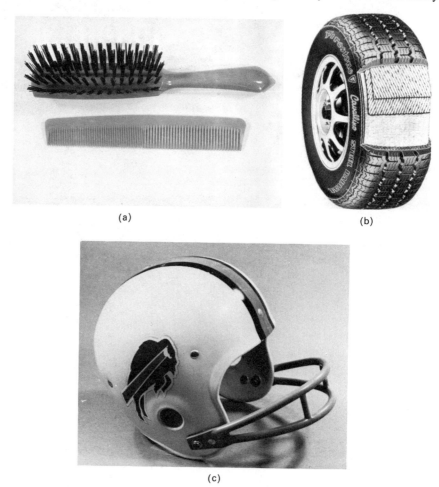

(a)

(b)

(c)

Figure 1-3 Examples of polymeric materials: (a) polyethylene hairbrush (with nylon bristles) and comb set; (b) rubber automobile tire with polyester cord body; (c) polycarbonate protective football helmet. (b—courtesy of Firestone Tire and Rubber Company, Akron, Ohio; c—courtesy of J. R. Senick.)

the synthetics, such as polyester (Dacron) and polyamide (nylon), which are popular textile materials used in the form of fibers. Also, vinyls, polyethylene, polystyrene, polyurethane, and various types of rubber are commonly used synthetic polymers.

Demand for polymeric engineering materials has increased substantially in recent years. This market is due to such factors as availability, economics, convenience, and technical advantages over other types of engineering materials for certain applications. Hydrocarbon polymers, by their very nature, are less dense than metals or ceramics, and although this may impair their utilization in certain structural applications, it does qualify them as attractive candidates for applications requiring *lightweight* materials. Additionally, polymers typically resist atmospheric and many other forms of corrosion; therefore, they can eliminate many unsightly or potentially dangerous corrosion problems that would ordinarily occur with metals. Furthermore, some polymers display good compatibility with human tissue. This particular feature, coupled with their corrosion resistance, makes these polymers excellent materials for surgical implantation and other biomedical applications as indicated in Figure 1-4.

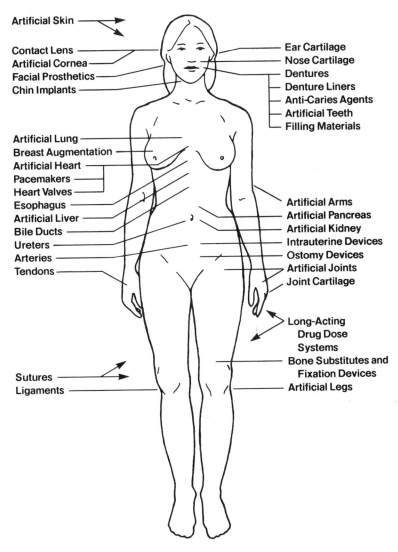

Figure 1-4 Areas of application where engineering materials have been implanted in humans. (From *SRI Bulletin*, Vol. 32, No. 1, Winter 1979–1980, Southern Research Institute, Birmingham, Ala.)

Finally, polymers exhibit excellent resistance to the conduction of electrical current. This characteristic makes them important in the fabrication of electronic and electrical devices.

Composites

Composite engineering materials generally consist of *combinations* of the other materials that we have discussed. Such combinations include metal or ceramic fibers dispersed in a polymeric matrix, ceramic fibers in a metal matrix, polymeric fibers in a polymer matrix, and particles of one material distributed throughout another. This class of engineering materials illustrates that although many dissimilarities may exist between different materials, they frequently can be utilized in conjunction to produce a material with unique properties and behavior—behavior not displayed by the individual constituents themselves.

A common example of a composite material is glass-fiber-reinforced plastic, or fiberglass, used for applications such as boat hulls, canopies, and storage tanks. Individually, the glass fibers are very strong but are quite susceptible to moisture attack and mechanical abrasion. Correspondingly, the polymeric matrix is comparatively weak but resists moisture penetration and protects the glass fibers from abrasion or mechanical damage. Combined properly, these two constituents act in concert to produce a new material which is very strong in the direction of fiber alignment and is relatively lightweight.

Due to their unique properties and the fact that these properties may be tailored to satisfy a certain set of requirements, composites are rapidly becoming a recognized class of engineering materials. Typical composites are illustrated in Figure 1-5. These

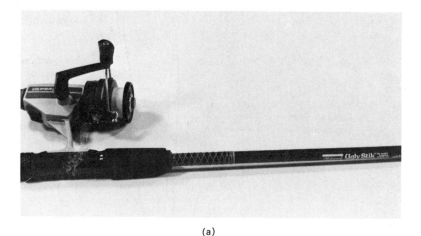

(a)

Figure 1-5 Typical applications for composite materials: (a) graphite fiber reinforced fishing rod; (b) glass fiber reinforced plastic tubing; (c) dacron fiber reinforced polyurethane ventricles (arrow) in JARVIK-7tm artificial heart.* (c—courtesy of Dr. R. K. Jarvik.)
*trademark of Kloff Medical Inc., Salt Lake City, Utah.

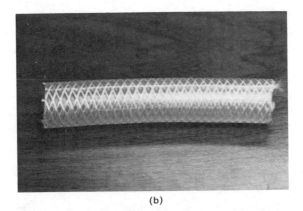

(b)

(c)

Figure 1-5 (Cont.)

materials are generally considered in situations or designs where one specific category of engineering material is insufficient.

INFLUENCE OF STRUCTURE ON PROPERTIES

Certain mechanical and physical properties of engineering materials are strongly dependent on structure. These properties, as you will discover throughout the course of this text, include density, thermal and electrical conductivity, strength, ductility, and hardness.

The concept of structure, however, may be somewhat more difficult to comprehend. In the context of engineering materials, the term "structure" is applied to the various geometrical or chemical arrangements that represent specific conditions in a material. Furthermore, the structures that play an important role in the behavior of engineering materials range, dimensionally, from extremely small values (on the order of atomic dimensions) to relatively large values (buildings and assemblies). A size comparison of structures relevant to materials is depicted in Figure 1-6. A brief

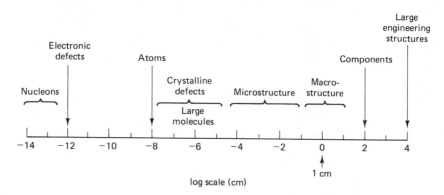

Figure 1-6 Dimensional comparison between certain quantities and structures that are relevant to engineering materials. Scale is logarithmic.

explanation of the pertinent structures that affect the properties and behavior of materials may be helpful at this point.

- *Electronic structure:* the configuration of electrons associated with an atom or group of atoms. We distinguish this from *inner* atomic structure (i.e., the nucleus) because most reactions between engineering materials can be interpreted in terms of changes in the electronic configuration surrounding the atom core.
- *Molecular structure:* the spatial arrangement of small groups of atoms which are strongly joined together within the group or molecule. Similar molecules, however, are joined in comparatively weaker fashion and often exhibit a random pattern.
- *Crystal structure:* atoms situated in an orderly, repetitive arrangement which occupies space. The arrangement of atoms in a crystalline manner produces solids with many familiar geometric shapes such as cubes, prisms, pyramids, and so on.
- *Microstructure:* refers to the physical appearance of the internal structure of engineering materials. This type of structure results from various combinations of chemical elements in the solid state, and is revealed under the microscope in specimens that have been prepared for microscopy by polishing and etching.[3]

[3] Etching is the process of applying an appropriate chemical reagent to a ground and polished specimen. Various constituents in the material react with the etching reagent at different rates, thereby producing different degrees of attack. The result is a surface that displays contrast under the microscope and may be used to identify the chemical compositions that produced the feature observed.

- *Macrostructure:* refers to the appearance of the internal structure, similar in principle to microstructure, but is generally resolved and viewed *without* the aid of a microscope. This type of structure also results from combinations of chemical elements and very often is enhanced by mechanical working (deformation) and/or thermal treatments of the material.

With the exception of radioactive materials and materials in nuclear reactors, the structure within an atom nucleus does not usually change significantly in a material with a specific chemical composition. However, the configuration of electrons associated with the atoms in a solid does change significantly, especially the electrons farthest from the nucleus. These electrons and the corresponding electronic structure exert a major influence on the mechanical and physical properties of the material. For example, the interaction between outer electrons (valence electrons) of adjacent atoms forms the basis for joining or bonding atoms together in solids. Also, certain materials, such as semiconductors, can have their electrical characteristics appreciably altered by the addition of foreign atoms that contain more valence electrons, or fewer valence electrons, than those of their host atoms.

If we move slightly up the scale shown in Figure 1-6, both molecular structure and crystal structure exert a strong influence on material behavior. Physical properties such as density, electrical conductivity (or conversely, resistivity), thermal conductivity, thermal expansion, and magnetic behavior are affected by the arrangement of atoms in a solid. Such structural configurations also affect strength, hardness, and ductility from the mechanical standpoint. Furthermore, the deformation behavior of a material, that is, its response to applied loads and forces, is affected strongly by molecular or crystal structure. For example, if the atoms in a molecule have the capability of forming different geometrical or structural arrangements without a change in chemical composition, we have an *isomer.* In a crystalline material, the ability to transform from one type of geometrical configuration to another is called *polymorphism.* In both cases, the mechanical and physical behavior of the materials are different (sometimes vastly different) depending on the structure displayed.

Correspondingly, the properties of engineering materials are intimately linked to how closely the molecular or crystal structures resemble a *perfect* structure. Simplistically, a perfect structure means that no atoms or ions are missing from their appointed positions and that the respective structure exhibits no geometrical deviations from perfection. In other words, no faults or defects can be present. Naturally, such an ideal situation is virtually impossible to achieve in the most closely controlled experiments, let alone in materials produced on a bulk scale for commercial applications. As you will learn, a host of geometrical imperfections are possible in crystalline and molecular structures, and these defects exert a profound influence on how the materials behave. Such imperfections also provide reasonable explanations for the differences in behavior observed between theoretical predictions and actual measurements.

Finally, the microstructures and macrostructures displayed by engineering materials have a major effect on their resultant properties. These structures can be very complex, depending on a number of factors, including chemical composition and its uniformity in the material, the degree (amount) of mechanical working that a particular product receives (if it is not simply cast as a liquid and allowed to "freeze" into its final shape), and certain thermal treatments or processes that a material may receive in order to achieve the desired properties. These factors synergistically affect the micro/macrostructure of an engineering material and correspondingly, its properties. Therefore, to predict and control the behavior of a material, we must understand the basic structure–property relationships and the factors that influence them. Accordingly, this understanding can be utilized to control the structure and hence the final properties of a material in service.

MATERIALS UTILIZATION

Very few engineering materials can be utilized without some refinement or processing. Although the processing of materials is a very important subject, it is also complex and lengthy. Therefore, we will not attempt a comprehensive treatment of these topics. Instead, certain processes that are intimately related to the use and performance of materials will be explained in conjunction with the appropriate material or materials.

Selection and Design

The selection of engineering materials and the design of a component or structure are inseparable factors in our modern technological world. It would be nice if there were "optimum" materials stored on the shelves of warehouses just waiting for applications. Unfortunately, this is not the case. There are no optimum materials as such. Rather, a material is optimized when it is considered relative to the parameters that the designer feels are important. For example, let us consider something as simple as a drinking vessel. If economics is a primary factor, one can use an ordinary paper cup to hold liquids. However, such a cup would not be suitable for a formal dinner. Similarly, a fine china teacup would be inappropriate on a camping trip. This concept can be extended to an example dealing with more sophisticated systems. For instance, certain aluminum alloys may be used for applications requiring moderate strength and light weight. But if elevated temperatures are a prime consideration, aluminum may be inadequate because of its relatively low melting temperature. In the same fashion, a ceramic material may be considered for an elevated-temperature application, because of its high melting point. However, if thermal shock (sudden temperature changes) is a consideration, the ceramic material may be inappropriate. Or a polymeric material (plastic) may be considered for an application requiring atmospheric corrosion resistance; but certain polymers are seriously degraded by exposure to sunlight and may therefore be inappropriate for outside applications.

The point of our example is that for a particular design or part, one must necessarily consider *all* the functional requirements in addition to economic factors. Then the designer must select the material that has the maximum number of positive attributes. In the event that a material selection cannot satisfy all the design requirements, one must also consider additional factors, such as combinations of materials, protective systems, or perhaps a change in design itself. Nevertheless, the net result of this analysis should be a component or system that will satisfy design requirements, operate safely and reliably, fulfill its intended function, and be available at a reasonable cost.

Analysis of Service Failures

In addition to periodic inspections in service, one method, particularly important to engineers, of determining whether the material selection and design has been properly conducted is to perform comprehensive analyses of any failures that might occur. Failures, although unfortunate, sometimes happen. This is due in part to the large number of synergistic factors involved in the design, production, and utilization of engineering components: for example, material defects, variation in properties, inadequate design, inadequately controlled processing, poor workmanship, insufficient maintenance, adverse or unexpected service conditions, and abuse or negligence. Although it may be time-consuming in many circumstances to study a part or system that has failed, the information to be gained from such a *postmortem* is very valuable and must not be wasted.

Proper application of failure analysis information can provide a valuable adjunct to the total engineering input for a product design. These studies can point out design flaws, material defects and limitations, fabrication defects, and incorrect usage of the product. These data are perhaps the last link on the product chain, as illustrated in Figure 1-7, and should be routed back to the design/engineering stage of product development. Such a procedure inevitably improves the overall reliability, safety, and usefulness of a product.

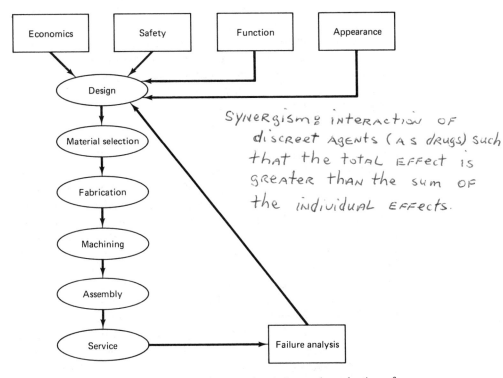

SyNeRgism & interaction of discreet Agents (as drugs) such that the total effect is greater than the sum of the individual effects.

Figure 1-7 Relationship of failure analysis to the design and production of typical engineering components.

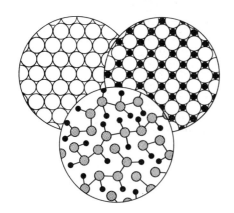

CHAPTER TWO:
Review of Atomic Theory
and Molecular Structure

ATOMIC STRUCTURE

The mechanical and physical properties of engineering materials depend very strongly on the nature and behavior of the atoms that constitute them. Therefore, a review of atomic theory may be helpful at this point in the study of materials. Our treatment of this particular subject will be somewhat abridged, because the student has very likely been introduced to the basics of atomic theory in other courses, such as general chemistry and introductory physics.

Universally, atoms are recognized as the basic constituents of matter, joining together in a collagenous manner to form crystalline or amorphous materials. This assemblage of atoms may be in the form of gases, liquids, or solids. Specific differences between atoms, and the way they behave mechanically, physically, and chemically, are due principally to their individual atomic or electronic structures.

Structurally, the core of an atom—its nucleus—is composed of positively charged particles known as *protons* and uncharged particles called *neutrons*. The number of protons is also known as the *atomic number* (Z) and determines the identity of the element. The sum of the protons and neutrons in the nucleus determines the *atomic weight*. Certain physical properties (e.g., atomic mass, density, atomic radius) are listed for selected elements, inside the back cover.

The nucleus is surrounded by the electrons associated with a particular atom. Electrons are relatively small charged particles, with a mass equal to 1/1836 of the mass of a proton. Their charge is equal in magnitude, but opposite in sign, to the charge on the proton. Also, the number of electrons in an atom equals the number of protons. Typically, atoms are on the order of 1 angstrom unit (Å) in diameter. Since 1 Å = 10^{-10} m, we are dealing with an extremely small quantity when one considers individual atoms.

Although the tendency is to model atoms conveniently as discrete, uniform, hard spheres, they are in fact rather diffuse, nonuniform, deformable bodies. Our present concept of atomic structure depicts the electrons associated with an atom as a "gaseous" cloud with poorly defined boundaries. This cloud of electrons exhibits greater density toward the nucleus, which is positively charged. A schematic example of a hydrogen atom displaying such characteristics is shown in Figure 2-1.

The orbital electrons or electron cloud of an atom can be altered or disturbed

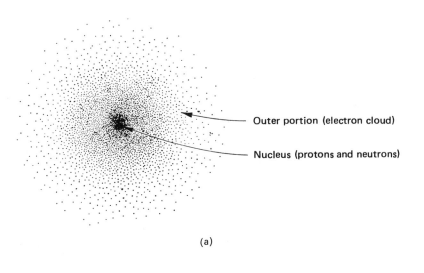

Outer portion (electron cloud)

Nucleus (protons and neutrons)

(a)

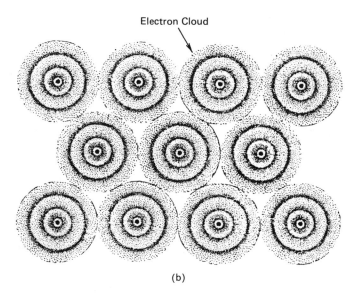

Electron Cloud

(b)

Figure 2-1 Schematic representations of the probability density cross section of atoms: (a) hydrogen atom; (b) lattice of atoms. (Adapted from J. V. Quagliano, *Chemistry*, Prentice-Hall, Inc., Englewood Cliffs, N.J., 1964, p. 34.)

by electrical, magnetic, and mechanical forces. Such alterations or disturbances to the electronic structure of the atoms contribute to a sizable portion of the observed behavior of bulk engineering materials: for instance, their electrical and thermal conductivity, magnetic properties, and corrosion resistance. Throughout this book, the student will constantly be reminded that the *macro* or bulk behavior and properties of engineering materials are closely related to the *micro* or atomic structure of the building blocks that constitute them. These two regimes, the atomistic level and the bulk material, may seem worlds apart dimensionally, but as we will see, they are inseparable with respect to mechanical and physical behavior.

The Periodic Table

The quantum theory, developed by Niels Bohr, predicted definite patterns among atoms and these patterns were eventually responsible for the classification of chemical elements into the *periodic system*. In this system, elements are arranged in tabular form so as to display, in rows and columns, those elements exhibiting similar chemical and physical properties. The periodic table is shown inside the front cover.

In this form, the elements are arranged in order of increasing atomic number in horizontal rows, called *series*, and in vertical columns, called *groups*. Two important features apply to this table:

1. The ionization[1] energy (energy necessary to remove electrons from the outer shell) associated with the elements usually increases from left to right in a row.
2. The ionization energy always decreases down the columns.

Also, trends in properties tend to be more uniform in columns than in rows.

Generically, the periodic table can be separated into two types of elements: *metals* and *nonmetals*. We have delineated this separation by the heavy line which extends in steplike fashion from boron (B) to astatine (At). Elements situated to the left of this line exhibit metallic characteristics, and those to the right exhibit nonmetallic behavior. Metals are elements with comparatively low ionization energies and they are located at the left of rows and bottom of columns. Nonmetals, on the other hand, have higher ionization energies and are located in the upper right-hand portion of the table, as indicated.

The transition from metals to nonmetals, or better yet, the transition from metallic to nonmetallic behavior, is not nearly as distinct as our boundary implies. A gradual change in behavior is actually observed, and such transition is duly reflected in the properties of elements in this region. Indeed, borderline elements are often referred to as *metalloids*: nonmetallic elements that exhibit some metallic characteristics.

---

[1] *Ionization* refers to the process whereby an atom gains or losses valence electrons and thus becomes an *ion*.

Electron Configuration

From the electronic standpoint, the periodic table is arranged according to the way in which additional electrons occur in elements of increasing atomic number. The electrons surrounding the nucleus of an atom do not all have the same energy level. The energy of the electrons becomes progressively greater as their distance from the nucleus increases. Thus it is convenient to separate the electrons into shells containing different energy characteristics. However, the electrons in a specific shell possess approximately the same energy. These shells may also be called the electron configuration of the atom. For example, the first, or lowest shell contains a maximum of 2 electrons. The second shell contains a maximum of 8 electrons, the third shell a maximum of 18, the fourth a maximum of 32, and so on. In other words, the maximum number of electrons in a given shell can be expressed as

$$2n^2 \qquad \text{where } n = 1, 2, 3, \ldots, 7$$

The letter n is called the *principal quantum number* of the shell and represents the energy level of the electrons in that shell.

In addition to the principal quantum numbers, the main energy levels can be further classified into subshells. The number of subshells in a given main energy level is equal to the principal quantum number. These subshells are called *orbitals* and designated s, p, d, and f. An example of electron configurations, or the shell-filling sequence, is demonstrated in Table 2-1 for the elements in the periodic table.

Often, it is convenient to express the electron configuration of an element in an abbreviated manner. One such method of shorthand notation consists of specifying the principal quantum numbers (n) and the subshell or orbitals (s, p, d, f) associated with a particular type of atom. Then the respective number of electrons in each orbit is expressed in exponential fashion. For example, the electron configurations of hydrogen (H), oxygen (O), aluminum (Al), and iron (Fe) may be represented as follows:

H	$1s^1$
O	$1s^2, 2s^2, 2p^4$
Al	$1s^2, 2s^2, 2p^6, 3s^2, 3p^1$
Fe	$1s^2, 2s^2, 2p^6, 3s^2, 3p^6, 3d^6, 4s^2$

Note the direct correspondence of this notation with the presentation in Table 2-1.

The exclusion principle. Expanding our discussion of electronic configurations, the exclusion principle, developed by Wolfgang Pauli, maintains that no two electrons can be in states of motion described by identical sets of quantum numbers. We have just seen the electronic configuration or filling sequence for the elements. However, this presentation does not reveal that two electrons in equivalent orbitals must have opposing spin states. Such a condition of opposing spins is illustrated in Table 2-2 by using arrows to represent the relative spin direction of electrons in their respective orbitals.

TABLE 2-1 ELECTRON CONFIGURATIONS

Atomic number	Element	K 1 s	L 2 s	L 2 p	M 3 s	M 3 p	M 3 d	N 4 s	N 4 p	N 4 d	N 4 f	O 5 s	O 5 p	O 5 d	O 5 f	P 6 s	P 6 p	P 6 d	Q 7
1	H	1																	
2	He	2																	
3	Li	2	1																
4	Be	2	2																
5	B	2	2	1															
6	C	2	2	2															
7	N	2	2	3															
8	O	2	2	4															
9	F	2	2	5															
10	Ne	2	2	6															
11	Na	2	2	6	1														
12	Mg	2	2	6	2														
13	Al	2	2	6	2	1													
14	Si	2	2	6	2	2													
15	P	2	2	6	2	3													
16	S	2	2	6	2	4													
17	Cl	2	2	6	2	5													
18	Ar	2	2	6	2	6													
19	K	2	2	6	2	6		1											
20	Ca	2	2	6	2	6		2											
21	Sc	2	2	6	2	6	1	2											
22	Ti	2	2	6	2	6	2	2											
23	V	2	2	6	2	6	3	2											
24	Cr	2	2	6	2	6	5a	1											
25	Mn	2	2	6	2	6	5	2											
26	Fe	2	2	6	2	6	6	2											
27	Co	2	2	6	2	6	7	2											
28	Ni	2	2	6	2	6	8	2											
54	Xe	2	2	6	2	6	10	2	6	10		2	6						
55	Cs	2	2	6	2	6	10	2	6	10		2	6			1			
56	Ba	2	2	6	2	6	10	2	6	10		2	6			2			
57	La	2	2	6	2	6	10	2	6	10		2	6	1		2			
58	Ce	2	2	6	2	6	10	2	6	10	2a	2	6			2			
59	Pr	2	2	6	2	6	10	2	6	10	3	2	6			2			
60	Nd	2	2	6	2	6	10	2	6	10	4	2	6			2			
61	Pm	2	2	6	2	6	10	2	6	10	5	2	6			2			
62	Sm	2	2	6	2	6	10	2	6	10	6	2	6			2			
63	Eu	2	2	6	2	6	10	2	6	10	7	2	6			2			
64	Gd	2	2	6	2	6	10	2	6	10	7	2	6	1		2			
65	Tb	2	2	6	2	6	10	2	6	10	9a	2	6			2			
66	Dy	2	2	6	2	6	10	2	6	10	10	2	6			2			
67	Ho	2	2	6	2	6	10	2	6	10	11	2	6			2			
68	Er	2	2	6	2	6	10	2	6	10	12	2	6			2			
69	Tm	2	2	6	2	6	10	2	6	10	13	2	6			2			
70	Yb	2	2	6	2	6	10	2	6	10	14	2	6			2			
71	Lu	2	2	6	2	6	10	2	6	10	14	2	6	1		2			
72	Hf	2	2	6	2	6	10	2	6	10	14	2	6	2		2			
73	Ta	2	2	6	2	6	10	2	6	10	14	2	6	3		2			
74	W	2	2	6	2	6	10	2	6	10	14	2	6	4		2			
75	Re	2	2	6	2	6	10	2	6	10	14	2	6	5		2			
76	Os	2	2	6	2	6	10	2	6	10	14	2	6	6		2			
77	Ir	2	2	6	2	6	10	2	6	10	14	2	6	9a		0			
78	Pt	2	2	6	2	6	10	2	6	10	14	2	6	9		1			
79	Au	2	2	6	2	6	10	2	6	10	14	2	6	10		1			
80	Hg	2	2	6	2	6	10	2	6	10	14	2	6	10		2			
81	Tl	2	2	6	2	6	10	2	6	10	14	2	6	10		2	1		
82	Pb	2	2	6	2	6	10	2	6	10	14	2	6	10		2	2		
83	Bi	2	2	6	2	6	10	2	6	10	14	2	6	10		2	3		
84	Po	2	2	6	2	6	10	2	6	10	14	2	6	10		2	4		

TABLE 2-1 ELECTRON CONFIGURATIONS (Cont.)

Z	El	1s	2s 2p	3s 3p 3d	4s 4p 4d	5s 5p
29	Cu	2	2 6	2 6 10ᵃ	1	
30	Zn	2	2 6	2 6 10	2	
31	Ga	2	2 6	2 6 10	2 1	
32	Ge	2	2 6	2 6 10	2 2	
33	As	2	2 6	2 6 10	2 3	
34	Se	2	2 6	2 6 10	2 4	
35	Br	2	2 6	2 6 10	2 5	
36	Kr	2	2 6	2 6 10	2 6	
37	Rb	2	2 6	2 6 10	2 6	1
38	Sr	2	2 6	2 6 10	2 6	2
39	Y	2	2 6	2 6 10	2 6 1	2
40	Zr	2	2 6	2 6 10	2 6 2	2
41	Cb	2	2 6	2 6 10	2 6 4ᵃ	1
42	Mo	2	2 6	2 6 10	2 6 5	1
43	Tc	2	2 6	2 6 10	2 6 6	1
44	Ru	2	2 6	2 6 10	2 6 7	1
45	Rh	2	2 6	2 6 10	2 6 8	1
46	Pd	2	2 6	2 6 10	2 6 10ᵃ	
47	Ag	2	2 6	2 6 10	2 6 10	1
48	Cd	2	2 6	2 6 10	2 6 10	2
49	In	2	2 6	2 6 10	2 6 10	2 1
50	Sn	2	2 6	2 6 10	2 6 10	2 2
51	Sb	2	2 6	2 6 10	2 6 10	2 3
52	Te	2	2 6	2 6 10	2 6 10	2 4
53	I	2	2 6	2 6 10	2 6 10	2 5

Z	El	1s	2s 2p	3s 3p 3d	4s 4p 4d 4f	5s 5p 5d 5f	6s 6p 6d	7s
85	At	2	2 6	2 6 10	2 6 10 14	2 6 10	2 5	
86	Rn	2	2 6	2 6 10	2 6 10 14	2 6 10	2 6	
87	Fr	2	2 6	2 6 10	2 6 10 14	2 6 10	2 6	1
88	Ra	2	2 6	2 6 10	2 6 10 14	2 6 10	2 6	2
89	Ac	2	2 6	2 6 10	2 6 10 14	2 6 10	2 6 1	2
90	Th	2	2 6	2 6 10	2 6 10 14	2 6 10	2 6 2	2
91	Pa	2	2 6	2 6 10	2 6 10 14	2 6 10 2ᵃ	2 6 1	2
92	U	2	2 6	2 6 10	2 6 10 14	2 6 10 3	2 6 1	2
93	Np	2	2 6	2 6 10	2 6 10 14	2 6 10 4	2 6 1	2
94	Pu	2	2 6	2 6 10	2 6 10 14	2 6 10 6	2 6	2
95	Am	2	2 6	2 6 10	2 6 10 14	2 6 10 7	2 6	2
96	Cm	2	2 6	2 6 10	2 6 10 14	2 6 10 7	2 6 1	2
97	Bk	2	2 6	2 6 10	2 6 10 14	2 6 10 8	2 6 1	2
98	Cf	2	2 6	2 6 10	2 6 10 14	2 6 10 10	2 6	2
99	Es	2	2 6	2 6 10	2 6 10 14	2 6 10 11	2 6	2
100	Fm	2	2 6	2 6 10	2 6 10 14	2 6 10 12	2 6	2
101	Md	2	2 6	2 6 10	2 6 10 14	2 6 10 13	2 6	2
102	No	2	2 6	2 6 10	2 6 10 14	2 6 10 14	2 6	2
103	Lr	2	2 6	2 6 10	2 6 10 14	2 6 10 14	2 6 1	2
104	Rf	2	2 6	2 6 10	2 6 10 14	2 6 10 14	2 6 2	2
105	Ha	2	2 6	2 6 10	2 6 10 14	2 6 10 14	2 6 3	2

ᵃNote irregularity.

TABLE 2-2 GROUND-STATE ORBITAL ELECTRON CONFIGURATIONS OF THE FIRST 18 ELEMENTS

Element	Atomic number	Electronic notation	Spin alignment
H	1	$1s^1$	[↑]
He	2	$1s^2$	[↑↓]
Li	3	$1s^2\ 2s^1$	[↑↓] [↑]
Be	4	$1s^2\ 2s^2$	[↑↓] [↑↓]
B	5	$1s^2\ 2s^2\ 2p^1$	[↑↓] [↑↓] [↑]
C	6	$1s^2\ 2s^2\ 2p^2$	[↑↓] [↑↓] [↑][↑]
N	7	$1s^2\ 2s^2\ 2p^3$	[↑↓] [↑↓] [↑][↑][↑]
O	8	$1s^2\ 2s^2\ 2p^4$	[↑↓] [↑↓] [↑↓][↑][↑]
F	9	$1s^2\ 2s^2\ 2p^5$	[↑↓] [↑↓] [↑↓][↑↓][↑]
Ne	10	$1s^2\ 2s^2\ 2p^6$	[↑↓] [↑↓] [↑↓][↑↓][↑↓]
Na	11	$1s^2\ 2s^2\ 2p^6\ 3s^1$	[↑↓] [↑↓] [↑↓][↑↓][↑↓] [↑]
Mg	12	$1s^2\ 2s^2\ 2p^6\ 3s^2$	[↑↓] [↑↓] [↑↓][↑↓][↑↓] [↑↓]
Al	13	$1s^2\ 2s^2\ 2p^6\ 3s^2\ 3p^1$	[↑↓] [↑↓] [↑↓][↑↓][↑↓] [↑↓] [↑]
Si	14	$1s^2\ 2s^2\ 2p^6\ 3s^2\ 3p^2$	[↑↓] [↑↓] [↑↓][↑↓][↑↓] [↑↓] [↑][↑]
P	15	$1s^2\ 2s^2\ 2p^6\ 3s^2\ 3p^3$	[↑↓] [↑↓] [↑↓][↑↓][↑↓] [↑↓] [↑][↑][↑]
S	16	$1s^2\ 2s^2\ 2p^6\ 3s^2\ 3p^4$	[↑↓] [↑↓] [↑↓][↑↓][↑↓] [↑↓] [↑↓][↑][↑]
Cl	17	$1s^2\ 2s^2\ 2p^6\ 3s^2\ 3p^5$	[↑↓] [↑↓] [↑↓][↑↓][↑↓] [↑↓] [↑↓][↑↓][↑]
Ar	18	$1s^2\ 2s^2\ 2p^6\ 3s^2\ 3p^6$	[↑↓] [↑↓] [↑↓][↑↓][↑↓] [↑↓] [↑↓][↑↓][↑↓]

In addition to depicting opposing electron spins, Table 2-2 also shows that when several electrons occupy several orbitals in the same energy level, such as the $2p$ orbitals in nitrogen, each electron has the same spin orientation but occupies a different orbital. Under these circumstances, the atom is at its lowest-energy state. More energy is required to place two electrons in the same orbital with opposing spins than is required to put them in different orbitals of equivalent energy with their spins aligned. This precept is called *Hund's rule*.

Valence Electrons

The outermost shell of electrons in an atom is known as the valency group. The properties of the elements and thus the behavior of engineering materials composed of the elements are closely related to the valence electrons. Recall that the innermost electrons are tightly bound to the nucleus, while the outermost electrons are more loosely bound. This factor contributes significantly to the behavior of the valence electrons and thus to the properties of the elemental species.

Most elements exhibit either electropositive or electronegative valence behavior. In other words, an element has either a positive or a negative valence number, unless it is inert. Inert elements are those which have a very stable electronic configuration and are reluctant to react chemically with other elements. The inert gases, Ne, Ar, Kr, Xe, and Rn, all show eight electrons in their valence shell.

Elements that have unfilled valence shells and tend to lose electrons (ionize) readily are termed *electropositive*. Because of their unfilled outer shells, it is easier for these elements (known as metals) to lose one or more of these "extra" electrons and become positively charged than it is to gain additional electrons. For example, iron (Fe) has 26 electrons distributed in the four principal quantum shells as shown in Table 2-1. Iron has 2 valence electrons because two electrons go to the fourth shell before the third shell is filled. Therefore, iron exhibits the tendency to give up these two electrons and become positively charged. Similarly, aluminum (Al) has 13 electrons distributed in the K, L, and M shells. Its valence shell (M) contains three, while the next innermost shell (L) contains eight. Note that according to the $2n^2$ rule, this is the maximum number of electrons that the L shell may hold. Therefore, if Al gives up its three valence electrons, it presents a very stable electronic arrangement. As you will see in Chapter 13, aluminum does readily give up these outer electrons and become positively charged. We can express these ionization processes in iron and aluminum as follows:

$$Fe^0 \longrightarrow Fe^{2+} + 2e^-$$

$$Al^0 \longrightarrow Al^{3+} + 3e^-$$

Elements that do not lose electrons readily are known as nonmetals. If an element tends to gain electrons during a chemical reaction (compound formation), it is referred to as *electronegative*. For example, oxygen (O) contains six electrons in its valence shell. If oxygen adds two electrons to this shell, it will be filled, resulting in a very stable electronic structure. Therefore, oxygen commonly exhibits a valency of -2. This type of reaction can be expressed as follows:

$$O^0 + 2e^- \longrightarrow O^{2-}$$

These relatively simple expressions of atoms losing or gaining electrons in the valence shell are much more important than they may seem. Such reactions are, by definition, *oxidation* (loss of electrons) and *reduction* (gain of electrons), and as the student will see in later chapters, these processes form the basis for many important properties and behavioral changes in engineering materials. Table 2-3 lists some elements according to the valence or oxidation number, they commonly exhibit. The student is encouraged to compare these valency numbers and their signs with the electron configuration for the respective elements given in Table 2-1.

TABLE 2-3 COMMON VALENCY OF SELECTED ELEMENTS

Electropositive		Electronegative	
Hydrogen	+1	Nitrogen	−3
Boron	+3	Oxygen	−2
Sodium	+1	Sulfur	−2
Magnesium	+2	Chlorine	−1
Aluminum	+3	Fluorine	−1
Silicon	+4		
Phosphorus	+5		

Band Theory

Thus far in our discussion of electronic structure we have considered individual atoms. However, when atoms are brought together in a solid, electronic *band* structures are formed. As the distance between the atoms decreases, their valence electrons begin to interact and the energy states (quantum states) subdivide into multiple energy levels which can accommodate all the electrons that formerly occupied single quantum levels. This subdivision is actually a requirement of the *exclusion principle*, which requires that no more than two electrons can occupy the same energy state at the same time—and those two electrons must have opposite spins. The valence level in a material containing N atoms subdivides into N energy levels. For example, the s level, which can contain two electrons, becomes the s band, with the capacity for $2N$ electrons. Correspondingly, the p level can contain $6N$ electrons, and so on.

The subdivided energy levels contain an array of very closely spaced electronic energy levels. These regions of permissible energy states are called *allowed bands*, and the energy gaps separating them are called *forbidden bands*. The electronic energy levels for magnesium (Mg) are illustrated in Figure 2-2.

In Chapter 10 you will see how the band theory of electrons accounts for the difference in electrical conductivity among metals, and the difference in electrical behavior between conductors, semiconductors, and insulators. Furthermore, band

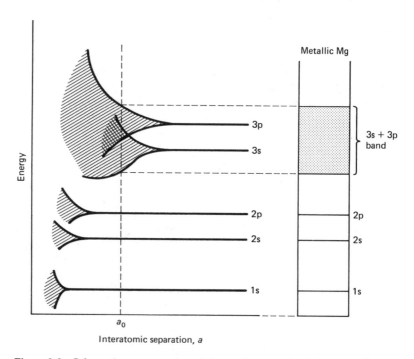

Figure 2-2 Schematic representation of electronic energy levels in magnesium as a function of interatomic separation. Equilibrium separation is denoted by a_0.

theory provides an explanation for the electronic behavior of certain semiconductor materials which are crucial to the operation of such familiar applications as your digital wristwatch or clock, calculator, television, stereo set, and computer.

FORCES BETWEEN ATOMS

The mechanical and physical properties of engineering materials depend strongly on the nature of the interatomic forces that exist between atoms. Now we will examine the manner in which these forces operate between the atoms and how the potential energy of the atoms depends on the distance between them. Figure 2-3 illustrates schematically the basic nature of this relationship. Examination of this diagram shows that where the atoms, represented by the dark circles, are separated by a large distance (case 4), there is very little force interaction between them. This is demonstrated by the small value of force for this spacing. As the atoms separate even further (infinite distance), the interactive force approaches zero, as does the potential energy.

Case 3 depicts the atoms separated by a relatively smaller distance. At this value of interatomic spacing (a), a large attractive force exists, tending to pull the atoms together. However, as the atoms approach closer, an interatomic spacing is eventually reached where the attractive forces are balanced by the forces of repulsion. In other words, the sum of the forces is equal to zero, and the potential energy of the atom pair is a *minimum*. This position is the most stable configuration for the atoms and is called the *equilibrium separation* (a_0). This condition is shown in case 2.

If the atoms approach closer than the equilibrium separation (a_0), large repulsive forces develop and the potential energy of their interaction rises sharply. Such a condition is shown in case 1.

For the purpose of modeling and many calculations, we sometimes schematize the equilibrium spacing between atoms as shown in Figure 2-4. For convenience, the atoms are represented by hard spheres just making contact with each other. The radius of the sphere corresponds to the equilibrium radius (r_0) of the atom. The equilibrium spacing (a_0) corresponds to the distance between the centers of the spheres. In addition to the specific atoms involved, this separation also depends on other factors, such as crystal, or molecular structure and temperature. Nevertheless, this model is a reasonable approximation of the equilibrium separation or the distance of closest approach, in most cases.

We have stated above that certain factors affect the equilibrium separation between atoms. These factors also include:

1. Temperature
2. Valency
3. Coordination number (number of next nearest neighbors)

The equilibrium spacing between atoms necessarily changes with temperature because the overall size of the atoms changes as they are heated or cooled. Changes in valency also affect the size of the atom. For example, if an iron atom ($r_0 = 1.241$ Å) loses two valence electrons and is ionized to Fe^{2+}, its ionic radius equals 0.74 Å. Numerical

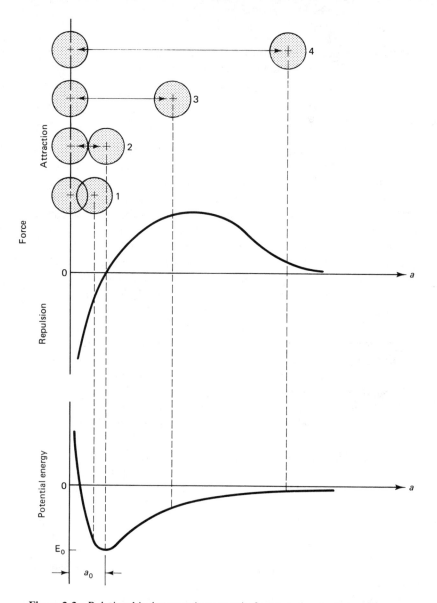

Figure 2-3 Relationship between interatomic forces and potential energy as a function of interatomic separation. Equilibrium separation a_0 corresponds to minimum energy (E_0).

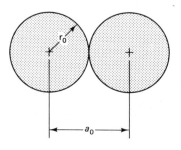

Figure 2-4 Schematic model of the equilibrium separation (or distance of closest approach) between two like atoms or ions r_0, equilibrium radius; a_0, equilibrium separation.

values of ionic radii for certain elements are presented inside the back cover. Additionally, changes in coordination number, that is, the number of next nearest neighboring atoms in a material, will affect the equilibrium separation between the atoms. For instance, if an Fe atom is directly surrounded by eight neighboring Fe atoms, $r_0 = 1.241$ Å. But if the coordination number increases to 12, $r_0 = 1.269$ Å.

Balance of Atomic Forces

The energy associated with the interatomic bond between atoms depends strongly on the distance or separation between them. As we schematized in Figure 2-3, the potential energy is at a minimum value for the equilibrium separation (a_0). This energy (E_0) represents the bonding energy between the atoms. If we attempt to change the interatomic separation in either direction (increase or decrease the distance), a force is required. This situation is analogous to connecting spheres with identical coil springs, as shown in Figure 2-5. At equilibrium the spheres are separated by an equal

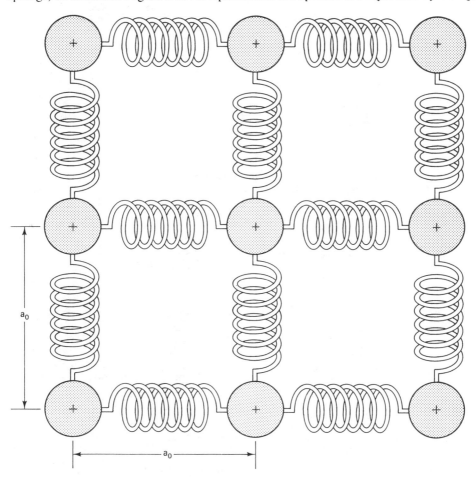

Figure 2-5 Mechanical analogy of interatomic forces. Equilibrium separation corresponds to a_0.

distance a_0. To bring two spheres closer, we must apply a force and compress the spring between them. Similarly, to pull two spheres apart, we must stretch the spring, which requires a certain force. In the case of real atoms situated at their equilibrium positions, any change in the interatomic distance *increases* the value of potential energy, as Figure 2-6 demonstrates.

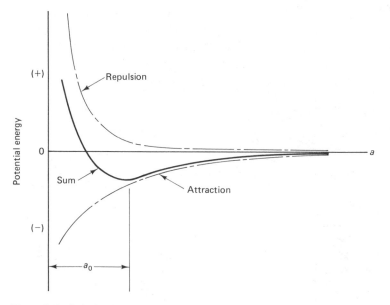

Figure 2-6 Relationship between the potential energy of interacting atoms and their separation.

We can derive an expression for the bonding energy between atoms by utilizing Coulomb's law of force between point charges. Treating the atoms as point charges, this law may be expressed as

$$F = \frac{q_1 q_2}{r^2} \tag{2-1}$$

where F = force between the charged particles

q_1, q_2 = charge associated with particles 1 and 2

r = distance separating the charged particles

Attractive forces. In our case, we may express the forces of attraction as follows:

$$F_A = \frac{1}{4\pi\epsilon_0} \frac{(Z_1)q(Z_2)q}{a^2} \tag{2-2}$$

where Z = number of charges at each point (the valence)

q = value of electronic charge (0.16×10^{-18} C)

a = interatomic separation

ϵ_0 = permittivity in vacuum (8.9×10^{-12} C/V-m)

Note that for monovalent ions ($Z = 1$), expression (2-2) reduces to

$$F_A = \frac{k_0 q^2}{a^2}$$

where $k_0 = \dfrac{1}{4\pi\epsilon_0}$

Repulsive forces. The forces of repulsion may be expressed as follows:

$$F_R = \frac{-nb}{a^{n+1}} \tag{2-3}$$

where n and b are empirical constants (n is equal to approximately 9 in ionic solids). Comparing equations (2-2) and (2-3) we see that the attractive and repulsive forces are proportional to a in the case of ionic solids as follows:

$$F_A \propto \frac{1}{a^2} \qquad F_R \propto \frac{1}{a^{10}}$$

Therefore, as the interatomic separation (a) increases to relatively large distances, the forces of attraction dominate. However, as the interatomic distance decreases below the value of a_0 (equilibrium separation), the repulsive forces dominate.

Bonding Energy

Since work or energy is equal to the product of force times distance, the interactive (potential) energy between neighboring atoms can be mathematically expressed as

$$E = \int (F_A + F_R)\, da \tag{2-4}$$

This integral or summation of the energy over the distance da will be evaluated from $a \rightarrow \infty$, where $E = 0$, to smaller values of interatomic separation (a). Thus we may write

$$E = \int_{\infty}^{a} \frac{(-k_0 Z_1 Z_2 q^2)\, da}{a^2} - \frac{nb\, da}{a^{n+1}}$$

Rearranging the integral in order to simplify evaluation, we have

$$E = \int_{\infty}^{a} (-k_0 Z_1 Z_2 q^2)\frac{da}{a^2} - \int_{\infty}^{a} \frac{nb\, da}{a^{n+1}}$$

Since the quantities k_0, Z_1, Z_2, q^2, n, and b are not functions (dependent) of the variable a, they can be treated as constants and removed from the integration process. The expression now becomes

$$E = (-k_0 Z_1 Z_2 q^2) \int_{\infty}^{a} \frac{da}{a^2} - nb \int_{\infty}^{a} \frac{da}{a^{n+1}}$$

In the form above, the student should now recognize the simple integral of the form

$$x^n\, dx = \frac{x^{n+1}}{n+1}$$

In the event that the student is not familiar with integration techniques, the solution of this integral as shown above can be found in tables of integrals[2] under elementary forms.

Our expression for energy is then integrated to the following result:

$$E = \left[-k_0 Z_1 Z_2 q^2 \frac{-1}{a} - nb \frac{-1}{na^n} \right]_\infty^a$$

By simplifying we obtain

$$E = \left[\frac{k_0 Z_1 Z_2 q^2}{a} + \frac{b}{a^n} \right]_\infty^a$$

Our initial premise stated that the interactive energy (E) was equal to zero when a approached an infinite distance. Therefore, substitution of the lower limit ($a = \infty$) produces a value of $E = 0$. Substitution of the upper limit (a) yields the final expression for energy as follows:

$$E = \frac{k_0 Z_1 Z_2 q^2}{a} + \frac{b}{a^n} \qquad (2\text{-}5)$$

The expression derived above for bonding or interactive energy between atoms is the basis for determining values of energy for various chemical bonds and for defining values of equilibrium spacing (a_0). This is accomplished by elementary calculus. We can determine a_0 because the first derivative of the energy function above with respect to separation is equal to zero when the energy is a minimum. Stated mathematically, we have

$$\frac{dE}{da} = 0$$

when $E = E_{min}$ = equilibrium
$\quad a = a_0$

The potential energy between atoms was shown schematically in Figure 2-6. The sum of the repulsive and attractive energies shows that the energy is a minimum value (slope of the energy curve = 0) for an interatomic spacing equal to the equilibrium separation (a_0).

INTERATOMIC BONDING

Since engineering materials are usually solids, it is important to understand the forces of attraction that hold these solids together. In other words, why do we obtain continuous aggregates of atoms rather than discontinuous or diffuse arrangements of these constituents? Typically, the forces that hold solids together are very strong. For example, to break down the bonds in diamond by melting, we must supply enough heat energy to raise the temperature of the solid to above 3500°C (6332°F). Or consider yet another aspect of bonding in engineering materials. Theoretically, to break atomic bonds, that is, to fracture a metal simply by separating adjacent atoms, tremendous

[2]Tables of integrals are given in *CRC Standard Mathematical Tables* and *CRC Handbook of Chemistry and Physics*, CRC Press, Boca Raton, Fla.

forces would have to be applied to the material. How large would such forces have to be? Well, for instance, a round steel bar 25.4 mm (1 in.) in diameter would have to be pulled with a force of about 9 million newtons (2 million pounds). Not impressed? You should be; this force amounts approximately to the weight of 400 automobiles (at 5000 lb each). In Chapter 4 we demonstrate how this theoretical value of *cohesive strength* (force holding atoms together) is derived and you will also see why your intuition is correct; engineering materials ordinarily "break" under forces *considerably* lower than their theoretical cohesive strength.

Interatomic attractions are caused by the electronic configuration and structural characteristics of atoms. The binding forces between atoms in a solid are attributed to four types of bonding mechanisms: *ionic, covalent, metallic,* and *van der Waals forces.* We will concentrate on the first three types, since the van der Waals is a comparatively weak bonding force and primarily involves inert-gas solids.

Ionic Bonding

This type of atomic bonding results from the mutual attraction of positive (+) and negative (−) charges. How do these charges originate? Elements such as sodium (Na) and calcium (Ca), which contain one or two electrons in their valence shells, tend to release these valence electrons easily and thus become positively charged. When an atom loses (or gains) electrons, consequently acquiring an electrical charge, we say that it has been ionized. This process is illustrated schematically in Figure 2-7, where the nucleus is represented by a large dark circle surrounded by the appropriate number of electrons (small dark circles) in their respective shells.

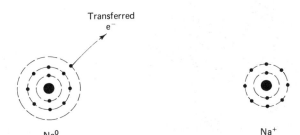

Figure 2-7 Schematic representation of the ionization of sodium (Na) to Na^+.

In this example, a sodium atom (Na^0) containing one electron in its valence shell releases this electron (e^-) and becomes a positively charged ion (Na^+). This process may also be represented in the following manner:

$$Na^0 \longrightarrow Na^+ + e^-$$

Note that by doing this, the sodium atom has achieved a more stable electron configuration. Its outer shell now contains eight electrons, resembling Ne. This positively charged ion can now be attracted by a negatively charged ion.

Chlorine (Cl) and oxygen (O) readily add electrons to their outermost shells. Inspection of Table 2-3 shows that these atoms have a valency of −1 and −2, respectively. This means that chlorine will readily accept one electron in its outer shell and oxygen will accept two. Note the effect that this process has on the stability of their electron configuration. By adding these electrons, the atoms become negatively charged ions. This reaction can be expressed as follows:

$$Cl^0 + e^- \longrightarrow Cl^-$$
$$O^0 + 2e^- \longrightarrow O^{2-}$$

A schematic example of this process is shown in Figure 2-8. The negatively charged

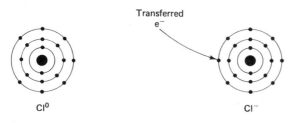

Transferred
e^-

Cl^0 Cl^-

Figure 2-8 Schematic representation of the ionization of chlorine (Cl) to Cl^-.

ion is now attracted to a positive ion, thus forming the basis for *ionic bonding*. Both types of ions have achieved a more stable electron configuration by the *transfer* of valence electrons, and in so doing, they have been mutually attracted. Compounds that form ionic bonds include sodium chloride (NaCl), calcium chloride ($CaCl_2$), aluminum oxide (Al_2O_3), and magnesium oxide (MgO). The ionic bond for a typical compound is illustrated schematically in Figure 2-9.

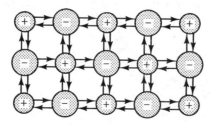

Figure 2-9 Schematic illustration of mutually attracted ions in an ionic compound.

The Covalent Bond

As we have emphasized several times already, the electron structure of an atom is especially stable if it contains eight electrons (an octet) in its outermost shell. Sometimes, an atom will *share* valence electrons with a neighboring atom in order to satisfy such a stable configuration. This sharing of electrons produces strong forces of attraction between the atoms involved and is termed *covalent bonding*. The best known example of this type of bonding is the hydrogen molecule (H_2). This bond is illustrated schematically in Figure 2-10, where the nucleus is represented by the elemental symbol (H) and the electrons, by small dark circles in orbit around the nuclei.

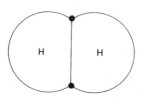

H   H

Figure 2-10 Schematic representation of covalent bonding in the hydrogen molecule (H_2).

According to the Heisenberg uncertainty principle, it is impossible to determine accurately the position and momentum of even one electron in orbit around an atom. However, the probable positions of electrons can be estimated by means of wave functions (i.e., the Schrödinger wave equation). Based on some probability distribution for the electrons involved, the covalent bond may be considered as a union of negatively charged electrons between the positively charged nuclei. Examples of covalent bonding include the oxygen and nitrogen molecules. These bonds are illustrated schematically in Figure 2-11. To simplify the schematic representation of cova-

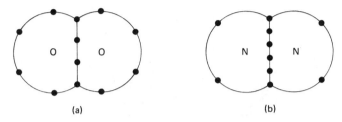

Figure 2-11 Examples of covalent bonding: (a) oxygen (O_2); (b) nitrogen (N_2).

lent bonding, only the valence shells are included in these sketches. The chemical symbols represent the nucleus plus all extra-nuclear electrons except the valence electrons. Note that whereas both the individual oxygen and nitrogen atoms originally contained fewer than eight valence electrons, the covalently bonded atoms exhibit the stable electronic structure of eight electrons in their outer shells.

Polyatomic arrangements of atoms are also common examples of the covalent bonding mechanism. Although the subject of polyatomic molecules will be discussed in more detail later in this chapter, the methane molecule (CH_4) is presented in Figure 2-12 along with a representation of a covalently bonded solid. These examples serve to illustrate the mutual sharing of valence electrons by atoms in order to achieve greater electronic stability.

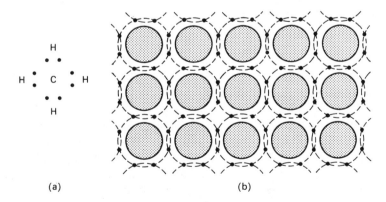

Figure 2-12 Schematic representation of covalent bonding. (a) Lewis electron-dot representation of CH_4. (b) Orbital shell representation of a typical covalently bonded material. Large shaded circles represent atoms exclusive of valence shell, while small dark circles are shared valence electrons.

In summary, the covalent bond produces very strong attractive forces between atoms. This is demonstrated most dramatically by the formation of diamond, the hardest substance found in nature. Diamond, which has an extremely high melting temperature (greater than 3500°C), is composed entirely of covalently bonded carbon atoms, as illustrated in Figure 2-13. The bond energy and thermal stability of some covalently bonded materials are presented in Table 2-4.

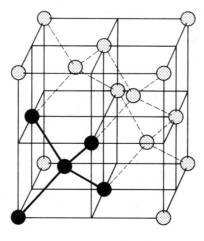

Figure 2-13 Space lattice representation of the diamond cubic structure. Dark outlines show tetrahedral bond.

TABLE 2-4 BOND ENERGY AND THERMAL STABILITY OF SOME COVALENTLY BONDED MATERIALS

Material	Bond Energy (kcal/mole)	Melting Temperature (°C)
Diamond	170	3500
Ge	75	958
Si	84	1420
SiC	283	2700

The Metallic Bond

Earlier we pointed out that certain elements tend to lose valence electrons readily and that such elements are known as metals. If there are only a few valence electrons associated with an atom, they may be removed rather easily while the balance of electrons (the inner shells) are held firmly to the nucleus. This process results in a positively charged *kernel* of an atom, the kernel being the nucleus plus extra-nuclear electrons (with the exception of the valence shell) and the electrons dissociated from that particular atom. Essentially, we can have a positive ion core (the kernel) and a number of "free" electrons. Such an atomistic condition is illustrated schematically in Figure 2-14.

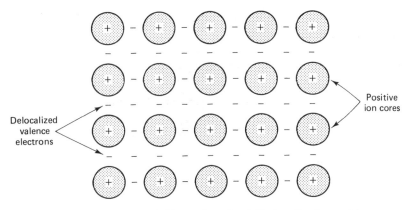

Figure 2-14 Schematic illustration of metallic bonding in crystals.

The de-localized valence electrons in this metallic structure, then, are free to drift about and associate with various positive ion cores. For this reason, they constitute what is frequently described as an *electron cloud* or *gas*. It is this combination of positively charged cores and the surrounding electron cloud which produces the attractive forces of the *metallic bond*. Furthermore, the detached valence electrons or electron cloud are responsible for such metallic characteristics as:

1. High thermal conductivity
2. High electrical conductivity
3. Opaqueness to light

Van der Waals Forces

Van der Waals interatomic attraction is relatively weak and therefore is not considered extremely important with respect to engineering materials. However, sometimes van der Waals forces are the only forces that operate between atoms; thus we will address them briefly.

In the inert gases, such as helium (He), the outer shell (1s) with two electrons is complete. Other elements, such as neon (Ne), argon (Ar), and krypton (Kr), contain filled levels in their valence shells (in addition to filled levels beneath the valence shell), and thus are very chemically stable. Accordingly, none of the primary bonding mechanisms discussed previously will be effective, since ionic, covalent, and metallic bonds all require adjustments in valence electrons and act on atoms with incomplete valence shells. Thus the atoms of the inert gases have little attraction for one another or for other atoms. Only at extremely low temperature, when thermal vibrations are greatly reduced, do these gases condense. This condensation or attraction between the inert-gas atoms demonstrates that there are weak interatomic forces pulling the atoms together. These forces, which consist of a weak coulombic interaction, are the van der Waals forces.

Summary

In summary, we have discussed and presented schematic examples of three primary types of atomic bonding mechanisms: ionic, covalent, and metallic. All of these are relatively strong and are responsible for the forces that hold most engineering materials together in solid form. The fourth type mentioned, van der Waals forces, are relatively weak and are not significant with regards to engineering materials. Table 2-5 presents a synopsis of the various interatomic bonding forces.

TABLE 2-5 SUMMARY OF INTERATOMIC BONDING PROPERTIES

Property	Ionic	Covalent	Metallic	Van der Waals
Energy origin	Coulombic interaction	Sharing of valence electrons	Dissociation of valence electrons	Weak coulombic interaction
Relative strength	Strong, hard crystals	Strong, hard crystals	Variable strength, gliding common	Weak, soft crystals
Thermal	Fairly high melting point, low coefficient of expansion	High melting point, low coefficient of expansion	Variable melting point	Low melting point, large coefficient of expansion
Electrical	Moderate insulators, conductors in liquid state	Insulators in solid and liquid state	Conduction by electron transport	Insulators
Optical and magnetic	Absorption and other properties those of individual ions	High refractive index	Opaque, lustrous	Properties those of individual molecules
Examples	NaCl, MgO, Al_2O_3	Diamond, CH_4, O_2	Fe, Cu, Ti, Cr	Ar, Ne, Kr

MOLECULAR STRUCTURE

In addition to crystalline structures, which exhibit regular repeating patterns of atoms in 3-dimensional space, there is another important class of atomic arrangements—molecular structures. Just as the properties of crystalline materials depend strongly on their specific crystal structure, the properties and behavior of "molecular materials" depend heavily on their particular molecular configuration. The molecular structure is often referred to as one that exists between the crystalline and the amorphous state.

Molecules

Molecules may be defined as a limited number of atoms which are strongly (chemically) bonded together, but whose bonds with other similar groups of atoms are relatively weak. These groups of atoms, which possess no net charge, act as a unit because the intramolecular attractions are strong, primary bonds, whereas the intermolecular adhesion (between molecules) actually operates between the electrons associated with individual molecules. Such attractive forces between molecules are similar to van der Waals forces.

The common examples of molecules include the elemental diatomic gases, such as H_2, O_2, N_2, and such compounds as H_2O, CO_2, CCl_4, and HNO_3, as well as the thousands of hydrocarbons. The existence of a stable polyatomic species, whether elemental or compound, suggests that the combined structure has a lower energy state than do the individual atoms. Within each of these molecules the atoms are usually covalently bonded, although ionic bonds are not uncommon. The fundamental reason for bonding between two atoms is that the potential energy of the system is at the lowest state when an electron is relatively near both atomic nuclei at the same time. As we shall see, this is the basis of covalent bond formation. The sharing of the electron permits it to spend more time in space where its potential energy is low, thereby reducing the total energy of the system.

In this sense, ionic bonding and covalent bonding are similar in that they both reduce the potential energy of the system. In ionic bonding, the redistribution of the electron density of the system is accomplished by means of electron transfer, while in covalent bonding, it is accomplished by the sharing of electrons.

Although in the diatomic molecules, such as H_2, O_2, N_2, the sharing of the electrons is equal, it should be realized that many chemical bonds are characterized by unequal sharing and in fact, there is a range of bond properties extending from complete ionic to complete covalent bonding.

Molecular Bonding

It is relatively easy to comprehend the nature of the ionic bond as described earlier in this chapter. The essential feature of such a bond is that electron transfer creates oppositely charged ions which thereby produces a coulombic attraction and results in a crystal formation. It is less easy to see how atoms of similar electron configuration such as those of hydrogen, oxygen, and nitrogen can form strong bonds which result in the formation of the diatomic molecules, H_2, O_2, and N_2. The formation and stability of these symmetric molecules is based on an equal *sharing* of valence electrons, thus the term "covalent bonding."

To understand the basis of covalent bonding, we should examine the distribution of valence electrons in a single atom compared to the distribution in the combined molecular state. If we examine the electron density of the hydrogen-ion molecule (H_2), as shown schematically in Figure 2-15, one can see that forming the bond transfers some of the energy from the regions outside both nuclei to the regions adjacent to and between the nuclei.

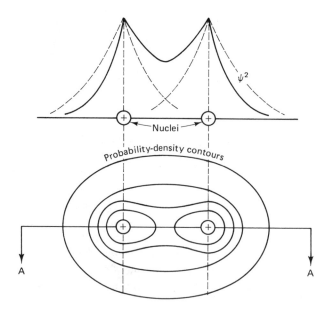

Figure 2-15 Electron sharing in the H_2 molecule. The dashed line represents the electron probability for two unbonded atoms, while the solid line represents the electron density for two bonded atoms.

Since the potential energy of the electron is a function of the distance from the nuclei as stated by the following expression:

$$E = -q^2\left(\frac{1}{r_A} + \frac{1}{r_B}\right)$$

where E = potential energy
q = value of electronic charge
r_A = distance from nucleus A
r_B = distance from nucleus B

It follows, then, that the electron lowers its energy by getting closer to the second nucleus. The total energy of the system decreases as the two atoms move together. This results in an attractive or bonding force.

Number of bonds. The number of covalent bonds around an atom depends on the number of electrons in the outer, or valence shell. Except for hydrogen and helium, an atom tends to form bonds until it is surrounded by eight electrons. This behavior is frequently referred to as the "octet rule" and may be expressed as follows:

$$N = 8 - G$$

where N = number of bonds

G = group number of the atom within the periodic table

This rule, as with so many others, has a number of exceptions that serve to limit its value. For example, if we examine the chlorides of phosphorus, phosphorus trichloride (PCl_3) and phosphorus pentachloride (PCl_5), we see that the formation of PCl_3 obeys the octet rule. In the case of PCl_5, however, the phosphorus atom is surrounded by 10 electrons, the reason being that the phosphorus atom can have two orbital configurations, the first having three half-filled orbitals, which permits the formation of PCl_3, and the second which has five half-filled orbitals, which permits the formation of PCl_5.

Bond energy. The significant factor in bond formation is that the sharing of an electron by two nuclei produces a decrease in the total energy of the system. When the shared electron occupies a bonding orbital, the energy of the system can be calculated as a function of the internuclear distance. This is plotted for H_2 in Figure 2-16.

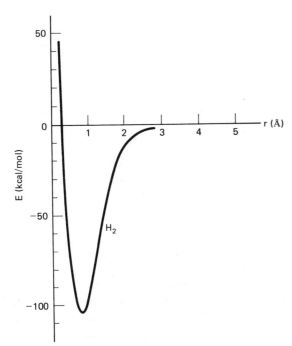

Figure 2-16 Total energy of the system as a function of internuclear distance for the H_2 molecule.

The minimum energy corresponds to the most stable *configuration* position and occurs at the equilibrium bond distance. The depth of the energy minimum corrected for the vibrational energy ($\frac{1}{2}h\nu$) at the zero point is roughly equivalent to the bond dissociation energy. Table 2-6 lists the bond energies at 25°C for a number of diatomic and polyatomic molecules. These values are the energy necessary to break one mole (Avogadro's number) of bonds. The magnitude of this energy is essentially the same for forming

TABLE 2-6 MOLECULAR BOND LENGTH AND ENERGY (25°C)

Bond	Bond length (Å)	Bond energy (kcal/mol)
C—C	1.54	83
C=C	1.33	146
C≡C	1.20	200
C—H	1.10	99
C—N	1.47	73
C=N	1.31	147
C—O	1.43	86
C—F	1.38	116
C—Cl	1.77	81
O—H	1.03	111
O—O	1.50	35
O—Si	1.50	90
N—H	1.03	93
N—O	1.20	53
N—N	1.10	39
N—Cl	1.80	46
F—F	1.42	37

bonds, also. However, the sign is negative for the formation of bonds (energy released) and positive for rupturing bonds (energy is required).

Electron-dot representation. A simple but effective method for representing the disposition of electrons is the use of electron-dot structures referred to previously. For example, the equation

$$4H\bullet + \bullet \overset{\bullet}{\underset{\bullet}{C}} \bullet \longrightarrow H\overset{\overset{\displaystyle H}{\bullet\bullet}}{\underset{\underset{\displaystyle H}{\bullet\bullet}}{\bullet C \bullet}} H$$

depicts the electrons which take part in the electron-pair bond and shows that the carbon atom and the hydrogen atoms each contribute one electron to the shared pair.

Similarly, if we describe the reaction of hydrogen and nitrogen as follows:

$$3H\bullet + \bullet \overset{\bullet}{\underset{\bullet}{N}}\colon \longrightarrow H\overset{\overset{\displaystyle H}{\bullet\bullet}}{\underset{\underset{\displaystyle H}{}}{\bullet N \bullet}}\colon$$

we can see that three covalent bonds are formed and one pair of valence electrons of nitrogen is not directly involved in the bonding.

A similar method for depicting the presence of covalent bonds is the use of lines to represent the bonded pairs. Methane (CH_4) would be described as follows:

$$H\text{—}\underset{\underset{H}{|}}{\overset{\overset{H}{|}}{C}}\text{—}H$$

It should be remembered that although these methods are convenient from the stand-point of graphical presentation, they are not precise geometrical models for the representative structures. Another important limitation is the fact that graphic representations are planar and lack the three-dimensional character often present in many molecules.

Multiple bonding. Up to this point we have discussed only the formation of a single covalent bond between atoms. However, in some molecules, two or three pairs of electrons combine to provide the bond to hold the atoms together. Ethylene (C_2H_4) is an example of a molecular structure that exhibits *double* bonding:

$$\overset{H\quad H}{\underset{H\quad H}{C\:\vdots\vdots\:C}}\qquad \overset{H}{\underset{H}{{>}}}C=C\overset{H}{\underset{H}{{<}}}$$

Hydrocarbon molecules such as ethylene which have in their structure the $C=C$ double bond are referred to as "unsaturated" since they are capable of combining with additional hydrogen atoms or atoms of other atomic species.

Similarly, it is possible to describe the existence of triple-bonded structures where three pairs of electrons serve to hold the atoms together. Acetylene (C_2H_2) is an example of a structure that exhibits a triple bond, depicted as follows:

$$H\text{—}C\equiv C\text{—}H$$

It is interesting and perhaps not surprising to note, as shown in Table 2-6, that as the bond between atoms becomes increasingly unsaturated, it becomes shorter and displays a higher bond strength.

Bond angles. The bond angles between atoms existent in a molecule represent the structure having the minimum energy condition. These angles arise from the fact that for varying reasons electrons try to avoid each other in atomic space. One reason is simply a matter of similar charge. Since they both possess a negative charge, they exert a repulsive force as they approach each other. The other principal reason deals with the low probability of two electrons with the same spin being in close proximity (exclusion principle). It is these characteristics that hold the key to molecular geometry. Since all electrons try to avoid each other because of like charge and since there is a low probability of electrons with the same spin being in the same location, then if one electron is located at a particular location in atomic space, any other electron is located somewhere else. These factors may be used to explain the geometry of molecular structures.

If we examine the structure of magnesium chloride ($MgCl_2$), we find that it has a linear molecular arrangement:

$$Cl—Mg—Cl$$

the reason being that this arrangement places the two valence electrons of the bonding pairs on opposite sides of the magnesium atom. The linear (180°) configuration is therefore the lowest-energy arrangement when there are two pairs of bonding electrons. Similarly, if there are three pairs of bonding electrons, the configuration of the bond around the central atom would be at an angle of 120° and indeed would probably be in the same plane, since this structure would place the electrons farthest apart. All this is preliminary to our discussion of the carbon atom. Let us examine the methane (CH_4) molecule. There are four pairs of valence electrons forming bonds around the carbon atom. The configuration that provides the best separation of the bonding pairs is that of a regular tetrahedron, as shown in Figure 2-17, with angles of 109.5°.

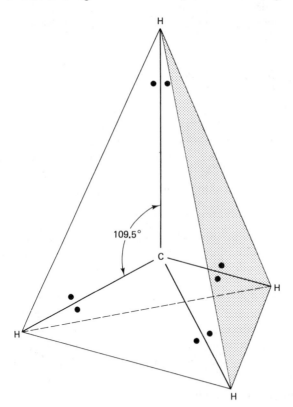

Figure 2-17 Schematic representation of the methane molecule. Each hydrogen atom lies at the apex of a regular tetrahedron forming a bond angle of 109.5°. This condition results in minimum electron repulsion (dark dots).

Resonance bonds. The covalent bonding we have described to this point illustrates how a pair of electrons holds two atoms together; however, there are cases where a pair of electrons provides the binding energy to hold a number of atoms together.

If we examine the structure of ozone (O_3), we see that there are two bonding arrangements:

The electronic distribution is such that the electrons oscillate between the nuclei and hold the atoms together. This fluidity of the electrons between the nuclei forms the basis for the concept of resonance.

Resonance occurs frequently in organic molecules. In the aromatic hydrocarbons such as benzene, for example, benzene (C_6H_6) has a hexagonal planar structure in which the bonds between the carbon atoms are in resonance as follows:

which can also be written

It should be noted that the resonance concept is useful in explaining not only structure, but also the chemical properties of compounds.

Molecular Arrangements

The linking of one atom to another by covalent bonding can continue and thereby form very large and complex molecular structures. Indeed, these complex molecular arrangements form the basis of the thousands of hydrocarbon compounds. In some, the carbon atoms are arranged in chains such as illustrated by the hexane molecule (C_6H_{14}):

In others, they are arranged in rings, as can be seen from an examination of the cyclohexane molecule (C_6H_{12}):

Table 2-7 describes the molecular arrangement and formulas for the various classes of hydrocarbons.

TABLE 2-7 CLASSES OF HYDROCARBONS

A. Carbon atoms in chains (aliphatic hydrocarbons)
    1. *Saturated*
       The alkanes: general formula, C_nH_{2n+2}; characterized by a single bond between the carbon atoms
    2. *Unsaturated*
       a. The alkenes: general formula, C_nH_{2n}; characterized by a double bond between two carbon atoms
       b. The alkynes: general formula, C_nH_{2n-2}; characterized by a triple bond between two carbon atoms

B. Carbon atoms in rings
    1. Aromatic hydrocarbons: those structures based on the benzene molecule (e.g., phenol)
    2. Polynuclear hydrocarbons: structures having two or more rings subdivided into (a) those with individual but connected rings (e.g., biphenyl), and (b) those with condensed rings (e.g., napthalene)
    3. Alicyclic hydrocarbons: cyclic compounds of an aliphatic nature; may have some unsaturated bonds (e.g., cyclopropane)

Isomers. Molecules of the same chemical composition can sometimes exhibit more than one atomic arrangement. These variations in structure, with the same chemical composition, are called isomers. The difference in molecular structure affects the properties of the molecule as illustrated in the pentane molecule. Note that all the molecules can be expressed by the formula C_5H_{12}.

H H H H H
| | | | |
H—C—C—C—C—C—H
| | | | |
H H H H H

Normal Pentane bp 36.2°C

H H H H
| | | |
H—C—C—C—C—H
| | | |
H | H H
H—C—H
|
H

Isopentane bp 28.0°C

H
|
H—C—H
H | H
| | |
H—C—C—C—H
| | |
H | H
H—C—H
|
H

Neopentane bp 9.5°C

STUDY PROBLEMS

2.1. Which electrons associated with a particular atom have the greatest energy: the electrons in the $1s$ shell or the electrons in the $3s$ shell? Explain your answer.

2.2. Determine the maximum number of electrons that can be contained in a shell with principal quantum number equal to 4 (the N shell).
Answer: 32

2.3. Using the abbreviated quantum number notation, write the electronic configurations of the following elements.
(a) Helium
(b) Sodium
(c) Silicon
(d) Calcium
(e) Germanium
(f) Molybdenum
(g) Silver

2.4. The abbreviated electronic configuration is given below for two elements. Identify these elements and comment on their common electronic characteristics that differentiate them from other elements.
Element 1: $1s^2, 2s^2, 2p^6, 3s^2, 3p^6$
Element 2: $1s^2, 2s^2, 2p^6, 3s^2, 3p^6, 3d^{10}, 4s^2, 4p^6$

2.5. Explain what is meant by the terms "electropositive" and "electronegative."

2.6. Based on their respective valence configurations and position in the periodic table, classify the following elements as either electropositive or electronegative. Compare your answers with the data inside the back cover.
(a) Aluminum
(b) Argon
(c) Iron
(d) Magnesium
(e) Oxygen
(f) Chlorine
(g) Silicon
(h) Tungsten
(i) Nitrogen
(j) Carbon

2.7. Write an expression that will describe the typical ionization process in the following elements.
(a) F
(b) Be
(c) Ag
(d) Cu
(e) N

2.8. What is meant by oxidation? Reduction? What effects do these processes have on the valence of an atom?

2.9. If a 1-cm³ sample of nickel contains 8.64×10^{22} atoms, how many electrons can be accommodated in the energy bands of the valence level?

2.10. Define the following terms.
  (a) Equilibrium separation   (b) Coordination number
  (c) Cohesive strength

2.11. Write an expression that expresses the ionization reactions in the following elements.
  (a) Calcium   (b) Aluminum
  (c) Copper   (d) Fluorine

2.12. Sketch the electronic configuration of the ions produced in problem 2.11 (see Figures 2-7 through 2-9). Identify the respective principal quantum shells for each ion.

2.13. If the interionic distance (a) between calcium and oxygen in CaO is 2.11 Å, what is the magnitude of the attractive forces developed between a pair of ions?
  Answer: $F_A = 2.07 \times 10^{-8}$ J/m

2.14. The attractive force between a pair of ions in KCl is determined to be 2.96×10^{-9} J/m (2.96×10^{-9} V-C/m). What is the interionic separation (a) for this condition?

2.15. Using the Lewis electron-dot method (Figure 2-12), sketch the covalent bonding configuration for the following molecules.
  (a) HF   (b) F_2
  (c) HCl

2.16. What unique properties do metals display? Explain why the metallic bond is largely responsible for the physical characteristics of metals.

2.17. Compare the relative differences in strength, electrical properties, and optical properties that result from ionic, covalent, and metallic bonding.

2.18. Chloromethane has a structure similar to methane except that the structure is distorted by the presence of a chlorine atom resulting in a polar molecule. If the angle between the carbon and hydrogen atoms is 110°, calculate the H—C—Cl angle.

2.19. Calculate the energy change that results when 100 g of vinyl chloride (C_2H_3Cl) is polymerized to form polyvinyl chloride. Is this energy absorbed or given off?

2.20. Molecular structures can vary from the very simple to those possessing multiple covalent bonds. These varying bonds have different bond strengths which affect the melting points of the materials involved. Several common organic materials are listed below. Draw their structures and arrange them in order of their expected melting points. Explain your rationale for the ranking.
  (a) C_2H_4, ethylene
  (b) C_2H_3Cl, vinyl chloride
  (c) $(C_2H_4)_n$, polymerized ethylene molecule
  (d) C_2H_6, ethane

2.21. Analysis of a chlorinated hydrocarbon reveals that it has the following composition: 14 weight percent (w/o) carbon, 2.1 w/o hydrogen, and 83.9 w/o chlorine. What is the most likely compound with this analysis?

2.22. Butane (C_4H_{10}) exists in various isometric forms.
  (a) Sketch the isomers possible and list them in order of the expected boiling points.
  (b) What do you believe is the basis for the variation in physical properties?

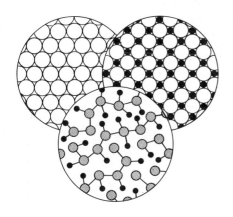

CHAPTER THREE:
Elements of Crystalline Structures

TYPES OF CRYSTAL STRUCTURES

(MAJORITY)

The preponderance of engineering and structural materials, with the exception of most glasses and polymers, are crystalline in nature. That is, they exhibit a repetitive or periodic, orderly arrangement of atoms (or ions) bonded together in three-dimensions. Although there are seven crystal systems as shown in Table 3-1, we will be concerned primarily with the cubic and the hexagonal systems, because they comprise the bulk of commercially important crystalline engineering materials.

TABLE 3-1 CRYSTAL SYSTEMS

Type of system	Length of axes	Axial angles
Cubic	$a_1 = a_2 = a_3$	All $= 90°$
Tetragonal	$a_1 = a_2 \neq a_3$	All $= 90°$
Orthorhombic	$a \neq b \neq c$	All $= 90°$
Monoclinic	$a \neq b \neq c$	$2 \gtrless = 90°, 1 \gtrless \neq 90°$
Triclinic	$a \neq b \neq c$	All different, none $= 90°$
Hexagonal	$a_1 = a_2 = a_3 \neq c$	Angles $= 90°$ and $120°$
Rhombohedral	$a_1 = a_2 = a_3$	All equal, but not $90°$

The spatial arrangement of equivalent points in a crystal system is called a *space lattice*. In a crystalline solid, these points are the positions of atoms on a space lattice. When this arrangement comprises the smallest repeating network of sites in the crystal, it is called a *unit cell*. As demonstrated in Figure 3-1, 14 space lattices may be constructed from the seven crystal systems. In other words, only 14 ways exist in

45

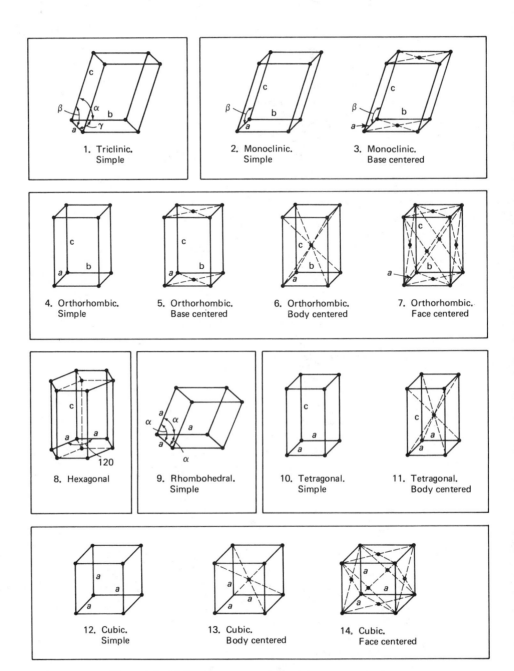

Figure 3-1 Unit cell representations of the 14 space lattices.

which points (or atoms located at these points) may be spatially arranged so that each site has similar surroundings. The coordinates of certain lattice positions in a cubic unit cell are shown in terms of the unit cell edge (a), also called the unit cell parameter, in Figure 3-2. A rather unique example of atoms arranged in a crystalline

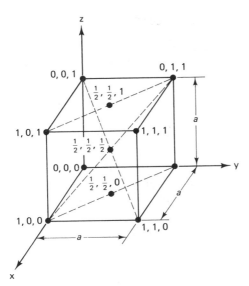

Figure 3-2 Selected lattice positions in Cartesian coordinates, for a cubic unit cell.

lattice is illustrated by Figure 3-3. This photograph shows the arrangement of atoms in the tip of a platinum needle as imaged in the *field-ion microscope*. In this device, helium atoms interact with the atoms in the lattice and become positively charged ions. These particles then travel in straight lines and impinge on a photographic film, yielding an image of the atomic arrangement in the metal.

Crystalline engineering materials that you are probably familiar with (e.g., steel,

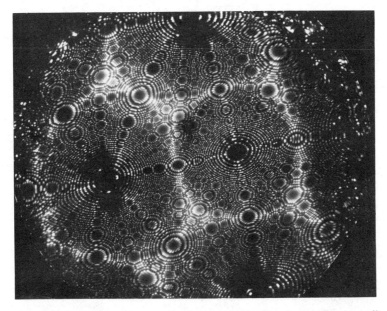

Figure 3-3 Crystalline arrangement of atoms in the tip of a platinum needle. Light spots indicate the positions of atoms in the crystal lattice (magnification: about 770,000 ×). (Courtesy of R. W. Newman, Georgia Institute of Technology)

aluminum, cement, porcelain, and diamond) consist of atoms arranged in various three-dimensional space lattices.

Simple Cubic

Although simple cubic arrangements of atoms do not ordinarily exist in engineering materials, they are an excellent example to introduce the subject of crystal structure because of their simplicity. Once the student comprehends the details and relationships in this elementary structure, we will move into the more complex but realistic systems of atomic arrangement.

Basically, the simple cubic crystal consists of a "square box" with the centers of atoms positioned on the eight corners. Even though this is actually a spatial arrangement, we can envision the crystal as a cube with definable sides for the purpose of our discussion. The eight atoms located on the corners are positioned so that the distance between their centers is *a*—the *lattice constant* or *lattice parameter*. Such a crystal is illustrated in Figure 3-4. This model displays corner atoms relatively far apart for the sake of clarity at this point in our study.

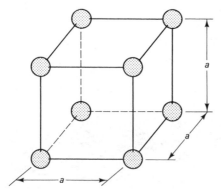

Figure 3-4 Simple cubic unit cell. (*a*) is the lattice parameter.

Further inspection of our simple cubic model reveals that even though we have associated the eight atoms with one cube—the *unit cell*—only a portion of each atom is actually contained within that particular unit cell! How much of a corner atom belongs to the unit cell? If we picture a section through an array of cubic unit cells as illustrated in Figure 3-5, it becomes clear that the atoms are shared equally among neighboring unit cells. Take, for example, the atom denoted with numbers. Its four quadrants shown are shared equally by the four adjacent cubes, as indicated by the numbers 1 to 4. But isn't this view a cross section through the atoms in the plane of the paper? Therefore, the same condition, that is, four equally shared quadrants of the atom, exists on the other side of this plane (out of the page). Thus each atom on the corners of this cubic cell is shared equally by *eight* cubes or unit cells. In other words, any individual unit cell contains one-eighth of an atom in each of the eight corners. We can now state that, in effect, one atom is associated with the simple cubic unit cell, or the unit cell contains one atom. This relation can be expressed as follows:

$$(\tfrac{1}{8} \text{ atom per corner})(8 \text{ corner atoms}) = 1 \text{ atom/unit cell}$$

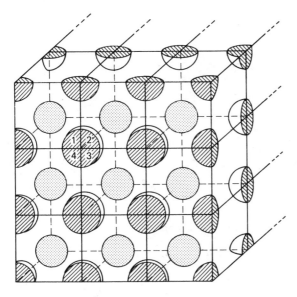

Figure 3-5 Three-dimensional illustration of cubic unit cells.

The student will soon realize that this relationship regarding corner atoms holds for all the cubic crystal structures.

Atomic packing factor. It is generally accepted that nature favors arrangements and states of matter that tend to minimize energy levels (potential energy). We may therefore question how efficient the simple cubic type of crystalline arrangement really is. To evaluate the efficiency of the crystal structure or how efficiently the atoms are arranged, we will compute the ratio of atomic volume contained within the unit cell to the total volume of the unit cell. This ratio, known as the *atomic packing factor* (APF) is expressed as follows:

$$\text{APF} = \frac{\text{volume of atomic material}}{\text{volume of unit cell}} \tag{3-1}$$

For example, let us calculate the APF for the simple cubic crystal structure in Figure 3-4. Previously, we determined that this unit cell contains the equivalent of one atom (one-eighth contribution from each corner). Since our model of the atom is a sphere, we can compute its volume as follows:

$$V_s = \tfrac{4}{3}\pi r^3$$

where r is the radius of the sphere. As illustrated in Figure 3-6, the atoms tend to approach each other, with a being the distance between their centers. Thus we can express r in terms of a as follows:

$$a = 2r$$

or

$$r = \frac{a}{2}$$

Types of Crystal Structures

49

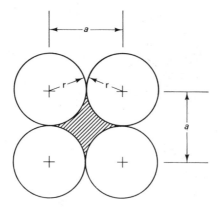

Figure 3-6 Plane view of one face of the cubic unit cell. Shaded area represents interstitial space between atoms.

The expression for volume in terms of a becomes

$$V_s = \frac{4\pi}{3}\left(\frac{a}{2}\right)^3 = \frac{4\pi}{3}\frac{a^3}{8}$$

which reduces to

$$V_s = \frac{\pi a^3}{6}$$

Substituting this relationship for volume in equation (3-1), we obtain the following result:

$$\text{APF} = \frac{(1 \text{ atom/unit cell})(\pi a^3)}{(6)(a^3)}$$

$$= 0.52 \quad \text{"simple cubic crystal"}$$

Since this value of APF is the ratio of utilized or occupied space to the total space available in the unit cell, in the simple cubic case above only 52% of the unit cell is actually occupied by atoms (mass). Such inefficiency results from the interstitial space (unoccupied) produced by the arrangement of atoms as depicted in Figure 3-6. The student is encouraged to experiment with solid spheres such as Ping-Pong or tennis balls, arranging them in simple cubic fashion. Observe the space in the central region of the "unit cell" you have made. This is analogous to the interstitial space in the crystal structure. A structure that essentially does not utilize 48% of its available space is very inefficient, and this is the primary reason such a crystalline arrangement is avoided by real engineering materials.

Coordination number. Another indication of atomic packing efficiency is the number of nearest neighbors an atom has. The number of these neighboring atoms is called the coordination number (CN). The higher the CN, the more efficient the atomic packing arrangement in a crystal and the more stable the structure.

As an example of coordination number, let us consider the number of nearest neighbors for an atom in the simple cubic structure. Figure 3-7 shows this structure with the dark corner atom under consideration. The nearest neighbors are shaded to distinguish them from other atoms in the lattice. There are six nearest neighbors, three on the corners of the unit cell shown and three associated with other unit cells, as

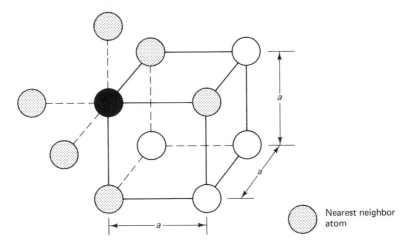

Nearest neighbor atom

Figure 3-7 Schematic illustration of the simple cubic structure showing nearest neighbor atoms—coordination number.

indicated by the dashed lines. The CN for the simple cubic structure therefore equals six.

Body-Centered Cubic

The body-centered type of crystal structure is just slightly more complex than the simple cubic. As the term implies, an atom is centered in the cubic structure equidistant from the eight corner atoms. This central atom is completely contained within the unit cell boundaries, hence the name "body-centered." Schematically, this crystal arrangement is shown in Figure 3-8. Again to assist perception of this model, we have positioned the atoms relatively far apart in Figure 3-8(a). The lattice parameter is a for all sides of the cube.

How many atoms per unit cell does the body-centered cubic (bcc) crystal structure contain? We know from the previous example (simple cubic) that the corner atoms contribute one-eighth of their volume and that there are eight corner atoms. Furthermore, the body-centered atom is completely contained within the unit cell. The number of atoms per unit cell can therefore be stated as follows:

$$\text{Corners:} \quad \tfrac{1}{8} \times 8 \text{ atoms} = 1 \text{ atom}$$

$$\text{Center:} \quad 1 \text{ atom} = \underline{1 \text{ atom}}$$

$$\text{Total} = 2 \text{ atoms/unit cell}$$

What about the coordination number for this crystal structure? Examination of the arrangement in Figure 3-8 shows that the body-centered atom has eight nearest neighbors (the corner atoms). Thus CN = 8 for bcc crystals. Why can we state this unequivocally? Although it is not completely evident from our illustration of one bcc unit cell, remember that each atom in the crystal is surrounded by an identical arrangement of atoms in the repeating space lattice.

Types of Crystal Structures

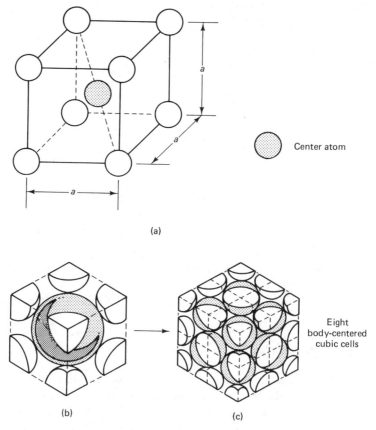

Center atom

(a)

(b)　　　　　(c)

Eight
body-centered
cubic cells

Figure 3-8 Schematic models of the body-centered cubic structure (bcc): (a) open arrangement; (b) atoms approaching each other (distance of closest approach); (c) space lattice arrangement for bcc.

Recall also that CN is an indication of atomic packing efficiency. Since CN = 8 is greater than that for the simple cubic (CN = 6), we can expect the APF for bcc to be greater also. To determine the APF for bcc crystals we must establish a relationship between the lattice parameter (a) and the atomic radius (r). Although we obtained such a relation for the simple cubic, it does not apply to the bcc, because a body-centered atom is present and the unit cell dimensions are changed. This feature can be appreciated from the representation shown in Figure 3-8(b).

The application of some trigonometry to the bcc crystal will assist the development of a relationship between a and r. Referring to Figure 3-9(a), construct two diagonal lines in this unit cell: one across the bottom, labeled B, and the other from one corner to the opposite, labeled C. The bottom diagonal is expressed as follows:

$$B = \sqrt{a^2 + a^2}$$
$$= \sqrt{2a^2}$$
$$= \sqrt{2}\, a$$

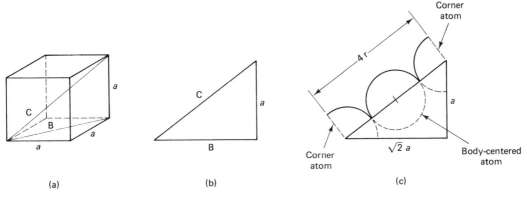

Figure 3-9 Trigonometric representation of a main diagonal through the bcc unit cell.

The value of B, in turn, becomes the bottom "leg" of the triangle formed by the lines B, C, and a. For the purpose of simplifying this analysis, we can remove the subject triangle from the unit cell as shown in Figure 3-9(b). Now we can determine the body diagonal (C) in terms of a as follows:

$$C = \sqrt{B^2 + a^2}$$

Substituting for B gives us

$$C = \sqrt{(\sqrt{2}\,a)^2 + a^2}$$

Simplifying we have

$$C = \sqrt{2a^2 + a^2}$$
$$= \sqrt{3a^2} = \sqrt{3}\,a \tag{3-2}$$

The result of our calculation is the body diagonal C in terms of the lattice parameter a. But in order to compute the APF we need a in terms of r, the atomic radius. Consider the body diagonal C in the same manner as Figure 3-9(b). If we position atoms along this line and allow them to approach, as they do in the real crystal, we can relate the atomic radius r to a as shown in Figure 3-9(c). This schematic demonstrates that the central atom contributes $2r$ while the corner atoms each contribute r, for a total of $4r$. We may now express this relationship as follows:

$$C = 4r = \sqrt{3}\,a \qquad \text{from equation} \tag{3-2}$$

or

$$a_{bcc} = \frac{4r}{\sqrt{3}} \tag{3-3}$$

Now we are in a position to determine the packing efficiency for the bcc structure. Recalling from equation (3-1) that the APF is expressed as follows:

$$APF = \frac{\text{volume of atoms}}{\text{volume of unit cell}}$$

Types of Crystal Structures

for bcc crystal structures,

$$\text{APF} = \frac{(2 \text{ atoms/unit cell})(\frac{4}{3}\pi r^3)}{a^3}$$

Substituting for r in terms of a from equation (3-3), we have

$$\text{APF} = \frac{(2) \frac{4}{3}\pi(\sqrt{3}\ a/4)^3}{a^3}$$

This expression is simplified to

$$\text{APF} = \frac{(8\pi)3\sqrt{3}\ a^3}{(3)(64)a^3}$$

which yields the following value:

$$\text{APF}_{bcc} = 0.68$$

The foregoing calculation of APF for body-centered crystals demonstrates that this system is considerably more efficient than the simple cubic in its packing arrangement. Of the available space in the unit cell, 68% is occupied in the body-centered structure compared to 52% in the simple cubic structure. In spite of this increase in packing efficiency, the body-centered cubic structure still contains a significant volume of unoccupied space. The unoccupied space in bcc crystals is quite evident as previously shown in Figure 3-8 and results from the interstices created between adjacent atoms in the lattice. Not only is this factor important with respect to the APF, but as we will see in later topics, the interstitial space has a significant effect on such properties as crystalline defects, alloying, and diffusion in the solid state.

Certain elements that exhibit the body-centered cubic crystal structure are listed inside the back cover, along with their respective atomic radius values (r).

Face-Centered Cubic

Still another type of cubic crystal arrangement is the face-centered cubic (fcc). It differs from the previous crystal structures that we have discussed in that each face of the cube has an atom positioned on it. A unit cell of the fcc system is shown schematically in Figure 3-10. Again, the cellular arrangement is intentionally spread apart so that the location of the atoms is clear. The lattice parameter or constant is indicated by a.

Inspection of this unit cell reveals that in addition to the eight corner atoms shared by the cell, six face-centered atoms are shared between this cell and its immediate neighbors. Half of any face atom is in one unit cell, the other half in an adjoining cell. Therefore, we can determine the number of atoms associated with the unit cell of this system as follows:

$$\text{Corners:} \quad 8 \text{ atoms} \times \tfrac{1}{8} = 1 \text{ atom}$$
$$\text{Faces:} \quad 6 \text{ atoms} \times \tfrac{1}{2} = 3 \text{ atoms}$$
$$\text{Total} = 4 \text{ atoms/unit cell}$$

Notice that the number of atoms per unit cell has doubled from that of the bcc system, which contained the equivalent of two atoms per unit cell. Certainly, this is an indication of more efficient atomic packing in the lattice.

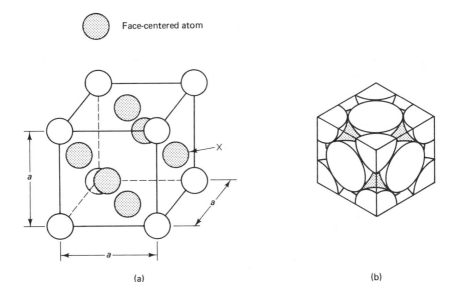

Face-centered atom

(a)   (b)

Figure 3-10 Fcc unit cell: (a) atoms separated for clarity; (b) atoms situated at distance of closest approach.

The coordination number for the fcc system can be determined from examination of Figure 3-10. For example, consider the face-centered atom marked by an $\times$. It is directly surrounded by four corner atoms in the same plane of the cube. In addition, it is adjacent to the four face-centered atoms in the unit cell pictured (the exception is the opposite face atom). So far this yields eight nearest neighbors. But the adjoining unit cell to the right (not pictured) also contributes four face-centered atoms to the coordination of our principal atom ($\times$). Thus the sum of next nearest neighboring atoms is 12. We can now state that the coordination number for fcc = 12. Again, because of identical surroundings, all atoms in this crystal structure have the same coordination.

Remember that CN indicates the efficiency of atomic packing in the crystal structure. A CN of 12, the highest number we have determined yet, implies very efficient packing of atoms in a crystal. In fact, the fcc system is commonly referred to as a *close-packed* structure. Such a close-packing arrangement is indicated in Figure 3-11, where three layers or planes of atoms are stacked on each other. Inspection of

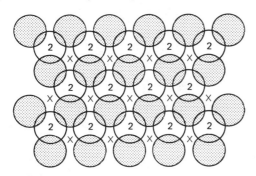

Figure 3-11 Top view of stacking sequence in fcc structure. Alternating planes of atoms consist of shaded atoms in first plane, number 2 atoms in second plane, and $\times$'s locate the centers of atoms in the third plane.

Types of Crystal Structures

55

this scheme shows that the atoms in any one plane lie in the interstices of the atoms in adjacent planes.

The close-packing condition therefore contributes to a greater density of atoms in the crystal structure. We can demonstrate this fact quantitatively by computing the APF for fcc.

First, let us relate the lattice parameter (a) to the radius of an atom in this structure. Figure 3-12(a) displays the fcc unit cell from a planar view. Examination of this

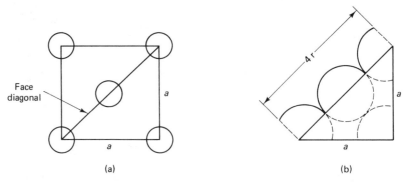

Figure 3-12 Trigonometric relationships in the fcc unit cell: (a) schematic view of one face; (b) atoms located at distance of closest approach.

structure shows that a face diagonal intersects one face-centered atom along its diameter. Now if we allow these atoms to approach and make contact, as illustrated in Figure 3-12(b), a relationship between a and r is developed as follows:

$$4r = \sqrt{a^2 + a^2}$$
$$= \sqrt{2a^2}$$
$$= \sqrt{2}\, a$$

Solving for a, we have

$$a_{fcc} = \frac{4r}{\sqrt{2}}$$

Utilizing this relationship and equation (3-1), we obtain the following expression:

$$\text{APF}_{fcc} = \frac{(4 \text{ atoms/unit cell})(\frac{4}{3}\pi r^3)}{a^3}$$

where $r = \sqrt{2}\,a/4$. Thus

$$\text{APF}_{fcc} = \frac{(4)(4\pi)}{3a^3}\left[\frac{\sqrt{2}\, a}{4}\right]^3$$

$$= \frac{(16\pi)(2\sqrt{2}\, a^3)}{192a^3}$$

$$= 0.74$$

This result confirms our previous information concerning the packing efficiency of this crystal system: 74% of the available space in a unit cell is occupied by atoms. Certain properties of some elements that form face-centered cubic crystals are listed inside the back cover.

Hexagonal Close-Packed

The third primary type of crystal structure we will discuss is the hexagonal close-packed system. As the term "close-packed" implies, this crystal structure displays a high-density packing arrangement of atoms, similar to the fcc. However, this system differs from the fcc arrangement in that the atoms in a third layer or plane lie directly over the atoms in the first plane, as shown in Figure 3-13.

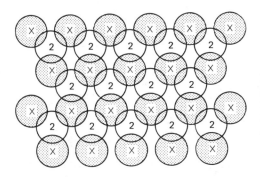

Figure 3-13 Top view of stacking sequence in hcp structure. Atoms in first plane are shaded. Number 2 atoms located in second plane. Position of atoms in the third plane denoted by ×'s. Note that atoms in third plane are directly over those in first plane.

The hexagonal close-packed (hcp) unit cell is illustrated in Figure 3-14. Inspection of this crystal reveals seven atoms in each basal plane, plus three atoms positioned between the basal planes on axes 120° apart. These central atoms lie between the interstices of the basal plane atoms as shown by the ×'s.

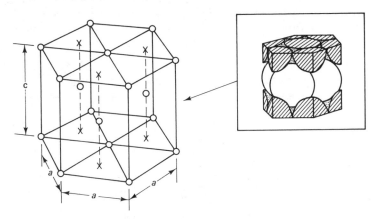

Figure 3-14 Hcp structure unit cell. Atoms in hard sphere model of unit cell (inset).

The coordination number for hcp crystals is the same as fcc and equals 12. We can verify this number of nearest neighbors by considering the central atom in a basal plane. This atom is surrounded by six atoms in its own basal plane, plus three atoms located in each adjacent parallel plane, for a total of 12 nearest neighbors. The atomic packing factor (APF) for hcp is also the same as the fcc system and equals 0.74. Incidentally, these two crystal structures are referred to as close-packed because they are the two ways in which identical spheres can be arranged with the greatest possible density and still exhibit periodicity. The number of atoms per unit cell can be determined if we consider a portion of the unit cell (actually one-third), as shown in Figure 3-15. This parallelepiped unit cell arrangement contains the following atoms:

$$1 \text{ atom inside} = 1$$

$$\text{corner atoms} = \underline{1}$$

$$\text{total} = \overline{2} \text{ atoms per unit}$$

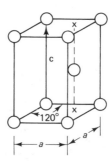

Figure 3-15 Schematic of the unit cell in the hexagonal system (one-third of the hcp structure).

But this is one-third of the hcp unit cell, so the total number of atoms associated with the hcp unit cell equals six.

A relationship exists between the lattice parameters c and a in the hcp structure. For spherical atoms that make contact in this unit cell, the ratio c/a equals 1.633. However, measurements of this parameter show values ranging from approximately 1.586 (Be) to 1.886 (Cd). This observation indicates that the hcp atoms are not perfect spheres, but ellipsoidal in shape. Some c/a values are presented in Table 3-2.

TABLE 3-2 LATTICE DATA: hcp CRYSTAL STRUCTURES

Metal	$c/2a$	c/a	
Cd	0.943	1.886	Prolate
Zn	0.928	1.856	spheroids
"Sphere"	0.816	1.633	
Mg	0.812	1.624	
Zr	0.795	1.590	Slightly
Ti	0.794	1.588	compressed
Be	0.793	1.586	

Selected elements that form hexagonal close-packed crystal structures, together with their properties, are also included inside the back cover.

Certain information regarding the principal crystal systems that we have analyzed will surface consistently in our studies of engineering materials. It will be beneficial, therefore, to summarize the important parameters and values concerning the two cubic systems, body-centered and face-centered, in addition to those related to the hexagonal close-packed structure. These data are presented in Table 3-3.

TABLE 3-3 SUMMARY OF CERTAIN CRYSTAL STRUCTURES DATA

System	Lattice parameter relationship	Number of atoms per unit cell	APF	CN
bcc	$a = \dfrac{4r}{\sqrt{3}}$	2	0.68	8
fcc	$a = \dfrac{4r}{\sqrt{2}}$	4	0.74	12
hcp	$a = 2r$ $c/a = 1.63$	6	0.74	12

Polymorphic Crystals

Some materials exhibit more than one type of crystal structure depending primarily on temperature and sometimes on pressure, and severe deformation. These materials are said to be *polymorphic (allotropic)*. Such a transformation of the crystalline structure without any change in chemical composition can occur because in certain temperature ranges, one particular arrangement of atoms is more stable than another. These transformations take place in many elements and materials involving many different types of crystal structures. Table 3-4 lists some examples of polymorphic metals and their temperature ranges.

TABLE 3-4 EXAMPLES OF POLYMORPHIC ELEMENTS

	Temperature	
	°C	°F
IRON		
Body-centered cubic α	< 906	< 1663
Face-centered cubic γ	906–1403	1663–2557
Body-centered cubic δ	1403–1535*	2557–2795
COBALT		
Close-packed hexagonal	< 420	< 788
Face-centered cubic	420–1495*	788–2723
TIN		
Cubic α (grey)	< 13	56
Tetragonal β (white)	13–232*	56–450
MANGANESE		
Cubic, 58 atoms per unit cell α	< 678	< 1252
Cubic, 20 atoms per unit cell β	678–1100	1252–2012
Tetragonal γ	1100–1138	2012–2080
Tetragonal δ	1138–1245*	2080–2273
CHROMIUM		
Close-packed hexagonal β	< 20	< 68
Body-centered cubic α	20–1799 ± 50	68–3270 ± 90
TITANIUM		
Close-packed hexagonal α	< 879	< 1615
Body-centered cubic β	879–1727*	1615–3140

*Melting point

EFFECTS ON DENSITY

Now that the student has become familiar with the crystal structures that comprise most engineering materials, we can apply this background to the determination of weight density of these metals. *Density* is defined as weight per unit volume; therefore, in the crystals just studied, we may conveniently apply this concept to the unit cell. The density of a material in this case would be: weight of atoms per unit cell divided by the volume of that particular unit cell. Such a definition of density implies perfect crystals with no imperfections, in other words, a perfect space lattice. Unfortunately, this condition is very difficult to obtain in most engineering metals. As you will see in Chapter 4, the periodicity of the lattice is frequently interrupted by crystalline imperfections which take up space or volume but add no mass. Therefore, the theoretical density based on our knowledge of crystal structure will be slightly greater than the actual density.

To perform our density calculation, we must determine the weight of atoms in a unit cell. Simply, this is the number of atoms per unit cell (which we know) times the weight per atom. What is the weight of an atom? We can solve this dilemma because the weight in grams of 6.02×10^{23} atoms of any element is its gram atomic weight. It is convenient to use atomic mass units (amu) in this type of calculation because the mass of an atom is so small. There are 6.02×10^{23} amu per gram of any element, or we can express the weight of an element as its atomic weight in grams divided by Avogadro's number. For example, carbon 12 has 12.01 amu per atom or 12.01 g per 6.02×10^{23} atoms. The number 6.02×10^{23} is called *Avogadro's number* (A_0).

We can express our density relationship as follows:

$$D = \frac{(\text{no. atoms/unit cell})(\text{atomic weight/Avogadro's no.})}{\text{volume of unit cell}} \tag{3-4}$$

Example 3-1

Copper (atomic number 29) has a fcc crystal structure and its atomic weight equals 63.54. If the approximate atomic radius of Cu = 1.278 Å, determine its weight density.

Solution

$$D = \frac{(\text{no. atoms/unit cell})(\text{atomic weight}/A_0)}{\text{volume of unit cell}}$$

$$D_{Cu} = \frac{(4 \text{ atoms})(63.54 \text{ amu/atom})}{(a^3)(6.02 \times 10^{23} \text{ amu/g})}$$

We previously related a_{fcc}, the lattice parameter, to the atomic radius as follows:

$$a_{fcc} = \frac{4r}{\sqrt{2}}$$

So for Cu this value is

$$a_{Cu} = \frac{(4)(1.278)\text{Å}}{\sqrt{2}}$$

$$= 3.61\text{Å} = 3.61 \times 10^{-8} \text{ cm}$$

Therefore,

$$a_{Cu}^3 = (3.61 \times 10^{-8} \text{ cm})^3$$

$$= 47.2 \times 10^{-24} \text{ cm}^3$$

Handwritten margin note (left side, rotated):

English $1 \overset{\circ}{A} = 10^{-8}$ cm

metric nanometer $NM = 10^{-9}$ m

Substituting this result in our expression for density, we obtain

$$D_{Cu} = \frac{(4 \text{ atoms})(63.54 \text{ amu/atom})}{(47.2 \times 10^{-24} \text{ cm}^3)(6.02 \times 10^{23} \text{ amu/g})}$$

$$= 8.94 \text{ g/cm}^3 \quad Ans.$$

Comment: In the interests of comprehension, we have included the units for all quantities in the calculation. The student should note that the units cancel appropriately, leaving g/cm³ or weight per unit volume. This answer may be compared with the value listed inside the back cover.

We may also determine the density for a hcp structure by considering the unit cell pictured in Figure 3-15. Although this is not the complete hcp structure commonly illustrated (Figure 3-14), as we stated earlier, this arrangement can also represent the hcp packing on a more elementary level. An example calculation may help to demonstrate this aspect.

Example 3-2

Titanium (Ti) exhibits a hcp crystal structure and has atomic weight of 47.90. Based on Figure 3-15, the "unit cell" contains two atoms and has a volume $= a^2 c \sin \theta$, where $\theta = 120°$. Calculate the weight density for Ti.

Solution First, let us determine the volume of our parallelepiped unit cell:

$$V = a^2 c \sin \theta$$

where a = lattice parameter = 2.950 Å
c = height of cell = 4.686 Å
$\sin \theta = \sin 120° = 0.866$
$V = (2.950 \text{ Å})^2(4.686 \text{ Å})(0.866)$
$= 35.3 \text{ Å}^3 = 35.3 \times 10^{-24} \text{ cm}^3$

Substituting this result and the information given above into equation (3-4), we obtain

$$D_{Ti} = \frac{(2 \text{ atoms})(47.90 \text{ amu/atom})}{(35.3 \times 10^{-24} \text{ cm}^3)(6.02 \times 10^{23} \text{ amu/g})}$$

$$= 4.51 \text{ g/cm}^3 \quad Ans.$$

As we pointed out, the density value obtained for Ti was calculated based on the rhombic arrangement shown in Figure 3-15. This representation of the hcp unit cell comprises one-third of the structure pictured in Figure 3-14. Will the answer above hold for the hexagonal close-packed structure, which contains six atoms per unit cell? Consider the information just stated. The hexagonal close-packed unit cell has a volume three times greater than the parallelepiped. It also contains three times as many atoms. What effect does this have on the density computation?

CRYSTAL PLANES

In earlier discussions of crystal structures we mentioned the process of building up unit cells in 3-dimensions as atoms joined the crystal. These "building blocks" stack side by side, repeating in space until their progress is impeded or a lack of atoms halts the process. Because of their symmetry and repetitive nature, crystalline materials contain planes of atoms which can be defined by Cartesian coordinates. Identification of planes in certain crystal structures and the atoms associated with these planes can assist our understanding of the influence atoms and crystal structure exert on mechanical or physical properties in engineering materials.

Identification of Planes

Ready identification of the planes in a crystal is accomplished by the use of *Miller indices*. By definition, the Miller indices of a plane are the reciprocal values of the plane's intercepts on the x, y, and z axes. These intercepts are measured in terms of unit cell distances from the Cartesian origin, wherever it is located. The identifying numbers of a plane, its Miller indices, are expressed in parentheses, as (hkl), where h, k, and l are the reciprocals of the intercepts on the x, y, and z axes, respectively, cleared of fractions. Families of crystal planes may also be designated by the use of Miller indices. In this sense, planes that belong to the same family are identical except for their orientation with respect to the Cartesian axes. Planes belonging to the same family are expressed in braces as follows: $\{hkl\}$, to differentiate them from a single plane.

An example will help illustrate these points. Consider the cubic unit cell pictured in Figure 3-16 without atoms for the sake of clarity. Let us describe in terms of Miller indices the planes designated a, b, and c. Note that these planes are mutually orthogonal; that is, they are oriented 90° with respect to each other. Although we have shown these planes as the faces of a unit cube, in reality each plane extends outward in two perpendicular directions within that plane. For instance, the plane labeled c extends outward in both the y and z directions. Furthermore, if a plane is parallel to one of the Cartesian axes, we define its intercept to be infinity (∞). Thus the Miller indices of our subject cube can be written as follows:

Plane a:

$$h = \frac{1}{x \text{ intercept}} = \frac{1}{\infty} = 0$$

$$k = \frac{1}{y \text{ intercept}} = \frac{1}{\infty} = 0$$

$$l = \frac{1}{z \text{ intercept}} = \frac{1}{1} = 1$$

Therefore, we can express plane a as the (001) plane.

Plane b:

$$h = \frac{1}{x \text{ intercept}} = \frac{1}{\infty} = 0$$

$$k = \frac{1}{y \text{ intercept}} = \frac{1}{1} = 1$$

$$l = \frac{1}{z \text{ intercept}} = \frac{1}{\infty} = 0$$

Plane b can therefore be written as the (010) plane.

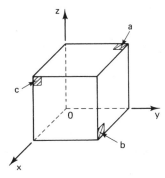

Figure 3-16 Cubic unit cell representation showing three mutually orthogonal planes (*a*, *b*, *c*) in the Cartesian coordinate system.

Plane *c*:

$$h = \frac{1}{x \text{ intercept}} = \frac{1}{1} = 1$$

$$k = \frac{1}{y \text{ intercept}} = \frac{1}{\infty} = 0$$

$$l = \frac{1}{z \text{ intercept}} = \frac{1}{\infty} = 0$$

So plane *c* is described as the (100) plane.

We have numerically described three of the six faces of the cubic unit cell in terms of the Miller indices of their planes in space. Furthermore, these planes are identical, as are the remaining three unidentified faces. Thus we can state that the faces of the cubic unit cell belong to the {100}, that is, the 100 family. From Figure 3-16 we can see that the {100} family of planes consists of:

	a	*b*	*c*
Front	(001)	(010)	(100)
Back	(00$\bar{1}$)	(0$\bar{1}$0)	($\bar{1}$00)

The identification of planes coinciding with the Cartesian axes is accomplished by a unit translation of the origin.

The concept of crystal-plane identification is beneficial because we can now discuss and communicate information regarding these planes without confusion as to their location and orientation.

As this point it is worth emphasizing that the location of the origin is arbitrary and that identical, parallel planes may be expressed by the same Miller indices. For example, in Figure 3-16, plane *a*—the (001) plane—is parallel and identical to the top face of every unit cell stacked directly on the subject cube (e.g., see Figure 3-23). Such planes can be reduced to the (001) designation, that is, the lowest set of integers for these indices.

Since this is perhaps a new concept for the student, some examples of planes with respect to a cubic system may be beneficial. Figure 3-17 illustrates various planes

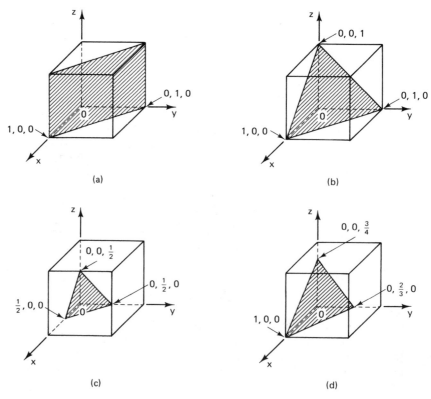

Figure 3-17 Unit cells in the cubic system illustrating various planes: (a) the (110); (b) the (111); (c) the (222); (d) the (698).

with their corresponding Miller indices (*hkl*). Another example of planes in the cubic system is shown in Figure 3-18. This series of planes from the {110} demonstrates two important points: (1) negative intercepts result in negative Miller indices (shown by bars over the indices); and (2) a plane passing through the origin cannot be designated by Miller indices. This is not a significant problem, since placement of the origin is arbitrary and as the figure indicates, a unit translation of the origin allows us to circumvent this problem.

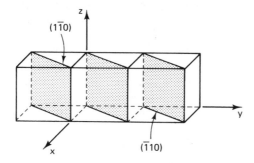

Figure 3-18 Example of parallel planes in the cubic system.

Hexagonal indices. The identification of crystal planes in the hexagonal system can be based on the three axes a_1, a_2, and c, or on four axes a_1, a_2, a_3, and c, where the three planar axes are located 60° apart, rather than 120° as in the former case. Regardless of which technique is employed, the reciprocals of the intercepts are determined and reduced to the smallest integers. The indices are expressed as ($hkil$) in the case of four intercepts, where the first three integers are related by the following expression:

$$i = -(h + k)$$

An example of crystal planes in the hcp structure is illustrated in Figure 3-19. Note that the origin is placed in the center of the bottom basal plane in this structure and that the basal planes have the indices (0001).

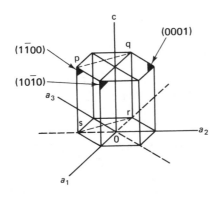

Figure 3-19 Example of crystal planes in a hexagonal unit cell.

Example 3-3

Determine the Miller indices of the shaded planes illustrated in Figure 3-17.

Solution The Miller indices of the planes in question are determined as follows:

	Intercept	Reciprocal	Miller indices
(a)	$x = 1$	1	(110)
	$y = 1$	1	
	$z = \infty$	0	
(b)	$x = 1$	1	(111)
	$y = 1$	1	
	$z = 1$	1	
(c)	$x = \frac{1}{2}$	2	(222)≡ (111)
	$y = \frac{1}{2}$	2	
	$z = \frac{1}{2}$	2	
(d)	$x = 1$	1	(698)
	$y = \frac{2}{3}$	$\frac{3}{2}$	
	$z = \frac{2}{4}$	$\frac{4}{3}$	

The set of indices for part (d) results when the reciprocal intercepts are cleared of fractions.

Crystal Planes and Deformation

The subject of crystallographic planes is pertinent to an understanding of deformation and fracture processes in engineering materials. Our representations of crystal planes and their identification with respect to Cartesian coordinates will now enable us to examine briefly the role of planes in deformation, from an elementary standpoint.

Certain planes in crystal structures are more important than others in terms of their capacity to displace or deform permanently. *Permanent deformation* (also called *plastic deformation*) simply means that the atoms of a material are permanently displaced from their equilibrium lattice positions. Details of such deformation will be discussed in later chapters. The parallel planes that contain the greatest density of atoms in a crystal tend to displace or move with respect to each other more easily than do other planes in the crystal. This circumstance occurs because the planes of greatest atomic density are also the planes with the *greatest interplanar spacing*. Interplanar spacing is the distance between parallel planes, as one can observe in Figure 3-18. Since these planes (greatest density of atoms) are relatively farther apart, the interatomic attractions between these planes are weaker than other planes; therefore, they may be displaced with comparatively less effort.

In the simplest fashion, displacement of atomic planes relative to one another is analogous to the sliding of cards over each other in a playing deck, as illustrated schematically in Figure 3-20. Alternatively, this movement in the crystal can be represented by the schematic models shown in Figure 3-21. Here the planes of atoms are portrayed as spheres with interplanar spacing (*d*). In case (a), the close-packed planes,

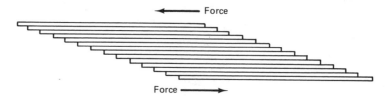

Figure 3-20 Schematic illustration of deformation or displacement of crystal planes.

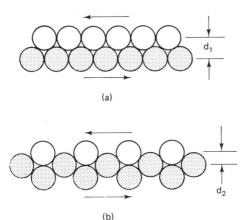

(a)

(b)

Figure 3-21 Example of displacement along close-packed planes.

atoms in the plane are closer together (greater density of atoms) and the interplanar spacing (d_1) is relatively large. The planes in case (b) are not close-packed, the atoms in the planes are farther apart, but the interplanar spacing is smaller compared to (a). By inspection, one can easily see that $d_2 < d_1$, and the planes in (a) can slide or "slip" past each other, as indicated by the arrows, with less difficulty or force than do planes in (b). The plane, identified by Miller indices, along which this displacement occurs is called the *slip plane*.

In the fcc and the hcp systems, the slip plane is the plane that is most densely packed, that is the (111) and the (0001) planes, respectively. In the bcc system, slip can occur on several planes, again those which are most densely packed with atoms. A summary of slip planes and directions are given in Table 6-3 for certain materials and the crystal structures of interest.

CRYSTAL DIRECTIONS

In a manner somewhat similar to that of identifying planes in a crystal, we can also identify directions in these structures. Crystal directions can be visualized or represented by arrows extending from an origin 0,0,0 and passing through points in the lattice, these points being measured in terms of unit cell dimensions. An example of crystal directions along the axes of Cartesian coordinates (x, y, z) is shown in Figure 3-22.

In this example, an arrow extends along the x axis one unit cell distance from the origin. This direction is denoted as the [100]. Notice that we use brackets [ ] for crystal directions to differentiate them from planes. Also note that reciprocals of the intercepts are *not* used. Similarly, the arrow along the y axis extending one unit cell distance is labeled the [010] and along z, the [001]. As this example shows, if the arrow extended any multiple of the unit cell dimensions along any of the three axes, the values could be reduced to those given by [100], [010], or [001]. Therefore, the lowest combination of integers is used. For instance, the [300] is the same direction as the [100], the [020] is equivalent to the [010], and so on.

As an example to familiarize the student with the concept of crystal directions, let us determine the [111] in a cubic crystal system. If we use the definition previously

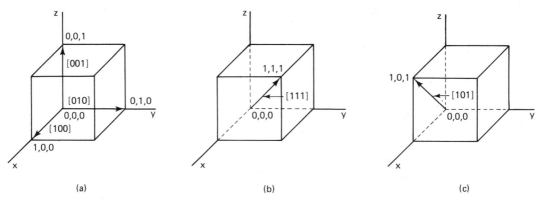

Figure 3-22 Crystal directions in the cubic system: (a) directions along the unit cell (Cartesian) axes; (b) ray showing the [111]; (c) ray showing the [101].

stated, the [111] is direction extending from the origin one unit cell distance in x, one unit cell distance in y, and one unit cell distance in z. This direction is shown in Figure 3-22(b) and consists of an arrow extending from the origin 0,0,0 on a diagonal through the body of the unit cell to a point located one unit cell distance from each axis. What is the [101] in this same unit cell? Well, the unit cell distances are one unit in the x direction and one unit in the z direction. Therefore, the arrow representing the [101] is a diagonal in the xz plane and is shown in Figure 3-22(c).

In a similar manner we can determine the [021]. This direction contains components along the y and z axes as illustrated in Figure 3-23(a). This example serves to illustrate that crystal directions denoted by integers greater than 1 can be determined

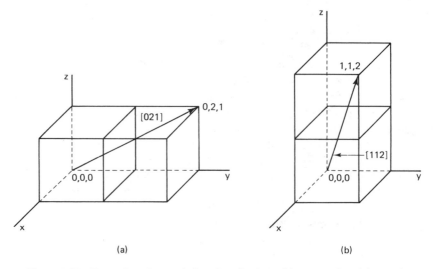

Figure 3-23 Examples of crystal directions in the cubic system involving multiple unit cells: (a) the [021]; (b) the [112].

by constructing multiple unit cells with respect to the appropriate axes. The unit cell intercept does not necessarily occur within a single unit cell, nor does it have to lie along the axes. This situation is demonstrated by the [112] shown in Figure 3-23(b). Note that in this example the arrow denoting the [112] intersects the first unit cell in the center of its top face, as indicated by the dashed circle.

Crystal directions that are negative with respect to the established coordinates are denoted in the same manner as crystal planes with negative intercepts, that is, by a bar over the intercepts. For instance, an arrow along the -y axis would be labeled [0$\bar{1}$0]. Equivalent directions or families of directions are expressed by angular brackets $\langle\ \rangle$.

An interesting feature of crystal directions is that in the cubic system, a crystal direction is perpendicular to a crystal plane having the same indices. For example, the [111] previously illustrated is oriented normal to the (111). The student can easily verify this fact by constructing these two crystal features in a single cubic unit cell. However, this relationship does not generally hold true for other crystal systems.

The crystal directions in the hexagonal system are determined essentially in the same manner as the cubic system. This is accomplished using the four-axis notation [a_1, a_2, a_3, c] or a simpler three-axis notation [a_1, a_2, c]. Figure 3-24 illustrates some crystal directions in a hexagonal unit cell the using three-axis notation.

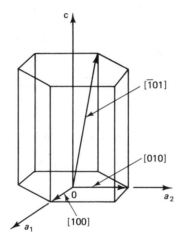

Figure 3-24 Some crystal directions in the hexagonal system using a three-axis notation [a_1, a_2, c].

INTERPLANAR RELATIONSHIPS

The subject of spacing between planes in a crystalline material is very important, as we noted earlier in the chapter when discussing the deformation behavior of crystals. It is worthwhile, therefore, to discuss briefly the method for determining the separation between crystal planes and to present the relationships between d, the interplanar spacing, and the lattice parameters of certain crystal structures. Such information can be utilized to determine experimentally the lattice constants for crystal structures, and other pertinent atomic information.

Bragg's Law

This concept involves the diffraction (reflection) of an X-ray beam or an electron beam[1] by a crystal when certain geometrical conditions are met. The process is illustrated schematically in Figure 3-25, where two parallel planes (*A* and *B*) containing atoms are separated by distance *d*. In this diagram, the radiation (waves) is indicated by rays

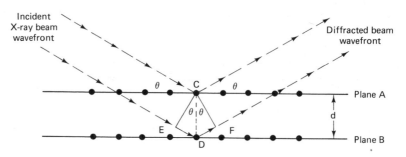

Figure 3-25 Schematic representation of X-ray diffraction process.

striking and leaving a crystal surface at an angle θ. The angle between the incident wavefront and the plane of atoms involved has to equal the angle of the diffracted wavefront. If these angles are not equal, the scattered radiation is not in phase and cancels by interference. The requirement for reinforcement or strengthening of the waves from adjacent parallel planes is met when the difference in length of the path for X-rays reflected from adjacent planes is equal to an integral number of wavelengths (λ) of the incident radiation. Figure 3-25 portrays the distance *EDF* as one wavelength or a multiple of it. Closer examination of this illustration shows that the distance *ED* equals *DF*, which also equals the quantity ($d \sin \theta$). When all the X-rays are in phase as they leave the crystal and are detected, the following condition exists:

$$n\frac{\lambda}{2} = d \sin \theta$$

Where *n* represents an integer number of wavelengths: 1, 2, 3, ... Simplified, this expression is written as follows and is commonly termed *Bragg's law*:

$$n\lambda = 2d \sin \theta \tag{3-5}$$

In practice, the value of *n* is typically taken to be 1, because the diffracted beam is usually weaker when more than one wavefront is present.

Experimentally there are three methods used for diffraction measurement in crystals:

1. *Laue method:* variable λ and fixed θ
2. *Rotating crystal method:* fixed λ and variable θ
3. *Powder method:* fixed λ and variable θ

---

[1]An electron beam can be diffracted by a crystalline material according to the same laws as those for electromagnetic radiation (X-rays) provided that the wavelength (λ) is expressed by the de Broglie relationship: $\lambda = h/p$, where h = Planck's constant and p = momentum or mass times velocity. This ability of electrons to be diffracted demonstrates their wavelike characteristics.

These techniques all utilize the Bragg condition expressed in equation (3-5). For example, in the powder method, diffraction lines are imaged on a strip of film positioned in a camera as illustrated in Figure 3-26. The distance of these lines from 0° or 180° is measured and subsequently related to θ in the Bragg expression.

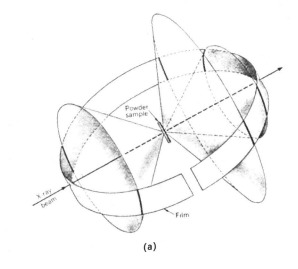

(a)

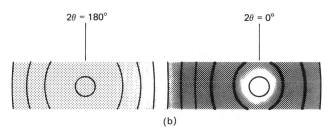

(b)

Figure 3-26 Schematic example of a diffraction experiment using the powder method: (a) diffraction cones from a powder specimen shown intercepting the film; (b) powder photograph showing diffraction lines for tungsten. (From L. V. Azaroff, *Elements of X-ray Crystallography*, McGraw-Hill Book Company, New York, 1968, p. 462.)

Distance between Planes

Using plane geometry, we can express the distance between parallel planes in the cubic system, such as the (110) and the (220), as illustrated in Figure 3-27. The distance between the (110) is determined as

$$d_{110} = \tfrac{1}{2}\sqrt{a^2 + a^2}$$
$$= \tfrac{1}{2}\sqrt{2a^2}$$
$$= \tfrac{1}{2}\sqrt{2}\, a$$

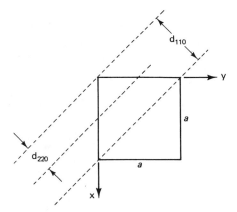

Figure 3-27 Top view of parallel planes (dashed) in a cubic cell.

This result can be simplified further by multiplying the numerator and denominator by $\sqrt{2}$ as follows:

$$d_{110} = \frac{\sqrt{2}}{\sqrt{2}}(\tfrac{1}{2}\sqrt{2}\,a)$$

Therefore,

$$d_{110} = \frac{2a}{2\sqrt{2}}$$

which yields

$$d_{110} = \frac{a}{\sqrt{2}}$$

Similarly, the distance between the (220) is determined to be

$$d_{220} = \tfrac{1}{4}\sqrt{a^2 + a^2}$$

and yields

$$d_{220} = \frac{a}{2\sqrt{2}}$$

In the cubic and hexagonal crystal systems, the distance between parallel planes is given as follows:

$$\text{Cubic:} \quad d_{hkl} = \frac{a}{\sqrt{h^2 + k^2 + l^2}}$$

$$\text{Hexagonal:} \quad d_{hkl} = \sqrt{\frac{3}{4}\frac{a^2}{h^2 + hk + k^2} + \frac{c^2}{l^2}} \tag{3-6}$$

where a = lattice parameter
  hkl = indices of the planes
  c = hexagonal unit cell height

The student is encouraged to verify this relationship by substituting in equation (3-6) the cubic indices for the (110) and the (220) that we analyzed above. Do the results agree?

Example 3-5

Unit cells for the bcc and fcc structures are illustrated in Figure 3-28. Parallel (011) are shown in the fcc. Determine the interplanar separation (d) in terms of the lattice parameter (a) for these planes.

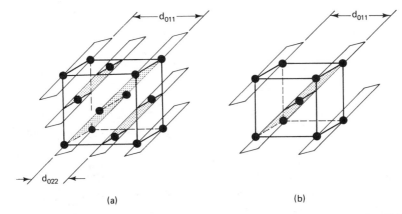

(a) (b)

Figure 3-28 Parallel planes in the bcc and fcc structure: (a) fcc showing {011} and {022}; (b) bcc showing {011}.

Solution Using equation (3-6), for the fcc structure we find

$$d_{011} = \frac{a}{\sqrt{0 + 1^2 + 1^2}} = \frac{a}{\sqrt{2}}$$

$$= 0.707a \quad \text{Ans.}$$

$$d_{022} = \frac{a}{\sqrt{0 + 2^2 + 2^2}} = \frac{a}{\sqrt{8}}$$

$$= 0.353a \quad \text{Ans.}$$

d_{011} for the bcc structure is also $0.707a$. However, recall that the value of a for the two structures is different (Table 3-3).

Example 3-6

A diffraction experiment is performed on a sample of bcc iron using CuK_α radiation. What value of θ will produce first-order ($n = 1$) diffraction lines for the (121)?

Solution

$$\text{For } CuK_\alpha, \quad \lambda = 1.54 \text{ Å}$$

From Table 3-3 we can find that $a = 2.87$ Å for bcc iron. Then utilizing equation (3-6) we can determine d_{121} as follows:

$$d_{121} = \frac{a}{\sqrt{(1)^2 + (2)^2 + (1)^2}}$$

$$= \frac{2.87 \text{ Å}}{\sqrt{6}}$$

$$= 1.2 \text{ Å}$$

Interplanar Relationships

Substituting this result in equation (3-5) and solving for θ, we obtain

$$n\lambda = 2d \sin \theta$$

$$\sin \theta = \frac{(1)(1.54 \text{ Å})}{(2)(1.2 \text{ Å})}$$

$$= 0.642$$

$$\theta = \sin^{-1} (0.642)$$

$$= 40° \qquad Ans.$$

Summary

Although our discussion of diffraction and interplanar spacing has been elementary, a number of important relationships are introduced. First, the interplanar spacing (d) is related to a, the lattice constant or parameter. Second, the interplanar spacing is also related to the wavelength (λ) of incident X-radiation or electron beam and the angle of diffraction (or reflection) (θ), by Bragg's law. Thus, by experimentation with a known wavelength and measurements of θ, the interplanar spacing (d) can be obtained. Such a determination then allows estimates of a, the lattice parameter, to be made and furthermore, since a is related to r, the atomic radius (see Table 3-3), estimates of this parameter can also be obtained.

Diffraction techniques also provide an accurate means for distinguishing between crystalline or amorphous structures, and can be utilized to determine the type of crystal structure from single crystals of a particular material. An example of this is shown in Figure 3-29, where a thin film (roughly 500 Å thick) of an amorphous

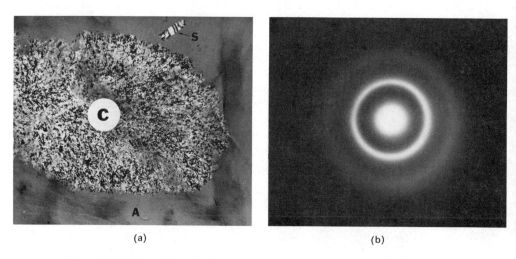

(a)   (b)

Figure 3-29 Examples of electron diffraction patterns in the Ni–P system. (a) Thin-film transmission electron micrograph of NiP at 26,000×. A, amorphous region; C, polycrystalline region; S, single crystal of NiP. (b) Diffraction pattern from region (A) shows relatively broad diffuse rings indicating randomly arranged atoms.

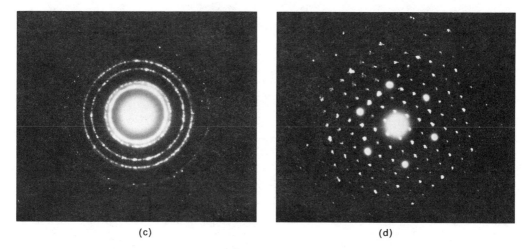

(c)   (d)

Figure 3-29 (Cont.) (c) Pattern from region (C) shows sharp narrow rings associated with diffraction by periodically arranged atoms (planes) in a polycrystalline material. (d) Diffraction pattern from area (S) displaying discrete diffraction spots indicative of a single crystal (i.e., no grain boundaries). Note, in particular, the hexagonal arrangement of spots corroborating the hexagonal crystal structure of NiP. (Courtesy of L. J. McNamara and P. J. Cote.)

nickel (Ni)–phosphorus (P) alloy has been crystallized by heating to about 300°C (572°F) momentarily with an electron beam.

Example 3-7

A diffraction experiment is performed on a sample of fcc copper, using CrK_α radiation ($\lambda = 2.29$ Å). The diffraction pattern yields first-order lines ($n = 1$) for the (111) at $\theta = 33.4°$. (a) Determine the lattice parameter (a_0) for this material, and (b) estimate the equilibrium atomic radius (r_0) for copper.

Solution (a) Using Bragg's law [equation (3-5)], we find the interplanar spacing to be

$$d = \frac{\lambda}{2 \sin \theta}$$

$$= \frac{2.29 \text{ Å}}{(2)(\sin 33.4°)}$$

$$= \frac{2.29 \text{ Å}}{(2)(0.55)}$$

$$= 2.08 \text{ Å}$$

The interplanar spacing is related to a in cubic structures by equation (3-6). Therefore,

$$d_{111} = \frac{a}{\sqrt{1^2 + 1^2 + 1^2}}$$

$$= \frac{a}{\sqrt{3}}$$

Interplanar Relationships

Substituting the value of d from the diffraction experiment, we obtain

$$\frac{a}{\sqrt{3}} = 2.08 \text{ Å}$$

$$a = 2.08\sqrt{3} \text{ Å}$$

$$a_{cu} = 3.60 \text{ Å} \qquad Ans.$$

(b) In fcc structures the lattice parameter (a) is related to the atomic radius (see Table 3-3) as follows:

$$a_{fcc} = \frac{4r}{\sqrt{2}}$$

Thus

$$r_{cu} = \frac{\sqrt{2}\,a_{cu}}{4}$$

$$= \frac{\sqrt{(2)}(3.60 \text{ Å})}{4}$$

$$= 1.27 \text{ Å} \qquad Ans.$$

Check result (b) with the property data inside the back cover.

STUDY PROBLEMS

3.1. (a) Calculate the distance between adjacent *corner* atoms in a unit cell for sodium (Na).
(b) What is the largest-diameter sphere that will fit into the unoccupied space (along the unit cell edge) between the atoms in part (a)?

3.2. Nickel forms a fcc structure and has an atomic radius of 1.246 Å. Calculate the volume of its unit cell.

3.3. X-ray diffraction measurements show that the lattice parameter of pure iron is 2.89 Å in the bcc structure and 3.66 Å in the fcc structure at the transformation temperature (910°C). Determine the atomic radius in each case. Calculate the percentage increase in radius due to this transformation.
Answer: $(\Delta r/r) = 3.2\%$

3.4. (a) Calculate the lattice parameter for fcc aluminum.
(b) Determine the diameter of the "hole" produced in the center of a unit cell consisting of aluminum atoms arranged in fcc structure.

3.5. Given the following portion of a hexagonal close-packed (hcp) structure, determine the ratio of c/a (the interbasal plane distance to the lattice parameter) for identically sized, perfect spheres.

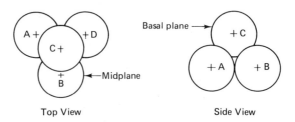

Top View   Side View

3.6. Iron experiences an allotropic transformation at 910°C (1670°F) (see Table 3-4). Calculate the difference that this transformation produces in the volume of the unit cell when iron is heated just above this temperature.
Answer: -0.85%

3.7. Determine the difference in theoretical density that occurs between iron atoms which form a bcc structure and an fcc structure. For the purpose of calculations, assume that temperature is not a factor in this case.

3.8. Given that the APF for hcp structures is 0.74, calculate the theoretical density for magnesium ($r = 1.610$ Å).

3.9. (a) Calculate the number of atoms in a wire 1 cm in diameter by 10 cm long made from pure copper (density $= 8.94$ g/cm³).
  (b) How many valence electrons would this wire contain?
Answer: (a) 6.65×10^{23} atoms

3.10. Beryllium exhibits a hcp structure and atomic radius of 1.140 Å. If $c/a = 1.586$,
  (a) Determine its unit cell volume. [*Hint:* The basal plane in hcp is comprised of six equilateral triangles, and $a = 2r$ (see Figure 3-14).
  (b) Calculate the density of Be.
Answer: (a) 4.88×10^{-23} cm³

3.11. Calculate the density of lead.

3.12. Based on the data inside the back cover, calculate the density of zirconium. (*Hint:* See Example 3-2.)

3.13. Given the following intercepts in Cartesian coordinates, determine the Miller indices for the planes in question.
  (a) $x = 1, y = 1, z = \frac{1}{2}$
  (b) $x = \frac{3}{4}, y = \infty, z = 1$
  (c) $x = 2, y = \frac{1}{3}, z = 1$
  (d) $x = \infty, y = -1, z = 1$
  (e) $x = -1, y = \frac{1}{4}, z = 1$
  (f) $x = 1, y = \infty, z = \frac{2}{3}$

3.14. Sketch planes in the cubic structure with the following Miller indices as they would appear in a unit cell. Also, label the intersections of the planes with the Cartesian axes, with their appropriate point locations.
  (a) (121)
  (b) (333)
  (c) (401)

3.15. (a) Identify the plane in the below unit cell with its proper Miller indices.
  (b) To what family of planes does it belong?

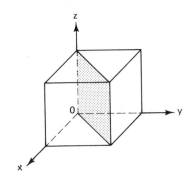

3.16. Sketch a hexagonal unit cell showing the a_1, a_2, a_3, and c axes. In this unit cell, identify (shade) the following crystal planes.

(a) $(11\bar{2}0)$

(b) $(1\bar{2}10)$

(c) $(1\bar{1}02)$

3.17. (a) Explain what is meant by the term *slip* in regards to crystalline materials.

(b) Why can slip occur comparatively easy on certain planes in crystalline materials?

(c) What are the Miller indices of the planes of easiest slip in the hcp system?

3.18. Determine the density (planar) on any face plane {100} of aluminum. Express your answer in atoms/cm².

Answer: $D_{100} = 1.22 \times 10^{15}$ atoms/cm²

3.19. Determine the planar density for aluminum on the (111). How does this compare with the (100)?

3.20. Compare the planar densities (atoms/cm²) of the (100) and (110) in molybdenum. Based on this determination, which of the two planes would you suspect are planes of easiest slip?

3.21. (a) Show the following crystal directions in the cubic system: [012], [110], [$\bar{1}$01].

(b) What are the coordinates of the intersection of these rays with the unit cell?

3.22. Identify the crystal directions which pass through the following coordinate points in a unit cell of the cubic system.

(a) $\frac{1}{2}, \frac{1}{2}, \frac{1}{2}$

(b) $\frac{1}{2}, 1, 1$

(c) $1, \frac{1}{2}, 0$

3.23. Show the following crystal directions in the hexagonal system using a three-axis notation for the indices.

(a) [110]   (b) [$\bar{1}\bar{1}$1]   (c) [210]

(d) [$\bar{1}\bar{2}$0]   (e) [$\bar{1}$21]

3.24. An X-ray diffraction experiment is performed using chromium K_α radiation ($\lambda = 2.29$ Å). If $n = 1$ and we obtain a value of 35° for the diffraction angle (θ), determine the distance between the diffraction planes.

3.25. Calculate the diffraction angle (θ) for the {110} type reflection in α-iron at room temperature, if $n = 1$ and the wavelength of the radiation is 1.79 Å (CoK_α).

Answer: 26.2°

3.26. A specimen of aluminum is analyzed by X-ray diffraction using CuK_α radiation ($\lambda = 1.54$ Å). The resulting pattern for the {100} reflection reveals a diffraction angle (θ) of 11°. What is the lattice parameter for this material?

3.27. A copper wire has been permanently deformed by the application of an external load. An X-ray diffraction experiment is performed on a sample of the wire using CrK_α radiation ($\lambda = 2.29$ Å). This analysis shows the diffraction angle (θ) to be 28.6° for the {111}-type reflection. Calculate the percentage change in the lattice parameter of the copper as compared to the equilibrium value. Was the wire elongated or shortened? Speculate on the type of loading that could produce this condition.

3.28. Cobalt K_α radiation ($\lambda = 1.79$ Å) is used to analyze a silver sample. What is the angle of diffraction for second-order ($n = 2$) diffraction lines from the {111}?

Answer: 49.6°

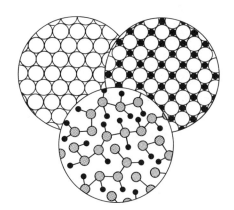

CHAPTER FOUR:
Imperfections in
Crystalline Materials

In our discussion and analyses of crystalline materials thus far, we have considered only perfect crystals. By "perfect crystals" we mean structures than contain atoms in all lattice positions that should be occupied and also contain no irregularities in long-range periodicity, that is, outside the immediate unit cell.

However, as the student should suspect by this point in the study of engineering materials, real crystals occurring in nature and in engineering materials deviate from perfect periodicity and contain numerous imperfections or flaws. There is a special class of engineering materials referred to as *single crystals* (contain no intercrystalline boundaries) that approach crystalline perfection, and such materials have been developed for use in very special applications such as certain semiconductors and magnetic materials used for information processing and storage, and certain blades in jet aircraft turbines. However, these materials are difficult and costly to produce, therefore, their use in routine commercial applications is presently limited.

Even though crystalline defects exist on an atomic scale, in other words, their individual proportions are extremely small, there can be extremely large numbers of such defects in ordinary crystals. For this reason, and the very nature of the imperfections themselves, they are responsible for a substantial part of the observed mechanical and physical behavior of engineering materials. Thus an introduction to these defects is in order before continuing on to the study of diffusion, solid solutions, alloying, solidification, and corrosion.

POINT DEFECTS

This category of crystalline imperfections is referred to as point defects because they involve single sites in the space lattice. Although these imperfections in the space lattice necessarily cover some finite area or volume, their extent is limited to a localized disturbance. Therefore, they are treated as though they behave like a point.

Vacancies

One important point defect is referred to as a *vacancy*. Such an imperfection is simply a vacant lattice site, where an atom or an ion is missing from its normal position in the crystal, as illustrated schematically in Figure 4-1(a).

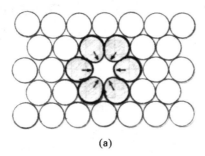

(a)

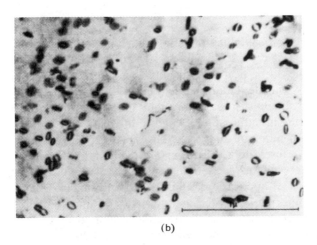

(b)

Figure 4-1 Examples of vacancies in crystalline materials: (a) schematic of a vacancy in a close-packed plane; (b) transmission electron micrograph of vacancy "loops" in aluminum. Typical size of a loop is approximately 5.2×10^{-8} m. (520 Å). (From E. Ruedl, P. Delavignette, and S. Amelinckx, *J. Nucl. Mater.* 6, 1962, p. 46.)

Aggregates of this type defect can sometimes be imaged by conducting transmission electron microscopy (TEM)[1] on very thin films of materials (up to about 1000 Å thick). For example, vacancy "loops" or rings produced by the congregation of individual vacancies in aluminum are shown in Figure 4-1(b). In this particular instance,

[1] TEM essentially consists of directing a beam of electrons onto a very thin film of the material to be characterized and imaging the transmitted beam. The intensity of the transmitted beam or signal depends on such factors as absorption, diffraction, and crystalline imperfections. Thus we can distinguish or resolve defects which have a significant effect on the transmitted intensity.

the aluminum foil was quenched in iced brine from 560°C, thus locking-in the high-temperature arrangement of vacancies.

Vacancies can occur during crystallization of the materials, by plastic or permanent deformation of the solid crystals, and by high-energy radiation. In pure metals, small numbers of vacancies are produced by thermal excitation. Such an event can be thought of as the process of moving an atom from within the crystal structure to a surface, with heat supplying the energy necessary to accomplish this task. At equilibrium, the fraction of lattice sites that are vacant, for a given temperature, can be expressed as

$$\frac{n}{N} = e^{-E_s/kT} \tag{4-1}$$

where n = equilibrium number of vacancies
N = total number of lattice sites
E_s = energy required to move atom from interior to surface
k = Boltzmann's constant
T = absolute temperature

We may now analyze how temperature affects vacancy formation by thermal excitation. By taking the natural logarithm of both sides of this equation, we obtain

$$\ln \frac{n}{N} = \ln e^{-E_s/kT}$$

From the logarithmic relationship,

$$\ln n - \ln N = -\frac{E_s}{kT}$$

$$\ln n = \ln N - \frac{E_s}{kT} \tag{4-2}$$

The final expression above reveals that as T increases, the quantity E_s/kT decreases. Since this value is deducted from the natural logarithm of the total number of lattice sites, as it decreases the natural logarithm of the equilibrium number of vacancies increases. In other words, as temperature increases, the number of vacancies in a crystalline material also increases. This is an important finding, since it influences solid-state processes such as diffusion and semiconduction.

The relationship in equation (4-1) can also be expressed in terms of molar quantities, making it more useful in determining equilibrium vacancy concentrations at particular temperatures. If we multiply equation (4-1) by Avogadro's number[2] (A_0) as shown, the following expression results:

$$\frac{n}{N} = e^{-E_s A_0/kTA_0}$$

$$= e^{-Q/RT} \tag{4-3}$$

where Q = activation energy to form one mole of vacancies (cal/mol)
R = gas constant = 2 cal/mol-°K

[2] Avogadro's number = 6.02×10^{23} mol^{-1}.

Example 4-1

Determine the equilibrium concentration of vacancies in copper at (a) room temperature (27°C) and (b) a temperature (1080°C) just under the melting point. Q for copper is approximately 2×10^4 cal/mol.

Solution Utilizing equation (4-3) and the absolute temperatures

$$T = 27°C + 273°C = 300°K$$

$$T = 1080°C + 273°C = 1353°K$$

we obtain the following relationships:

    (a) Room temperature:

$$\frac{n}{N} = e^{-2 \times 10^4/(2)(300)}$$

$$= e^{-33.3} = 3.4 \times 10^{-15}$$

One vacancy for $\simeq 3 \times 10^{14}$ atoms! *Ans.*

    (b) Approximate melting temperature:

$$\frac{n}{N} = e^{-2 \times 10^4/(2)(1353)}$$

$$= e^{-7.4} = 6.1 \times 10^{-4}$$

One vacancy for $\simeq 1636$ atoms! *Ans.*

Comment: At this stage, the student may not appreciate this enormous difference in the vacancy concentration. However, in Chapter 5 when we study diffusion processes, this disparity will surely become more meaningful.

Vacancies can also occur in other forms, such as *di-vacancies*, which is simply a situation where two neighboring atoms are missing; *Schottky defects*, ion-pair vacancies which occur in crystals with ionic bonding; and *Frenkel defects*, an interstitial vacancy created by a displaced ion. Let us consider the latter two crystalline imperfections, which are important in ceramics, in more detail.

Schottky defect. This imperfection is produced by the migration of an ion pair (positive and negative) to a position on a free surface or an intercrystalline boundary. The net result is an ion-pair vacancy, found in materials such as ceramics, which must maintain a charge balance. This process involves a relatively low "activation energy" and therefore can occur as a result of thermally induced vibrations in the lattice. A Schottky-type defect is illustrated for NaCl in Figure 4-2.

Frenkel defect. A Frenkel defect is formed by the displacement of an ion from its normal lattice site into an interstitial position in the crystal. As a result, a material containing Frenkel defects has both interstitial ions, causing localized expansion and lattice vacancies. The formation of this type of defect requires considerably more "activation energy" than the Schottky defect, since interstitial space is very

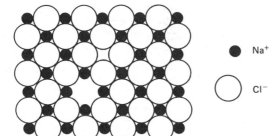

Figure 4-2 Example of an ion-pair vacancy (Schottky defect) in NaCl.

limited, and therefore they are not produced by thermal vibrations in the lattice. Frenkel defects are produced in quantity only by high-energy particles such as those encountered in nuclear reactors. Indeed, these defects are one form of *radiation damage* in reactor components. A Frenkel-type defect is illustrated in Figure 4-3.

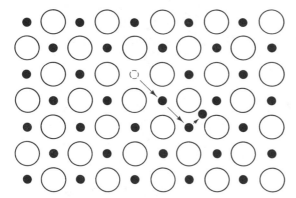

Figure 4-3 Ion displacement to an interstitial site (Frenkel defect).

Overall, vacancies play an important role in many metallurgical and chemical processes performed on crystalline engineering materials. These defects all have one feature in common—empty lattice sites! This unoccupied space plus the localized lattice distortion accompanying the defect influence such properties as electrical conduction and semiconduction in ionic materials, and solid-state movement of atoms (or ions), as we will see in Chapter 5.

In addition, vacancies can interact with the other forms of crystalline defects that will be described later in this chapter. Such interactions help explain many different mechanical and physical processes.

Interstitial Impurity Atom

Another important type of point defect that occurs in crystalline materials is referred to as an *interstitial defect* or *interstitial impurity atom*. This imperfection in the lattice simply amounts to a smaller atom occupying an interstitial site in the lattice of relatively larger atoms. The student will recall that interstices are the unoccupied spaces produced by the arrangement of atoms (spherical models) in the crystal structure. A plane view of an interstice was given previously in Figure 3-6 (see the shaded region). An interstitial impurity atom or defect is shown in Figure 4-4. Interstitial impurity atoms produce a local disturbance in the crystal lattice, depending primarily on their size, the size of the "host" atoms, valency effects, and the structure of the host lattice.

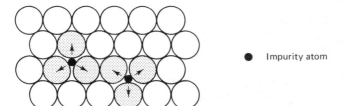

● Impurity atom

Figure 4-4 Schematic showing interstitial impurity atoms in the lattice.

Consider for the moment the cubic unit cell structures depicted previously in Figures 3-8 and 3-10. This representation helps us to appreciate the interstitial spaces produced as a result of these particular atomic arrangements. Even though all the interstitial sites are not clearly defined, there are both *fourfold* and *sixfold* sites in these structures. By fourfold (tetrahedral) we mean that the interstitial site exists among four neighboring atoms; similarly a sixfold (octahedral) site is produced by six neighboring atoms. The interstitial void spaces formed by these arrangements in the fcc structure are depicted in Figure 4-5. Moreover, the *average* number of interstitial sites associated with a unit cell in the systems of interest are summarized in Table 4-1.

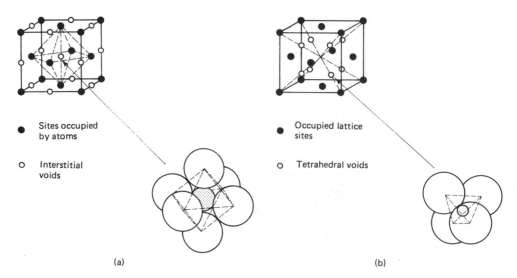

● Sites occupied by atoms

○ Interstitial voids

● Occupied lattice sites

○ Tetrahedral voids

(a)

(b)

Figure 4-5 Interstitial voids pictured in the fcc structure. (a) Octahedral interstices; inset shows an impurity atom located within the interstice produced by six neighboring atoms. (b) Tetrahedral interstices; inset shows an impurity atom in the interstice produced by four neighboring atoms. (Corner plus three adjacent face-centered atoms.)

TABLE 4-1 INTERSTITIAL SITES IN CERTAIN CRYSTAL STRUCTURES

| Structure | Interstitial sites | |
	Tetrahedral	Octahedral
bcc	12	6
fcc	8	4
hcp[a]	12	6

[a]Hexagonal close-packed structures exhibit tetrahedral and octahedral interstices in the same ratio of sites to atoms as fcc.

Substitutional Impurity Atom

This third category of point defect consists of impurity atoms occupying lattice positions in the crystal. Essentially, the impurity atom is a different chemical element (atomic species) from the rest of the atoms in the lattice. We may also refer to the normal atoms in the lattice as host atoms or the *solvent* species. Such a concept places the impurity atoms in the category of the *solute* species. The substitutional impurity defect is shown schematically in Figure 4-6, where the impurity atom is larger than its hosts. However, this is not always the case, and substitutional impurities may be smaller than their hosts.

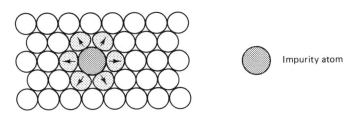

Impurity atom

Figure 4-6 Substitutional impurity atom (larger than host atoms).

Obviously, this is a simplistic representation, but it serves to illustrate the concept of a substitutional impurity atom serving as a point defect. Since different atoms exhibit different chemical or electronic behavior, there is a certain incompatibility between the impurity and its neighboring hosts. Depending on the particular geometrical and chemical differential, a localized disturbance is created in the crystal lattice. In turn, this perturbation in the lattice affects the mechanical and physical properties of the crystal. When such crystalline imperfections are considered en masse, they make a substantial contribution to the observed behavior of bulk engineering materials. In spite of the connotation of "defect," we shall soon see that this condition of impurity atoms can be extremely beneficial to certain classes of engineering materials and therefore is purposely contrived. The mechanisms of strengthening by adding substitutional and interstitial atoms of different chemical species is discussed in Chapter 6.

LINE DEFECTS

This category of crystalline imperfections consists of defects which are considerably larger than the point defects just discussed in that they may extend many atomic distances through the crystal. Such imperfections are often referred to as *line* defects, and constitute a crystalline irregularity extending along a line threading through adjacent lattice sites.

The Dislocation

A dislocation, as the term literally implies, is an interruption in the periodicity of a lattice, which puts the crystal out of order. This type of line defect can be created during crystallization, when there is a mismatch in crystallographic orientation, or a break in periodicity of the growing crystal, and also during deformation of the crystals in a solid engineering material, when they are subjected to forces that *permanently* change their shape. As you will see in Chapter 6, this type of deformation is also referred to as *plastic deformation*. Two types of dislocations can be present in crystals, and they are referred to as *edge* and *screw* dislocations.

Edge type. The formation of an edge dislocation consists of either introducing or eliminating an extra row (plane) of atoms in the crystal lattice. An edge dislocation can be represented schematically in cross section as illustrated in Figure 4-7.

Examination of the sketch shows that these "extra" atoms are situated above a specific plane in the crystal referred to as the *slip plane*. The slip plane, as you will see later in this section, is the plane on which displacement of the atoms occurs under

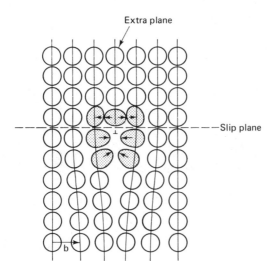

Figure 4-7 Cross section through a positive edge dislocation. Shaded atoms indicate lattice disturbance—compressive above slip plane, tensile below slip plane. Burger's vector (*b*) denotes the lattice displacement associated with dislocation.

certain conditions. Thus the connotation of "slipping." The edge dislocation is indicated by the symbol $\perp$ and the atoms in proximity to this defect are shaded to draw the students' attention to this disturbance. The extra atoms "wedged in" above the dislocation produce a compressive effect in the lattice. Correspondingly, the atoms in the vicinity of the dislocation below the slip plane also adjust to this defect, thereby producing a tensile or stretching effect in this region of the lattice. The dislocation pictured in Figure 4-7 is a *positive* edge dislocation. A similar condition is just as likely to be produced if the extra plane is situated below the slip plane. This condition is termed a *negative* edge dislocation. To picture such a situation, simply turn Figure 4-7 upside down. You will note that the compressive and tensile regions are just reversed with respect to the slip-plane orientation. The negative edge dislocation is sometimes denoted by the symbol $\top$. Consider for the moment what would happen if two dislocations, one positive and one negative, met along the same slip plane. What would be the result? Of course, they would annihilate each other and the atomic registry of the crystal would be restored!

Note that the disturbance caused by the dislocation is localized (shaded atoms) in our representation. This is true, but remember that we are really discussing a three-dimensional imperfection, which extends in and out of the plane of the page. Edge dislocations are illustrated in Figure 4-8. The spheres in Figure 4-8(a) represent atomic positions in a space lattice. Dislocation lines may terminate at the surface of a crystal and at grain boundaries in polycrystalline materials, but they can never end within a crystal. Therefore, dislocations form closed loops, join other dislocations, or intersect surfaces.

The atomic displacement produced by a dislocation can be expressed by its *Burger's vector*. This vector (b) is shown in Figure 4-7 and is oriented normal to the line of the edge dislocation. The magnitude of b is determined by tracing a path around the dislocation (Burger's circuit). The vector needed to close the loop is the Burger's vector.

Screw type. Another type of dislocation in crystalline materials is the *screw dislocation*. This defect differs from the edge type in that the displacement of atoms is parallel to the line of the dislocation rather than normal to it. Examples of screw dislocations are presented in Figure 4-9. The periodic disturbance in this case is concentrated along the heavy dark line and the lattice is displaced one Burger's vector (b). Note that if we trace the Burger's circuit for a screw dislocation, it describes a helical path. In addition, the following features pertain to this defect:

1. The dislocation line lies in the slip plane, but is parallel to the slip plane.
2. The dislocation line moves in the slip plane in a direction normal to the slip direction (as defined by the Burger's vector).

In summary, it is important to point out that while we have simplistically described two different types of dislocations, in real crystals there can be combinations of edge and screw dislocations. Furthermore, the passage of a dislocation line through the crystal displaces the portions of the crystal separated by the slip plane, by one

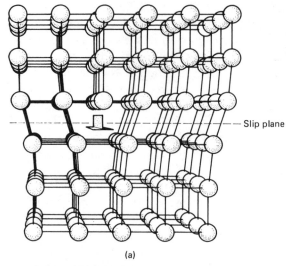

(a)

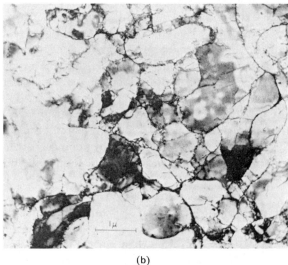

(b)

Figure 4-8 Examples of an edge dislocation: (a) schematic three-dimensional representation of an edge dislocation ($\perp$) in a crystal; (b) edge dislocations forming a "substructure" in aluminum as imaged by thin-film techniques in the transmission electron microscope (TEM). Dislocation density in this example is approximately 10^{10} cm^{-2}. (From P. B. Hirsch et al., "Direct Observation of the Arrangement and Motion of Dislocations in Aluminum," in *Dislocations and Mechanical Properties of Crystals*, John Wiley & Sons, Inc., New York, 1957, p. 93.)

Burger's vector relative to one another. The next section will reveal just how important this fact is.

Dislocations and Deformation

Now that we have been properly introduced to dislocations in crystalline materials, these concepts will help explain some properties and observed behavior of engineering materials.

Typically, the number of dislocations in a crystalline solid is extremely large, unless special precautions have been exercised to prevent or reduce their numbers.

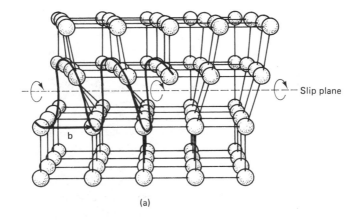

(a)

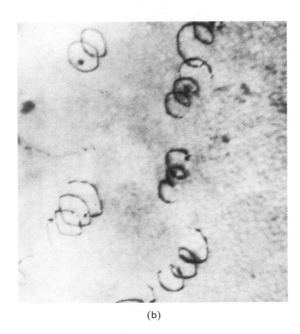

(b)

Figure 4-9 Examples of a screw dislocation. (a) Three-dimensional representation of the screw dislocation in a crystal. The dislocation line (dark) describes a helix around (S). (b) Helical dislocations containing a screw component as seen by TEM in an aluminum–5% magnesium alloy. (Courtesy of G. Thomas, University of California at Berkeley)

The number of dislocations in a unit volume is defined as the total length of dislocation lines (l) per unit volume. This quantity is also referred to as the dislocation density N and equals l/V. The dislocation density of a crystal is usually expressed in units of cm^{-2}, because it conveniently indicates the number of intersections over a unit area. For example, an annealed metal (one that has been heated and slowly cooled) can typically have dislocation densities on the order of 10^6 to 10^8 dislocation lines/cm². Heavily cold-worked metals (severely deformed at room temperature) can exhibit as many as 10^{12} dislocation lines/cm².

As we stated in the section "Interatomic Bonding" in Chapter 2, in order to pull atoms apart and actually break their bonds, *very* high forces are required. Early determinations of this ideal strength of crystals indicated that about 3 to 10% elastic deformation[3] should occur in crystals before slip or displacement of the material begins. However, engineers traditionally prefer to deal in terms of forces or *stress*. Although the concept of stress is discussed in considerable detail in Chapter 6, at this point we can simply define stress as: force per unit area. Stress is usually stated in units of pascals (newtons/m²) or pounds per square inch (psi). Do you recall our hypothetical example of 400 automobiles suspended on a steel bar in Chapter 2? *Theoretically*, the stress required to break the atomic bonds in the steel amounts to approximately 14 billion pascals (2 million psi). If this value of theoretical strength sounds large, be assured that it is. How was this determination arrived at? Well, based on the concepts of atomic bonding we have previously studied and some elementary calculus, we can derive an expression for this "ideal" strength in crystals. The ideal strength of crystals is also known as the *theoretical cohesive strength*.

Theoretical cohesive strength.   Based on the atomic models and structural arrangements presented earlier in this book (refer to Figure 2-3), we may express the cohesive or binding force between atoms as a function of the distance separating them. The diagram shown in Figure 4-10 represents the summation curve of the attractive and repulsive forces between atoms. Because of the shape of this

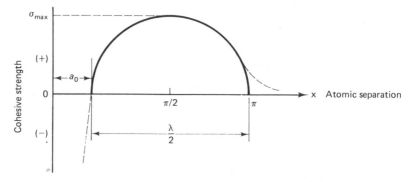

Figure 4-10   Schematic representation of cohesive forces between atoms.

curve, we may approximate the cohesive strength (σ) by a portion of the sine function. Half of the sine curve ($\lambda/2$) represents the cohesive force between two atoms. Mathematically, this relationship can be expressed as follows:

$$\sigma = \sigma_{max} \sin \frac{2\pi}{\lambda} x \qquad (4\text{-}4)$$

As the diagram illustrates, when $x = a_0$, the two atoms are separated by the equilibrium distance and the sine function $= 0$; therefore, $\sigma = 0$. If the atoms are gradually pulled apart, the cohesive strength increases to a maximum value σ_{max}. This

[3]Elastic deformation is displacement of the atoms from their equilibrium positions in the crystal, which is recovered when the forces producing such deformation are removed (see Chapter 6).

maximum occurs at $x = \pi/2$. σ_{max} is the *theoretical cohesive strength* between the two atoms. Once this strength is exceeded, or the atoms are separated farther than $x = \pi/2$, the force required to pull them farther apart gradually decreases. Thus our objective is to obtain an expression for σ_{max} in terms of some known parameters.

First, the work per unit area done during fracture, that is, breaking the atoms apart, is actually the area under the curve (from 0 to π). Work or energy is equal to force times displacement. In our case, this energy per unit area can be expressed as the integral of σ times displacement, or simply the integral of equation (4-4) with respect to x. Let us perform this integration as follows:

$$U = \int_0^{\lambda/2} \sigma_{max} \sin \frac{2\pi}{\lambda} x \, dx \tag{4-5}$$

where U is the energy involved in the fracture process (area under curve). Since σ_{max} is not a function of x, we can rearrange the expression:

$$U = \sigma_{max} \int_0^{\lambda/2} \sin \frac{2\pi}{\lambda} x \, dx \tag{4-6}$$

This integral is found in a table of integrals under forms involving trigonometric functions and takes the form

$$\int \sin A \, x \, dx$$

where $A = 2\pi/\lambda$. The solution of this integral is as follows:

$$\int \sin Ax \, dx = -\frac{1}{A} \cos Ax$$

Thus equation (4-6) becomes

$$U = \sigma_{max} \left[-\frac{\lambda}{2\pi} \cos \frac{2\pi}{\lambda} x \right]_0^{\lambda/2 = \pi}$$

Substituting the limits in this equation, we obtain

$$U = \sigma_{max} \left[-\frac{\lambda}{2\pi} \cos \frac{2\pi}{\lambda}(\pi) + \frac{\lambda}{2\pi} \cos \frac{2\pi}{\lambda}(0) \right]$$

Therefore, our expression reduces to[4]

$$U = \sigma_{max} \left[-\frac{\lambda}{2\pi}(-1) + \frac{\lambda}{2\pi}(1) \right]$$

which yields

$$U = \frac{\lambda}{\pi} \sigma_{max}$$

The energy per unit area required to produce a new surface (such as breaking atoms apart) is often designated as γ. If we make the assumption that all the work

[4]The student is referred to any elementary text on trigonometric functions to find: $\cos \pi = -1$ and $\cos 0 = 1$.

involved in fracture goes into the creation of two new surfaces (as it would in an ideal brittle solid), we can state

$$2\gamma = \frac{\lambda}{\pi} \sigma_{max}$$

or, solving for the theoretical cohesive strength,

$$\sigma_{max} = \frac{2\pi\gamma}{\lambda} \qquad (4\text{-}7)$$

The second part of our derivation consists of obtaining λ in known terms. In the initial stages of our diagram, the displacement or deformation is elastic (not permanent), so *Hooke's law*[5] can be applied to express the cohesive strength as follows:

$$\sigma = Ee$$

where E = Young's modulus (modulus of elasticity)
e = engineering strain

In our example, the strain ($e = \Delta l/l_0$) equals the displacement in x divided by the equilibrium separation:

$$e = \frac{\Delta x}{a_0} = \frac{x}{a_0}$$

So we can write

$$\sigma = \frac{Ex}{a_0} \qquad (4\text{-}8)$$

We can eliminate λ from equation (4-7) if we differentiate equations (4-4) and (4-8) with respect to x and then set the resultant expressions equal. First, differentiate equation (4-4) as follows:[6]

$$\sigma = \sigma_{max} \sin \frac{2\pi}{\lambda} x$$

$$\frac{d\sigma}{dx} = \sigma_{max} \frac{2\pi}{\lambda} \cos \frac{2\pi}{\lambda} x$$

If we assume that the strains involved are very small, then for small values of x, $\cos (2\pi/\lambda)x \simeq 1$. Thus our expression is

$$\frac{d\sigma}{dx} = \sigma_{max} \frac{2\pi}{\lambda} \qquad (4\text{-}9)$$

Now differentiate equation (4-8) as follows:[7]

$$\sigma = \frac{Ex}{a_0}$$

$$\frac{d\sigma}{dx} = \frac{E}{a_0} \qquad (4\text{-}10)$$

---

[5]See Chapter 6 for an explanation of E and σ.
[6]$d(\sin Ax)/dx = A \cos Ax$.
[7]$d(Ax)/dx = A$.

Thus we have two expressions for $d\sigma/dx$, equations (4-9) and (4-10). By equating these relationships, we can eliminate λ from equation (4-7) and finally solve for σ_{max} in known terms. From the two differentials,

$$\sigma_{max}\frac{2\pi}{\lambda} = \frac{E}{a_0}$$

Solving for λ, we have

$$\lambda = \frac{a_0\sigma_{max}2\pi}{E}$$

Substituting this value in equation (4-7), we obtain

$$\sigma_{max} = \frac{2\pi\gamma E}{a_0\sigma_{max}(2\pi)}$$

which simplifies to

$$\sigma_{max}^2 = \frac{\gamma E}{a_0}$$

or

$$\sigma_{max} = \left(\frac{\gamma E}{a_0}\right)^{1/2} \qquad (4\text{-}11)$$

Our derivation shows that the theoretical cohesive strength (σ_{max}) is a function of the energy or work to create new surfaces during fracture (γ), the elastic modulus (E), and the equilibrium separation between the atoms (a_0). As an exercise to familiarize the student with this concept and the units that are involved, let us consider the following example.

Example 4-2

Given the following typical data for a brittle material, determine the theoretical cohesive strength (σ_{max}). Express the answer in both SI and English units.

$$\gamma = 1 \text{ J/m}^2$$
$$a_0 = 3 \text{ Å}$$
$$E = 100 \text{ GPa}$$

Solution   Utilizing equation (4-11) and the appropriate conversion factors, we obtain

$$\sigma_{max} = \left(\frac{\gamma E}{a_0}\right)^{1/2}$$

$$= \left[\frac{(1 \text{ N-m/m}^2)(100 \times 10^9 \text{ N/m}^2)}{3 \times 10^{-10} \text{ m}}\right]^{1/2}$$

$$= \left(\frac{100 \times 10^9 \text{ N}^2/\text{m}^4}{3 \times 10^{-10}}\right)^{1/2}$$

$$= (33.3 \times 10^{19} \text{ N}^2/\text{m}^4)^{1/2}$$

$$= 1.8 \times 10^{10} \text{ N/m}^2$$

$$= 18,000 \text{ MPa} \ (2.6 \times 10^6 \text{ psi}) \qquad \textit{Ans.}$$

If the determination for σ_{max} in Example 4-2 is expressed as a fraction of the elastic modulus, we find that

$$\sigma_{max} \approx E/6$$

Estimates of σ_{max} range from $E/4$ to $E/15$, with a convenient value being in the neighborhood of $E/10$. This approximation supports our previous assertion of 2 million psi for the approximate theoretical cohesive strength in steels. But how does this theoretical strength compare to the actual strength behavior of steels? For instance, a typical tensile test of high-strength steel in the laboratory demonstrates that the ultimate tensile strength (maximum strength exhibited by a material prior to failure) ranges from 100,000 to 200,000 psi (690 to 1380 MPa). The predicted theoretical cohesive strength, then, is roughly 10 to 20 times greater. A comparison of the theoretical strength with the typical strength of bulk materials is presented in Table 4-2 for several metals and ceramics.

TABLE 4-2 COMPARISON OF THEORETICAL COHESIVE STRENGTH WITH TYPICAL STRENGTH OF SELECTED BULK MATERIALS

Material	Theoretical strength, σ_{max}		Bulk strength	
	psi	MPa	psi	MPa
Iron (carbon steel)	2,000,000	13,800	100,000	690
Copper	2,000,000	13,800	70,000	483
Glass	1,200,000	8,280	60,000	414
Graphite	4,000,000	27,600	20,000	138
Alumina	5,200,000	35,880	80,000	552

An explanation for such an obvious discrepancy has been formulated on the models of dislocation motion through a crystalline substance. Apparently, the passage of a dislocation through the lattice requires far less applied force than that required for rupturing atomic bonds (σ_{max}). Why is this so? Consider the schematic representation of the edge dislocation in Figure 4-7. If a shear stress (τ) is applied to this crystal along the slip plane, the atoms just ahead of the dislocation resist its movement because it tends to push them out of their lattice sites. However, the atoms behind the dislocation tend to push it forward so they can achieve new stable lattice sites. The dislocation is in a state of "quasi-equilibrium," being acted on by both positive and negative forces along the slip plane. In theory the resistance to its passage is approximately zero! The movement of a dislocation through a crystal is illustrated schematically in Figure 4-11.

Don't forget that there are extremely high numbers of these defects in typical engineering materials. Dislocation densities on the order of 10^6 to 10^8 dislocation lines/cm^2 are not uncommon. So, in spite of their individual, atomistically small

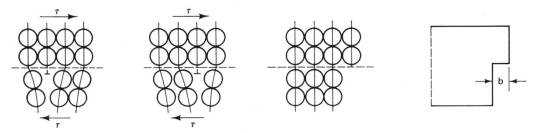

Figure 4-11 Example of dislocation movement through a crystal (unit slip).

nature, these crystalline imperfections can have an enormous effect on the deformation behavior of the crystals because of their huge numbers and their unique property of relatively easy motion. As the schematic illustrates in Figure 4-11, once the dislocation has passed through the crystal, the surface is displaced by one atomic distance. Scale up this process by the quantities of dislocations that routinely exist in engineering materials and you can appreciate their influence on permanent (plastic) deformation at seemingly low applied forces.

Sir Nevill Mott has called the passage of a dislocation across a slip plane analogous to the movement of a wrinkle or a rumple in a heavy rug. A large heavy rug can present formidable resistance to being pulled across a floor. However, if one places a wrinkle along the edge and progressively pushes the wrinkle through the length of the rug, surprisingly little effort is required. After the wrinkle has passed through the length of the rug, similar to a wave, the entire rug is displaced by the dimension of the wrinkle. This process is shown schematically in Figure 4-12. The student may find

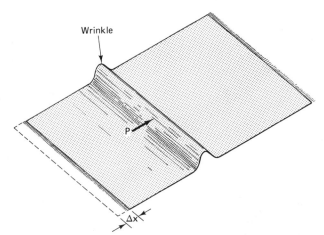

Figure 4-12 Schematic analogy of dislocation motion and resultant displacement, using a large carpet.

this analogy humorous, but it serves to illustrate dislocation motion very neatly. The relatively small force or stress required to move the dislocation is represented by P, the force on the wrinkle, and the eventual rug displacement Δx represents the subsequent atomic displacement.

Line Defects

Summary

In summarizing our brief discussion of dislocations and their effect on deformation, the important points to remember are:

1. Dislocation models offer a reasonable explanation of the huge discrepancy between the theoretical or ideal strength of crystalline materials and the actual strengths observed.
2. Dislocation densities on the order of 10^6 to 10^8 dislocation lines/cm^2 typically exist in commercial, crystalline engineering materials.
3. Dislocations cannot end abruptly within crystals. They either form closed loops or intersect some type of surface.

SURFACE DEFECTS

There are a number of crystalline imperfections that exhibit the features and dimensions of a surface. These defects may be broadly classified as follows: *grain boundaries, interphase interfaces, domain boundaries, stacking faults,* and *twin boundaries.* In this section we discuss these types in a basic manner.

Grain Boundaries

Internally, crystalline engineering materials such as metals and ceramics may be composed of just one structural arrangement (i.e., bcc, fcc, hcp) at a particular tem-

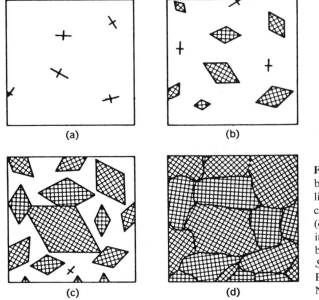

(a)   (b)

(c)   (d)

Figure 4-13 Formation of grain boundaries during solidification of a liquid metal: (a)–(c) nucleation of crystals—grain growth; (d) solidification completed. Dark, irregular, lines represent grain boundaries. (From C. Smith, *The Science of Engineering Materials*, Prentice-Hall, Inc., Englewood Cliffs, N.J., 1977, p. 101.)

perature. During crystallization, the aggregates of crystals (unit cells) take on various orientations, as illustrated in Figure 4-13. In this schematic representation, the heavy dark lines are the boundaries between grains. Within an individual grain, the unit cells have the same spatial orientation. These unit cells then repeat, enlarging the space lattice until they impinge on another grain. Such a crystallization process results in a *polycrystalline* material. The details of this *freezing* process (solidification) are discussed in Chapter 7.

Large-angle boundaries. Boundaries that separate adjacent grains differing in orientation by more than a few degrees are known as large-angle grain boundaries. The junction between two such grains is very likely to be separated by a thin layer (in the range 2 to 10 atoms) within which the atomic arrangement undergoes a gradual transition from conformity (registry) with the lattice of one grain into conformity with that of the other. An example of a large-angle grain boundary is shown in Figure 4-14.

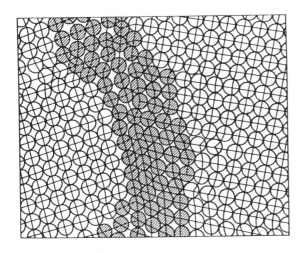

Figure 4-14 Large-angle grain boundary. Shaded "atoms" constitute grain boundary region. (From C. Smith, *The Science of Engineering Materials*, Prentice-Hall, Inc., Englewood Cliffs, N.J., 1977, p. 103.)

The shape of a grain, then, is influenced by the orientation of the unit cells and the presence of surrounding grains. In any specific grain, the atoms are arranged in a preferred orientation, characterized by the unit cell structure. This arrangement repeats periodically until a boundary is formed with an adjacent grain. The boundary is actually a transition zone from one preferred crystal orientation to another.

Small-angle boundaries. Boundaries that differ in orientation by less than a few degrees are referred to as small-angle grain boundaries. Sometimes these small-angle boundaries occur as subgrain boundaries within a single grain. In such cases, the grain boundary has been described as an array of dislocations, as illustrated in

Surface Defects

97

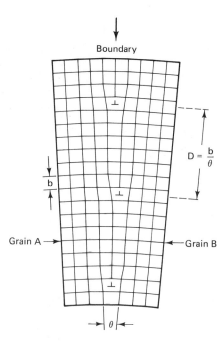

Figure 4-15 Low-angle grain boundary formed by an aligned dislocation array (arrow denotes boundary line).

Figure 4-15. The existence of extra "half-planes" of atoms produces a boundary which is highly regular except for the distortion associated with the dislocation. In this schematic, $D = b/\theta$ is an estimate of the spacing between dislocations for small-angle mismatch.

In summarizing our brief discussion of grain boundaries, these imperfections are often conveniently displayed as two-dimensional features, as shown in Figure 4-16. However, in reality, grain boundaries are surfaces and thus should be treated as three-dimensional quantities. Figure 4-17 demonstrates this aspect of grain boundaries in a polycrystalline material.

Since the atoms are obviously packed less efficiently in the grain boundary regions, these atoms exhibit higher energy than that of the atoms located within the grains. Furthermore, the grain boundary regions contain more void space than the grain interior, due to their irregularities. This feature has very important implications with respect to both mechanical and physical properties of the bulk engineering materials produced by these crystals. For instance, the grain boundaries can act as a *sink* for impurity or tramp elements, thus lowering their resistance to fracture. Also, since the grain boundary region is the last portion of the liquid to solidify (as the growing grains impinge on each other; see Figure 4-13), low-melting-temperature phases can be entrapped along these boundaries. The mechanism for this behavior is explained fully in Chapter 7. However, it is important to recognize that the entrapment of any "second" phases at the grain boundaries generally weakens these regions and increases their susceptibility to *intergranular failure*. Such failure may be referred to as intergranular attack (IGA) or environmental attack, and includes hydrogen embrittlement

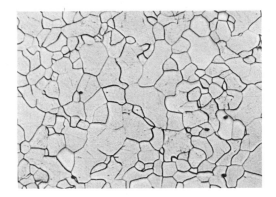

Figure 4-16 Microstructure of polycrystalline low-carbon steel (60 ×). (Courtesy of T. V. Brassard, Benét Laboratories.)

Figure 4-17 SEM micrograph illustrating three-dimensional nature of polycrystalline TiN (340 ×). (Courtesy of L. J. McNamara, Benét Laboratories.)

and stress corrosion in certain alloys, as we discuss in Chapter 17. Finally, the grain boundaries may provide ready avenues for diffusion (migration of atoms and ions) in the solid state, as we will see in Chapter 5.

Grain size. The grain size in an engineering material generally refers to the type of microstructural features illustrated in Figure 4-16. In single-phase materials the grains simply differ from one another in crystallographic orientation as depicted in Figure 4-13(d). However, in multiphase materials, one phase can predominate and the grains of this phase may be contiguous. In the latter case, the grain size of the material usually refers to the contiguous or matrix phase.

In the preceding section we indicated that grain boundary regions can adversely affect the mechanical and physical properties of crystalline engineering materials such as metals and ceramics. Therefore, it is important to control grain size, thus the amount of grain boundary region, during the various processes that are routinely performed in the production of these materials. Such control then requires that we be able to measure and specify grain size.

One method that is commonly applied to the problem of estimating grain size is known as the *intercept analysis*. In contrast to counting the actual number of grains in a specific area (i.e., the field of view in a microscope) and relating this count to the number of grains in an attendant volume, the intercept method relates the number of grain boundary intersections on a *test* line of known length to an equivalent ASTM[8] grain size number (g_e). Mathematically, the grain size number can be expressed as follows:

$$g_e = -10.00 + 6.64 \log_{10} \frac{NM}{C} \tag{4-12}$$

[8]ASTM, American Society for Testing and Materials.

where N is the number of intersections on a test figure (the line can be curved) of perimeter length C, viewed at a magnification of M. To facilitate grain size determinations, this relationship can also be expressed by the nomograph (for a 10-cm test circle) shown in Figure 4-18.

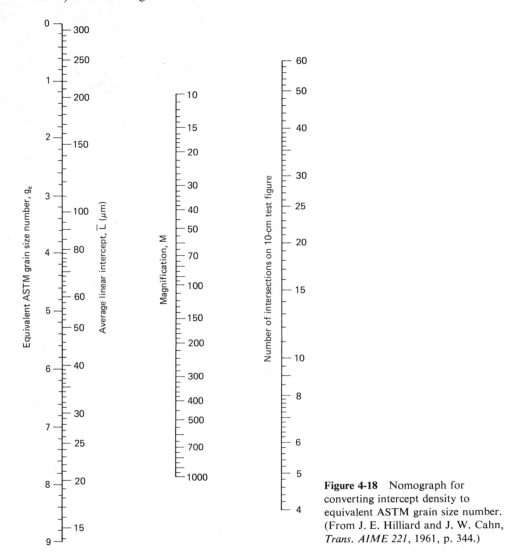

Figure 4-18 Nomograph for converting intercept density to equivalent ASTM grain size number. (From J. E. Hilliard and J. W. Cahn, *Trans. AIME 221*, 1961, p. 344.)

The ASTM grain size number indicates the relative size of the grains and is widely used throughout industry as a convention for expressing grain size. Note in Figure 4-18 that as the grain size number (g_e) increases, the value of the average linear intercept (the average chord length of intercept in the grains) decreases. Thus, as the actual size of the grains decreases, the ASTM grain size number increases.

Example 4-3

During the examination of a polycrystalline material a 20-cm test circle is applied to the microstructure at a magnification of $500\times$. Thirteen intersections of grain boundaries with the test circle are counted. Determine the equivalent ASTM grain size number (g_e).

Solution Utilizing equation (4-12), we can estimate the ASTM grain size number as follows:

$$g_e = -10.00 + 6.64 \log_{10} \frac{(13)(500)}{20}$$

$$= -10.00 + 6.64 \log_{10} 325$$

$$= -10.00 + 6.64(2.5)$$

$$= -10.00 + 16.7 = 6.7$$

$$= 6.7 \quad Ans.$$

Stacking Faults

Another type of crystalline defect which exhibits surface dimensions is the stacking fault. This imperfection consists of a disruption in the orderly sequence of stacking of close-packed planes in close-packed crystal structures, such as fcc and hcp. Two representations of the stacking fault in fcc structures are illustrated in Figure 4-19. An *intrinsic* fault, one in which part of a layer (plane) of atoms is missing from the

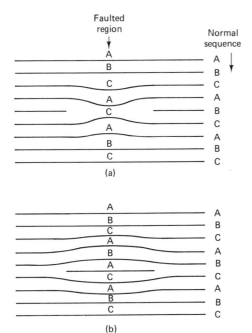

Figure 4-19 Schematic representation of stacking faults in fcc structure: (a) intrinsic fault; (b) extrinsic fault.

natural stacking sequence of the lattice, is shown in part (a). An *extrinsic* fault, consisting of an extra partial layer between planes, is illustrated in part (b).

Stacking faults may be produced by two different mechanisms. First, if the growth of the crystalline structure during solidification is disturbed and a new layer starts incorrectly, it may be incorporated into the structure due to rapid freezing. Such a sequential imperfection is termed a *growth fault*. A second mechanism involves the dissociation of a dislocation into two partial dislocations, producing a configuration known as an *extended* dislocation. The stacking in the region between the two partial dislocations is disrupted (A-B is converted to A-C) and this condition is termed a *deformation fault*.

Other Surface Imperfections

There are several other types of surface defects produced in crystalline materials. These imperfections are also important with respect to the mechanical and physical behavior. However, we will introduce them only briefly at this point, and treat them in more detail in later chapters where they are appropriate.

For example, *twin boundaries* are formed under certain conditions when a highly ordered type of deformation known as *twinning* occurs. Twinning involves the movement of each consecutive atomic plane in the twinned region, a fixed fraction of an interatomic distance with respect to its neighbor (see Figure 6-20). Twinning occurs spontaneously in most fcc materials at elevated temperatures (*annealing twins*), presumably nucleated at stacking faults. In the case of hcp and bcc materials, *mechanical* or *deformation* twins form under the action of applied forces. The subject of twinning is discussed in Chapter 6 with regard to plastic deformation.

Another important type of surface defect in engineering materials is the *domain boundary*. These imperfections are associated with certain regions within grains where ordering of the atoms occurs (see Figure 6-11). Ordering refers to the non-random arrangement of dissimilar atoms in a particular region of the crystal (domain). At the domain boundary like atoms face each other, while inside the domain each atom is surrounded by atoms of the other species.

Finally, since many commercial engineering materials are *polyphase* rather than pure metals, pure compounds, or single-phase materials, an interface exists between the phases. Such a surface imperfection is referred to as an *interphase interface*. The interface formed between two phases in a material is usually noncoherent (crystallographically incompatible) under equilibrium conditions.[9] Such surface imperfections are important from the standpoint of strength and hardening in certain materials.

SUMMARY

Although we have partially summarized certain sections in this chapter, it will be worthwhile to point out that many "interactions" are possible between the various

---

[9]Equilibrium conditions in this context implies sufficient time for a reaction to reach a steady-state condition.

crystalline defects just discussed. Very often these interactions can have a pronounced effect on both the structure-insensitive and the structure-sensitive properties, as listed in Table 4-3. Structure-insensitive properties depend on the crystalline relationships and atomic behavior we have discussed in Chapters 2 through 4, whereas structure-sensitive properties primarily depend on the thermal and mechanical history of the material. But in the latter case, crystalline imperfections can account for the discrepancies between experiment or actual values, and theory.

TABLE 4-3 PROPERTY CLASSIFICATIONS

Characteristic	Material property	
	Structure insensitive	Structure sensitive
Mechanical	Elastic modulus Density	Elastic limit Hardness Ductility Ultimate tensile strength
Thermal	Melting temperature Thermal conductivity Thermal expansion Specific heat	
Optical	Reflectivity Emissivity	
Electrical	Electrical conductivity Thermoelectric properties Electrochemical potential	Semiconduction Superconductivity
Magnetic	Saturation magnetization Diamagnetism Paramagnetism	Induction Permeability Magnetostriction Hysteresis loss
Nuclear	Nuclear cross section	Radiation damage

The importance of understanding the various types of defects and how they interact with each other and with the "defect-free" lattice will become more and more evident as we progress. A fundamental background in both crystal structure and crystalline defects is essential in interpreting, analyzing, and predicting the behavior of metallic and ceramic engineering materials.

STUDY PROBLEMS

4.1. Based on the relationship between vacancies (n) and temperature (T) given in equation (4-1), sketch a plot showing the effect of temperature on vacancy concentration in a material with N atoms.

4.2. A solid cylindrical bar of silicon 2 cm in diameter by 20 cm long contains 3.1×10^2 vacancies/cm³ at 300°C and 2.5×10^9 vacancies/cm³ at 600°C. Construct a plot of ln n versus $1/T$ and determine the activation energy for vacancy formation in this material.
Answer: 2.28 eV

4.3. Based on the results of problem 4.2, calculate the number of atoms (N) contained in the bar of silicon.

4.4. Calculate the equilibrium number of vacancies that will occur in copper at 300°C. The activation energy for vacancy formation in copper is approximately 2×10^4 cal/mol of vacancies.
Answer: 1 vacancy for 3.8×10^7 atoms

4.5. Defect concentration in ionic materials may be determined from: $n/N = \exp(-\Delta g/ 2kT)$, where Δg is the energy of defect formation. If Δg for Frankel defects in silver bromide (AgBr) is 2 eV, determine the equilibrium concentration of defects at 1000°C.

4.6. At 25°C the equilibrium concentration of vacancies in gold is 1.5×10^6 vacancies/cm³. Determine the activation energy in eV for vacancy formation in this material. [*Hint:* Use equation (4-1).]
Answer: $E_s = 1$ eV

4.7. How many tetrahedral interstices exist in a fcc unit cell? How many octahedral interstices? Explain how you arrived at your answers.

4.8. Sketch a bcc unit cell (see Figure 3-8) and show the location of tetrahedral interstices in this structure. How many are there?

4.9. Repeat problem 4.8 for the octahedral interstices in a bcc unit cell. How many sites are there?

4.10. The solubility of substitutional impurity atoms may also be expressed by the relationship given in equation (4-1). If the energy associated with introducing each impurity atom (n) into a crystalline substance is 1.25 eV, determine the temperature at which the equilibrium composition would be 0.05 atomic percent impurities.
Answer: 1639°C

4.11. The following figure represents an edge dislocation in a crystal. Show by the appropriate number of sketches the movement of this defect through the crystal. How many unit slip movements are necessary?

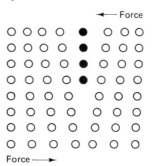

4.12. An annealed copper wire 6.5 mm in diameter is drawn through a series of dies, at room temperature, until it is 2 mm in diameter. Estimate the change in the number of dislocations that a cross section of this wire may contain.

4.13. Estimate the dislocation density of the material shown in Figure 4-8(b).

4.14. If the equilibrium separation between ions in alumina is 2 Å and the elastic modulus is 53 million psi, determine the energy (γ) associated with producing new surfaces during fracture based on the data in Table 4-1. Express your answer in SI units.

4.15. Rank the following engineering materials according to an estimate of their theoretical fracture strength (σ_{max}). Use a value of $E/10$ for your estimate. (*Hint:* Values of E for various materials may be found in Appendix 1 and certain tables in pertinent chapters throughout this text.)

  (a) Copper   **(b)** Graphite (high strength)

  (c) SiC (dense)   **(d)** Aluminum

  (e) TiC   **(f)** Tungsten

  (g) Beryllium   **(h)** Pyrex glass

Express your answer in both SI and English units.

4.16. Comment on the difference between grain boundary regions and the interior of crystalline grains. Can you think of a mechanism whereby the grain boundaries can actually move through a crystalline material?

4.17. Determine the distance (D) between dislocations in a small-angle grain boundary if the angle of mismatch is 2° and the Burger's vector (b) associated with the dislocations is 2.5 Å.

4.18. A polished sample of polycrystalline brass (alloy of copper and zinc) is microscopically examined at $100\times$ with a 10-cm test circle. If eight intersections of the test circle with grain boundaries are noted, calculate the ASTM grain size number (g_e).
Answer: $g_e = 2.6$

4.19. Using a 10-cm test circle, determine the ASTM grain size number for the carbon steel shown in Figure 4-16.

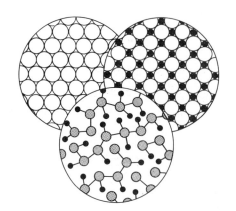

CHAPTER FIVE:
Material Transport in
Engineering Solids

Solid-state diffusion is the process whereby atoms (and ions) migrate or move through the lattice of crystalline structures and other solid nonmetallic materials. Diffusion in solids often has been analogously compared to the erratic *Brownian movement* of particles in suspensions, as illustrated in Figure 5-1. This example depicts the particles (or atoms) in ceaseless, random motion.

Basically, the movement of atoms in solid engineering materials is related to chemical composition, temperature, and crystallographic defects. When diffusion takes place in a solid solution, the net movement of solute atoms is away from regions where they are most highly concentrated. The result is an improvement in chemical homogeneity and the composition of the solid solution becomes more uniform. This initial condition of nonuniformity (i.e., segregation of solute) is explained in Chapter 7 and is shown in Figure 5-2. Interestingly, the nonuniform distribution of alloying elements or *concentration gradient*, as it is often referred to, provides the driving force for diffusion.

Diffusion also occurs in relatively pure metals, where atoms of the same species exchange lattice positions within the crystalline solid. Such a process is known as *self-diffusion* and several mechanisms are proposed for this type of atomic movement.

Many important chemical and material-related processes are dependent on or at least related to diffusion. These include sintering (coalescence of powder particles at elevated temperature), precipitation, grain growth, oxidation and corrosion, creep (time-dependent, elevated-temperature deformation) and various other *thermally* activated processes. The thermal treatments may be applied to ferrous alloys (steels) in order to soften, homogenize, or harden these materials. Or they may be appropriate for nonferrous alloys such as aluminum, copper, and the nickel-based superalloys, where the treatments consist of solutionizing and precipitation hardening (aging).

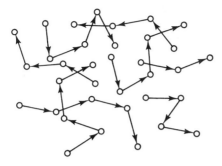

Figure 5-1 Brownian movement of suspended particles.

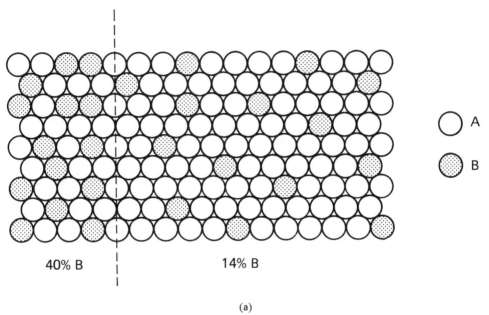

40% B   14% B

(a)

(b)

Figure 5-2 (a) Atomistic representation of solute segregation (%B) in a solid solution; (b) microstructure showing nonuniform distribution of chemical elements in a steel forging (15×). Dark bands are rich in solute.

Furthermore, the production of certain semiconductor devices and micro-electronic circuits utilizes diffusion techniques to introduce impurity atoms (e.g., phosphorous or boron) selectively to the lattice of semiconductor materials such as silicon and germanium, thereby changing the electronic characteristics of these materials. Our study of diffusion will encompass the fundamental mechanisms whereby atoms move throughout solids and the factors that affect these movements.

MECHANISMS OF DIFFUSION

Vacancy Mechanism

In simple fashion, the vacancy mechanism may be thought of as the movement of an atom into an adjacent vacant lattice site. On the other hand, it is just as appropriate to view this exchange as the process of vacancy movement to an adjacent lattice site. Such a procedure is shown in Figure 5-3, which is a view of a single plane of atoms.

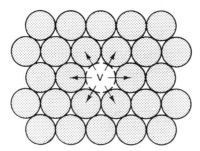

Figure 5-3 Possible vacancy movement in a close-packed plane.

Note that in this planar arrangement the vacancy can move to any one of the six neighboring atom positions shown. Furthermore, from our previous examination of crystal structures we know that the coordination number is the number of nearest neighbors. Therefore, a vacancy can exchange positions with any one of eight neighboring atoms in the bcc structure or with 12 neighboring atoms in fcc and hcp.

Vacancies, you will recall, were discussed as point defects in Chapter 4. There we determined that the number of vacancies increased as temperature increased. For instance, just below the melting point of copper one vacancy exists for roughly 1600 atoms. If this ratio, 1 : 1600, does not sound very impressive, consider the following illustrative example.

Example 5-1

Assume that the composition of a U.S. penny [Figure 5-4(a)] is 100% copper (actually it is 95% Cu–5% Zn). If the diameter of this coin is 1.9 cm, determine the approximate number of vacancies in a (100) plane parallel to the faces at a temperature of 1080°C.

Solution The (100) plane in fcc copper contains the equivalent of two atoms per unit cell face, as shown in Figure 5-4(b). From Table 3-3, the lattice parameter of Cu is

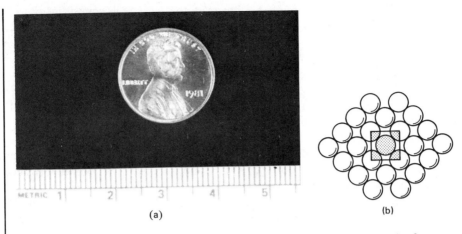

(a) (b)

Figure 5-4 (a) U.S. penny; (b) schematic illustration of (100) plane in fcc structure (unit cell face darkly outlined).

$$a_{fcc} = \frac{4r}{\sqrt{2}}$$

The radius of copper, $r_{cu} = 1.278$ Å. Therefore,

$$a = \frac{(4)(1.278 \text{ Å})}{\sqrt{2}}$$

$$= 3.61 \times 10^{-8} \text{ cm}$$

$$a^2 = 13 \times 10^{-16} \text{ cm}^2$$

Thus the number of atoms/cm^2 on this plane, (100), is

$$N_A = \frac{2 \text{ atoms}}{13 \times 10^{-16} \text{ cm}^2}$$

$$= 1.54 \times 10^{15} \text{ atoms/cm}^2$$

The area of the coin face is

$$A = \frac{\pi d^2}{4} = \frac{\pi (1.9 \text{ cm})^2}{4}$$

$$= 2.83 \text{ cm}^2$$

Therefore, the total number of atoms in a (100) parallel to the face is

$$N_T = (1.54 \times 10^{15} \text{ atoms/cm}^2)(2.83 \text{ cm}^2)$$

$$= 4.4 \times 10^{15} \text{ atoms}$$

This figure represents the approximate total number of lattice sites. If one vacancy exists for about 1600 atoms at this temperature, the total number of vacancies (N_V) in this plane is given by

$$N_V = (4.4 \times 10^{15} \text{ atoms})\left(\frac{1 \text{ vacancy}}{1600 \text{ atoms}}\right)$$

$$= 2.8 \times 10^{12} \text{ vacancies} \qquad Ans.$$

Mechanisms of Diffusion

The results of Example 5-1 illustrate that very high concentrations of vacancies can exist in crystalline solids. Perhaps this amount, 2.8×10^{12}, is more meaningful if expressed as 2.8 *trillion* vacancies, in essence, just on the face of a small coin. Clearly, vacant lattice sites by virtue of their numbers and possible jump directions are instrumental in the process of diffusion.

The actual process of vacancy diffusion occurs when the diffusing atom has sufficient energy to leave its equilibrium lattice position and *jump* into the adjacent vacant site. How often does such a jumping procedure take place? It has been estimated that in most solid metals near their melting temperature, each atom changes its lattice position about 10^8 times per second. If one considers this in terms of vibrational frequency of an atom,[1] the atom jumps to a new position roughly once in 10,000 to 100,000 oscillations.

Such a process can be pictured analogously by the sphere on an elastic band, as in Figure 5-5. As the amplitude of the sphere's vibration increases, its displacement

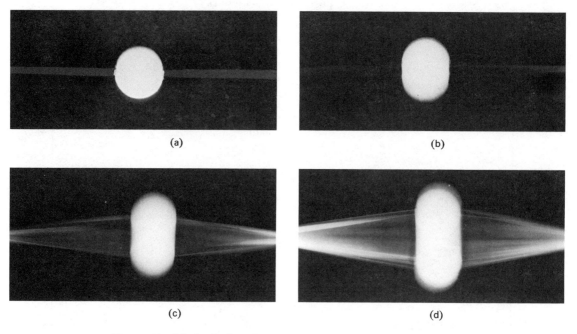

(a)

(b)

(c)

(d)

Figure 5-5 Mechanical analogy of atomic vibration: (a)–(d): vibrational amplitude increasing.

about an equilibrium position also increases. This mechanical analogy infers that the probability of an atom jumping to a new site is greater if the vibrational frequency is greater and that this event is strongly influenced by temperature.

The vacancy mechanism of diffusion can be schematized as shown in Figure

[1] The vibrational frequency (Debye frequency) of atoms in a solid metal near the melting point is 10^{12} to 10^{13} sec^{-1}.

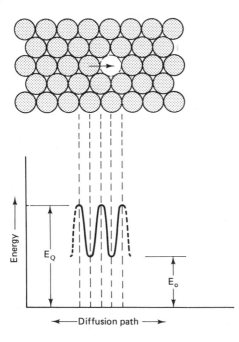

Figure 5-6 Schematic of vacancy mechanism and the potential energy associated with it. E_0, energy associated with equilibrium position; E_Q, energy barrier (activation energy).

5-6. This portrayal illustrates that for an atom to move into an adjacent vacancy it must overcome an energy barrier. The barrier is produced by the neighboring atoms in the vicinity of the diffusing atom and the vacancy. The diffusing atom must overcome its bonds and push by the atoms bordering the vacancy in order to reach the next equilibrium position in the lattice. This process corresponds to the hills and valleys of the energy curve depicted in Figure 5-6. For this event to occur, energy must be supplied to the atom. Where does the necessary energy come from? It is usually supplied in the form of heat. In the introduction to this chapter we cited a number of chemical and materials processes which were related to diffusion. Notice that typically they are thermally related processes.

Diffusion by the vacancy mechanism has been shown to be the principal mechanism of atom transport in fcc materials. Furthermore, this particular process has also been observed in many bcc and hcp metals and ceramic compounds.

Interstitial Mechanism

The interstitial method of diffusion consists of moving an atom from one interstitial site to another in the structure. Interstitial lattice sites were described in Chapters 3 and 4 and interstitial solid solutions are discussed in Chapter 6. According to the Hume–Rothery rules stated in Chapter 6, interstitial atoms are usually smaller than their host atoms: diameter of solute < 0.59 diameter of solvent. The number of possible interstitial sites depends on crystal structure, and as we discussed in Chapter 4, there are fourfold (tetrahedral) sites and sixfold (octahedral) sites in the cubic and hexagonal systems (see Table 4-1). For example, six potential octahedral jump sites are indicated for an interstitial atom in bcc in Figure 5-7.

Mechanisms of Diffusion

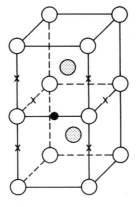

Figure 5-7 Bcc structure containing an interstitial atom (dark circle) located in an octahedral (6-f) site. The potential 6-f interstitial jump sites contained in this sketch are indicated by ×'s.

The interstitial mechanism of diffusion is illustrated in Figures 5-8 and 5-9 together with a portrayal of the energy associated with this particular process. In this model the diffusing atom must squeeze past the adjacent solvent atoms. Depending on their relative sizes, the structure, and the path the atom takes, this jump may require a considerable distortion of the host lattice. It is this displacement of the solvent lattice that comprises the bulk of the energy barrier shown in Figure 5-9. As the size of the interstitial increases relative to its host, the distortion associated with its jump also increases. Correspondingly, the energy required for this type of diffusion rises and it becomes more difficult for diffusion to take place.

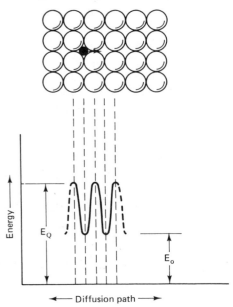

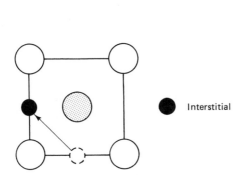

Figure 5-8 Schematic example of an interstitial jump in bcc (center atom is shaded).

Figure 5-9 Interstitial diffusion mechanism and the potential energy associated with it. E_0, energy associated with equilibrium position; E_Q, energy barrier (activation energy).

Ring Diffusion

This mechanism employs direct exchange of atoms on the lattice, as illustrated in Figure 5-10. The two-ring process is more difficult, especially in close-packed structures, due to the amount of distortion it produces. Multiatomic rings do not involve such large lattice distortions and therefore the three-ring and four-ring models are more energetically favorable.

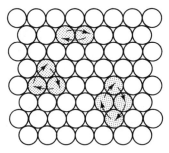

Figure 5-10 Ring diffusion models showing two-ring, three-ring, and four-ring exchanges. Diffusing atoms are shaded.

SELF-DIFFUSION

Ordinarily, no net diffusion of atoms is recognized in pure or relatively pure metals, although the atoms may exchange positions. This is because atom movements are random and all the atoms are physically and chemically identical. The same condition arises with the solvent species in dilute solid solutions. However, by utilizing radioactive isotopes[2] (tracers) of an element, this movement, termed *self-diffusion*, may be detected.

For example, gold has an atomic weight of 197. If a radioactive isotope of gold (^{195}Au) is electroplated on the surface of a normal gold specimen, depending on temperature and time, there will be a progressive migration of the ^{195}Au into the normal gold lattice. Such movement can be experimentally monitored by detecting the radioactive tracer as it proceeds throughout the specimen. Self-diffusion of ^{195}Au in normal gold is illustrated schematically in Figure 5-11. Examination of this figure shows that in part (a) all the tracer atoms (^{195}Au) are initially concentrated at the left surface. As time progresses however, the tracer, which is chemically identical to the regular atoms, moves through the lattice as depicted by parts (b) and (c). This process lowers the concentration of ^{195}Au at the surface and correspondingly increases the tracer fraction in the bulk solid. Eventually, the concentration of ^{195}Au reaches an equilibrium distribution in the gold lattice as shown in part (d). An examination of the diffusion models we examined previously shows that the tracer movement can be attributed to vacancy and ring mechanisms.

---

[2] Atoms of the same element with different masses are called *isotopes*. Isotopes of an element all have the same atomic number, because they have the same number of protons. But they have different atomic weights because they exhibit a different number of neutrons.

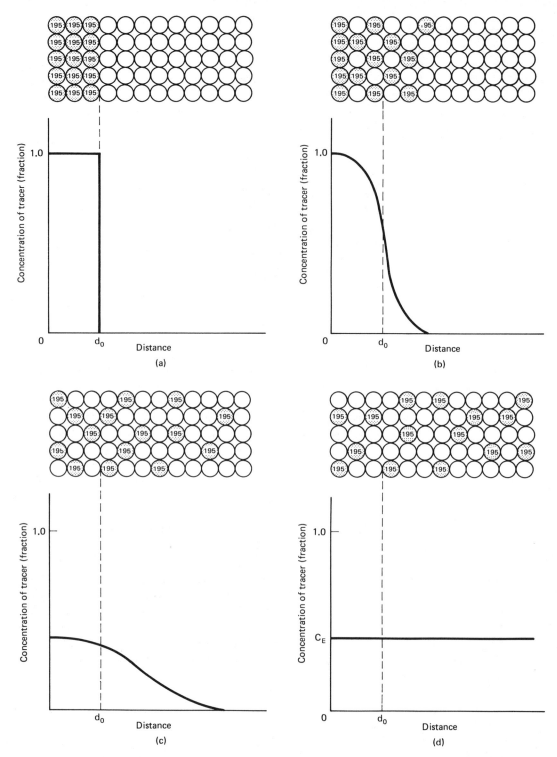

Figure 5-11 Schematic representation of self-diffusion, ^{195}Au into normal Au: (a) time equals zero (t_0); (b) time $t_1 > t_0$; (c) time $t_2 > t_1$; (d) time $t \gg t_0$, C_E is equilibrium concentration.

MATHEMATICAL ANALYSIS OF DIFFUSION

Many chemical and metallurgical processes, including diffusion, can be quantitatively analyzed by describing the rate at which the process occurs. In mathematical terms, very often the *Arrhenius equation* describes the relationship between the time rate of a particular process or reaction and temperature. This expression takes the following form:

$$\text{process rate} = Ae^{-Q/RT} \tag{5-1}$$

where A = constant (independent of temperature)
Q = activation energy for the process (cal/mol)
R = gas constant (2 cal/mol-°K)
T = absolute temperature

For instance, if equation (5-1) is applied to the movement of an atom into a neighboring vacancy, the number of jumps it makes per second can be estimated by the following equation:

$$r_v = Ae^{-Q_m/RT} \tag{5-2}$$

In this form, r_v is the jump rate, the number of times per second an atom and a vacancy exchange positions, and Q_m is the activation energy or energy barrier for such a process. Let us examine the movement of vacancies in copper by the following illustrative example.

Example 5-2

Compare the jump rate (r_v) for copper at temperatures of (a) 27°C and (b) 1080°C if $A_{cu} \simeq 10^{15}$ and $Q_m = 29,000$ cal/mol.

Solution (a) For temperature = 27°C, the absolute temperature is

$$27°C + 273°C = 300°K$$

Therefore,

$$r_v = (10^{15})e^{-29,000/(2)(300)}$$

$$= 1.1 \times 10^{-6} \text{ jumps/sec} \qquad \textit{Ans.}$$

(b) Similarly, for temperature = 1080°C, the absolute temperature equals 1353°K. Thus

$$r_v = (10^{15})e^{-29,000/(2)(1353)}$$

$$= 2.2 \times 10^{10} \text{ jumps/sec} \qquad \textit{Ans.}$$

Examination of the results of this simple example shows that at 1080°C (melting point of Cu = 1084.5°C), a mole[3] of copper exhibits a vacancy jump rate of approxi-

---

[3]The mole is that quantity of matter which contains 6.02×10^{23} atoms (Avogadro's number). Therefore, a mole of copper is 1 gram atomic weight or 63.54 g.

mately 20 billion jumps per second. But at room temperature, a time interval of about 10.5 days elapses between jumps. Clearly, this is an astonishing difference, but it succinctly emphasizes the important relationship between temperature and the diffusion of atoms by the vacancy mechanism in solid engineering alloys.

Steady-State Diffusion

Previously in this chapter we referred to the segregation of solute atoms as being the result of a concentration gradient (Figure 5-2). In this context a concentration gradient simply means that the solute concentration changes with distance. For example, consider the adjacent atomic planes separated by one interatomic distance (a) in a volume element of unit cross-sectional area as illustrated in Figure 5-12. The concen-

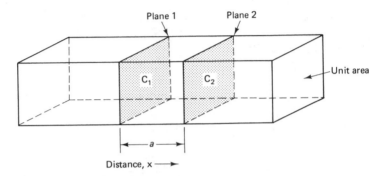

Figure 5-12 Rectangular volume of unit cross-sectional area containing adjacent atomic planes.

tration of solute atoms on plane 1 (C_1) is greater than on plane 2 (C_2). Therefore, we can state the respective jump rates (J) as follows:

$$J_{1,2} = \frac{nac_1}{2} \tag{5-3}$$

$$J_{2,1} = \frac{nac_2}{2} \tag{5-4}$$

$J_{1,2}$ indicates the number of solute atoms jumping from plane 1 to plane 2 per second, and $J_{2,1}$ indicates the solute jump rate from plane 2 to plane 1. In equations (5-3) and (5-4), n is the number of jumps per solute atoms (per second) in either direction, out of either plane. The net flux or movement of solute atoms (J) can be stated as

$$J = J_{1,2} - J_{2,1}$$

$$= \frac{nac_1}{2} - \frac{nac_2}{2}$$

$$= \frac{na}{2}(c_1 - c_2) \tag{5-5}$$

However, the concentration gradient (dc/dx) can be expressed as $c_1 - c_2/a$ in this situation. Thus

$$\frac{c_1 - c_2}{a} = \frac{dc}{dx}$$

$$c_1 - c_2 = a\frac{dc}{dx}$$

Substituting this result in equation (5-5), we find

$$J = \frac{na^2}{2}\frac{dc}{dx} \tag{5-6}$$

Equation (5-6) is usually rewritten as follows:

$$J = -D\frac{dc}{dx} \tag{5-7}$$

where $D = na^2/2$. This expression for diffusion [equation (5-7)], also referred to as the *diffusion equation*, is Fick's first law, and equates the flux of atoms (J) to the concentration gradient dc/dx. D, the proportionality constant in this equation, is the *coefficient of diffusion*. This parameter is also referred to as the *diffusivity* and depends on the nature of the solute and solvent, crystal structure, and temperature. Typical units for the terms in equation (5-7) are as follows:

$J = $ atoms/cm² sec

$-D = $ cm²/sec (negative sign indicates movement from higher to lower concentration)

$\frac{dc}{dx} = $ atoms/cm³-cm

In other words, the flux is the movement of mass past a unit area in unit time, under the influence of a uniform concentration gradient. These conditions represent *steady-state* diffusion.

As an exercise to familiarize the student with steady-state diffusion, let us apply Fick's first law to the movement of solute atoms in the presence of a concentration gradient.

Example 5-3

A sample of fcc iron (austenitic) contains electroplated nickel at the surface. The initial concentration of Ni in the surface is 1.5 a/o.[4] At a depth of 5 mm the concentration of Ni is 0.8 a/o. If the temperature of the sample is held at 1000°C, determine the flux of Ni atoms from the surface. D for this diffusion couple is $\simeq 3 \times 10^{-12}$ cm²/sec at this temperature.

[4]a/o denotes atomic percent as opposed to w/o or weight percent.

Solution The mass density of fcc iron is determined as follows:

$$\text{mass density} = \frac{\text{mass}}{\text{unit volume}}$$

or four atoms per unit cell where the unit cell volume is a^3.

$$a_{Fe} = \frac{4r_{Fe}}{\sqrt{2}} = \frac{(4)(1.269 \text{ Å})}{\sqrt{2}} = 3.59 \text{ Å}$$

$$= 3.59 \times 10^{-8} \text{ cm}$$

Therefore,

$$\text{mass density} = \frac{4 \text{ atoms}}{a^3}$$

$$\text{density} = 8.64 \times 10^{22} \text{ atoms/cm}^3$$

The concentration gradient can now be expressed as follows:

$$\left(\frac{dc}{dx}\right)_{Ni} = \frac{(0.0150 - 0.008)(8.64 \times 10^{22} \text{ atoms/cm}^3)}{0.5 \text{ cm}}$$

$$= 1.21 \times 10^{21} \text{ atoms/cm}^3\text{-cm}$$

Substituting this result in equation (5-7), we finally obtain the flux:

$$J_{Ni} = (-3 \times 10^{-12} \text{ cm}^2/\text{sec})(1.2 \times 10^{21} \text{ atoms/cm}^3\text{-cm})$$

$$= -3.6 \times 10^{9} \text{ atoms/cm}^2\text{-sec} \quad \textit{Ans.}$$

Non-Steady-State Diffusion

In the case of non-steady-state diffusion, we may expand Fick's first law by assuming the parallel planes in Figure 5-12 are separated by a distance (dx). According to equation (5-7), the flux of solute atoms into the volume element (described by the unit area) across plane 1 is

$$J_1 = -D\left(\frac{dc}{dx}\right)_1$$

Correspondingly, the flux of solute out of the volume element across plane 2 is

$$J_2 = -D\left(\frac{dc}{dx}\right)_2$$

Therefore, the net flux through the volume element is

$$J = J_1 - J_2$$

$$= -D\left(\frac{dc}{dx}\right)_1 - \left[-D\left(\frac{dc}{dx}\right)_2\right]$$

$$= D\left[\left(\frac{dc}{dx}\right)_2 - \left(\frac{dc}{dx}\right)_1\right] \tag{5-8}$$

where $(dc/dx)_1$ is the concentration gradient at plane 1 and $(dc/dx)_2$ is the concentration gradient at plane 2. But this latter gradient may also be expressed as follows:

$$\left(\frac{dc}{dx}\right)_2 = \left(\frac{dc}{dx}\right)_1 + dx\left(\frac{d^2c}{dx^2}\right)$$

where (d^2c/dx^2) is the rate of change in the concentration gradient with distance (x). Now we may rewrite equation (5-8) as follows:

$$J = D\left[\left(\frac{dc}{dx^1}\right) + dx\left(\frac{d^2c}{dx^2}\right) - \left(\frac{dc}{dx^1}\right)\right]$$

which reduces to

$$J = D\frac{d^2c}{dx^2}\,dx$$

or

$$\frac{J}{dx} = D\frac{d^2c}{dx^2} \tag{5-9}$$

But

$$\frac{J}{dx} = \frac{\text{atoms}}{\text{cm}^3\text{-sec}} = \frac{dc}{dt} \tag{5-10}$$

where dc/dt is the change in volume concentration with respect to time. Substituting this result in equation (5-9), we may finally write

$$\frac{dc}{dt} = D\frac{d^2c}{dx^2} \tag{5-11}$$

which means that the time rate of concentration change is proportional to the rate of change of the concentration gradient. This relationship is known as *Fick's second law* of diffusion.

A very useful solution of equation (5-11) can be expressed as

$$\frac{C_s - C_x}{C_s - C_0} = \text{erf}\left(\frac{x}{2\sqrt{Dt}}\right) \tag{5-12}$$

where C_s is the concentration at the surface, C_x is the concentration at some depth (x) from the surface at time t, and C_0 is the initial concentration of the diffusing species in the material (solvent). You will recall that *erf* is the error function, which is found in standard math tables. However, if we are interested in a specific composition (C_x), this expression may be reduced to

$$\frac{x}{\sqrt{Dt}} = \text{constant} \tag{5-13}$$

Therefore, at a given temperature the depth (x) of the composition (C_x) varies with the square root of diffusivity and time. The quantity $\sqrt{Dt}$ can be used to estimate the distance over which appreciable diffusion will take place. For example, if C_x is equal to $\frac{1}{2}C_s$, and $C_0 = 0$, equation (5-12) reduces to

$$\text{erf}\left(\frac{x}{2\sqrt{Dt}}\right) = 0.5$$

From standard mathematical tables,[5]

$$\frac{x}{2\sqrt{Dt}} = 0.5$$

or

$$x = \sqrt{Dt}$$

Thus, for estimation purposes, we can simply relate diffusion distance and time for a particular temperature.

Diffusion Coefficients

The proportionality coefficient (D) in the diffusion equation is influenced by several factors, including the nature of the solute and solvent atoms, bond strength, and the crystal structure. Furthermore, this coefficient is a function of temperature. Diffusion coefficients for some selected metals are shown in Table 5-1 for a temperature of 500°C.

TABLE 5-1 APPROXIMATE DIFFUSION COEFFICIENTS FOR CERTAIN METALS AT 500°C

Solute	Solvent	D (cm²/sec)
Ag	Ag	1.3×10^{-13}
	Al	5.5×10^{-6}
	Sn	1.7×10^{-1}
Au	Au	3.4×10^{-14}
	Ag	1.0×10^{-13}
	Bi	1.9×10^{-1}
	Pb	1.3×10^{-1}
	Sn	1.9×10^{-1}
C	Fe (bcc)	6.3×10^{-8}
	Fe (fcc)	5.0×10^{-11}
Cu	Cu	7.9×10^{-16}
	Al	5.0×10^{-10}
Fe	Fe (bcc)	6.7×10^{-17}
	Fe (fcc)	3.1×10^{-20}
Mg	Mg	8.3×10^{-10}
	Al	5.5×10^{-6}
Mn	Fe (fcc)	2.5×10^{-20}
Ni	Ni	5.2×10^{-20}
	Fe (fcc)	1.0×10^{-19}
Pb	Pb	7.9×10^{-8}
	Sn	1.3×10^{-1}
Si	Al	7.0×10^{-6}
Zn	Zn	8.2×10^{-8}
	Al	1.0×10^{-5}
	Cu	6.3×10^{-13}

[5] *CRC Standard Mathematical Tables*, CRC Press, Boca Raton, Fla.

The dependence of D on temperature can be expressed by the mathematical relationship presented in equation (5-1). The diffusivity (D) is given by

$$D = D_0 e^{-Q/RT} \qquad (5\text{-}14)$$

where D_0 = diffusion constant (independent of temperature) (cm²/sec)
Q = activation energy for diffusion (cal/mole)
R = gas constant (2 cal/mol-°K)
T = absolute temperature (°K)

The temperature dependence of this relationship becomes clearer if we take the natural logarithim of equation (5-14) and rearrange as follows:

$$\ln D = \ln D_0 e^{-Q/RT}$$
$$= \ln D_0 + \ln e^{-Q/RT}$$

Since the $\ln e^x = x$, we may simplify this expression to

$$\ln D = \ln D_0 - \frac{Q}{RT} \qquad (5\text{-}15)$$

The student may recognize the form of equation (5-15) to be a straight line ($y = mx + b$) where

$$\ln D = y = \text{dependent variable}$$

$$\ln D_0 = b = \text{intercept}$$

$$-\frac{Q}{R} = m = \text{slope}$$

$$\frac{1}{T} = x = \text{independent variable}$$

Experimental data for diffusivity generally take the form illustrated in Figure 5-13,

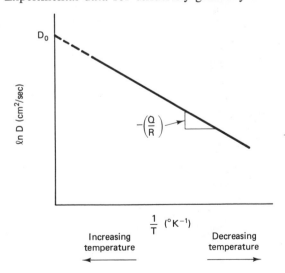

Figure 5-13 Diffusivity (D)–temperature (T) relationship.

and D_0 and Q are empirically determined from this plot. Values of D_0 and Q are given in Table 5-2 for self-diffusion and for selected diffusion couples. These data plus the relationship presented in equation (5-14) can be utilized to calculate D as a function of temperature. The temperature dependence of D is shown in Figure 5-14 for several systems of interest.

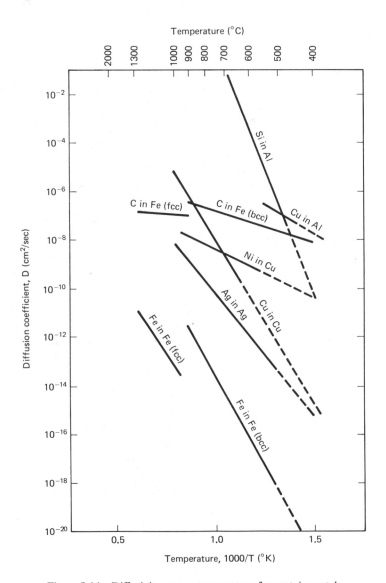

Figure 5-14 Diffusivity versus temperature for certain metals.

TABLE 5-2 DIFFUSION CONSTANTS FOR SELECTED METALS

Solute	Solvent	D_0 (cm^2/sec)	Q (cal/mol)
Al	Al	1.71	34,000
Cu	Cu	0.6	49,000
	Al	2.0	33,900
C	Fe (bcc)	0.01	18,100
	Fe (fcc)	0.21	33,800
Au	Au	0.2	45,000
Fe	Fe (bcc)	5.8	59,700
	Fe (fcc)	0.6	67,900
Mg	Mg	1.0	32,000
Mn	Fe (fcc)	0.4	67,500
Ni	Ni	3.36	69,800
	Fe (fcc)	0.5	66,000
Ag	Ag	0.9	45,000
Zn	Zn	0.4	23,800
As	Ge	6.0	57,500
P	Ge	2.0	57,500
B	Si	10.5	85,000
Al	Si	8.0	80,000
As	Si	0.32	82,000

Example 5-4

Based on the information in Table 5-2, determine the diffusion coefficient (D) for carbon in fcc iron at (a) 500°C and (b) 1000°C.

Solution Examination of Table 5-2 shows that D_0 for C in fcc iron is 0.21 cm^2/sec and $Q = 33,800$ cal/mol. Substitute this information in equation (5-14) for the appropriate temperatures.

(a) For $T = 500$°C:

$$T = 500°C = 773°K$$

$$D_C = (0.21 \text{ cm}^2/\text{sec})e^{-33,800/(2)(773)}$$

$$= 6.7 \times 10^{-11} \text{ cm}^2/\text{sec} \quad Ans.$$

(b) For $T = 1000$°C:

$$T = 1000°C = 1273°K$$

$$D_C = (0.21 \text{ cm}^2/\text{sec})e^{-33,800(2)(1273)}$$

$$= 4 \times 10^{-7} \text{ cm}^2/\text{sec} \quad Ans.$$

Comparison of the results shows that the diffusivity of carbon in fcc iron increases by about 10^4 between 500 and 1000°C.

Mathematical Analysis of Diffusion

Example 5-5

Consider the problem where aluminum is diffused into pure silicon to make a semiconductor material. How long should the silicon be heated at 1300°C in contact with the aluminum such that the aluminum concentration at 0.01 mm below the surface is equal to half the concentration of aluminum at the immediate surface?

Solution From Table 5-2 we see that D_0 for Al in Si is 8.0 cm²/sec and Q equals 80,000 cal/mol. Therefore,

$$D = D_0 e^{-Q/RT}$$

$$= \left(8.0\, \frac{cm^2}{sec}\right) \exp\left[-\frac{80,000\ cal/mol}{(2\ cal/mol\text{-}°K)(1573°K)}\right]$$

$$= 7.2 \times 10^{-11}\ cm^2/sec$$

Utilizing this value and the relationship $x = \sqrt{Dt}$ for this particular situation, we have

$$t = \frac{x^2}{D} = \frac{(0.001\ cm)^2}{7.2 \times 10^{-11}\ cm^2/sec}$$

$$= \frac{1 \times 10^{-6}}{7.2 \times 10^{-11}}\ sec$$

$$= 13,888\ sec\ or\ 3.9\ hr \qquad Ans.$$

DIFFUSION IN GRAIN BOUNDARIES

So far our discussion of diffusion has been concerned with the movement of atoms through the lattice or within grains. This type of diffusion is often referred to as *volume diffusion*. However, we have previously studied the crystallography of engineering materials and realize that where the aggregates of unit cells impinge on each other, *grain boundaries* are formed. Due to the energy associated with the atoms at the grain boundaries and the crystallographic mismatch between adjacent grains, which can result in a high vacancy concentration, these regions generally display high diffusivity compared to that of the lattice.

Self-Diffusion in Grain Boundaries

Analysis of self-diffusion in grain boundaries is possible by conducting radioactive tracer experiments on materials with varying grain size. In fact, measurements of D have been made on metals in the polycrystalline condition versus a single crystal (no internal grain boundaries). The results of this comparison for self-diffusion in silver are shown in Figure 5-15. In this example, the diffusivity of Ag is the same for both samples at temperatures above approximately 700°C. But below this temperature, the values of D in the polycrystalline sample (D_{app}) are greater than those of the single crystal (D_L). Notice also that the activation energy for lattice diffusion ($Q = 45,950$) is almost twice the value for that of grain boundary diffusion ($Q = 26,400$). This is a clear indication that diffusion along the grain boundaries is more energetically favorable.

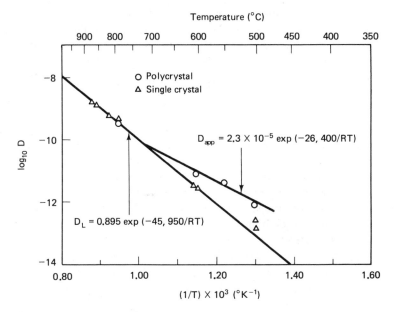

Figure 5-15 Diffusivity for single-crystal and polycrystalline silver. (From *Atom Movements*, American Society for Metals, Metals Park, Ohio, 1951, p. 141.)

Example 5-6

As an exercise, determine the difference in diffusivity (D) between polycrystalline Ag and single crystal Ag at 500°C.

Solution   Utilizing equation (5-14) and the empirical information in Figure 5-15, we obtain

$$\text{Single crystal:}\quad D_L = 0.9e^{-46,000/(2)(773)}$$
$$= 1 \times 10^{-13}\ \text{cm}^2/\text{sec}$$
$$\text{Polycrystal:}\quad D_{\text{app}} = (2.3 \times 10^{-5})e^{-26,400/(2)(773)}$$
$$= 8.8 \times 10^{-13}\ \text{cm}^2/\text{sec}$$

Therefore, the difference in diffusivity amounts to approximately a factor of 10:

$$D_L = 10^{-13}\ \text{cm}^2/\text{sec}$$
$$D_{\text{app}} \simeq 10^{-12}\ \text{cm}^2/\text{sec} \qquad \textit{Ans.}$$

In other words, at this temperature, diffusion takes place at a rate 10 times greater in the material containing the grain boundaries.

Solute Diffusion along Grain Boundaries

In many instances, diffusion of solute elements along the grain boundaries of a solvent also takes place at a greater rate than solute diffusion in the solvent lattice. For example, the diffusivity of thorium (Th) in tungsten (W) has been examined in considerable detail because concentration differences are relatively easy to detect (by emissivity measurements) in this system. Diffusion of Th along the grain boundaries of W can be compared with lattice diffusion, respectively, by the following expressions:

$$D_{\text{G.B.}} = 0.74e^{-90,000/RT}$$

$$D_L = 1.0e^{-120,000/RT}$$

For such a comparison we can easily see that the activation energy (Q) for solute diffusion in the solvent lattice is significantly greater (by one-third, in fact) than diffusion along the grain boundaries in this situation.

As was the case for self-diffusion in grain boundaries, solute diffusion along the grain boundaries of the solvent tends to be more important at relatively low temperatures. Therefore, such high-diffusivity paths can be a problem in metals and ceramics that exhibit relatively low lattice diffusivity. For instance, diffusion of solute species and foreign atoms along grain boundaries in a solvent has been associated with intergranular attack or degradation of a host of engineering materials. In general, such forms of grain boundary deterioration include corrosion, embrittlement, and fracture.

Grain Boundary Migration

Finally, in this discussion of grain boundary diffusion we should recognize that grain boundaries *themselves* can move or migrate. Therefore, when one discusses grain boundary diffusion, a distinction must be made whether diffusion along the boundaries is involved or whether movement of the grain boundaries is actually occurring.

Simplistically, the migration of grain boundaries can be envisioned as a net flow of atoms or ions from the surface of one grain to the surface of an adjacent grain. In other words, atoms simply jump across the boundary region and occupy positions on the other side. Consequently, the grain boundary moves in a direction opposite to the net flow of atoms. This process is pictured in Figure 5-16. Such a transport mechanism involves individual atom movements on a relatively small scale and does *not* necessarily depend on concentration gradients. Essentially, the driving force for grain boundary migration in an engineering material depends primarily on; surface tension on the grain boundary, stored energy resulting from permanent deformation, the structure of the boundary region, and temperature.

As your study of engineering materials progresses, the role of grain boundary migration will become more apparent, especially in understanding both certain types of mechanical behavior and the fundamentals of materials processing to develop desirable properties. For example, deformation and failure of structural materials in elevated temperature applications often involves grain boundary movement. The consolidation of metal or ceramic powders to produce dense, solid parts also involves grain boundary migration, in a process called *sintering*. In general, grain growth or the enlargement of grains in a polycrystalline material is the direct result of grain boundary migration.

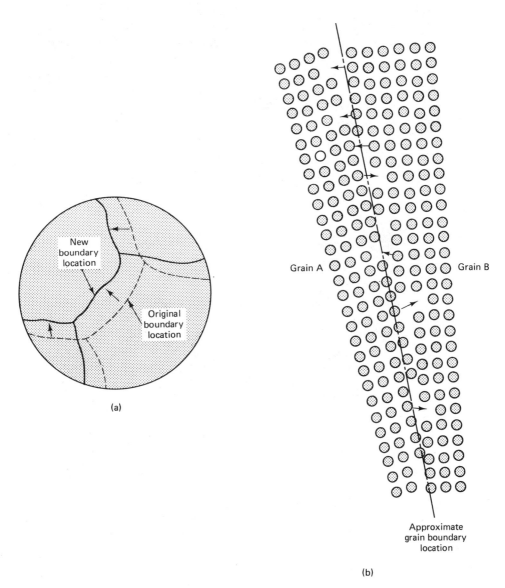

Figure 5-16 Schematic illustration of grain boundary diffusion: (a) migration of a grain boundary toward its center of curvature; (b) random atom jumps across a grain boundary into vacant lattice sites.

DIFFUSION IN NONMETALLIC SOLIDS

In amorphous or noncrystalline solids, the diffusion mechanism depends on the nature of atomic bonding and the structural aspects of the solid. In Chapter 2 we briefly discussed the bonds in molecular structures. Recall that the atoms in a molecule are covalently bonded (strong attractive *intramolecular* forces), whereas the bonds between molecules (*intermolecular*) are comparatively weak. Polymeric materials can exhibit diffusion by the movement of molecules rather than individual atoms. For example, strain energy in deformed plastics is relieved by this type of diffusion mechanism.

Diffusion in Glass

The movement of gaseous atoms through glass is very often presented in the form of *permeability* data rather than diffusivity. Permeability is the rate of diffusion of a pressurized gas through a porous substance. For example, the permeability (K) of helium (He) for various glasses is shown in Figure 5-17. Not only do these data reflect

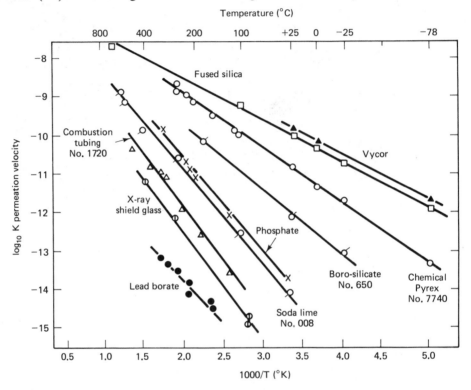

Figure 5-17 Permeation velocity (K) of helium (He) through various glasses. (From F. J. Norton, *J. Amer. Ceramic Society 36*, 1953, p. 90)

the temperature dependence of K, but they also demonstrate that significant differences in permeability are exhibited by certain glasses at a given temperature. This aspect of diffusion in glass may be quite important for high-vacuum and high-pressure applications. In such cases it may be necessary to consider the more impermeable types of glass.

Although we will discuss the structure of glass in Chapter 15, it is worthwhile to point out that certain elements are added to the various glasses in order to modify their properties. These elements are referred to as modifying cations and include Na, K, Ca, and Mg. In one sense, these modifiers can block the holes or openings in the glass networks (random arrays of molecules), thus causing differences in permeability. But they may also become diffusing species themselves. Indeed, modifying cations may migrate through the cavities in the glass structure. As the amount of these cations increases, the activation energy for diffusion decreases and D increases.

Diffusion in Ceramics

Diffusion in ceramic materials may occur by the vacancy and interstitial mechanisms that we discussed for metals and alloys. However, in the case of ceramics, the diffusing species are ions, which carry a charge, so the presence of Schottky and Frenkel defects can heavily influence diffusivity in these materials. Diffusion in ceramics may also occur by an *interstitialcy* mechanism, which differs from the Frenkel disorder in that an interstitial ion moves into a *lattice* site by bumping the lattice ion to an interstitial site. Overall, the movement of ions in and out of ceramics is heavily influenced by temperature, impurities, defects, and high-diffusivity paths such as grain boundaries. The temperature dependence of D, and the wide variation in diffusivity in various ceramic materials, are shown in Figure 5-18.

The electrical conductivity of ionic crystals at high temperatures is due principally to the diffusion of ions. The conductivity (σ) may be related to the diffusivity (D) in ionic solids by a form of the Nernst–Einstein equation as follows:

$$\sigma = \frac{NZ^2q^2D}{kT} \qquad (5\text{-}16)$$

where N = number of ions/m³
$\quad Z$ = valence of ion
$\quad q$ = charge on an ion
$\quad k$ = Boltzmann's constant
$\quad T$ = absolute temperature

A comparison of the diffusivity of Na^+ in NaCl, made by experimental radioactive tracer techniques and calculated from conductivity measurements utilizing the Nernst–Einstein relationship, is shown in Figure 5-19. In addition to the excellent agreement between experimental and analytical techniques, these data demonstrate the temperature dependence of both diffusivity and electrical conductivity in ionic solids. As temperature increases, conductivity increases because diffusion is a thermally activated process. In contrast, the electrical conductivity in metallic materials decreases with increasing temperature. Recall the difference in atomic bonding between the two types of crystalline solids. Certainly, the charge carriers in metals are the conduction (de-localized) electrons. Thus, as temperature increases, the mean free path between collisions decreases, due to the increased vibration (amplitude and frequency) of atoms on the metal lattice. Such a decrease in the conduction path results in decreased electrical conductivity (greater resistivity). This aspect of diffusion controlled conductivity in ceramics will surface again in Chapter 10, when we examine the electrical properties of ionic materials.

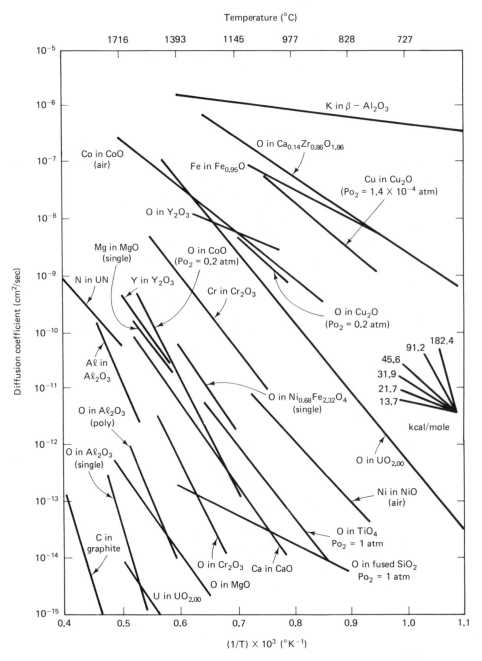

Figure 5-18 Diffusion coefficients in selected ceramic systems. (From W. D. Kingery, H. K. Bowen, and D. R. Uhlmann, *Introduction to Ceramics*, John Wiley & Sons, Inc., New York, 1960, p. 240.)

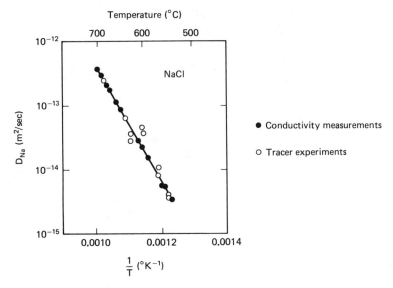

Figure 5-19 Diffusivity of Na in NaCl at high temperatures. (From C. A. Wert and R. H. Thompson, *Physics of Solids*, McGraw-Hill Book Company, New York, 1970, p. 210.)

STUDY PROBLEMS

5.1. Explain what is meant by concentration gradient. How do concentration gradients affect diffusion? Are they necessary for diffusion to occur?

5.2. List some crystalline imperfections which you think might influence diffusion in solids. Why?

5.3. In a certain material at 500°C, one atom out of 10^9 has the energy required to jump into an adjacent vacancy. However, at 1000°C, one atom out of 10^6 has sufficient energy to accomplish this. If the fraction of atoms (n/N) possessing this energy can be described as follows:

$$\frac{n}{N} = Ae^{-E/kT}$$

where n is the number of atoms with sufficient energy to jump, N is the total number of atoms, and A is a proportionality constant, determine the energy (E) required for this process.
Answer: 1.2 eV

5.4. How much energy would be required for the process described in problem 5.3, in terms of a mole of the material? Express your answer in cal/mol.

5.5. In the material analyzed in problem 5.3, at what temperature would the fraction of atoms with sufficient energy to jump be equal to 10^{-4}?

5.6. (a) An impurity atom is located in the octahedral interstice produced in the center of a fcc unit cell. How many possible jump directions does this atom have, presuming that it will only jump into another octahedral interstice?

(b) Sketch atoms on the (100) plane in adjacent unit cells in the fcc system. If an interstitial atom is located in an octahedral interstice on this plane, show by arrows the possible jumps that it can make in the (100) plane.

5.7. The following sketch of a fcc unit cell shows a face-center vacancy located at 1/2, 1, 1/2. Indicate which atoms have to be displaced in order for an adjoining face atom (x) located at 1, 1/2, 1/2 to jump into the vacancy.

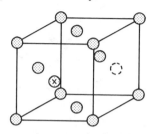

5.8. The diffusion coefficient (D_0) for arsenic in silicon is 0.32 cm²/sec and the activation energy for diffusion (Q) is 82,000 cal/mol. Determine the diffusivity of As in Si at 500°C and 1000°C.

5.9. A 0.5-in. (12.7-mm)-diameter bar of aluminum contains 0.25 a/o copper at its surface. The concentration of copper at the center of the bar is 0.10 a/o. At 500°C what is the flux of copper toward the center of the bar? The lattice parameter of Al is 4.05 Å.
Answer: $J = -3.56 \times 10^{10}$ atoms/cm² · sec

5.10. The surface of steel can be selectively hardened by increasing its carbon content (this process is called *carburizing*). If the surface of a steel part has a carbon concentration of 40 a/o, and the carbon content is 10 a/o at a depth of 0.5 mm (0.020 in.), what is the flux of carbon atoms between these two locations at 1000°C? At 1000°C, Fe is fcc and a is approximately 3.6 Å.

5.11. If the depth (x) of a carburized layer on a steel part is proportional to $(Dt)^{1/2}$, how much longer will it take to produce a carburized "case" 1.0 mm deep at 500°C compared to 1000°C?
Answer: 36.3 hr

5.12. In certain cases, the quantity $(Dt)^{1/2}$ can be used to estimate the depth (x) below a surface where the concentration of the diffusing species will be one-half of the surface concentration. If aluminum (Al) is diffused into silicon at 1200°C until the concentration at 0.01 mm is half the surface concentration, how much time will elapse? ($D = 8 \times 10^{-12}$ cm²/sec.) (*Hint*: See Example 5-5.)

5.13. A certain material penetrates an average depth of 10 μm into Al_2O_3 in 1 hr. What is an estimate of the diffusion coefficient of this material in alumina? Does this value of D correspond to any diffusion couple involving alumina in Figure 5-18.

5.14. In order to produce a certain semiconductor, phosphorus (P) is diffused into germanium (Ge). This process is called *doping*. If the diffusion coefficient (D_0) of P in Ge is 2.0 cm²/sec and the activation energy is 57,500 cal/mol, determine the diffusivity (D) of P in Ge at 800°C.

5.15. The semiconductor material in problem 5.14 is designed to operate most efficiently with a controlled concentration of phosphorus. If the P concentration at the surface of the germanium is 50 a/o and the concentration at a depth of 0.1 mm is zero, calculate the flux of P atoms between the surface and this depth at 800°C. Ge is diamond cubic and a is approximately 5.6 Å.

5.16. Estimate how long the process will take if the phosphorus in problem 5.15 is diffused into the germanium at 800°C until the concentration at 0.01 mm reaches half the concentration of the surface? (See Example 5-5.)

5.17. Which region of a polycrystalline material exhibits greater diffusivity, the crystalline lattice within grains or the grain boundary regions? Briefly discuss your answer.

5.18. Compare the diffusivity (D) at 500°C for self-diffusion in zinc through the lattice and along grain boundaries if D_0 and Q for lattice diffusion are 0.4 cm^2/sec and 23 kcal/mol, while these values are 0.14 cm^2/sec and 14k cal/mol for grain boundary diffusion, respectively.

5.19. Is grain-boundary-related diffusion more important at high or low temperatures?

5.20. Based on the experimental data in Figure 5-19, calculate the approximate activation energy for diffusion of sodium (Na) in NaCl at 600°C if $D_0 = 0.001$ m^2/sec. Express your answer in eV.

5.21. The electrical conductivity of NaCl is experimentally determined to be 10^{-5} (Ω-cm)$^{-1}$ at 600°C. If the number of Na$^+$ ions is approximately 9×10^{21} cm^{-3}, calculate the diffusivity of Na at this temperature. How well does this compare with Figure 5-19? Answer: $D = 3 \times 10^{-10}$ cm^2/sec

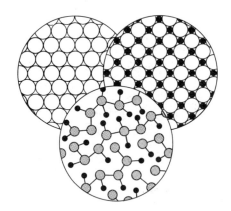

CHAPTER SIX:
Strengthening
Mechanisms
and Response to
Thermal Treatment

EFFECTS OF DEFORMATION AND STRESS

The study of engineering materials must necessarily examine the aspects of mechanical behavior or how materials respond to applied and residual forces. The importance of this topic cannot be overemphasized. Questions concerning the strength of materials (i.e., "the breaking stress in pieces of wood of different shapes") were raised as early in history as circa 300 B.C. (Aristotelian period) in an engineering article entitled *Mechanika*.[1] Unfortunately, these questions went unanswered until the time of Galileo (seventeenth century) and his successors. Today, the designer, the engineer, the builder, and virtually everyone involved from the conceptual stage to the final inspection of a component or structure is concerned about mechanical properties at some point. The characterization of mechanical behavior is essential to the *proper* selection and utilization of engineering materials.

In this chapter we introduce and explain the fundamental mechanical relationships exhibited by engineering materials. Materials in this context include polymers, ceramics, wood, concrete, and fiber-reinforced composites, in addition to metals and alloys. Furthermore, since mechanical properties and behavior are frequently structure sensitive; closely linked to crystal and molecular structure and to microstructure, we will accentuate these relationships whenever it is appropriate.

An understanding of the way materials react to stress is fundamentally necessary in order to understand the way in which materials are strengthened, formed, and utilized. We have separated the basic mechanical behavior of materials from the testing aspects presented later in Chapter 8, in order to introduce certain fundamental aspects of materials' behavior at this time.

[1] L. Sprague De Camp, *The Ancient Engineers*, Ballantine Books, New York, 1963, p. 126.

134

Deformation

When an engineering material is subjected to forces, such as those typically imposed by service loads, its atoms may be displaced from their equilibrium positions. In Chapter 2 we discussed the potential energy associated with atomic separation and showed schematically that the energy is at a minimum for equilibrium separation (see Figure 2-3). Thus any displacement from this position or separation (a_0) results in an energy increase. This requires work, which is supplied by the forces creating the displacement. Referring to this figure, we see that this premise holds whether the material is stretched, causing the atoms to separate, thus bringing attractive forces into play, or compressed, bringing the atoms closer together and causing repulsion.

Elastic deformation. The displacement of atoms from their equilibrium positions constitutes deformation. Such deformation is termed *elastic* if the atoms can resume their equilibrium positions when the imposed forces are released. Elastic deformation, then, is recoverable and indicates the relative resilience of a material. For example, a rubber band can be stretched quite far, yet snap back to its original dimensions upon being released. A slightly different manner of stating this concept of elasticity is that it is the property of a material to return to its initial form and dimensions after the deforming force is removed.

The process of elastic deformation is presented schematically in Figure 6-1(a). Here the atoms are represented as "hard" spheres on a lattice. When no forces are applied, they assume equilibrium separation (a_0). A relatively small tensile force tends to pull the atoms apart, producing elastic deformation (δ_E). Their separation is now slightly larger than a_0. However, when the force is released, the atoms resume their equilibrium positions and no deformation or displacement remains. The material is restored to its initial condition.

Plastic deformation. On the other hand, if an engineering material undergoes deformation which exceeds the elastic capability (elastic limit) to restore the atoms to their equilibrium positions, the deformation is permanent and termed *plastic*. Plastic deformation is nonrecoverable and leaves the atoms permanently displaced from their original positions when the forces are released. Deformation of materials may be entirely elastic, or elastic plus plastic. The total deformation may then consist of the combined elastic and plastic portions. In this case, removal of the load or forces producing the deformation results in recovery of the elastic portion, while the plastic portion remains. This process is illustrated in Figure 6-1(b).

Plastic deformation of engineering materials is permanent in that work or energy must be supplied to restore the atoms to their original equilibrium positions. For instance, the effects of this type of deformation can be alleviated by thermal treatments, when necessary, as discussed later in the section on annealing.

Strain

Engineering strain. We have just briefly examined the response of atoms to deformation by mechanical forces. If this concept of atomic displacement is extended to bulk engineering materials, we can define the deformation in terms of the

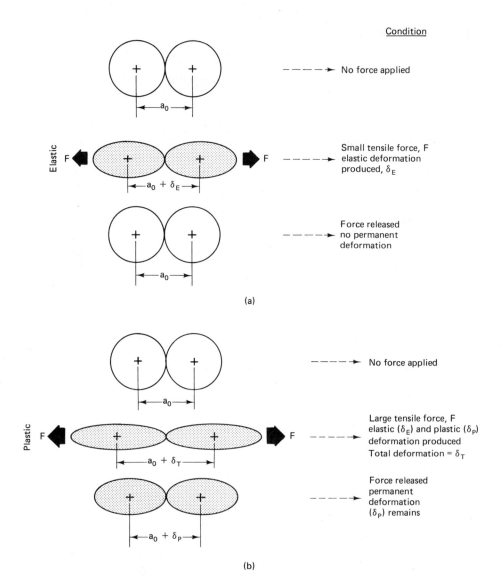

Figure 6-1 Schematic illustration of deformation in atoms: (a) elastic deformation; (b) elastic plus plastic.

original dimensions of the material under consideration: in other words, the ratio of the dimensional change to the original dimension. For example, consider a bar of length (l_0), as shown in Figure 6-2. Under the action of an applied load (P), this bar experiences deformation (δ) and elongates to a new length (l_f). The ratio of this change in length (Δl) to the original length (l_0) is the *average linear strain* (e) and can be expressed as follows:

$$e = \frac{\Delta l}{l_0} = \frac{l_f - l_0}{l_0} \tag{6-1}$$

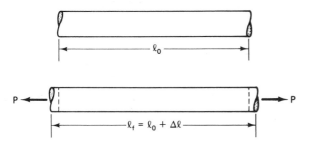

Figure 6-2 Deformation (elongation) of a round bar produced by axial load.

This quantity is referred to as the average linear strain because only the dimensional change in the axial direction is considered and it is considered over the entire length of the sample. In reality, because its volume remains constant, the bar diameter does decrease slightly, resulting in a decrease in the cross-sectional area, but for small strains this response is usually insignificant. Average strain (e) is commonly referred to as the *engineering strain* and expressed in units of in./in. or mm/mm. However, strain may also be treated as a dimensionless quantity, because these units cancel.

True strain. Since a material undergoing deformation is continuously changing its dimensions (i.e., length and width or diameter), a more precise definition of strain is given by the ratio of the change in dimensions to the instantaneous dimension. In the case of our round bar, this may be viewed as the change in length (dl) with respect to the instantaneous length (l) at any point in the process. This ratio is actually the *true strain* (ϵ) and is expressed as follows:

$$\epsilon = \frac{dl}{l} \tag{6-2}$$

A more specific expression for ϵ is obtained by placing some limits on equation (6-2). For example, if the initial length is l_0 and the final length is l_f, we can integrate this equation as follows:

$$\epsilon = \int_{l_0}^{l_f} \frac{dl}{l}$$

$$= \left[\ln l \right]_{l_0}^{l_f}$$

$$= (\ln l_f) - (\ln l_0)$$

$$= \ln \frac{l_f}{l_0} \tag{6-3}$$

The engineering strain (e) is related to the true strain (ϵ) and this relationship can be demonstrated as follows:

$$e = \frac{l_f - l_0}{l_0}$$

Simplifying to

$$e = \frac{l_f}{l_0} - 1$$

Effects of Deformation and Stress

solving for

$$\frac{l_f}{l_0} = e + 1$$

then substituting this result in equation (6-3), we obtain

$$\epsilon = \ln (e + 1) \tag{6-4}$$

Example 6-1

A metal rod 100 mm in length is pulled in tension to a length of 102 mm. (a) What is the value of the engineering strain produced in the bar? (b) What is the true strain?

Solution

     (a) Original length $l_0 = 100$ mm
           final length $l_f = 102$ mm
        difference $\Delta l = 2$ mm

From equation (6-1),

$$e = \frac{\Delta l}{l_0} = \frac{2 \text{ mm}}{100 \text{ mm}}$$

$$= 0.02 \text{ mm/mm} \quad Ans.$$

     (b) In this case we may use either equation (6-3) or (6-4), since we have sufficient information. Utilizing equation (6-4), the true strain is

$$\epsilon = \ln (e + 1)$$

$$= \ln (0.02 + 1)$$

$$= \ln 1.02$$

$$= 0.0198 \quad Ans.$$

This example illustrates that the two values of strain are essentially identical at relatively small strains (< 0.1). As strain increases, the discrepancy between the two values increases, with the true strain being the smaller.

Shear strain. In addition to linear strain, an engineering material can experience *shear strain* (γ). This type of strain is due to the displacement of parallel planes through a certain angle (θ), as shown in Figure 6-3. The shear strain is there-

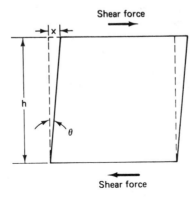

Shear force

Shear force

Figure 6-3 Shear displacement on parallel planes of a solid material.

fore defined as the ratio of the displacement (x) to the distance (h) between the planes, expressed as follows:

$$\gamma = \frac{x}{h} \qquad (6\text{-}5)$$

Since the ratio x/h is also the tangent of θ, and the angles involved are typically small (in the case of elastic strains), we may state

$$\gamma = \tan \theta \approx \theta \qquad \text{(in radians)} \qquad (6\text{-}6)$$

In a simple analogy, the student can simulate this concept of shear strain by pushing parallel to the top face of a soft eraser or rubber block, producing an angular displacement similar to Figure 6-3.

Poisson's ratio. Because of the constancy of volume, when a material is deformed in one direction there is a corresponding displacement or deformation in a direction perpendicular to it. For example, consider the bar in Figure 6-4. If an axial

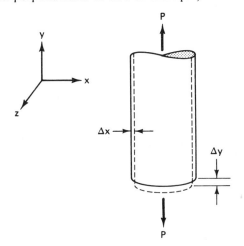

Figure 6-4 Deformation in x direction due to axial load (P) applied in y direction.

load (P) is applied, the bar elongates in the y direction. The ratio of the strain in the x direction to the strain in the y direction is termed *Poisson's ratio* (v) and expressed as follows:

$$\frac{e_x}{e_y} = v \qquad (6\text{-}7)$$

Since our example is a symmetrical shape and the strains have been assumed to be uniform, the strain in the z direction (e_z) equals e_x. So v is equivalent in the x and z directions. This relationship is important, because v is a constant for elastic strains and can be used to compute strains in directions other than the direction of the applied force.

Stress

The response of an engineering material to forces imposed on it has been discussed in terms of deformation and strain in the preceding sections. Now let us analyze the resistance of a material to deformation. Referring once more to Figure 2-3, one can

Effects of Deformation and Stress **139**

see by inspection that the net force on an atom is zero only for the equilibrium position. Displacement in either direction produces an increase in the forces (tensile or compressive) that oppose the deformation. This resistance is due to the interatomic attractive and repulsive forces that operate in a particular material, as discussed in Chapter 2.

Stress is the result of the internal response that a material exhibits when forces are imposed on it. To simplify matters at this point, we will assume that the force acts uniformly over a certain area. Then we can state that the internal response, the stress, is a *force per unit area.*

Consider the load (*P*) applied to the cylindrical bar in Figure 6-5. The bar re-

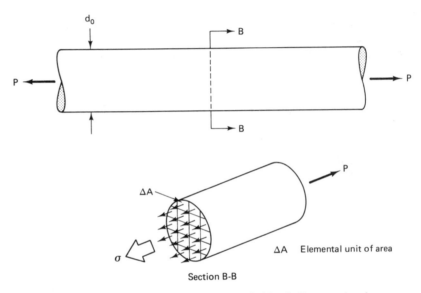

Figure 6-5   Response of cylindrical bar to applied load. Cross-sectional cutway shows uniform resistance balancing (*P*).

mains intact, indicating that the external force is balanced internally by a response of the material. If we section this bar at any particular location (normal to the axis of the applied load), *P* must be opposed by the stress (σ) produced in each elemental area (ΔA). If these increments of area become smaller and smaller, in the limit

$$P = \sigma\, dA \tag{6-8}$$

Since we previously assumed that the force is uniformly applied over the cross section, the summation of stress over the entire area may be expressed as follows:

$$P = \sigma \int_0^A dA$$

which yields

$$P = \sigma A$$

Therefore, the stress (σ) can be expressed as

$$\sigma = \frac{P}{A}$$

(6-9)

where σ = average stress
     P = load or force
     A = cross-sectional area over which the force acts

Stress is commonly denoted in units of lb/in.^{2} (psi), or in international units (SI) of pascals (Pa).[2] Since the pascal is a rather small value of stress, it is customary to express stress values in megapascals (MPa), where the prefix M stands for 10^6. Detailed information concerning SI units[3] and conversion factors is given inside the front and back covers.

Engineering stress. A common application of the stress concept is termed the engineering stress. This particular expression, which is used in many design calculations and analyses, is given as follows:

$$\sigma = \frac{P}{A_0}$$

(6-10)

The only modification to our basic stress equation is A_0, which represents the original area. Therefore, engineering stress treats cross-sectional area as a constant. Although this is not completely accurate, for elastic strains in the loading direction the corresponding changes in cross-sectional area are generally small. Furthermore, most designs and structures are based on service conditions in the elastic range; thus the engineering stress is a very useful parameter.

Example 6-2

Determine the engineering stress produced in the bar shown in Figure 6-5 if the original diameter (d_0) equals 20 mm (0.79 in.) and the load (P) is 4900 N (1104 lb). Express your answer in both English and SI units.

Solution From equation (6-10) the engineering stress is

$$\sigma = \frac{P}{A_0}$$

The area in this case is

$$A_0 = \frac{\pi(0.020 \text{ m})^2}{4} = 3.14 \times 10^{-4} \text{ m}^2$$

Therefore, the stress is

$$\sigma = \frac{4900 \text{ N}}{3.14 \times 10^{-4} \text{ m}^2} = 1.56 \times 10^7 \text{ Pa} = 15.6 \text{ MPa} \qquad \textit{Ans.}$$

---

[2] 1 Pa = 1 newton per square meter (N/m^2).

[3] For an explanation of SI units and rules concerning their use, see *SAE Handbook*, Part 1, Society of Automotive Engineers, Inc., Warrendale, Pa., 1983, p. J916; or the *International System of Units (SI)*, National Bureau of Standards Special Publication No. 330, July 1974.

Referring to inside the back cover for the appropriate conversion factors, the engineering stress is given as follows:

English: $\sigma = 2262 \text{ lb/in.}^2$ (2.26 ksi) *Ans.*

True stress. The concept of engineering stress treats the area under consideration as a constant (A_0). In reality, as we have implied in equation (6-7), however, the area does not remain constant, and in the case of an axially loaded tensile bar, gradually decreases as the stress (and corresponding strain) increases. This change is actually a manifestation of Poisson's ratio effect.

The *true stress* therefore can be expressed as follows:

$$\sigma_{tr} = \frac{P}{A_i} \qquad (6-11)$$

where A_i is the actual or instantaneous area over which the force is acting. Ordinarily, the true stress is larger than the engineering stress. However, in the elastic region, changes in area are usually inconsequential and the engineering stress is sufficiently accurate. When the elastic limit is exceeded and plastic strains come into play, the discrepancy between engineering strain and true strain becomes more significant. From a design standpoint, however, this condition (plastic deformation) is typically avoided in service.

The true stress (σ_{tr}) is related to the engineering stress (σ) in the following manner:[4]

$$\sigma_{tr} = \frac{P}{A_i} \frac{A_0}{A_0}$$

$$= \frac{P}{A_0} \frac{A_0}{A_i}$$

$$= \sigma \frac{A_0}{A_i}$$

In this type of analysis, the volume of material is constant (even though the dimensions may change). Therefore, the following relationship also applies:

initial volume (V_0) = instantaneous volume (V_i)

$$V_0 = V_i$$

$$A_0 l_0 = A_i l_i$$

Rearranging yields

$$\frac{A_0}{A_i} = \frac{l_i}{l_0}$$

Substituting above gives us

$$\sigma_{tr} = \sigma \frac{l_i}{l_0}$$

But from the section on strain [equation (6-4)], since at any instant in time l_i is equivalent to l_f, we can substitute for l_i/l_0 as follows:

$$\sigma_{tr} = \sigma(e + 1) \qquad (6-12)$$

Thus the true stress equals the engineering stress times the quantity engineering strain plus 1.

---

[4]Multiplying by the factor A_0/A_0 does not change the value of the equation.

Hooke's law.   The *elastic limit* of an engineering material is the highest stress that can be produced without experiencing any plastic (permanent) deformation. We alluded to the concept of elastic limit earlier in this chapter in the discussion of plastic deformation. In most materials for values of stress below the elastic limit, stress is proportional to strain as follows:

$$\sigma = Ee \tag{6-13}$$

This relationship is known as *Hooke's law*, and the proportionality constant (E) is the *modulus of elasticity* (Young's modulus).

Elastic modulus.   The elastic modulus (E) is a measure of the stiffness of an engineering material. Examination of equation (6-13) reveals that for a given stress, greater values of E result in smaller elastic strains, meaning that the higher the elastic modulus, the smaller the response of the structure to a particular stress. This parameter is important for design and analysis purposes, especially in computing the allowable displacements and deflections of engineering components or structures.

Values of E for selected engineering materials are listed in Table 6-1. Comparison of steel and titanium to aluminum, for example, shows that titanium is about $1\frac{1}{2}$ times stiffer, while steel is about 3 times as stiff. As we have stated, the implied design considerations are extremely important. Many structures are not strength limited but are modulus limited; that is, the structures do not require materials of higher strength but rather of higher modulus or stiffness. These structures will not fail if lower modulus materials are used, but the deflections might well be excessive. For example, consider the relative deflection (δ) of three identical-size cantilever beams produced from the three materials previously mentioned, for the same load (P), in Figure 6-6. One can see that the deflections are proportional to the elastic modulus, and although the low-modulus material may not fail, its use might be unsuitable in some applications. Consider, for example, the metal framework of a very tall skyscraper. Wind loads can produce deflections of the structure. With a low-modulus material, these deflections could make the building uninhabitable even though it would not topple.

Although the modulus of elasticity is a *structure-insensitive* property, it is influenced by temperature. As temperature is increased, E decreases, thereby reducing the stiffness of a material. This reduction in elastic behavior is due to an inverse relationship between the modulus and the interatomic or interionic distance in metals and ceramics, respectively. Therefore, as we increase temperature, the equilibrium separation (a_0) increases and E gradually decreases.

Modulus of rigidity.   Whereas E represents the modulus of elasticity in tension or compression, the *modulus of rigidity* (G) is the modulus of elasticity in shear. G is a measure of the shearing force necessary to produce a given, small amount of deformation. The student should recall the concept of shear strain (γ) discussed earlier in this chapter. Therefore, if we apply a Hooke's law relationship to the shear stress (τ), the following expression results:

$$\tau = G\gamma \tag{6-14}$$

Values of G, usually determined by torsion testing, are related to E by the following expression:

Effects of Deformation and Stress   **143**

$$G = \frac{E}{2(1 + v)} \tag{6-15}$$

where v is Poisson's ratio from equation (6-7). Typical values of these elastic constants (G, v) are also given in Table 6-1 for several common engineering materials.

TABLE 6-1 ELASTIC PROPERTIES FOR SELECTED ENGINEERING MATERIALS AT ROOM TEMPERATURE

Material	Elastic modulus, E		Shear modulus, G		Poisson's ratio
	10^6 psi	GPa	10^6 psi	GPa	
Metals					
Aluminum alloys	10.5	72.4	4.0	27.6	0.31
Copper alloys	16.0	110.4	6.0	41.4	0.33
Steels (plain carbon—low alloy)	30.0	207.0	11.0	75.9	0.33
Stainless steel (18-8)	28.0	193.2	9.5	65.6	0.28
Titanium	16.0	110.4	6.5	44.8	0.31
Tungsten	56.0	386.4	22.8	157.3	0.27
Ceramics					
Alumina porcelain (90–95% Al_2O_3)	53.0	365.7	—	—	—
Beryllia	45.0	310.5	—	—	—
Graphite	1.3	9.0	—	—	—
Silicon carbide	68.0	469.2	—	—	—
Titanium carbide	45.0	310.5	—	—	—
Tungsten carbide	77.5	534.8	31.8	—	0.22
Silica glass	10.5	72.4	—	—	—
Fused quartz	10.6	73.1	4.5	—	0.17
Pyrex glass	10.0	69.0	—	—	—
Mullite porcelain	10.0	69.0	—	—	—
Fireclay brick	14.0	96.6	—	—	—
Plastics					
Polycarbonate	0.35	2.4	—	—	—
Polyethylene	0.058–0.19	0.4–1.3	—	—	—
Polymethyl methacrylate	0.35–0.49	2.4–3.4	—	—	—
Polystyrene	0.39–0.61	2.7–4.2	—	—	—
Nylon 66	0.17	1.2	—	—	—
Other materials					
Concrete[a]	1.0–5.0	6.9–34.5	—	—	—
Common brick[a]	1.5–2.5	10.4–17.2	—	—	—
Hardwoods	1.2–1.8	8.3–12.4	—	—	—
Softwoods (Conifers)	1.0–1.6	6.9–11.0	—	—	—
Rubber	(150 psi)	0.001	—	—	—

[a]Not a constant for any range of loading.

Source: G. W. C. Kaye et al., Eds., *Tables of Physical Chemical Constants*, 14th ed., Longman Group Ltd., London, 1973, p. 31.

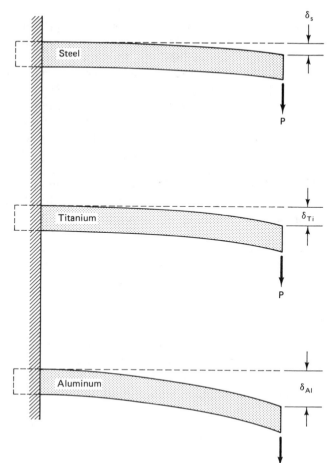

Figure 6-6 Deflections in three identical size cantilever beams for the same load (*P*). Note the relative differences in deflection (δ) for the respective materials.

Example 6-3

If the elastic modulus for medium-carbon steel is 208 GPa and Poisson's ratio is 0.26, calculate the shear modulus for steel.

Solution

$$G = \frac{E}{(2)(1 + v)}$$

$$= \frac{208 \text{ GPa}}{(2)(1 + 0.26)} = 82.5 \text{ GPa} \qquad \textit{Ans.}$$

Note that the modulus in shear is slightly less than half of the elastic modulus.

Effects of Deformation and Stress

SOLID SOLUTION STRENGTHENING

While relatively "pure" metals such as copper and aluminum, containing only small amounts of contaminants, are used commercially, metals of extremely high purity exist rarely in nature and are uncommon even in industry. To prepare such pure metals requires a great effort to eliminate the trace elements commonly found as impurities. We have previously considered the mixing of the atoms of one elemental species with another as a structural defect. However, the combining of certain elements in the proper proportions can yield beneficial changes in the mechanical and physical properties of many metallic and ceramic systems. This process of adding elemental species to a base metal is called *alloying*. Although alloying can have additional effects on the basic system, one effect is to produce a blend of the atoms in such a way that one species is dissolved in the other, in other words, a *solid solution*. The atoms of the host species are referred to as the *solvent* and those of the dissolved species as the *solute*. Solid solutions are also formed by combinations of many ceramic compounds.

Solid solutions fall in two distinct classes, which differ with respect to the position of the solute atom relative to the lattice site. These are (1) substitutional solid solutions and (2) interstitial solid solutions.

Substitutional Solid Solutions

In this type of solid solution, the solute atoms occupy lattice sites that would normally be occupied by the solvent atoms. These solute atoms can occupy random substitutional lattice sites as shown in Figure 6-7(a), forming a disordered solid solution as exemplified by the copper–zinc system. On the other hand, it is also possible for the

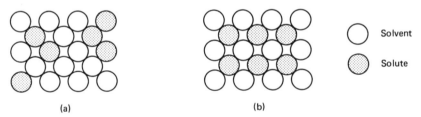

(a)   (b)

○ Solvent

◉ Solute

Figure 6-7 Schematic diagram illustrating the relative positions of the solvent and solute atoms in a substitutional solid solution: (a) disordered; (b) ordered.

solute atoms to occupy regular repeating lattice sites, thereby forming an ordered solid solution, as shown in Figure 6-7(b). In an ordered solid solution such as the gold–platinum system, the two species tend to form a statistically greater number of Au–Pt bonds than would occur on a random basis.

In the schematic examples shown above we have substituted solute atoms having the same diameter as the host atoms. In reality, the atomic diameters of the species would probably differ somewhat. As we will see, this difference in the size of the species affects the solubility of the species and their strengthening capability.

Solubility limits. Although some systems, such as cadmium–magnesium, exhibit complete solid solubility, that is, the two species are completely soluble in each other in all proportions, most systems exhibit only limited solid solubility at any temperature. Solute additions above the solubility limits result in the formation of a second phase. The solubility limit at any temperature can be determined from an examination of the phase diagram of the system as you will eventually see in our discussion of equilibrium phase relationships in Chapter 7.

There are several determinants governing the limits of solid solubility. These can be expressed as a series of rules which, although not absolute, can serve to provide guidance as to the factors favoring extensive substitutional solubility. These factors are as follows:

1. *Size:* Hume–Rothery first pointed out that extensive substitutional solid solubility occurs only if the relative difference between the atomic diameters of the two species was less than 15%. If the difference is greater than 15%, the solubility is usually quite limited.

2. *Crystal structure:* In order to achieve solubility, the solvent and solute atoms should crystallize in the same structural lattice form: face-centered cubic, for example.

3. *Valence:* The solute and solvent atoms should typically have the same valence in order for maximum solid solubility to occur.

4. *Chemical reactivity:* The two species should resemble each other chemically and lie relatively close in the electrochemical series. Chemical reaction between the species tends to promote the formation of stable compounds rather than solid solutions.

Example 6-4

Based on the Hume–Rothery rules, determine if copper will exhibit extensive substitutional solid solubility in silver.

Solution   From inside the back cover we find the following information:

$$\text{atomic diameter Cu} = 2.556 \text{ Å}$$
$$\text{atomic diameter Ag} = 2.888 \text{ Å}$$
$$\Delta d = 0.332 \text{ Å}$$
$$\text{difference} = \frac{0.332 \text{ Å}}{2.888 \text{ Å}} \times 100 = 11.5\%$$

Extensive solid solubility is predicted.   *Ans.*

Substitutional "alloying" can also occur in ceramic compounds. The requirements for such solid solutions include similar ion sizes and identical valence charges on the ions involved in the replacement reactions. For example, the substitution of a Be^{2+} ion for Mg^{2+} in MgO would not be favorable because a sizable difference exists between their sizes: radius of $Be^{2+} = 0.35$ Å, radius of $Mg^{2+} = 0.66$ Å (see inside back cover). It would also be difficult to replace Mn^{2+} in MnO with a Zr^{4+} ion even though their ionic radii are virtually the same, since a surplus of charge results from this substitution.

Interstitial Solid Solutions

In this type of solid solution, the solute atoms occupy positions in the interstices of the solvent lattice, as illustrated in Figure 6-8. This illustration points up one important

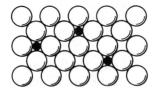

Figure 6-8 Schematic diagram illustrating the relative positions of the solvent and solute in an interstitial solid solution.

detail regarding interstitial solid solutions: namely, the diameter of the solute atom must be small relative to that of the solvent atom. Extensive interstitial solid solubility occurs only when the ratio of the apparent diameter of the solute atom to that of the solvent is less than 0.59. Several important atomic species readily forming interstitial solid solutions are carbon, nitrogen, boron, hydrogen, and oxygen. While carbon, nitrogen, and boron have important beneficial roles as alloying elements, hydrogen and oxygen can exert a devastating effect on the mechanical properties of many alloys. Interstitially dissolved species can exert an influence on the host element disproportionate to the concentration of the solute. One reason for this is that the interstitial atoms, because of their small size, may diffuse quite readily from one interstitial position to another and do not require a high concentration of vacancies for diffusion to occur.

Example 6-5

Determine, based on the Hume–Rothery rules, whether carbon will have significant interstitial solubility in iron.

Solution

$$\text{radius C} \simeq 0.77 \text{ Å (diameter C} \simeq 1.54 \text{ Å)}$$

$$\text{radius Fe} \simeq 1.24 \text{ Å (diameter Fe} \simeq 2.48 \text{ Å)}$$

$$\text{ratio } \frac{d_{\text{solute}}}{d_{\text{solvent}}} = \frac{1.54 \text{ Å}}{2.48 \text{ Å}} = 0.62$$

Since in order for extensive interstitial solubility to occur, the ratio $d_{\text{solute}}/d_{\text{solvent}} < 0.59$, carbon would not be expected to exhibit extensive interstitial solubility.   *Ans.*

As you will see in Chapter 12, this prediction is quite valid.

An additional factor in the formation of this type of solid solution is that the small atoms dissolve interstitially much more readily in the *transition metals* than in other metals. If you will recall from the periodic table, the transition metals include iron, chromium, nickel, cobalt, titanium, vanadium, molybdenum, and tungsten. A review of the electronic structure of these elements discloses that these metals have a peculiar electronic configuration: an incomplete electron shell inside the outer valence shell. This condition apparently contributes to their ability to accommodate the interstitial solute atoms so readily.

Solid Solution Strengthening

The addition of alloying elements can, by means of solid solutions, create remarkable changes in the mechanical (and also physical) properties of metals and ceramics. Generally, the strength and hardness of a material increases as the result of solid solution alloying and there may be attendant decreases in ductility. Typically, interstitial solute atoms have a relative strengthening effect of about three times the shear modulus (G) per unit concentration of the solute species. Substitutional solute atoms generally produce a relative strengthening effect of about $\frac{1}{10}G$, indicating the greater effect of the interstitial atoms in modifying material behavior.

The most important factors affecting this type of hardening are described below.

1. *Size factor:* Obviously, the size of the solute atom relative to that of the solvent atom will affect the degree to which straining will occur in the lattice adjacent to the site. As can be seen in Figure 6-9, the strains imposed on the lattice with an atom larger than the solvent atoms are largely compressive, while if the atom is smaller, the strain field tends to be tensile.

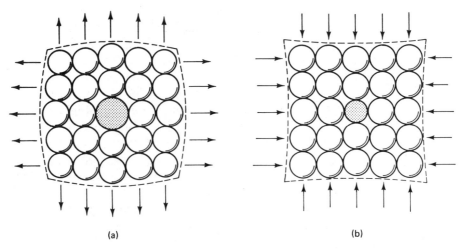

(a)   (b)

Figure 6-9   Effect of solute atom size on surrounding lattice: (a) solute atom larger than solvent atom; (b) solute atom smaller than solvent atom.

2. *Modulus factor:* It has been shown that the relative moduli of the two species also has an effect. For example, the shear stress per unit concentration of solute ($d\tau/dc$) for a series of alloying elements in copper varies linearly when plotted

against an elastic modulus parameter (ϵ_s) which takes into account the shear modulus (ϵ'_G) and the change in lattice parameter (ϵ_b) as shown in Figure 6-10.

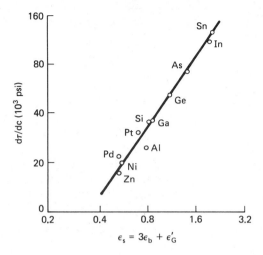

Figure 6-10 Relationship between shear stress per unit concentration of substitutional alloying element and the amount of misfit as measured by ϵ_s. (From H. W. Hayden, et. al., "The Structure and Properties of Materials", vol. 3, John Wiley and Sons, Inc. N.Y. 1965, p. 172.)

This parameter is expressed as

$$\epsilon_s = 3\epsilon_b + \epsilon'_G$$

where ϵ_b = change in the lattice parameter
ϵ'_G = shear modulus of the matrix

3. *Electronic interaction:* The compressive and tensile stresses associated with dislocations may produce a displacement of the conduction electrons, creating regions of positive and negative charge around the dislocation. This electronic dipole can interact with solute atoms having a different valence from the solvent atoms.

4. *Chemical interaction:* When a dislocation dissociates into partial dislocations, a stacking fault is formed. (Stacking faults are discussed in Chapter 4.) In a fcc structure, the faulted regions will have a hcp structure. This will result in an interaction between the dislocations and the solute atoms.

5. *Configurational interaction:* As we have described previously, solid solutions can occur in either an ordered form or in the disordered or random form. It is unusual for a solid solution to be truly random, and usually short-range order or clustering occurs. These conditions create domains, as shown in Figure 6-11, which retard the motion of dislocations.

As we have indicated several times in the foregoing sections, solid solution strengthening also occurs in ceramic materials. These ceramics, which are described fully in Chapter 15, are *ionic* compounds of metallic and nonmetallic elements: for example, aluminum oxide (Al_2O_3), magnesium oxide (MgO), silica (SiO_2), tungsten carbide (WC), boron nitride (BN), just to name a few.

Many of these ceramics consist of metal-oxides, arranged with oxygen ions (*anions*) in a close-packed structure and metal ions (*cations*) situated in available inter-

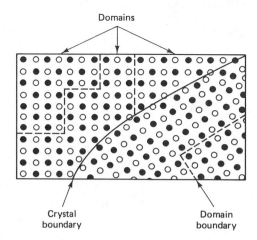

Domains

Figure 6-11 Domains and domain boundaries in an ordered solid solution.

Crystal boundary

Domain boundary

stitial spaces. Therefore, in addition to the importance of relative ion size, it is essential that charge balance be maintained in ceramic materials. Replacement of ions must occur with identically charged ions.

ADDITIONAL STRENGTHENING MECHANISMS

In addition to solid-solution strengthening there are several other mechanisms which are utilized to strengthen and harden metals and alloys. The nature of the mechanism differs from system to system, but the common effect is to produce changes in the mechanical behavior of the materials. A general listing of the mechanisms in common use is as follows: (1) strain hardening, (2) precipitation hardening, and (3) polymorphic transformation. All the mechanisms are not employed in all systems; however, a number of them are used in conjunction with each other to provide complementary improvements in the mechanical properties.

Strain Hardening

This hardening mechanism may be defined as the increased resistance to further plastic deformation. Strain hardening or cold working occurs when a metal or alloy is plastically deformed at temperatures below the *recrystallization* temperature of the alloy. This is the temperature at which stressed, plastically deformed grains begin to recrystallize into new stress-free grains. The cold working produces several effects. One is that it markedly increases the number of dislocations in the metal. A soft, annealed metal can have a dislocation density ranging from 10^6 to 10^8 dislocations/cm^2, and this number can increase to 10^{12} in a heavily cold-worked metal. The increase in dislocation density due to cold work in iron is revealed by transmission electron microscopy in Figure 6-12. As we have indicated in Chapter 4, each dislocation has a strain field associated with it; therefore, increasing the number of dislocations increases the total strain energy of the system. This can be observed from an examination of Figure 6-13, which shows the increase in the resolved shear stress as a function of dislocation concentration.

Strain hardening can be used in conjunction with solid-solution strengthening, since the strength increase for a solid solution alloy is usually greater than that for an

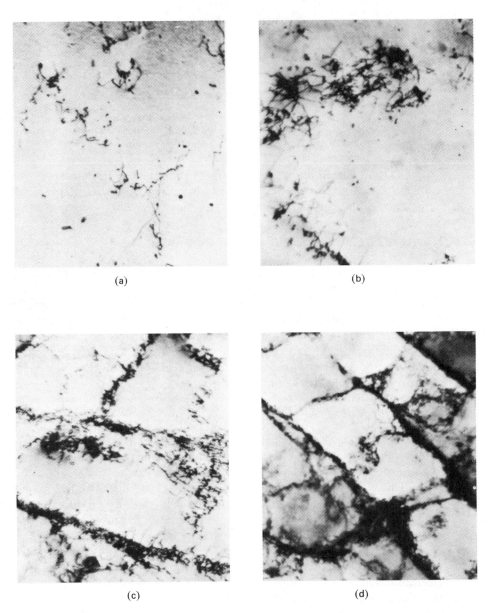

(a)

(b)

(c)

(d)

Figure 6-12 TEM micrograph of dislocation tangles in iron deformed at 25°C, magnification 20,000× : (a) 1% strain; (b) 3.5% strain; (c) 9% strain; (d) 20% strain. (From *Making, Shaping, and Heat Treating of Steel*, United States Steel Corp., Pittsburgh, Pa., 1964, p. 559.)

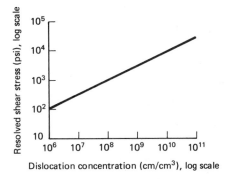

Figure 6-13 Increase in the critical resolved shear stress as a function of dislocation concentration. (From H. Wiedersick, "Hardening Mechanisms and Theory of Deformation," *J. Metals 16*, 1964, p. 425.)

unalloyed metal for the same degree of cold work. Other factors also influence the results; for example, the rate of strain hardening is generally lower for metals forming a hcp lattice than for those with a cubic structure. Ordered structures are also affected differently than disordered structures by cold working. As seen in Figure 6-14, a fully ordered alloy strain hardens more rapidly than does a disordered alloy of the same

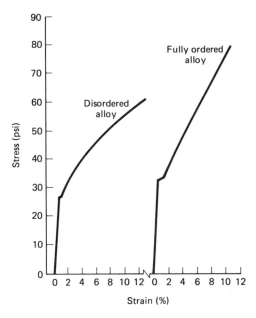

Figure 6-14 Effect of cold working on the mechanical properties of an ordered and disordered alloy. (From M. J. Marcinkowski and D. S. Miller, *Phil. Mag. 6*, 1961, p. 871.)

composition. This effect is attributed to the formation of domain boundaries in the ordered structure. These domains are formed between regions of differing order as shown in Figure 6-11 and act to retard the motion of dislocations, thereby raising the stress required for slip to occur.

The net effect of the cold working is to change the mechanical properties of the metal from those obtained in the annealed state. As can be seen from an examination of Figure 6-15, the strength parameters (yield strength and tensile strength) increase while the ductility parameters (percent reduction in area and percent elongation) decrease. Ductility is a measure of a material's ability to deform plastically (per-

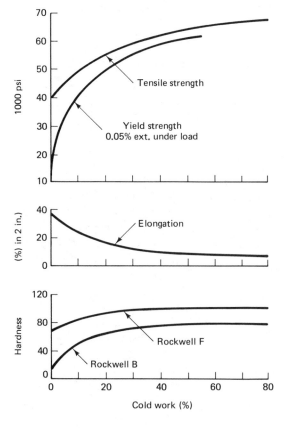

Figure 6-15 Change in mechanical properties of cupronickel (89 Cu–10 Ni–1 Fe) as a function of the degree of cold work. (From *Metals Handbook*, 8th ed., Vol. 1, American Society for Metals, Metals Park, Ohio, 1961, p. 1030.)

manently). Methods for determining ductility are explained in detail in Chapter 8. This figure also illustrates the use of the term "percent cold work," which is simply the ratio of the reduced cross-sectional area of the metal bar in its final size to the original cross-sectional area. Percent cold work may be expressed as follows:

$$\% \text{ cold work} = \frac{A_0 - A_f}{A_0} \times 100 \tag{6-16}$$

where A_f = final cross-sectional area
A_0 = original cross-sectional area

or for a round bar, simply

$$\% \text{ cold work} = \frac{d_0^2 - d_f^2}{d_0^2} \times 100$$

$$= \left(1 - \frac{d_f^2}{d_0^2}\right) \times 100 \tag{6-17}$$

where d_f = final diameter
d_0 = original diameter

As can be seen, the mechanical properties can be modified markedly, depending on the degree of cold work. This capability allows the properties to be tailored somewhat to suit various applications.

Precipitation Hardening

Early in the twentieth century it was observed that the strength and hardness of certain aluminum alloys increased upon aging at room temperature. It was later discovered that this modification in behavior was attributable to the formation of a very fine precipitated phase within the structure.

In several alloy systems, precipitation hardening can be more effective than other strengthening mechanisms, such as cold working or solid-solution strengthening. The alloys exhibiting this phenomenon are those binary or ternary alloys in which the solid solubility of one metal in another decreases as a function of decreasing temperature. The actual precipitate formation is discussed in Chapter 7.

This mechanism of precipitation hardening can be illustrated by the aluminum–copper system. Upon holding at the aging temperature, small clusters of copper atoms segregate on the {100} planes of the aluminum-rich solid solution. These clusters in their initial stages are called Guinier–Preston (GP) zones and were discovered using X-ray diffraction techniques. There is a strong tendency for these clustered atoms to attempt to remain in registry (coherent) with the matrix lattice, as shown in Figure 6-16(a). However, since the precipitated phase has different lattice parameters from

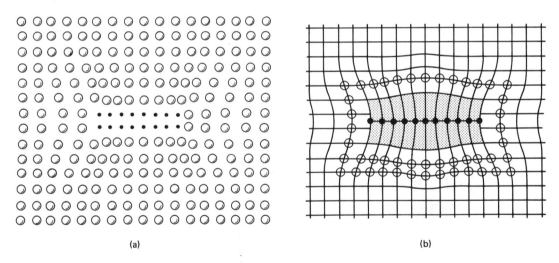

(a) (b)

Figure 6-16 Schematic illustration of GP zones: (a) GP-1 zone showing relative coherency of the precipitated atoms with those of the matrix; (b) GP-2 zone, edge view showing coherency on upper and lower surfaces, but not along the edges.

those of the matrix, strains in the matrix lattice are created. These strains interfere with dislocation motion and produce the observed hardening effect. In addition, these particles can produce dislocation pile-ups and then the yield strength of the system would be determined by the stress necessary to fracture or deform the particle.

Continued aging will produce a second transitional structure called the GP-2 zones. This is a disk (lozenge)-shaped precipitate which is coherent at its planar surfaces but is incoherent along the edges, as shown in Figure 6-16(b). A line of dislocations will be impeded and distorted by the precipitated particles in the manner shown in Figure 6-17. The dislocations cannot pass through the particles, but are forced

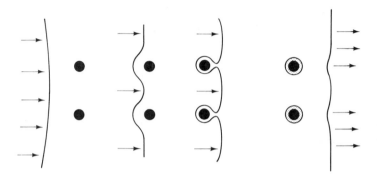

Figure 6-17 Creation of dislocation loops by the progress of a dislocation through precipitate particles. Shaded areas around particles indicate slipped regions.

between them, requiring additional stress to move them. Dislocation loops are created and left behind on the particles, creating an additional stress field. Further aging leads to the development of the $(CuAl_2)$ phase and a reduction in the number of particles, since the larger particles grow at the expense of the smaller neighboring ones. The final result is that upon overaging there is a decrease in the strength and hardness of the alloy.

Polymorphic Transformation

A polymorphic transformation may be defined as the change in the morphology of the lattice structure of metallic or ceramic systems which occurs upon heating or cooling (e.g., fcc to bcc). The system of principal importance exhibiting this type of behavior is the iron–carbon system, which is discussed in detail in Chapter 12. In this alloy system, a solution of carbon in face-centered cubic (fcc) iron transforms to a body-centered tetragonal (bct) structure upon cooling. This transformation is called the *martensite transformation*. In general, the martensite transformation functions as a strengthening mechanism by creating a fine substructure and increased strains in the lattice which retard and block dislocation movement, thereby requiring higher applied stresses to move them through the crystal structure. The phenomenon also has applications in other systems. Listed in Table 6-2 are a number of systems in which polymorphic transformation is known to occur. These types of transformations, which are characterized by the martensite transformation, are diffusionless reactions occurring by a process of lattice shearing. Such reactions reflect the propensity of the system to assume a crystal structure which more closely resembles the equilibrium condition than does the high-temperature structure.

TABLE 6-2 POLYMORPHIC TRANSFORMATIONS IN VARIOUS SYSTEMS

Structural change on cooling	System
fcc to hcp	Pure Co, Fe–Mn
fcc to bcc	Fe–Ni
fcc to bct	Fe–C, Fe–Ni–C, Fe–Cr–C, Fe–Mn–C
fcc to fct	In–Tl, Mn–Cu
bcc to hcp	Pure Li, Zr, Ti, Ti–Mo, Ti–Mn
bcc to fct	Cu–Zn, Cu–Sn
bcc to distorted hcp	Cu–Al
bcc to orthorhombic	Au–Cd
Tetragonal to orthorhombic	U–Cr

Source: Strengthening Mechanisms in Solids, American Society for Metals, Metals Park, Ohio, 1960.

PLASTIC DEFORMATION AND ANNEALING

Plastic Deformation

As stated earlier in this chapter, when a metallic material is permanently deformed in tension, torsion, or compression, the stress required to continue deformation increases continuously with strain. This process is termed *work hardening* or *strain hardening* and arises from the continued increase in dislocation density with plastic deformation. The plastic deformation of a metal system increases the internal stresses within the material that produce changes in the mechanical properties (i.e., strength and ductility) and in the physical properties (e.g., electrical conductivity and magnetic permeability).

Recall from our discussions of crystal imperfections in Chapter 4 that heavily cold-worked metals can have dislocation densities as high as 10^{12} dislocation lines/cm². As the mobile dislocations (those free to move) become entangled and pinned (i.e., their motion is impeded), an increasing magnitude of stress is required to generate new dislocations and then pass them through this "tangle." Thus the material becomes stronger and harder. Correspondingly, a decrease in ductility is observed. This process of plastic deformation is extremely important in the materials industry, for several reasons. By means of such manufacturing methods as rolling, extrusion, forging, and wire drawing, it permits the attainment of a shape or configuration that would not otherwise be possible. In addition, the properties are generally altered and may be adjusted by the degree of cold work to suit the design application.

We can represent various cold-working methods by the schematic rolling process illustrated in Figure 6-18. In the illustration presented, the initial material, prior to rolling, is a polycrystalline metal comprised of equiaxed stress-free crystals. After rolling, the plastically deformed (strain hardened) material contains highly stressed,

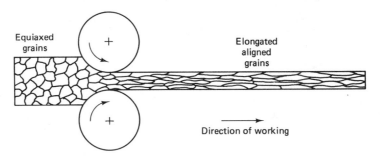

Figure 6-18 Schematic representation of cold working by a rolling operation. Note the change in grain morphology.

elongated crystals aligned in the direction of working. In general, the deformation produces a substantial difference in strength and ductility, when measured parallel to the direction of working (longitudinally oriented), compared to the properties measured perpendicular to the direction of working (transverse). This difference in properties with regard to direction is called *anisotropy* and the degree of anisotropy is dependent in part on the amount of deformation (percent cold work).

Anisotropy is related to several factors, such as the creation of a preferential crystallographic orientation (texturing), and the alignment of; nonmetallic inclusions, chemical segregation, and grain boundaries into planes and directions parallel to the direction of working. This topic is discussed in more detail in Chapter 8.

Slip. In an individual single crystal, the most important mechanism of plastic deformation is slip, the mechanism whereby one plane of atoms slides over another as described in Chapter 4. Slip occurs most easily on planes and directions that exhibit the densest atomic packing. On planes that are less densely packed, a higher shear stress is necessary to achieve slip. This occurs because slip is the result of dislocation motion, and dislocation motion occurs most readily with dislocations that have the shortest Burger's vector—those in the closest packed directions.

Slip initiates when the shear stress on the slip plane reaches a certain magnitude in the slip direction. This threshold value of shear stress is called the *critical resolved shear* (τ_{CR}). The critical value of resolved shear stress can be derived by examining a single crystal rod subjected to an axial tensile load (P), as illustrated in Figure 6-19. In this situation, the shear load (P_s) acts in the slip direction (in the slip plane) at an angle θ to the tensile load. The angle between the load (P_N) normal to the slip plane and the tensile axis is ϕ, and the area of the slip plane (A') is given by:

$$A' = \frac{A}{\cos \phi}$$

Therefore, the critical resolved shear stress (τ_{CR}) can be determined from:

$$\tau_{CR} = \frac{P_s}{A'} = \frac{P_s}{A/\cos \phi}$$

where

$$P_s = P \cos \theta$$

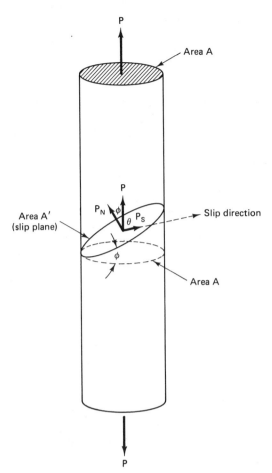

Figure 6-19 Diagram for determining critical resolved shear stress on a plane oriented at an angle (ϕ) to the direction of applied load (P).

So:

$$\tau_{CR} = \frac{P \cos \theta}{A/\cos \phi} = \frac{P}{A} \cos \theta \cos \phi \qquad (6\text{-}18)$$

Since P/A = engineering stress,

$$\tau_{CR} = \sigma \cos \theta \cos \phi \qquad (6\text{-}19)$$

The critical resolved shear stress is a maximum when the following condition is met: $\cos \theta = \cos \phi = 45°$. In this case, $\tau_{CR} = 0.5\sigma$. If the tensile load is applied normal to the slip plane, $\theta = 90°$, the resolved shear stress is zero. Also, if the tensile load is applied parallel to the slip plane, $\phi = 90°$, the resolved shear stress is zero. For both instances there is no shear component on the slip plane; therefore slip cannot occur.

Values of critical resolved shear stress vary considerably from material to material, and are influenced by factors such as crystal structure, temperature and chemical composition (alloy additions and impurities). Table 6-3 lists the planes and directions in which slip has been observed in a number of structures. Examination of this data discloses that although the number of slip systems varies with the structure, hcp structures tend to exhibit fewer slip systems.

Plastic Deformation and Annealing

TABLE 6-3 SLIP SYSTEMS IN VARIOUS CRYSTALS

Structure	Slip plane	Slip direction	Number of slip systems
FCC			
Cu, Al, Ni, Pb, Au, Ag, Fe, α-brass, . . .	111	110	4 × 3 = 12
BCC			
Fe, W, Mo, β-brass	110	111	6 × 2 = 12
Fe, Mo, W, Na	211	111	12 × 1 = 12
Fe, K	321	111	24 × 1 = 24
HCP			
Cd, Zn, Mg, Ti, Be, . . .	0001	1120	1 × 3 = 3
Ti	1010	1120	3 × 1 = 3
Ti, Mg	1011	1120	6 × 1 = 6
SODIUM CHLORIDE			
NaCl, AgCl, MgO LiF	110	110	6 × 1 = 6

Example 6-6

A single crystal specimen of dimensions 2 cm × 1 cm is subjected to an axial load of 10,000 kg in tension. Determine the critical resolved shear stress if slip is first observed on a plane oriented at 30° to the direction of tensile load, in a direction which makes an angle of 56° with the tensile axis.

Solution Based on Figure 6-19 and Equation (6-18) the critical resolved shear stress is

$$\tau_{CR} = \frac{P}{A} \cos \theta \cos \phi$$

$$= \frac{10,000 \text{ kg}}{(.02 \text{ m})(.01 \text{ m})} (\cos 56°)(\cos 60°)$$

$$= 5 \times 10^7 \text{ kg/m}^2 (0.559)(0.500)$$

$$= 1.40 \times 10^7 \text{ kg/m}^2$$

We can convert kg to N by multiplying 9.8, thus

$$\tau_{CR} = 1.4 \times 10^8 \frac{\text{N}}{\text{m}^2} \text{ or } 140 \times 10^6 \text{ Pa}$$

so

$$\tau_{CR} = 140 \text{ MPa} \qquad \textit{Ans.}$$

Deformation twinning. Twinning is a phenomenon which often occurs as the result of the applied stresses accompanying plastic deformation in metals. Shear and realignment of the lattice occur with the result that a portion of the crystal is a mirror image of the other. The effect of twinning on a crystal structure is schematically shown in Figure 6-20. The twin plane is defined as that plane which separates the sheared region from the original matrix. It is therefore the crystallographic plane of reflection. Twins may be produced as a result of mechanical deformation or as a result of nucleation during phase transformations. In either case, the event occurs by the uniform shear of successive planes of atoms in the original lattice. Like slip, it permits a change in the crystal shape with a negligible change in volume.

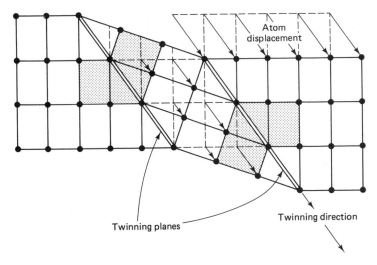

Figure 6-20 Crystallographic illustration of twinning.

Twinning differs from bulk slip in that the twinned region is a mirror image of the original lattice, whereas in slip, the slipped region has the same orientation as the original grain. Also, slip occurs by integral units of atomic displacement on the slip plane, whereas twinning involves successive planes of atoms moving successive fractional amounts.

Plastic deformation by twinning occurs primarily in bcc and hcp metals. In the bcc structure, twinning commonly occurs on the (112) plane, whereas in hcp, the common twinning plane is the (10$\bar{1}$2).

Twinning may also be produced as the result of annealing following plastic deformation. This type of twinning is referred to as *annealing twins* and occurs principally in fcc metals. In this particular crystal structure, the common twin plane is the (111). Figure 6-21 illustrates the microstructural appearance of twinning in copper alloys.

Annealing

The change in properties that occurs as the result of plastic deformation can be reversed by controlled heating. Historically, the fact that a strengthened metal softened and regained its ductility after heating above some particular temperature was dis-

Figure 6-21 Microstructure of brass (65 Cu–35 Zn) as shown at 70×. Structure consists of a single phase (α solid solution) with twins (arrows).

covered early in the utilization of metals. Modern metallography (examination and analysis of microstructures) has revealed that during this softening process, which we refer to as *annealing*, the deformed grains of the work-hardened metal were consumed by the nucleation and growth of new grains.

Briefly, the process of annealing generally refers to an elevated-temperature treatment used to soften metals and refine their grains. This treatment is ordinarily applied to metals in the mill after they have been shaped or formed by a mechanical working process such as rolling or forging. In this context, the deformed material is heated above its recrystallization temperature and the procedure is called *full annealing*.

Recrystallization. The recrystallization temperature may be defined as the lowest temperature at which stress-free equiaxed grains appear in the microstructure of a cold-worked metal. The temperature at which recrystallization occurs varies markedly with the metal system, but is approximately 0.3 to 0.5 of the melting point of the metal. Therefore, while lead recrystallizes at temperatures approximating room temperature, iron requires temperatures of 450°C (840°F) and above. The precise temperature at which recrystallization occurs is dependent on several general factors, all of which exert some influence on the recrystallization temperature. These factors can be listed as follows:

1. The grain size existent prior to the initiation of plastic deformation. The finer the initial grain size, the lower the recrystallization temperature for any system.
2. The degree of plastic deformation. The greater the degree of plastic working, the lower the recrystallization temperature. The effect of the amount of plastic deformation on the recrystallization temperature for several materials is shown in Figure 6-22.
3. The temperature at which plastic deformation occurs. When plastic deformation takes place at lower temperatures, the recrystallization temperature is reduced.

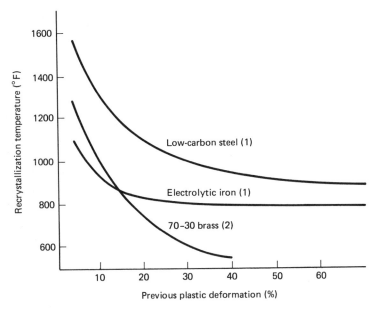

Figure 6-22 Effect of percent plastic deformation on the recrystallization temperature in iron, steel, and brass. [(1) After H. F. Kaiser and H. F. Taylor, *Trans. ASM 27*, 1939, p. 227; (2) after C. H. Mathewson and A. Phillips, *Trans. AIME 54*, 1916, p. 608.]

4. The duration of the heating cycle. The recrystallization temperature can be reduced by holding the worked metal at a lower temperature for a longer time. For example, 3 to 4 hr at a particular temperature might produce visible evidence of recrystallization, whereas 30 min might not.

The net effect of the factors listed above may be summarized by the following: In general, any factor that tends to increase the amount of residual stress in the structure (e.g., high degree of working, low working temperatures) will tend to reduce the temperature at which recrystallization will occur.

The effect of temperature on the mechanical and physical properties of a plastically deformed metal is illustrated in Figure 6-23. In the early stages of heating, or at temperatures below the recrystallization temperatures, the concentration of crystallographic defects is reduced without the movement of grain boundaries. This process, known as *recovery*, has a pronounced effect on the residual stress in the workpiece and on certain physical properties, such as electrical conductivity and magnetic permeability. However, at this stage, no change in the microstructure is discernible and no significant change in the mechanical properties is evident.

Recovery. The *recovery* process, which occurs at temperatures below the recrystallization temperature (T_R), consists of the increased movement and annihilation of point defects, plus the annihilation and redistribution of dislocations, re-

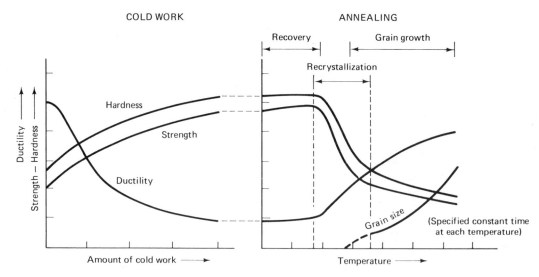

Figure 6-23 Schematic example showing the effects of cold work and annealing on mechanical properties.

sulting in a marked reduction in the internal stress of the grains. As the temperature increases to some temperature above T_R, nucleation of stress-free equiaxed grains occurs at various regions of high stress grain boundaries, triple points (grain boundary junctions), and slip planes. With additional heating, the nuclei continue to grow as stress-relieved equiaxed grains, at the expense of the plastically deformed grains. The process continues until the structure is entirely comprised of fine equiaxed grains. In the metalworking industry, the technique of cold working followed by recrystallization is of great importance as a method of achieving grain refinement, and as a method for achieving a degree of cross-sectional reduction that could not be achieved in a single cold-working operation.

Grain growth.   With continued heating at temperatures above the recrystallization temperature, the fine grains grow larger and coalesce until the structure consists of fewer but larger equiaxed grains. Accompanying this process is an attendant decrease in the strength parameters (yield strength, tensile strength) and hardness. It should be noted that the uniformity of the grain structure is largely dependent on the uniformity of the prior plastic deformation and of the chemical composition. If these factors are not uniform, the heavily worked areas or areas of nonuniform composition will nucleate earlier and a variable grain can result as illustrated in Figure 6-24.

Generally, large grains are deleterious to the mechanical behavior of metals and alloys. For example, the flow stress or yield stress (σ_y) of polycrystalline metals is related to the grain size by the Petch–Hall relationship as follows:

$$\sigma_y = \sigma_i + k_y d^{-1/2} \tag{6-20}$$

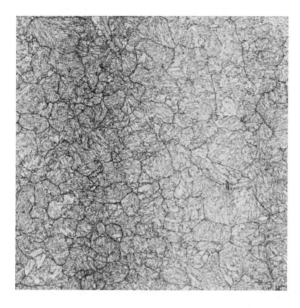

Figure 6-24 Microstructure of alloy steel showing duplex grain size due to alloy segregation ($425\times$). Dark band is solute rich.

where σ_i = friction stress opposing dislocation motion

k_y = constant associated with strengthening contributed by grain boundaries

d = average grain diameter

We can illustrate this expression in graphical form as illustrated in Figure 6-25. An examination of this figure reveals that as the grain size for a material decreases, the yield strength increases.

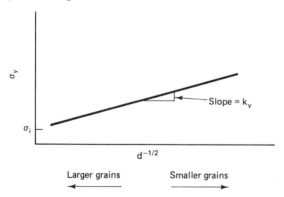

Figure 6-25 Graphical relationship between yield stress (σ_y) and grain size (d).

Summary

The overall process of annealing, that is, recovery, recrystallization, and grain growth, has many useful applications as an intermediate step in processing metallic materials before they achieve their final properties. Therefore, it will be beneficial to recapitulate the processes of plastic deformation and annealing with the effects they produce in a metal:

Plastic deformation (strain hardening)	Increases hardness
	Increases tensile and yield strength
	Increases residual stress
	Decreases ductility
	Decreases toughness

Plastic deformation
   (strain hardening)

Increases hardness
Increases tensile and yield strength
Increases residual stress
Decreases ductility
Decreases toughness

Annealing
   (elevated temperature
   treatment)
   Recovery period
   Recrystallization

Decreases residual stress
Decreases hardness
Decreases strength
Increases ductility
Increases toughness

Grain growth

Continues to decrease strength
Continues to increase ductility

To summarize the entire procedure of plastic deformation and annealing, the student is referred to Figure 6-26, where the sequence of microstructural behavior is presented in schematic form.

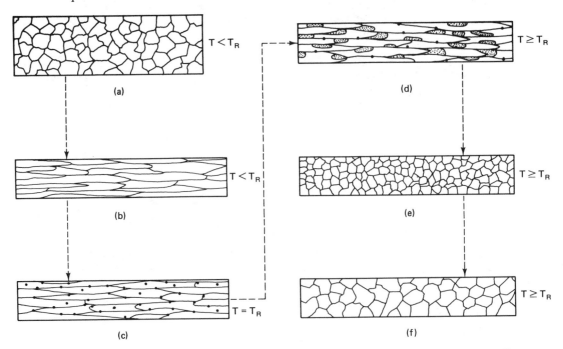

Figure 6-26 Schematic representation of plastic deformation followed by annealing: (a) equiaxed microstructure prior to deformation; (b) elongated highly stressed grains after plastic deformation; (c) recrystallization starts at T_R, new grains nucleate at grain boundaries and other nucleation sites in the deformed grains; (d) new "stress-free" grains grow with time at $T \geq T_R$; (e) annealing complete, structure consists of equiaxed stress free grains; (f) grain growth occurs if annealing is continued, resulting in larger equiaxed grains.

STUDY PROBLEMS

6.1. A 5-in.-long cylindrical steel bar is loaded in compression. After loading, the height of the bar is 4.98 in. Calculate the engineering strain in the bar.

6.2. As the load on the bar in problem 6.1 is increased, the bar continues to shorten until the total change in length is 0.085 in.

6.3. Each corner of a large structure is supported by a solid steel column at each of four corners. The cross-sectional area of each column is 5.0 in.². The strain produced in each column is 0.0012. Assuming that the modulus of elasticity for steel is 30×10^6, calculate the weight of the structure.
Answer: 720,000 lb.

6.4. Calculate the true strain for the bar conditions presented in problems 6.1 and 6.2.

6.5. A cube-shaped rubber engine mount 3 cm high is subjected to shear stresses when the vehicle stops. The angle of displacement is 11°.
  (a) What is the shear strain in the engine mount?
  (b) How far has the engine shifted forward?

6.6. What is the magnitude of the error in problem 6.5 if the displacement angle is expressed in radians and this is used to calculate strain?

6.7. A metal cube is subjected to a shear force of 90,000 N. If the cube has sides of 50 mm, what is the shear stress? If the angle of shear is 0.04°, what is the shear strain?
Answer: 36 MPa

6.8. For problem 6.7.
  (a) Calculate the shear modulus.
  (b) What would it be if the shear angle were increased by 11%?

6.9. During a cold-forming operation, an aluminum alloy rod, 18 cm long and 0.75 cm in diameter, is elongated by 1.5 cm and the diameter is reduced by 0.020 cm.
  (a) Calculate Poisson's ratio.
  (b) How does this value compare with the literature value?

6.10. A load of 91,300 N (20,500 lb) is supported by a steel bridge member 30 mm square.
  (a) What is the stress in the bar?
  (b) Calculate the strain in the bar if the modulus of elasticity for steel is 200 GPa (29.3×10^6 psi).
Answer: (a) 101.4 MPa

6.11. For problem 6.10, calculate the true stress obtained under the cited conditions and discuss the significance of true stress.

6.12. A power line comprised of individual copper wire conductors is strung from pole to pole. It is important that the stresses be kept in the elastic region. The wire diameter is 10 cm.
  (a) What is the maximum line tension permissible if the elastic limit is 100 MPa (14,450 psi)?
  (b) What is the strain at this load?

6.13. A metal bar $\frac{1}{2}$ in. (12.7 mm) in diameter is loaded in axial tension to 450 lb (1998 N). What is the shear stress produced on a plane oriented at 50° to the long axis of the bar?
Answer: 1133 psi

6.14. The critical resolved shear stress in a copper single crystal is 142 psi (0.98 MPa) on the (111) plane in the [10$\bar{1}$] direction. What is the engineering stress normal to the unit cell face (100) which will produce slip in this system?
Answer: 284 psi

6.15. An engineering stress (σ) of 62 MPa (9000 psi) normal to the unit cell face (100) in a single crystal of bcc metal produces slip on the (110) plane in the [$1\overline{1}\overline{1}$] direction. What is the critical resolved shear stress (CRSS) in this material? Compare your answer with the CRSS in problem 6.14.

6.16. Using the data presented inside the back cover, determine:
(a) If lead and tin would be expected to exhibit extensive solid solubility.
(b) What type of solubility would occur, if any?

6.17. Using the Hume–Rothery rules, determine whether you would expect hydrogen to exhibit a large degree of solubility in
(a) Steel (iron).
(b) Titanium.
(c) Are there any other factors that would influence solubility?

6.18. What are the principal methods employed to strengthen and harden metals and alloys?

6.19. A 89% Cu–10% Ni–1% Fe bar is cold worked (drawn) from 12 mm in diameter to a final size of 8 mm. Using Figure 6-15, determine the tensile strength, yield strength, and hardness for the final product.

6.20. Using Figure 6-14 as an example, establish the difference in strength between an ordered and a disordered alloy when both are subjected to a strain of 8%.

6.21. A cupronickel tubing section with an initial diameter of 1.75 in. and a wall thickness of 0.250 in. has the following properties: tensile strength 50,000, elongation 17%, hardness 65 Rockwell B. How much additional cold work must be given to the tube to raise the properties to the level specified: tensile strength 60,000 psi minimum, elongation 10% minimum, and hardness not greater than 78 Rockwell B?

6.22. Using the data presented and determined in problem 6.21, calculate the final cross-sectional area of the tubing if you assume that the reduction is uniform.

6.23. Show that for a round bar the relationship given in equation (6-17) is equivalent to that in equation (6-16).

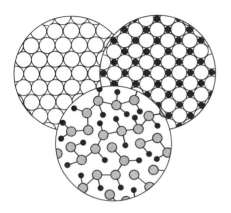

CHAPTER SEVEN:
Basic Relationships
in Single-Phase
and Multiphase
Materials

ALLOY SYSTEMS

An alloy system is basically a combination of two or more elements forming solid solutions within a specific range of *temperature, pressure,* and *composition.* The types (mechanisms) of solid solutions and their influence on hardening or strengthening was discussed in Chapter 6.

Alloy systems are classified according to the number of components or elements that constitute the system. For instance, two components, binary system; three components, ternary system; four components, quarternary system; and so on. In a particular alloy system, the elements may combine within a certain temperature range to form more than one homogeneous, coexisting portion. Each of these portions can have a different chemical composition and exhibit different properties. However, considered individually, each portion is chemically homogeneous throughout. A homogeneous (chemically uniform), physically distinct portion of an alloy system is called a *phase.* In alloys, phases are delineated by definite boundaries and their composition is either invariant (fixed) or varies in a continuous manner.

Chemical equilibrium may be defined as that condition where the potential energy of a given atom or molecule is the same in every part of the thermodynamic system under consideration. When there is more than one phase present in a system, the composition and amounts of the phases are fixed and no net change occurs under conditions of equilibrium. For example, liquid alloy may exist in equilibrium with a solid solution of the same composition. Analogously, equilibrium can be explained as that condition which exists in a closed container of water with a limited amount of air space. At constant temperature and pressure, the rate of evaporation of the liquid water will eventually equal the rate of condensation of the water vapor, as shown schematically in Figure 7-1. At this point, there is no net change in the amounts of the

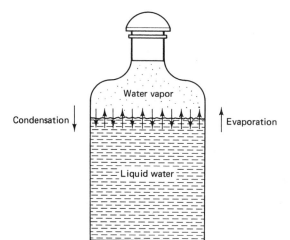

Figure 7-1 Condition of equilibrium between the liquid and vapor phases.

two phases; the system is at equilibrium. Such a condition can also be represented for constant temperature and pressure as follows:

$$H_2O_{liquid} \rightleftharpoons H_2O_{vapor}$$

The concept of two phases coexisting in a *solid solution* is illustrated schematically from an atomistic viewpoint in Figure 7-2. Although this example is highly schematic

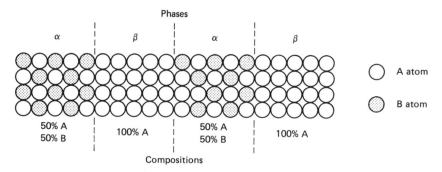

Figure 7-2 Two phases coexisting in a solid solution.

[atoms are uniformly dispersed in phase α,[1] and the phases (α and β) are uniformly separated], it suggests that two chemical species may coexist in the same materials as different phases. Does this representation meet the definition of a phase? Each phase is chemically homogeneous; phase α is a 50–50 solid solution of the hypothetical elements A and B, while phase β is pure A. Also, the phases are physically distinct from each other. This is true on both the atomic level and the microscopic level. Because of their differences in composition, the respective phases will exhibit different rates of dissolution during etching (attack) by the appropriate chemical reagents. An example of a multiphase solid solution is shown in Figure 7-3, where the phases have

[1] Beginning letters in the Greek alphabet are used to denote simple phases, and later letters are used for more complex phases.

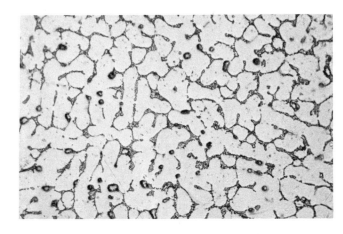

Figure 7-3 Microstructure of a two-phase aluminum alloy (300×).

been selectively contrasted. The distinction in appearance between these phases is obvious under the microscope.

The various phases in a multiphase material are not necessarily distributed in a completely uniform manner either. Phases may combine in a recognizable fashion to form what is termed a structural *constituent*. A constituent may be defined as any portion of an alloy which when viewed under the microscope appears as a definite unit of the microstructure. Figure 7-4 illustrates schematically a constituent consisting of the phases α and β. This representation shows that this constituent (No. 1) contains

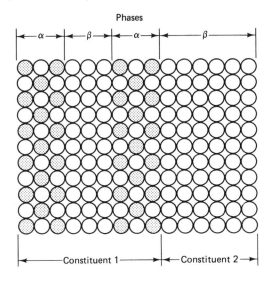

Figure 7-4 Atomistic representation of the microstructural constituents in a binary alloy system.

$\frac{2}{3}\alpha$ and $\frac{1}{3}\beta$ in an intimate association or mixture. Furthermore, this simplified model of a crystalline structure shows that one of the phases (β in this case) may exist by itself, in conjunction with the mixture. When this occurs, we term the *free* phase as a structural constituent also. In this example, the free β phase is identified as constituent No. 2. Figure 7-5 demonstrates that this type of crystalline structure does occur and is microscopically recognizable in many engineering materials.

Figure 7-5 Microstructure of bismuth–tin alloy (1150×). Bismuth appears dark, while the tin-rich solid solution is light. (Courtesy of T. V. Brassard.)

To avoid confusion later in this chapter, when we study the concepts of phase equilibria, let us summarize a few important terms with regards to alloys:

- Three *states of matter* are possible: solid, liquid and gas. Engineering materials generally are utilized in the solid or liquid state.

- An *alloy system* refers to the chemical elements that combine to produce the solid solution.

- A *component* of an alloy system is one of the chemical elements constituting the system. An alloy system can contain any number of components.

- A *phase* is a physically distinct (mechanically separable) portion of an alloy system delineated by definite boundaries, within which the chemical composition is either invariant or varies in a continuous manner. An alloy system may be comprised entirely of one phase, in either the liquid or the solid state. A system may also contain one or more physically distinct phases in the solid, as we have illustrated.

- A *constituent* of an alloy system is a recognizable, definite unit of the microstructure. In other words, we can identify a constituent under the microscope because of its distinct appearance, and it may consist of one or several phases.

THERMODYNAMICS OF ALLOYING

The formation of alloys and their resultant phases or constituents depends, in addition to composition, on relationships between pressure, volume, and temperature. Such relationships are usually developed for the various alloy systems under conditions of near equilibrium. We previously explained equilibrium conditions as that situation where no net change takes place in a system. A simple example of this condition is the freezing point of water; at 0°C (32°F), ice and water can exist in equilibrium. Although true equilibrium conditions are rarely achieved, actual conditions in alloy systems very often approach those of equilibrium. Therefore, approximations of the equilibrium condition can be useful in explaining or interpreting behavior in the real alloy systems.

Free Energy

Thermodynamically, an alloy system is in a state of equilibrium when the *free energy* of that system is at a minimum. The change in free energy (ΔF) associated with a chemical reaction is essentially a measure of a system's tendency to undergo that reaction. Consider the concept of free energy with respect to our example of ice and water (the H_2O system). The free energy of this system as a function of temperature is illustrated schematically in Figure 7-6. Examination of this relationship reveals the following:

- Below 0°C ice has the lower free energy and thus is the stable phase.
- Above 0°C water has the lowest free energy and therefore is the more stable phase.

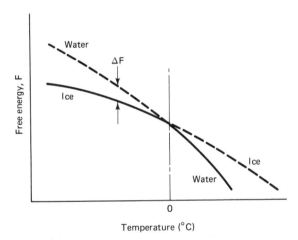

Figure 7-6 Free energy of H_2O as a function of temperature.

The solid curve represents the phase that actually exists at any temperature for this system. The change in free energy (ΔF) increases as temperature either increases or decreases from 0°C. Note that at 0°C the free energy associated with ice equals that of water; these two phases are in equilibrium at this temperature.

Perhaps some additional thoughts about this concept of free energy will be worthwhile. Free energy is a property of a system that can spontaneously decrease but cannot increase unless external energy or work is supplied. Refer to our schematic example; if water is supercooled (cooled below 0°C), ice may form spontaneously if the system is disturbed. For example, the student may have actually observed a similar spontaneous event when opening a bottle of carbonated beverage which has been cooled slightly below its freezing temperature but is still liquid (supercooled). Depend-

Thermodynamics of Alloying

ing on the composition of the beverage, the temperature and pressure of the system, ice crystals may quickly form and plug the neck of the bottle.

Furthermore, if a pure material can exist in either of two phases, the stable phase is that which has the lower free energy, because the free energy of the system would be increased if any of the material changed to the other phase. Finally, the free energies of two phases of a material are equal only at one temperature, the transformation temperature (e.g., solid and liquid at the melting point).

Phase Rule

A useful relationship between the number of phases, the number of components, and the number of variables in a system at equilibrium was developed by J. Williard Gibbs. This relationship is called the *Gibbs phase rule* and expressed as follows:

$$P + F = C + 2 \qquad (7\text{-}1)$$

where P = number of phases in equilibrium

F = degrees of freedom (number of variables that can be controlled, i.e., temperature, pressure, composition)

C = number of components in the system

In most cases, alloy systems are considered at constant pressure (1 atm),[2] and we can rewrite equation (7-1) as follows:

$$P + F = C + 1 \qquad (7\text{-}2)$$

For example, let us apply the phase rule to a pure metal. At the melting temperature (T_M), solid and liquid can coexist. Thus $P = 2$ and $C = 1$. Substituting this information in equation (7-2), we have

$$P + F = C + 1$$
$$2 + F = 1 + 1$$
$$F = 2 - 2$$
$$= 0$$

The phase rule predicts that there are no degrees of freedom in the system under these conditions. In other words, pure metals *must* exhibit one discrete value of T_M. Actually, commercially "pure" metals show a range of melting temperatures depending on their degree of contamination by other elements.

Example 7-1

Consider a binary alloy system of components A and B. Under what conditions may three phases coexist in equilibrium?

Solution Utilizing the phase rule [equation (7-2)], we obtain the following:

$$P + F = C + 1$$

---

[2] Exceptions to this include melting or processing in a vacuum and hot isostatic pressing (HIP) of metal or ceramic powders.

$$3 + F = 2 + 1$$
$$F = 3 - 3$$
$$= 0$$

The phase rule implies that three phases may only exist at a *fixed* temperature and composition in this system! *Ans.*

Later in this chapter you will see an example of this condition that occurs when a *eutectic* is formed.

In binary alloy systems, when two phases are in equilibrium, the phase rule predicts the following condition:

$$P + F = C + 1$$
$$2 + F = 2 + 1$$
$$F = 3 - 2$$
$$= 1$$

This result means that the state of the system can be established by specifying one of the variables, temperature or composition. Furthermore, a change in one of these variables produces a corresponding change in the other.

Examination of a *phase diagram* may assist the understanding of these concepts. Take, for example, the phase diagram of a "one-component" system, H_2O,[3] as illustrated schematically in Figure 7-7. In this diagram we can locate points in P–T space by specifying values of the two variables, pressure and temperature. One can easily see that if such a point lies above the line $\overline{BDC}$, the phase in existence is liquid, or water in this system. But what if the point determined by values of P and T falls directly

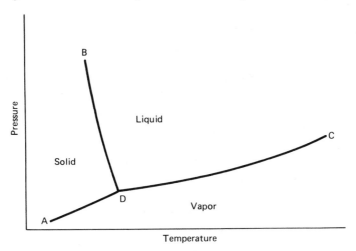

Figure 7-7 Schematic phase diagram for H_2O.

[3]Even though two elements (H and O) are involved, we may consider compounds like H_2O to be one-component systems by utilizing the definitions previously stated.

Thermodynamics of Alloying

on the line $\overline{DC}$? In this case, two phases coexist in equilibrium, liquid water and gaseous water vapor. Finally, if the specified temperature and pressure are the coordinates of point D, then all three phases—solid, liquid, and vapor—coexist in equilibrium. This condition is known as a *triple point*.

If we apply the phase rule to such a situation (triple point), what is the result? From equation (7-2),

$$P + F = C + 2$$

Solving for F, we obtain

$$F = C - P + 2$$

From Figure 7-6 we know that

$$C = 1 \text{ component (H}_2\text{O)}$$

$$P = 3 \text{ phases coexisting}$$

(both temperature and pressure are variables in this case). Therefore,

$$F = 1 - 3 + 2$$

$$= 0$$

A very interesting result: The number of degrees of freedom for this condition equals zero. Three-phase equilibrium is invariant; it exists at only one point (fixed temperature and pressure) for this system.

Before proceeding to the development of equilibrium phase diagrams for engineering alloys, let us briefly consider one more familiar example, a two-component system, which we know by experience exhibits some degree of solubility, *sugar* and *water*.

Example 7-2

The phases in equilibrium for the binary sugar–water system may be represented by the diagram shown in Figure 7-8. Based on this illustration and the information previously discussed in this chapter, let us address the following questions:

Question 1: How many phases exist at $T = 50°C$ and $C = 20\%$ sugar?

Answer: One phase—liquid solution composed of 20% sugar and 80% water.

Question 2: For $T = 100°C$ and $C = 40\%$ sugar, how many components exist?

Answer: Two components (sugar and water) exist for *any* combination of T and C in this system.

Question 3: For $T = 100°F$, $C = 20\%$ H$_2$O, how many phases exist?

Answer: Two phases exist—liquid solution and solid sugar.

Question 4: At any point along the solubility limit curve, how many phases are in equilibrium?

Answer: Two phases—liquid solution and solid sugar.

Question 5: Assuming constant pressure in this system, how many degrees of freedom are there for any point along this solubility line?

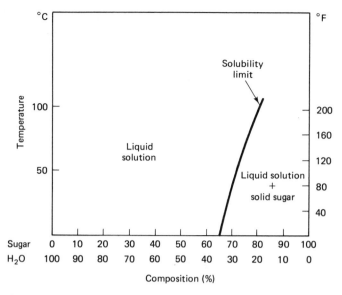

Figure 7-8 Phase diagram for a system composed of sugar and H_2O.

Answer: Using equation (7-1) and the following information:

$$C = 2 \text{ (sugar and water)}$$
$$P = 2 \text{ (from question 4)}$$
$$F = C + 1 - P$$
$$= 1$$

Therefore, the state of the system can be established by specifying either temperature or composition (of one component).

EQUILIBRIUM PHASE RELATIONSHIPS

In the basic study of alloy systems and ceramic compounds, the interrelation of phases for various temperatures and compositions will be useful in understanding and predicting material characteristics or behavior. As we previously emphasized, seldom is an alloy truly in a state of equilibrium, but equilibrium conditions often provide a reasonable approximation of actual alloy conditions. Therefore, equilibrium conditions may be applied as the limiting case.

Recall that we briefly mentioned equilibrium phase diagrams in the section "Alloy Systems." From an elementary standpoint, phase diagrams are considered the "road maps" of an alloy system. In this sense, they show the various "states" or phases in which an alloy exists depending on temperature and composition. Also, these diagrams delineate the conditions under which phases coexist in equilibrium (solubility lines).

In the following sections we will study principally three types of binary alloy phase diagrams which have considerable commercial applications as engineering materials. For convenience, these diagrams may be classified as:

- *Type I:* An alloy system that exhibits complete solubility in both the liquid and solid states. Examples of this type include Sb–Bi, Cu–Ni, and MgO–FeO.
- *Type II:* A system in which the components are completely soluble in the liquid state and insoluble in the solid state. This type is also known as a *eutectic alloy.* An example is the Bi–Cd system.
- *Type III:* A system in which the components are completely soluble in the liquid state, but only partially soluble in the solid state. Examples of this type include Pb–Sn, Ag–Cu, and Al–Si.

Although one glance at any handbook on phase diagrams will alert the student that phase diagrams are often more complex than those treated here, the principles presented in this section may be extended to the more complex diagrams.

Phase Diagram Construction

The construction or determination of equilibrium phase diagrams for metals and ceramics can be accomplished by several experimental techniques which employ thermocouples to detect changes in temperature within the alloy system being studied. Thermocouples, as you will learn in Chapter 10, are very precise temperature measuring devices consisting of dissimilar metal wires. Briefly, the electrical potential (voltage) across the junction of the dissimilar metal wires is extremely sensitive to temperature and thus may be utilized to observe reactions (e.g., phase changes) in the system that either absorb or liberate heat.

To better appreciate and understand phase diagrams, let us conduct a typical "cooling rate" experiment and use these "data" to generate an equilibrium phase diagram. In this experimental method, the components of the alloy system under investigation are combined in the desired proportions. For example, a hypothetical binary system of components A and B may be combined as 90% A–10% B, 80% A–20% B, 70% A–30% B, and so on. Each combination then represents an individual alloy of that system. The components of each alloy are melted together, allowed to mix thoroughly, then very slowly cooled to room temperature. Precise temperature measurements are continuously made by an immersion thermocouple, during the entire procedure. The very slow cooling process approximates the equilibrium cooling conditions for each alloy combination. The results of this experiment, temperature versus time data, are recorded and plotted for each compositional variation under examination. For example, the hypothetical binary system of components A and B may yield the schematic cooling rate curves illustrated in Figure 7-9.

Let us briefly consider the schematic cooling rates shown in Figure 7-9. All five "alloys" are heated to T_{liq}, where they are completely liquid, chemically homogeneous, and at a steady-state temperature. Then each alloy is very slowly cooled at constant rate of heat transfer. In case (a) for an alloy of 100% A (pure A, if we could really attain it), the cooling rate is constant for a brief period, then abruptly changes slope to a horizontal line (slope = 0) between points x and y. No change in temperature

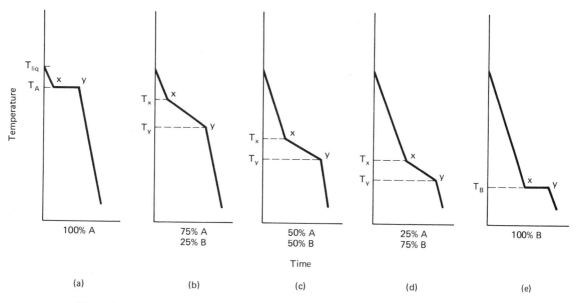

Figure 7-9 Schematic cooling curves for hypothetical alloy composed of A and B. T_A, melting-point, component A; T_{liq}, initial temperature of system (liquid); T_x, temperature of x inflection; T_y, temperature of y inflection; T_B, melting point, component B.

occurs for this period of time, even though heat continues to be removed from the system. In this situation, the two points x and y are equal in temperature. What causes this thermal behavior? Well, if we are continuously removing heat from the system but the temperature remains unchanged, heat energy must be supplied somewhere in the system. The *latent heat of fusion* is responsible for the observed behavior. Latent heat of fusion is heat energy liberated by the system when a solid is formed. This means that the liquid is solidifying at this temperature and that the freezing process occurs between x and y. Once solidification is completed, temperature begins to decrease once more, at a constant rate. This cooling process continues until the solid alloy reaches room temperature.

Notice that as the composition of the alloys changes, the temperatures associated with the initiation and completion of freezing also change. In our schematic example, T_x and T_y decrease as the percentage of component B increases. It is also interesting to note that freezing occurs over the temperature range T_x–T_y for mixtures of A and B as shown in Figure 7-9 b, c, d, rather than at a constant temperature. The final case (e) shows a cooling rate behavior similar to case (a), except that T_B occurs at a much lower point.

The phase diagram for the hypothetical alloy system of A and B can now be constructed from these data. The temperatures of interest, T_A and T_B, and the respective T_x and T_y data are plotted, corresponding to their approximate composition, as shown schematically in Figure 7-10. Although we have shown only four compositions, the accurate phase diagram determination of a real system would necessarily involve more compositional variations. However, the experimental procedure would generally be the same.

Equilibrium Phase Relationships

179

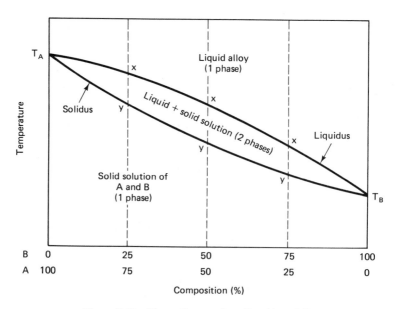

Figure 7-10 Phase diagram for alloy (A and B).

Systems Exhibiting Complete Solubility

The hypothetical phase diagram we have just constructed from components A and B is a type I system. Recall that this type exhibits complete solubility in the liquid and also in the solid state. Figure 7-10 serves as an example of the elementary details of phase diagrams. Consider the three regions outlined on this "map." The uppermost portion labeled liquid alloy is a liquid solution of A and B for all compositions. The central region contains liquid alloy plus a solid solution of the components. This is the region where solidification is occurring. The lower region is completely solid and consists of a solid solution of A and B atoms as a single phase.

An interesting feature of this phase diagram is the narrow central region containing both liquid and solid. The upper boundary of this region is the locus of T_x points, or the temperature at which freezing initiated for any composition. This curve is called the *liquidus line*. The lower boundary, the locus of T_y points, indicates the temperature where freezing is complete. This curve is referred to as the *solidus line*. By definition, then, the liquidus is the locus of temperatures above which all compositions are in the liquid state. Similarly, the solidus is the locus of temperatures below which all compositions are solid. In essence, then, the liquid–solid solution region of the phase diagram is the transition zone from the liquid state to the solid state. The question that arises is: Why is there a transition zone at all? Why doesn't the liquid simply freeze at one common temperature? Although we briefly touched on this topic during our discussion of the phase rule, the answer lies in the mechanisms of alloy solidification. The topic of solidification is discussed in detail later in this chapter. There we will see why freezing occurs over a range of temperatures for any system except a "pure" element.

Lever-arm technique. The lever-arm principle is an analytical technique for determining the proportions of solid and liquid coexisting in the region of the phase diagram containing both liquid and solid solutions. This technique is applicable at any temperature and composition. Furthermore, it can be used in any *two-phase* region of phase diagrams in general, as we will soon demonstrate. This principle is based on the simple mechanical analogy of a *balanced lever* as shown in Figure 7-11(a).

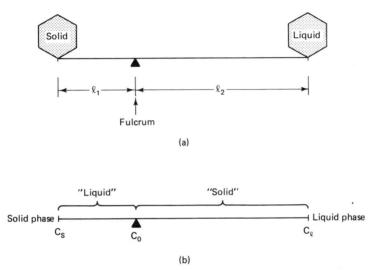

Figure 7-11 Example of lever-arm principle: (a) balanced lever; (b) inverse portrayal of phases.

The lever or "beam" is balanced when the quantity of solid on the left end times the distance l_1 equals the quantity of liquid on the right end times l_2.

We have conveniently placed the solid on the left end and liquid on the right end to correspond with the diagram in Figure 7-10. The entire length of our example beam is the "distance" (actually the difference in composition) from the solidus to the liquidus along a temperature isotherm (constant temperature). The fulcrum point represents the nominal composition (C_0) of the particular alloy under consideration. Thus in a real alloy system, we would know the values of l_1, l_2, and C_0.

Suppose that the fulcrum point (C_0) moves to the right in Figure 7-11(a). If the beam is to stay balanced, the amount of liquid must increase because its lever arm (l_2) is decreasing. Correspondingly, the amount of solid decreases because l_1 is increasing. In essence, a mass balance exists.

The lever arm is reproduced in Figure 7-11(b) in a slightly different manner. This time, the ends correspond to the compositions of the solid (C_s), the liquid (C_l), and the nominal composition (C_0) for a specific temperature. In addition, we have represented the difference between C_s and C_0 as the liquid phase, while the difference between C_0 and C_l represents the solid phase. Why is this seemingly inverse relationship valid? Notice what happens if the fulcrum (C_0) is moved to the left, closer to C_s. Correct! The portion of the beam termed "solid" increases in length. In other words, more solid phase is formed. Our inverse relationship (mass balance) illustrates that the portion of the mass that has formed a solid phase is the ratio of $C_l - C_0$ to the total mass $C_l - C_s$. Expressed as a percent, this portion is stated:

$$\% \text{ solid phase} = \frac{C_l - C_0}{C_l - C_s} \times 100 \tag{7-3}$$

Similarly, the amount of liquid phase remaining is expressed as

$$\% \text{ liquid phase} = \frac{C_0 - C_s}{C_l - C_s} \times 100 \tag{7-4}$$

In summary, the lever-arm technique just discussed is useful in determining the proportions of the phases present in two-phase regions of the phase diagrams. The important points to remember are:

- The ends of the lever arm represent the compositions at the intersection of an isotherm with the phase boundaries (liquidus and solidus in this case).
- The fulcrum is the nominal composition of the alloy under consideration.
- An *inverse* relationship exists between the quantities being calculated and the respective portions of the lever arm.

As an exercise, let us perform the appropriate phase determinations on an example of a type I phase diagram.

Example 7-3

Consider the phase diagram for the copper (Cu)–nickel (Ni) system shown in Figure 7-12. (a) For an alloy consisting of 70% Cu–30% Ni, determine the amount of solid solution (α phase) that has formed at a temperature of 1200°C. (b) How much liquid exists at this same temperature?

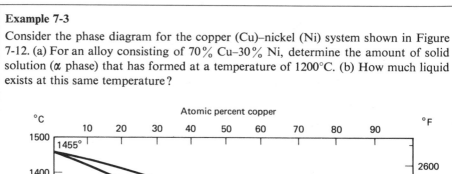

Figure 7-12   Phase diagram for copper–nickel system. (From *Metals Handbook*, 1948 ed., American Society for Metals, Metals Park, Ohio, 1948, p. 1198.)

Solution (a) At 1200°C we can sketch a lever arm similar to Figure 7-11(b), with the fulcrum at 70% Cu, and the ends intersecting the solidus and liquidus as follows:

α Solid solution ⊢————————▲————————⊣ Liquid
                        62   70 % Cu   78

Then, utilizing equations (7-3) and (7-4), the relative amounts of the desired phases are calculated as follows:

$$\% \text{ solid} = \frac{C_l - C_0}{C_l - C_s} \times 100$$

$$= \frac{78 - 70}{78 - 62} \times 100$$

$$= 50 \quad \textit{Ans.}$$

(b) The balance of mass (100% − % solid) must equal 50% liquid.   *Ans.*

Note that in this example the solid phase is labeled α. Alpha (α) is a *solid solution* of Cu and Ni atoms as illustrated in a hypothetical manner in Figure 7-2.

Eutectic Alloys

In this class of alloys, the components are completely soluble in the liquid state but insoluble in the solid state. In other words, the two components do not form a solid solution (one phase) over the entire composition range. The type II phase diagram is shown for the bismuth (Bi)–cadmium (Cd) system in Figure 7-13.

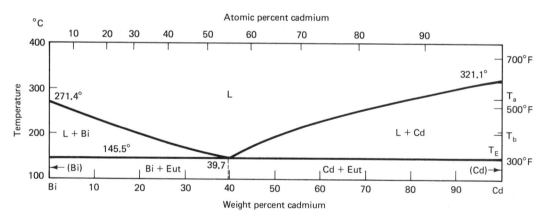

Figure 7-13 Equilibrium phase diagram for the bismuth (Bi)–cadmium (Cd) system. (From *Metals Handbook*, 8th ed., Vol. 8, American Society for Metals, Metals Park, Ohio, 1973, p. 272.)

Eutectic reaction. Inspection of this diagram reveals several interesting features not present in the type I diagrams. Below the liquidus a two-phase region exists for all compositions. But the solid phase above T_E is either pure Cd or pure Bi. The lowest temperature that the alloy is still liquid is referred to as the *eutectic point*. The eutectic reaction occurs in this type of alloy at a specific temperature and composition and can be represented by the following expression:

$$\text{liquid} \xrightleftharpoons[\text{heating}]{\text{cooling}} \text{solid 1} + \text{solid 2}$$

The phase rule predicts this three-phase equilibrium at the eutectic point (one composition–one temperature), as we illustrated in Example 7-1.

In this particular case, the solids are Bi and Cd in the ratio 60% Bi–40% Cd. Below the eutectic temperature (T_E) all compositions of these components are completely solid. However, because of the mutual insolubility of these two components, the product formed does not consist of a solid solution of Bi and Cd. Rather, the solid contains the eutectic mixture plus either free Cd or Bi, depending on C_0.

At this point it will be helpful to our understanding of eutectics to study some examples associated with the type II phase diagrams.

Example 7-4

Consider the phase diagram shown in Figure 7-13. What phases coexist at (a) $T = T_a$ and (b) $T = T_b$ for a nominal composition of 20% Bi (80% Cd)? (c) How much of each phase is present at T_b?

Solution (a) Inspection of the diagram shows that as the liquid alloy is cooled, at T_a, freezing initiates. The solid formed consists of pure Cd. Therefore, at this temperature two phases exist: liquid + solid Cd.   *Ans.*

Note that as temperature decreases between T_a and T_E, the composition of the liquid progressively becomes richer in Bi. Does this seem realistic? Certainly it should; if we are selectively removing Cd by solidification, the relative concentration of Bi in the liquid must increase.

(b) At T_b we may determine the quantity of each phase present by the following lever arm:

Liquid |———————————▲———————————| Solid Cd

53   80   100

$$\text{solid Cd} = \frac{80 - 53}{100 - 53} \times 100 = 57.4\% \quad \textit{Ans.}$$

Therefore, the % liquid $= 100\% - 57.4\% = 42.6\%$ *Ans.*

Comment: The student may note that we have used % Cd on the lever-arm scale while calculating % Cd. Since this is simply a proportional mass balance (with system totaling 100%), the substitution of the respective Bi concentrations on this scale will produce identical results.

Referring to Figure 7-13 again, as any composition of these two components (except the eutectic composition) is progressively cooled below the liquidus, the composition of the liquid in equilibrium with the solid formed becomes enriched with *one* of the components. At the eutectic temperature (T_E), the composition of the remaining liquid contains 60% Bi–40% Cd, regardless of what C_0 we started with! When this liquid phase freezes, it forms the eutectic mixture, which we can identify under the microscope. A typical eutectic microstructure from the Bi–Sn system was previously illustrated in Figure 7-5.

Below the eutectic temperature (T_E) the entire range of compositions are solid (Figure 7-13). The phases in existence are solid Cd and solid Bi. The microstructural constituents consist of eutectic plus free Bi or Cd. If one considers the exact eutectic composition as C_0, the microstructure would consist entirely of the eutectic mixture.

In summarizing the type II phase diagram, let us consider the following two examples with respect to the Bi–Cd system.

Example 7-5

Determine the amounts of the constituents coexisting at room temperature for the 80% Cd alloy in Figure 7-13.

Solution At room temperature (or any temperature below T_E),[4] the constituents are determined from the following lever arm:

Eutectic ├─────────▲─────────┤ Solid Cd

40   80   100
  % Cd

$$\text{solid Cd} = \frac{80 - 40}{100 - 40} \times 100$$

$$= 66.7\% \quad Ans.$$

Therefore,

$$\text{eutectic} = 100\% - 66.7\%$$

$$= 33.3\% \quad Ans.$$

Example 7-6

Determine the amounts of constituents in the structure of an 85% Bi alloy at room temperature (Figure 7-13).

Solution The lever arm for this situation is as follows:

Solid Bi ├─────────▲─────────┤ Eutectic

100   85   60
  % Bi

$$\text{eutectic} = \frac{100 - 85}{100 - 60} \times 100$$

$$= 37.5\% \quad Ans.$$

Thus

$$\% \text{ solid Bi} = 100\% - 37.5\%$$

$$= 62.5\% \quad Ans.$$

---

[4]Below the eutectic temperature, no other phase changes occur in this type of alloy.

Partial Solid Solubility

The third type of phase diagram we will analyze consists of components that are completely soluble in the liquid state, but only partially soluble in the solid state. In other words, this type of alloy exhibits *partial solid solubility*. Many commercially important alloy systems exhibit this type of solubility behavior as you will see in the next section, "Real Systems." Let us begin by examining a hypothetical binary alloy system of components A and B as illustrated in Figure 7-14.

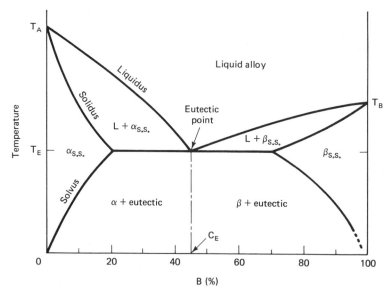

Figure 7-14 Phase diagram for a hypothetical binary system which exhibits partial solid solubility.

This diagram exhibits a noticeable difference from the previous phase diagrams we have studied. It will be worthwhile to emphasize three regions which compose the solid state of this type of alloy:

1. *Alpha* (α) *solid solution:* atoms of B (solute) dispersed in A (solvent)
2. *Beta* (β) *solid solution:* atoms of A (solute) dispersed in B (solvent)
3. *Eutectic:* mixture of α and β; the eutectic composition of this system is 45% B–55% A.

The essential difference in this type of alloy system is the solid solution which forms on either side of the eutectic region ($20 \leq \%\ B \leq 70$). The partial solubility characteristics of these components in the solid state results in the regions bounded by the solidus and *solvus* lines. The solvus indicates the maximum solubility of the components in their respective solid solutions (α and β).

In the eutectic region, the solid formed consists of one of the solid solutions (α or β) plus the eutectic mixture. Remember: the eutectic itself consists of α solid solution *plus* β solid solution. Sound confusing? Well, we can differentiate between the phases in the solid state by referring to the α or β that form prior to the eutectic reaction (T_E) as *proeutectic* α or β (sometimes called *free* α or β). This simply means that the solid solution which forms above the eutectic temperature is not *intimately* associated with the eutectic mixture. In this case it appears as a separate *constituent* in the microstructure. Correspondingly, we could refer to the α and β in the eutectic as eutectic α and eutectic β. Perhaps some visualization of these concepts as illustrated in Figure 7-15 will be helpful.

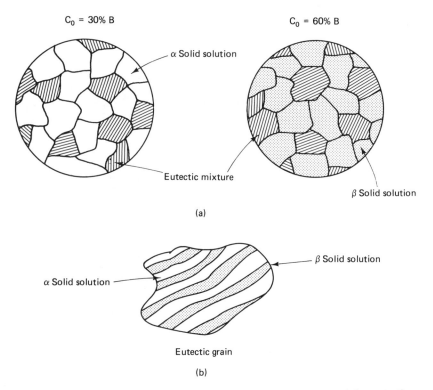

Figure 7-15 (a) Schematic illustrations of microstructures containing eutectic constituent; (b) eutectic constituent composed of α and β phases.

This schematic representation shows two microstructural arrangements in the solid state: one for a nominal composition of 30% B and the other for 60% B. In Figure 7-15(a), the light equiaxed grains consist of proeutectic α. The eutectic grains are depicted as alternating bands of light and dark phases. The structure of the 60% B alloy consists of slightly shaded, equiaxed grains composed of β solid solution, while the appearance of the eutectic mixture is unchanged. Admittedly, the portrayal of the eutectic is highly schematic, but it serves to illustrate the compositional nature of this *constituent*. A closer look at this feature is schematically shown in Figure 7-15(b).

As a final reminder, the eutectic grain depicted in Figure 7-15(b) consists of a "mechanical mixture" of α and β solid solutions. This constituent is referred to as a mechanical mixture because of the rather sharp demarcation between the alternating phases. Also recall that earlier in this chapter a structural constituent was defined as a recognizable association of phases. The eutectic qualifies for this definition, although

Equilibrium Phase Relationships

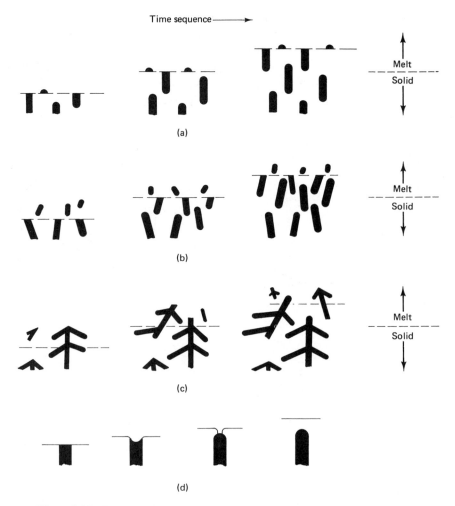

Time sequence ⟶

(a)

(b)

(c)

(d)

Figure 7-16 Illustrative examples of various eutectic morphologies: (a) lamellar eutectic; (b) nonlamellar eutectic; (c) dendritic eutectic; (d) closure of discontinuous plate. (From W. Rostoker and J. R. Dvorak, *Interpretation of Metallographic Structures*, Academic Press, Inc., New York, 1965, p. 69.)

its morphology does not necessarily have to be laminar. For example, consider the possible eutectic formations illustrated in Figure 7-16.

Eutectics often have mechanical and physical properties quite different from those of the individual phases that comprise them. Such properties are usually influenced by the morphology of the eutectic formation and in certain cases can be controlled by various solidification techniques. For example, directionally solidified eutectics are being applied to several applications which require strength in a preferred orientation. Buckets or blades for aircraft gas turbines are one particular component in which longitudinal "fiber" reinforcement may be developed during solidification by uniaxial alignment of the eutectic structure (see Figure 16-22). Examples of eutectic microstructures, including directional solidification, are shown in Figure 7-17.

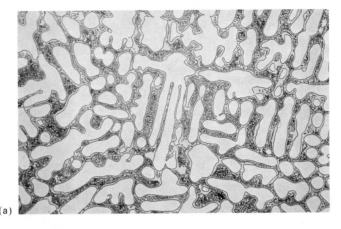

(a)

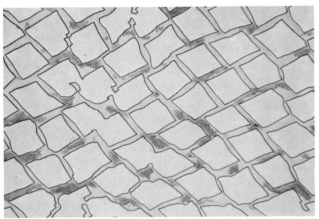

(b)

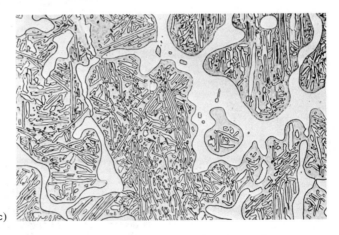

(c)

Figure 7-17 Microstructural examples of eutectic structures; (a) Ni–Ni$_2$B eutectic (gray region surrounding Ni-rich phase) at 380 $\times$; (b) transverse section of a directionally solidified CoAl–Co eutectic (Co, light rectangles; CoAl eutectic, gray matrix) at 1150 $\times$; (c) Cr–Ni eutectic in a 65 Cr–35 Ni alloy (light etching phase is Cr-rich solid solution, gray area is the Ni-rich solid solution) at 675 $\times$. (Courtesy of T. V. Brassard.)

189

As an exercise to familiarize the student with the alloy systems that exhibit partial solid solubility, let us work out an illustrative example.

Example 7-7

For the hypothetical alloy system shown in Figure 7-14, determine the amounts of *phases* and *constituents* that exist at room temperature for $C_0 = 35\%$ B.

Solution For this concentration of B, the required information can be determined from the following lever arm:

α Solid solution ├──────────▲──────────┤ Eutectic

20   35   45
%B

Therefore,

$$\% \, \alpha = \frac{45 - 35}{45 - 20} \times 100$$

$$\alpha = 40\% \ (\text{proeutectic } \alpha)$$

Correspondingly,

$$\% \text{ eutectic} = 100\% - 40\% = 60\%$$

Thus we have determined that the structure contains the following constituents:

$$\text{proeutectic } \alpha = 40\%$$

$$\text{eutectic} = 60\% \quad \textit{Ans.}$$

However, the computation above did not fully answer how much of each individual phase (α and β) exists in this structure. To do this, we must ascertain the amounts of α and β in the eutectic. This is accomplished with the following lever arm:

α ├──────────▲──────────┤ β

20   45   70
%B

Thus the $\% \, \alpha$ in the eutectic is

$$\% \, \alpha = \frac{70 - 45}{70 - 20} \times 100$$

$$\alpha_{\text{eut}} = 50\%$$

Therefore,

$$\beta_{\text{eut}} = 50\%$$

So the eutectic mixture consists of 50% α solid solution and 50% β solid solution, in whatever particular morphology this constituent prefers to solidify. The eutectic may appear as laminar platelet, nodules, dendrites, and so on. Nevertheless, its composition will always be 50% α–50% β (in this hypothetical system).

    Now to finish our determination of the amounts of phases in the structure for $C_0 = 35\%$ B. Recall that the eutectic comprised 60% of the structure (the mass); therefore, the relative amounts of α and β contained in this mass are as follows:

$$\alpha_{\text{eut}} = 0.60 \, (50\% \ \alpha \text{ in eutectic}) = 30\%$$

$$\beta_{\text{eut}} = 0.60 \, (50\% \ \beta \text{ in eutectic}) = 30\%$$

These percentages (30% α and 30% β) are now the portion of the entire mass which is associated in particular with the eutectic. To obtain the total amounts of each phase in the structure, simply sum the proeutectic phase and the eutectic phases as follows:

$$proeutectic\ \alpha = 40\%$$
$$eutectic\ \alpha = 30\%$$
$$total\ \alpha = \overline{70\%}\ \Big\}$$
$$eutectic\ \beta = 30\%\ \Big\}\ \ Ans.$$
$$total\ mass = \overline{100\%}\quad check$$

Real Systems

Up to this point we have considered a hypothetical system which exhibits partial solid solubility. Real type III systems appear slightly different, as illustrated by the Pb–Sn phase diagram in Figure 7-18. This is a popular commercial alloy used exten-

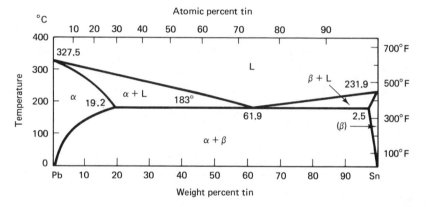

Figure 7-18 Lead–tin phase diagram. (From *Metals Handbook*, 8th ed., Vol. 8, American Society for Metals, Metals Park, Ohio, 1973, p. 330.)

sively in the vicinity of the eutectic composition for *soldering*[5] because of its relatively low melting temperature. The low-melting-temperature characteristics of solder accounts in part for its widespread utilization as a *joining* material for the electrical and plumbing industries.

In this specific case, the solid solutions of α and β can now be described in terms of the components Pb and Sn as follows:

- α *phase:* solid solution of Sn atoms (solute)
  dispersed in Pb atoms (solvent)
- β *phase:* solid solution of Pb atoms (solute)
  dispersed in Sn atoms (solvent)

[5]Soldering is the process of joining metals by "wetting" the metals to be joined with a molten filler metal at temperatures below 427°C (800°F). Above this temperature, the process is termed *brazing* and the molten metal is called *brazing filler metal*.

Two other points of interest on this diagram are particularly worthy of note. The point specified by $C = 19.2\%$ Sn and $T = 183°C$ indicates the maximum solubility of Sn in the solid solution. In other words, the α phase cannot contain more than 19.2% Sn in solid solution under equilibrium conditions. Similarly, the β phase cannot contain more than 2.5% Pb and this maximum solid solubility occurs at $T = 183°C$. Incidentally, the lines that bound the solid-solution regions α and β indicate the solubility of the respective components in these phases at any given temperature. Consider the α phase, for instance. Between 327 and 183°C, the solubility of Sn in the α solid solution increases. But from the eutectic temperature (183°C) to room temperature, this solubility decreases.

Such behavior (decreasing solid solubility) results in the precipitation of β (Sn-rich solid solution) from the α, as cooling progresses below 183°C in the Pb–Sn system under equilibrium conditions. However, under *nonequilibrium* conditions, this behavior is responsible for *age hardening* or *precipitation hardening* in some very important commercial alloys. The precipitation hardening reaction is discussed later in the section "Nonequilibrium Behavior."

The lead–tin system exhibits a eutectic reaction at 183°C and 61.9% Sn. As previously mentioned, solders composed of Pb–Sn alloys in the vicinity of the eutectic

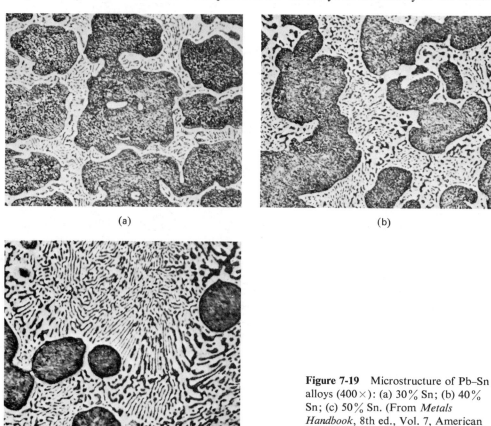

(a)

(b)

(c)

Figure 7-19 Microstructure of Pb–Sn alloys (400×): (a) 30% Sn; (b) 40% Sn; (c) 50% Sn. (From *Metals Handbook*, 8th ed., Vol. 7, American Society for Metals, Metals Park, Ohio, 1972, p. 302.)

composition are used in the electrical and plumbing industries to make connections and joints. The advantage of these alloys is, of course, their relatively low-melting-temperature range, which allows the solder to be melted on-site with a minimum of effort (equipment). Correspondingly, the liquid solder solidifies rapidly upon removal of the heat source enabling a rigid connection to be accomplished in a minimum of time.

The microstructures of three Pb–Sn alloys are shown in Figure 7-19. These structures are arranged in parts (a)–(c) according to increasing Sn concentration, so the student can observe the decrease in α solid solution (large, dark dendritic grains) and corresponding increase in eutectic (light matrix) as the eutectic composition is approached.

An exercise with a real alloy exhibiting partial solid solubility will reinforce the concepts discussed in this section.

Example 7-8

The phase diagram for the silver (Ag)–copper (Cu) system is given in Figure 7-20. For an alloy of $C_0 = 82\%$ Ag, determine the following: (a) At 900°C, what are the quantities of Ag and Cu in existence? (b) At 800°C, what are the amounts of the phases in existence? (c) What are the eutectic composition and temperature for this alloy system? (d) For the 82% Ag alloy, what phases exist at 400°C? (e) What is the amount of the

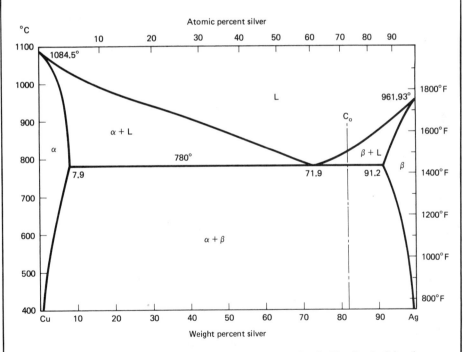

Figure 7-20 Silver–copper phase diagram. (From *Metals Handbook*, 8th ed., Vol. 8, American Society for Metals, Metals Park, Ohio, 1973, p. 253.)

Equilibrium Phase Relationships

193

eutectic constituent at 400°C? (f) What amounts of the individual phases coexist in this system at 400°C?

Solution (a) At 900°C, the 82% Ag alloy is in the liquid region. Therefore, one phase exists, consisting of 82% Ag and 18% Cu.   *Ans.*

(b) At 800°C, inspection of Figure 7-20 shows that two phases coexist: β solid solution (silver rich) plus liquid. The quantities are determined by the lever-arm technique, where this lever is specified by the intersections with the solidus and liquidus as follows:

Liquid |————————▲————————————| β Solid solution

78   82   92

% Ag

$$\beta \text{ solid solution} = \frac{82 - 78}{92 - 78} \times 100$$

$$\beta = 28.6\%$$

$$\% \text{ liquid} = 100 - 28.6$$

$$\text{liquid} = 71.4\% \quad \textit{Ans.}$$

(c) By inspection of the phase diagram we see that

$$T_E = 780°C \quad \textit{Ans.}$$

$$C_E = 71.9\% \text{ Ag} \quad \textit{Ans.}$$

(d) At 400°C, the phases in existence consist of α solid solution and β solid solution, regardless of how they are arranged (as constituents) in the microstructure. *Ans.*

(e) The percentage of eutectic at 400°C is determined by the lever-arm technique as follows:

Eutectic |————————————▲————————| β

71.9   82   91.2

% Ag

$$\text{eutectic} = \frac{91.2 - 82}{91.2 - 71.9} \times 100$$

$$= 47.7\% \quad \textit{Ans.}$$

(f) Since this is a two-phase region, we may determine the total amounts of the coexisting phases as follows:

α Solid solution |————————————————————▲————| β Solid solution

2   82   98

$$\alpha = \frac{98 - 82}{98 - 2} \times 100$$

$$= 16.7\%$$

Thus

$$\beta = 83.3\% \quad \textit{Ans.}$$

Before we complete our discussion of the phase diagrams for real binary systems, it should be pointed out that for simplicity and convenience, the more complex diagrams have been avoided. In all fairness, therefore, to prevent misleading the student, the phase diagrams for several of the more commercially important complex binary systems are presented in Figures 7-21 through 7-23. Not only do these diagrams illustrate the phase complexity of various real systems, but they also illustrate two important reactions we have not yet discussed: the *peritectic* and the *eutectoid* reactions.

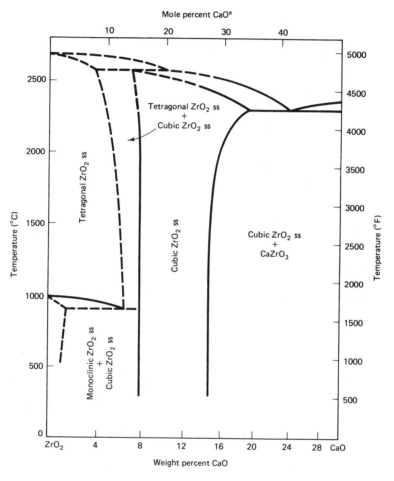

[a] Mixture of components A and B in the amounts $n_A + n_B = n_T$, where n is the number of moles. The mole percent is expressed as the mole fraction $\times$ 100; e.g. $(n_A/n_T) \times 100$ = mole percent A.

Figure 7-21 Equilibrium phase diagram for the CaO–ZrO₂ system. (From P. Duwez, F. Odell, and F. H. Brown, Jr., *J. Amer. Ceramic Soc. 35*, 1952, p. 109.)

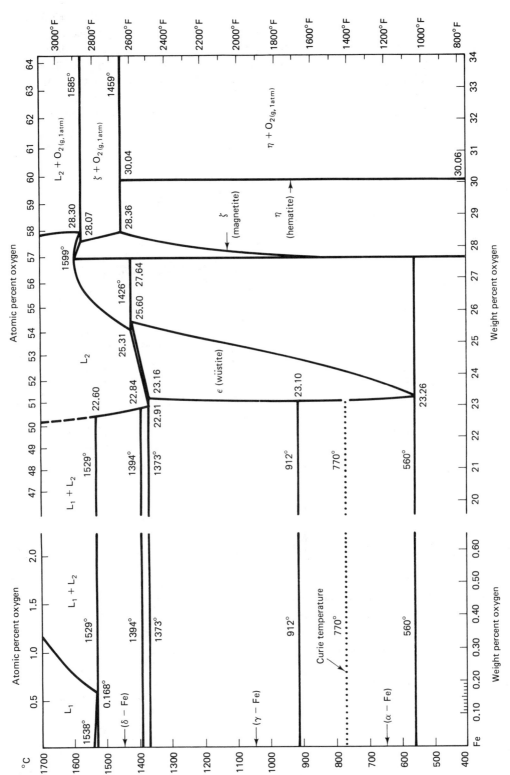

Figure 7-22 Iron–oxygen phase diagram. (From *Metals Handbook*, 8th ed., Vol. 8, American Society for Metals, Metals Park, Ohio, 1973, p. 304.)

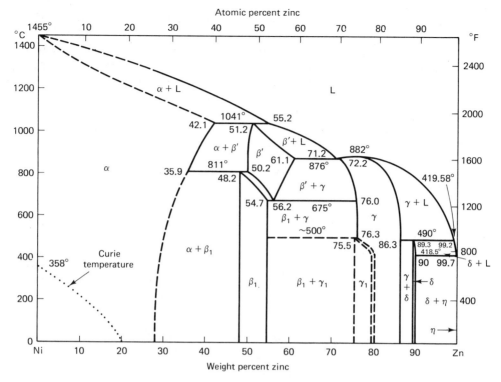

Figure 7-23 Equilibrium phase diagram for the nickel–zinc system. (From *Metals Handbook*, 8th ed., Vol. 8, American Society for Metals, Metals Park, Ohio, 1973, p. 326.)

Peritectic reactions. This reaction consists of the transformation of a solid phase plus liquid to a new solid phase. The reaction may be expressed as follows:

$$S_1 + L \; \underset{\text{heating}}{\overset{\text{cooling}}{\rightleftarrows}} \; S_2$$

In the case of the CaO–ZrO$_2$ system, shown in Figure 7-21, the peritectic occurs at approximately $T = 2575°C$ and % CaO = 7. S_1 is the tetragonal zirconia solid solution and S_2 is the cubic zirconia solid solution. A peritectic reaction is also observed in the Ni–Zn system at 1040°C and 52% Zn, as illustrated in Figure 7-23.

Eutectoid reactions. The eutectoid transformation is similar to the eutectic reaction we studied earlier in this section, except that it occurs entirely in the solid state. The eutectoid reaction consists of one solid phase transforming to two new solid phases and is represented as follows:

$$S_1 \; \underset{\text{heating}}{\overset{\text{cooling}}{\rightleftarrows}} \; S_2 + S_3$$

In the CaO–ZrO$_2$ system, solid 1 is the tetragonal zirconia phase, solid 2 is the monoclinic zirconia phase, and solid 3 is the cubic zirconia phase. This transformation occurs at approximately $T = 900°C$ and 6% CaO, as shown in Figure 7-21. Similarly, a eutectoid reaction occurs in the Ni–Zn system at 56% Zn and $675°C$.

Furthermore, as the preceding section demonstrates, binary systems of many important engineering materials may be composed of compounds rather than individual elements. Take, for example, the binary system formed between alumina (Al_2O_3) and silica (SiO_2). The *components* in this case are ionic compounds of aluminum–oxygen, and silicon–oxygen. The phase diagram for this system is shown in Figure 7-24. Several high-temperature applications are related to the phases presented in this

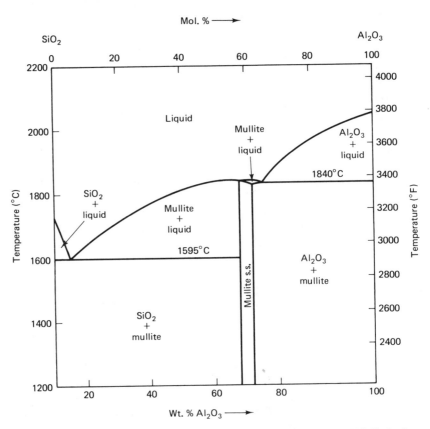

Figure 7-24 Phase diagram for the alumina–silica system. Mullite ($Al_6Si_2O_{13}$) is a solid solution of alumina and silica.

diagram, including refractory brick for lining furnaces, crucibles and ladles, fireclay tiles, ceramic laboratory ware and heat-resistant cookware. Simple inspection of this phase diagram reveals that the alumina-rich portion produces materials with relatively higher melting temperatures, thus demonstrating the usefulness of the diagram with respect to materials selection and design.

ATOMIC PERCENT VERSUS WEIGHT PERCENT

Thus far in our study of phase diagrams we have expressed the components in terms of weight percent (w/o). Although this is common practice, because the alloys are measured by weight, weight percent is not the limiting factor in determining solubilities. Rather, atomic percent of the solute species is a limiting criterion. The observant student may have already noticed that some of the example phase diagrams are also expressed in atomic percent (a/o). In some cases, alloys are designed on the basis of atomic percent rather than weight percent (e.g., high-temperature superalloys). The following examples will serve to illustrate the relationship between weight percent and atomic percent.

Example 7-9

A lightweight die-casting alloy consists of 92% aluminum and 8% magnesium by weight. What are the atomic percents of each component in this alloy?

Solution Based on a 100-g sample of this material, we can determine the a/o of each component as follows:

$$N_{Al} = \frac{(92 \text{ g})(6.02 \times 10^{23} \text{ amu/g})}{26.98 \text{ amu/atom}} = 20.53 \times 10^{23} \text{ atoms}$$

$$N_{Mg} = \frac{(8 \text{ g})(6.02 \times 10^{23} \text{ amu/g})}{24.31 \text{ amu/atom}} = 1.98 \times 10^{23} \text{ atoms}$$

The total number of atoms in our sample is

$$N_T = 20.53 \times 10^{23} + 1.98 \times 10^{23} = 22.51 \times 10^{23} \text{ atoms}$$

By taking the number of atoms for each component as a percentage of the total, we obtain the respective atomic percents as follows:

$$\text{Al:} \quad \frac{20.53 \times 10^{23} \times 100}{22.51 \times 10^{23}} = 91.2 \text{ a/o} \qquad Ans.$$

$$\text{Mg:} \quad \frac{1.98 \times 10^{23}}{22.51 \times 10^{23}} \times 100 = 8.8 \text{ a/o} \qquad Ans.$$

Example 7-10

Inconel is a wrought heat-resistant nickel-based alloy widely utilized in elevated-temperature applications. This alloy nominally contains 15.5% chromium, 1.0% iron, 0.04% carbon, and the balance nickel. Determine the atomic percents of these components.

Solution Based on 10,000 amu[6] of the alloy, we have

1550 amu Cr

100 amu Fe

4 amu C

8346 amu Ni

[6]amu, atomic mass unit. There are 6.02×10^{23} amu per gram.

Atomic percents:

Cr: $\dfrac{1550 \text{ amu}}{52.00 \text{ amu/atom}}$ = 29.81 atoms   or   17.1 a/o   *Ans.*

Fe: $\dfrac{100 \text{ amu}}{55.85 \text{ amu/atom}}$ = 1.79 atoms   or   1.0 a/o   *Ans.*

C: $\dfrac{4 \text{ amu}}{12.01 \text{ amu/atom}}$ = 0.33 atoms   or   0.2 a/o   *Ans.*

Ni: $\dfrac{8346 \text{ amu}}{58.71 \text{ amu/atom}}$ = 142.16 atoms   or   81.7 a/o   *Ans.*

     Total   174.09 atoms   100 a/o   *check*

MONOTECTIC ALLOYS

Another class of alloy systems which we will mention briefly are the *monotectics*. This term refers to alloys that exhibit *immiscibility* in the liquid state. Because at certain temperatures the components are unable to mix or blend with each other as one liquid phase, they form two separate liquid phases. A familiar example of this characteristic is an oil slick on top of water.

The importance of this type of alloy resides in their ability to produce a dispersion of droplets of one component in another. When certain monotectic alloys are cooled from the liquid state, the liquid phase with the higher melting point tends to

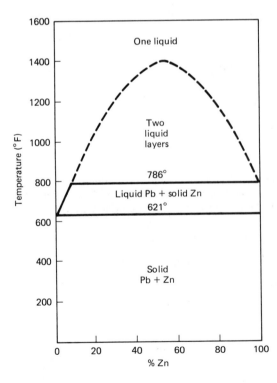

Figure 7-25 Lead–zinc phase diagram illustrating immiscibility in the liquid state.

form globules in the lower-melting-point liquid, due to surface tension. These globules then persist into the solid structure. Such a process is used to make bearing bronze, which consists of lead droplets dispersed throughout copper or zinc. In service, the Pb acts as a lubricant, thereby decreasing friction and wear. Similarly, Pb is added to certain steels to improve machineability. The lead is insoluble and forms droplets dispersed randomly in the steel. Then, during machining, these droplets smear over the cutting edge of the tool, reducing friction and localized heating.

The phase diagram for the Pb–Zn alloy system is shown in Figure 7-25. This diagram demonstrates that the immiscibility of these components is temperature dependent. If the temperature is raised high enough, Pb and Zn eventually become soluble in each other, forming one liquid phase.

MULTICOMPONENT ALLOYS

In the beginning of this chapter we stated that alloy systems are classified according to the number of components involved. Our discussions of phase diagrams concentrated on binary systems because they are relatively easy to understand and form the basis of many commercially important engineering materials. However, a large number of important alloys are multicomponent systems containing three (ternary) or more elements. In many cases, small percentages of certain elements are present in the alloys as impurities because it is impractical to remove them. But multicomponent alloys are often deliberately produced because the addition of the extra elements results in improved properties, such as increased toughness, greater corrosion resistance, or better machineability. For example, chromium is added to steel (Fe–C) to improve corrosion resistance. Furthermore, the addition of nickel and chromium to steel produces the familiar *stainless steels* used in many applications requiring corrosion resistance.

Ternary Phase Diagrams

The equilibrium phase diagrams for a ternary system is an extension of the principles we used to develop and interpret binary diagrams. However, because there is another variable to consider (third component), the ternary phase diagram is three-dimensional under conditions of constant pressure and can be represented by a polyhedron (triangular prism). The phase relationship of a hypothetical ternary system composed of components A, B, and C is portrayed in Figure 7-26(a). This three-dimensional diagram depicts a system which exhibits various degrees of solid solubility, depending on temperature and composition.

Typically, three-dimensional representations of ternary phase diagrams are difficult to analyze. Therefore, we may consider isothermal sections through the diagram, or sections of constant composition. For an isotherm, such a strategy allows us to examine an equilateral triangle with the three components at the corners. For example, the basal (room temperature) isothermal plane of our hypothetical system (A, B, C) is shown in Figure 7-26(b). The amount of each component in a specific ternary alloy at any given temperature (isotherm) is determined by constructing a line

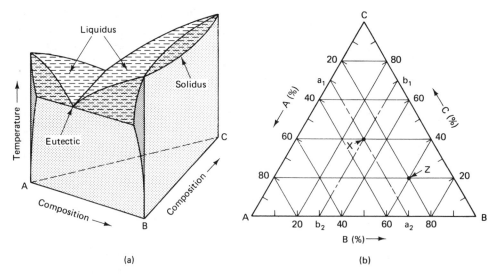

Figure 7-26 (a) Three-dimensional phase diagram representation of a hypothetical ternary system consisting of components A, B, and C. Note that compositions consisting of just A and B display partial solid solubility, while compositions of B and C exhibit complete solid solubility. (b) Isothermal section (room temperature) of the hypothetical system illustrated in part (a).

through the alloy (point) in question parallel to the side of the triangle opposite 100 % of the component. The intersection of this line with the appropriate components' axis is the composition of that component.

Example 7-11

Determine the concentration of A, B, and C for alloy X in Figure 7-26(b).

Solution   Construct lines parallel to the three axes, through the point X. From the respective intersections:

Component A: (line a_1–a_2) % A = 30 %   *Ans.*

Component B: (line b_1–b_2) % B = 30 %   *Ans.*

Component C: (existing line) % C = 40 %   *Ans.*

Although the isothermal section of ternary phase diagrams in general may contain numerous phases and phase combinations, the respective compositions are determined in the manner as illustrated in Example 7-11. For example, consider the isothermal section of the chromium–iron–nickel system (*stainless steels*) at 1400°C (2552°F) shown in Figure 7-27. Depending on composition, an alloy may be solid or liquid at this temperature. The solid solutions of the three components are identified as α and γ in this diagram and the liquid phase by L. Note in particular the triangular region containing all three phases ($\alpha + \gamma + L$). The ternary diagram predicts that the three phases exist in equilibrium for conditions of composition and temperature (constant in this case) in this region. Recall from our earlier discussion of eutectic

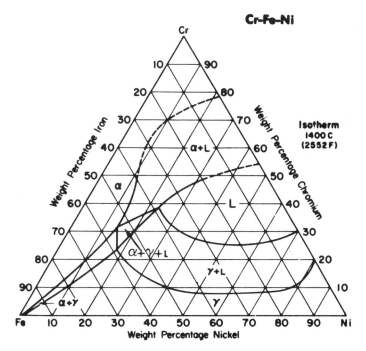

Figure 7-27 Isothermal section of the Cr–Fe–Ni phase diagram showing various phases in equilibrium at 1400°C. (From *Source Book on Stainless Steels*, American Society for Metals, Metals Park, Ohio, 1976, p. 400.)

alloys that a eutectic reaction consists of liquid transforming to two solid phases. Therefore, this portion of the ternary diagram is referred to as the *eutectic triangle*.

The complexities of many ternary diagrams and multiphase diagrams (quaternary and higher) may require considerable graphical effort and analysis. Therefore, such systems are potential candidates for analysis by interactive computer graphics if they can be analytically described or if they can be plotted.

NONEQUILIBRIUM BEHAVIOR

Solidification

As we have seen earlier in this chapter, the idealized pure metal is composed of a single component and there are only two phases present (liquid and solid); therefore, there are zero degrees of freedom. That is, the liquid and solid have the same concentration, and the temperature of the interface, which is invariant, is the melting point of the pure metal.

In solidifying, the atoms in the liquid must give up a part of the energy they possess and assume positions in the crystal structure of the pure metal, characteristic of the solid state. The difference in energy between the solid and liquid states corresponds to the *latent heat of fusion*.

Even the purest common structural metals, however, usually contain sufficient impurities to allow them to be considered alloys from the solidification standpoint. Alloys, by definition, must be multicomponent systems, and solidification usually occurs over a range of temperatures. The mechanics of solidification for alloys differs from that of pure metals in two ways:

1. Solidification occurs over a range of temperatures rather than at a constant temperature.
2. The solidification process requires a redistribution of the solute atoms and the dissipation of the latent heat of solidification (fusion).

The distribution coefficient.   The ratio of the composition of the solid (C_s) to the composition of the liquid (C_l) which it contacts and is in equilibrium with is called the *equilibrium distribution coefficient* (K_0). This coefficient is expressed as follows:

$$K_0 = \left(\frac{C_s}{C_l}\right)_T \tag{7-5}$$

For example, consider the equilibrium distribution coefficient at a given temperature (T_x) for the two cases illustrated in Figure 7-28. In part (a) of this figure, $K < 1$, and the result is solute rejection at the liquid–solid interface. When $K > 1$ as shown in Figure 7-28(b), rejection of the solvent species occurs.

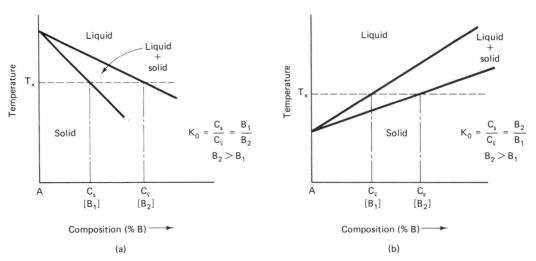

Figure 7-28   Effect of the distribution coefficient upon: (a) solute rejection, $K_0 < 1.0$; (b) solvent rejection, $K_0 > 1.0$.

In real systems, the distribution coefficient is not constant over the entire melting range, because the liquidus and solidus lines are not linear. However, over the melting temperature range for a particular alloy, the error introduced by selecting the distribution coefficient corresponding to the solidus temperature for that alloy is usually negligible.

Redistribution of Solute

It is advantageous to begin a discussion of solidification by considering variations in the process which are related to the form and degree of solute distribution. Upon solidifying, a redistribution of solute occurs, which depends on the boundary conditions. Three cases may be postulated:

1. Equilibrium is maintained in the system at all times.
2. Negligible diffusion in the solid with sufficient mechanical mixing in the liquid to render it completely homogeneous at all times.
3. Negligible diffusion in the solid and no mechanical mixing in the liquid; diffusion is the only method for modifying the composition of the liquid phase.

Case 1: This is the condition where equilibrium is maintained in the system at all times, and is illustrated in Figure 7-29. As the melt of mean composition C_0

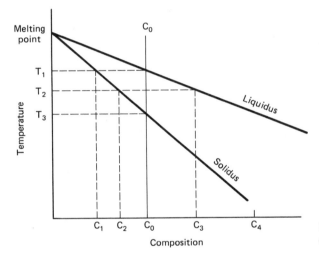

Figure 7-29 Solidification under equilibrium conditions.

loses heat, the temperature falls, and upon reaching the liquidus, solidification begins. The first solid to form is of composition C_1 at temperature T_1. Since for the case presented the composition of the solid, C_1, is less than that of the melt, C_0, in terms of solute concentration, the formation of a significant amount of solid must therefore be accompanied by rejection of the solute and a corresponding increase in the solute concentration in the liquid. However, since we have assumed the system to be in equilibrium, compositional adjustments must occur in the solid and in the liquid. For example, at any temperature T_2, the concentration of the solute in the solid has been made uniform throughout, by diffusion, and is given by composition C_2. At this same temperature, the concentration of solute in the liquid has been made uniform throughout, by mixing and diffusion, and is given by C_3.

The percentage of the solid and liquid in equilibrium at any temperature can be

Nonequilibrium Behavior

calculated using the lever law. At T_2,

$$\% \text{ of liquid} = \frac{C_0 - C_2}{C_3 - C_2} \times 100$$

$$\% \text{ of solid} = \frac{C_3 - C_0}{C_3 - C_2} \times 100$$

As the temperature is slowly lowered, solidification continues. The volume of the solid increases while the volume of liquid decreases, and the homogenizing processes (diffusion and mixing) continue to operate to keep each at the equilibrium composition for the temperature in question.

At T_3 the last liquid has disappeared and solidification is complete. The excess solute in the last infinitesimal quantity of liquid is uniformly distributed throughout the solid phase by diffusion, and the composition of the solid at this point is C_0, the nominal composition of the alloy.

Case 2: In this case there is negligible diffusion in the solid accompanied by sufficient mechanical mixing (stirring) in the liquid to render it completely homogeneous at all times. The nominal composition of the melt is C_0. Upon solidification, the first solid to form is C_1 at temperature T_1, as shown in Figure 7-30. At any tem-

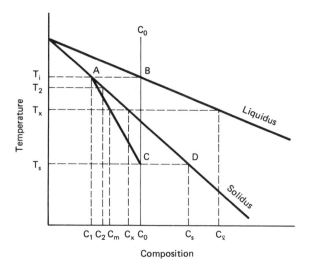

Figure 7-30 Solidification with mechanical mixing in liquid, negligible diffusion in solid.

perature T_2 below T_1, the composition of the forming solid is given by the intersection with the solidus C_2. Since by definition homogenization cannot occur, there is now a distinction from case 1. A concentration gradient exists between the last solid to form and the first, and each infinitely small layer of solid is at a slightly different composition than that of the previous one. The mean composition C_m of the entire solid at any temperature T_x may be expressed by $(C_x - C_1)/2$. Therefore, a line AC may be constructed so that $AB = CD$. This line describes the locus of points representing the mean composition of solid phase.

The composition of the liquid at temperature T_x is given by C_l. Complete mixing is assumed, and no concentration gradient exists in the liquid. The percentage of solid and liquid in equilibrium at any temperature T_x may again be calculated from the lever law. However, the mean composition line AC must be used rather than the solidus. At T_x,

$$\% \text{ of solid} = \frac{C_l - C_0}{C_l - C_m} \times 100$$

$$\% \text{ of liquid} = \frac{C_0 - C_m}{C_l - C_m} \times 100$$

At T_s solidification is complete. The temperature is defined by the coincidence of the mean solid composition line AC with the nominal composition C_0. At this temperature, the last remnants of liquid solidify, forming a solid of composition C_s, with the mean composition of the entire solid being the nominal composition C_0.

Case 3: This situation presumes that there is negligible diffusion in the solid and no mechanical mixing in the liquid. Diffusion is the only method for modifying the composition of the liquid phase. Given a melt of nominal composition C_0, upon cooling to T_i the composition of the first solid to form is C_i, as shown in Figure 7-31. Using the distribution coefficient, the composition $C_i = K_0 C_l$; therefore, $C_i = K_0 C_0$.

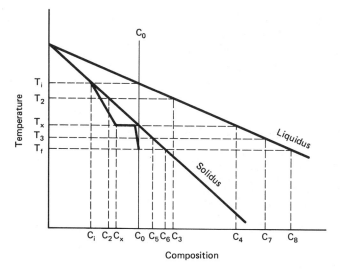

Figure 7-31 Solidification with no mechanical mixing in liquid, negligible diffusion in solid (for $K_0 = 0.5$).

The composition of this solid is significantly lower than the original melt (C_0); therefore, its formation must be accompanied by the rejection of solute, thereby enriching the liquid. Since the concentration of solute in the liquid can change only by diffusion, a concentration gradient is established in the liquid and the next solid to form is higher in solute than the previous one. Consequently, a concentration gradient is created in the solid as well. Solidification proceeds until the temperature falls to T_x, which marks the onset of dynamic equilibrium. As in case 2, the mean or average composition of the solid up to this point may be calculated by $(C_i + C_0)/2$.

At T_x the composition of the solid forming at the interface is equal to C_0, and that of the liquid is equal to C_4. This composition is considerably richer in solute than the bulk of the liquid. A steep concentration gradient exists on the liquid side of the interface, due to the lack of mechanical mixing. The mean composition shifts along the composition axis in the direction of the nominal composition. The steady-state stage is limited, however, by the physical confines of the system. As the volume of the solid formed increases, there is, of course, a corresponding decrease in the liquid. When this liquid diminishes to the point where only the enriched volume of liquid material preceding the interface is available, the terminal stage of solidification begins. This stage is marked by increased concentrations of solute in the solid, since the liquid from which solidification proceeds is heavily enriched. For example, at T_3, the composition of the solid formed is C_5, from liquid of composition C_7. Similarly, at T_f the composition of the final solid to form is C_6, from liquid of composition C_8. Upon total solidification, the mean composition is identical to the nominal composition, C_0.

Macrostructure

Any liquid under equilibrium conditions contains "clusters" of atoms which, on a microscale, resemble the structure of the solid phase. As the temperature of the liquid decreases, the growth of these clusters results in the formation of grain nuclei. These nucleating grains hold the key to the form the cast metal will take. If a relatively small mass of metal is cooled very rapidly, the structure will form primarily equiaxed grains; grains whose shape, while multifaceted, have essentially the same dimensions in all directions, as shown in Figure 7-32. However, if a large volume of metal is poured into a mold and cooled, the structure will consist of grains which nucleated at the mold wall and grew inward as heat flows from the liquid metal into the mold.

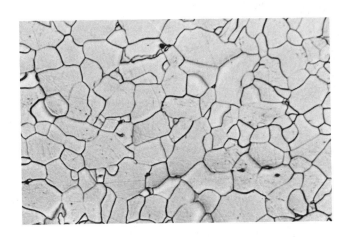

Figure 7-32 Equiaxed grains in an iron alloy ($65 \times$).

This inward growth results in the formation of elongated grains called *columnar grains*. A typical example of columnar grain formation is shown in Figure 7-33.

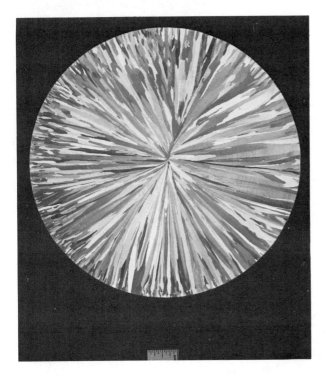

Figure 7-33   Columnar grains in centrifugally cast copper. Scale in inches. (Courtesy of N. Gendron.)

Dendritic Solidification

Solid solutions of one (or more) elements in another tend to produce a rather unique structure upon solidification. The structure formed by these alloys upon solidification is referred to as *dendritic*. Dendritic solidification is characterized by the formation of dendrites, a linear branched structure with arms parallel to specific crystallographic directions and branches that are spaced at fairly regular intervals. A typical dendrite is shown schematically in Figure 7-34(a), and actually in Figure 7-34(b). A cluster of dendrites is shown in Figure 7-35, indicating their Christmas tree-like appearance and random orientation.

All solidification, however, is not dendritic. It usually occurs only when the liquid melt is supercooled, and even under these conditions, only a minor proportion of the total liquid solidifies this way. The discussion of dendritic segregation may be clarified by the third solidification case (3) presented previously. Figure 7-36 illustrates the situation for dendrite solidification. The temperatures shown refer back to Figure 7-31.

In Figure 7-36(a), solidification begins at temperature T_i, initiating at the dendrite nucleus. The initial solid is lower in solute and the liquid is slightly enriched. At temperature T_2 [Figure 7-36(b)], the composition of the forming solid is C_2, which is

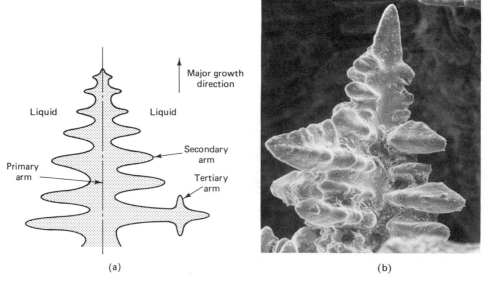

(a)   (b)

Figure 7-34 Appearance of a dendrite: (a) schematic; (b) actual dendrite in cast iron (130×).

Figure 7-35 Cluster of dendrites showing tree-like structure (15×).

higher in solute content; however, the solute concentration in the liquid at the interface is also increasing. Figure 7-36(c) shows the conditions that exist when dynamic equilibrium commences. The composition of the solid at any point along the curve from C_i to C_0, called the *initial transient*, is approximately equal to

$$C_s = C_0 \left\{ (1 - K_0)\left[1 - \exp\left(-K_0 \frac{R}{D} X \right) \right] + K_0 \right\} \qquad (7\text{-}6)$$

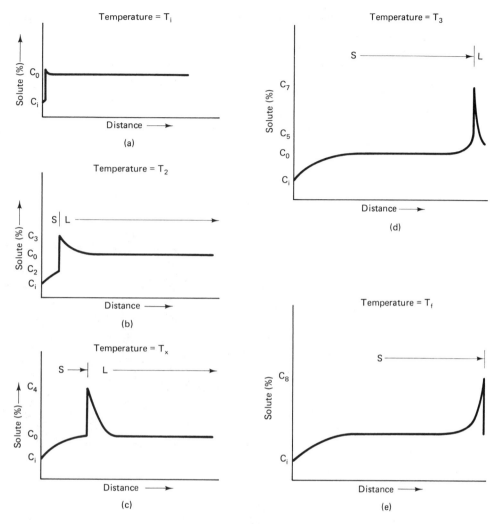

Figure 7-36 Schematic showing the redistribution of solute during dendrite solidification.

where: K_0 = distribution coefficient
D = diffusion coefficient in liquid
R = rate of growth of interface
X = diffusion distance (thickness of the boundary layer at interface)

It should be noted that the *rate of growth* is now a consideration, and can appreciably affect the distribution of solute.

Upon completion of the initial transient, the temperature remains constant at T_x and the solid of composition, C_0, forms from the liquid of composition, C_4, as long as such liquid is available. As the liquid interface proceeds ahead of the growing dendrite, it approaches the interfaces of competing dendrites. When the two concentration spikes impinge on one another, the solute content in the solid at the interface begins to increase, as shown in Figure 7-36(d). Finally, as the two interfaces meet, the conditions shown in Figure 7-36(e) exist. The last liquid of composition, C_8, reacts to form the solid of composition, C_8. The ultimate result is that the redistribution of solute atoms occurs with the composition of the dendrite core or centerline, differing from that at its surface (this condition is referred to as *coring*), the interdendritic region being very rich in rejected solute. A further consequence in real systems is that the last metal to solidify usually contains large numbers of inclusions (see following section on solidification defects) pushed ahead of the solidifying front, which become entrapped in the interdendritic region. Also, since this is the last metal to solidify, it is rich in phases having a low melting point. In steels these phases often consist of sulfides and phosphides, as illustrated in Figure 7-37.

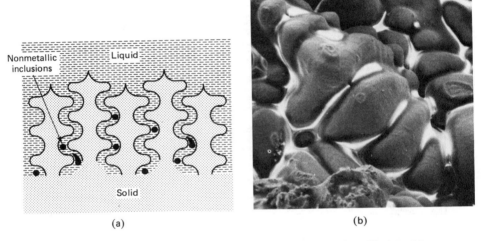

Figure 7-37 (a) Schematic illustration of second phase formation; (b) Actual interdendritic segregation viewed in the SEM. Light regions are rich in sulfur and phosphorous (200×).

This deviation from the nominal composition at various locations in the casting is called *chemical segregation* and is undesirable, since it can be reasonably expected that it produces variations in mechanical and physical properties as well as corrosion resistance, since these properties are all dependent on composition.

Rapid Solidification Processes (RSP)

Recent developments in solidification technology have led to processes which involve extremely high cooling rates ($> 10^3$ °C/sec). From a practical standpoint, these high cooling rates may be achieved by several methods currently used on a commercial basis. For instance, in one process, called *melt extraction*, a rapidly rotating disk is used to extract a wire-like filament from the molten surface of metal in a crucible, as shown in Figure 7-38(a). Another technique involves directing a stream of molten metal at a rapidly spinning disk or roll, as illustrated in Figure 7-38(b).

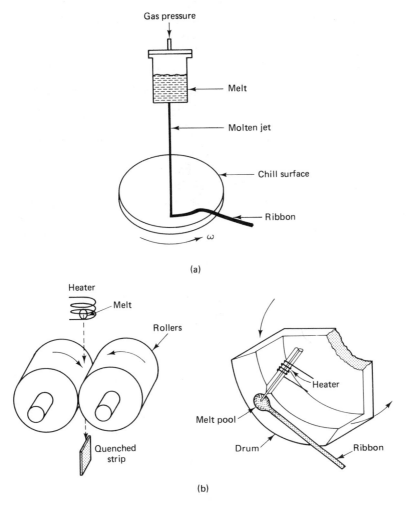

Figure 7-38 Schematic illustration of rapid solidification techniques: (a) chill block casting; (b) roller quenching and spin casting. (From *Metallic Glasses*, American Society for Metals, Metals Park, Ohio, 1978, pp. 38, 276.)

Similar cooling rates may also be obtained by utilizing *lasers* (or electron beams) to quickly heat the surface of a metal or ceramic material, which is subsequently "quenched" by the underlying material (at ambient temperature). Since the laser beam can be focused to concentrate its energy in a very small region, relatively thin surface layers can be melted. Therefore, in thermally conductive materials such as metals, extremely high cooling rates can be developed (on the order of 10^6 °C/sec).

The boundary conditions for rapid solidification are different from the three cases that we considered in our discussion of solute redistribution. The following conditions may apply during the rapid solidification process:

1. No mechanical mixing (stirring) occurs in the liquid.
2. No diffusion occurs in the liquid.
3. No diffusion occurs in the solid.

Because of the ultrahigh cooling rates, the usual solute redistribution no longer takes place and dissipation of the latent heat of fusion is quickly accomplished by conduction to the surrounding metal. In fact, the liquid (melt) may be supercooled if the rate of solidification is fast enough. When this happens, an amorphous (glassy) material[7] may be produced rather than a crystallized state.

Metal powders which are used commercially to produce parts close to *net shape* (parts that require little machining), may also be made by a variation of the rapid solidification process. In this case the process is referred to as *atomization*, and consists of directing jets of an appropriate gas (e.g., N_2) or liquid (e.g., H_2O) at a stream of the liquid metal or alloy. This process causes the stream to break into very small droplets which solidify in a rapid manner due to their diminutive size. One important benefit of atomization is that the chemical composition of the individual powder particles closely approaches that of the liquid since there is both little time and mass for any appreciable segregation. Correspondingly, components manufactured from such powders can exhibit a high degree of chemical homogeneity.

In general, the rapid solidification processes result in structures which can have many of the following characteristics:

- *Greater homogeneity:* The consequence of this characteristic is that detrimental effects associated with anisotropy may also be reduced.
- *Amorphousness:* In this noncrystalline state, many interesting changes in mechanical and physical behavior occur. These include improvement of corrosion resistance due to a lack of grain boundaries, increased resistance to fatigue crack initiation, and unexpected changes in magnetic properties, as will be discussed in Chapter 11.
- *Extremely fine grain:* This characteristic usually occurs when the cooling rate is not high enough to yield an amorphous condition. However, the grain size is sufficiently reduced compared to conventional processes as to create improved mechanical properties.
- *Greater induced supersaturation:* Alloys from nonmiscible systems, or systems that exhibit negligible solubility under equilibrium conditions, can be obtained. The properties of many such systems have yet to be explored and perhaps hold great potential.

[7]The condition of amorphous structure is discussed in Chapter 15 in the section "The Glass State."

Solidification Defects

Initially, most engineering alloys begin with melting and casting, regardless of whether the alloy is poured as a cast product or as an ingot which will ultimately be reshaped by mechanical working. Although many factors that influence solidification can be controlled adequately, certain problems may develop at this stage which subsequently result in defects in the final product.

Some solidification-related defects are common to both castings and ingots. Therefore, we shall briefly discuss the more important ones, including chemical segregation, nonmetallic inclusions, gas formation, and shrinkage.

Segregation. The phenomenon of chemical segregation, that is, the creation of regions in the solidified product which differ from the nominal chemical composition of the alloy, is entirely related to the fact that upon solidification, solute rejection occurs. This has been discussed in considerable detail in the preceding section and will not be repeated here. However, it should be reemphasized that the segregation produces regions of varying chemical composition with attendant variations in mechanical, chemical, and physical characteristics, as described in much more detail in Chapter 8 in the section "Anisotropy." Therefore, any belief that the nominal composition alone determines properties should be promptly discarded.[8] The greater the degree of segregation, the greater the likelihood of variation in properties. To achieve a uniform isotropic material, segregation must be effectively minimized or eliminated.

Nonmetallic inclusions. Nonmetallic inclusions are, as the name implies, included bodies of nonmetallic material which are entrapped or produced in the metal during solidification. The inclusions arise from two sources and are classified as follows:

- Indigenous inclusions are those that result from chemical reactions occurring in the melting of the alloy and from impurities in the raw materials. The reaction of manganese and sulfur in steel to form the MnS inclusions shown in Figure 7-39 is typical of the method of formation of these types of inclusions.
- Exogenous inclusions are those which occur as a result of particles of refractory materials eroding from furnace linings and transfer vessels and becoming entrapped in the solidified metal.

The potentially harmful nature of these inclusions depends on a number of factors, including:

- Chemical composition of inclusion
- Volume percentage
- Shape or morphology

[8]Unfortunately, the nominal composition is usually what is measured and reported when one specifies a chemical composition requirement.

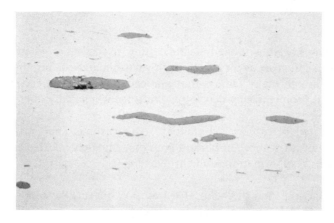

Figure 7-39 Photomicrograph of indigenous MnS inclusions in resulfurized steel (350×).

- Distribution in the matrix
- Orientation with respect to applied stresses
- Mechanical behavior of inclusions as compared to surrounding matrix

Depending on the factors above, then, nonmetallic inclusions can seriously affect the mechanical properties and performance of metallic and ceramic materials. The following properties are affected the most, especially in the direction transverse (perpendicular) to the direction of any mechanical working:

- Tensile ductility
- Impact energy
- Fatigue properties
- Fracture toughness

Shrinkage. This casting defect, which results in a network of voids within the interior of a casting, is caused by the contraction of the metal as it cools from the liquid and solidifies. As shown in Figure 7-40, this contraction occurs in three stages; as a result of the metal cooling while still liquid, as a result of the liquid-to-solid

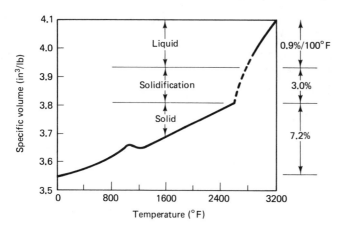

Figure 7-40 Volume contraction in steel as a function of temperature change.

transformation, and as a result of cooling in the solid state. Since the metal solidifies from the exterior first, the volume shrinkage that occurs results in a cavity on the interior unless the cavity is fed by a molten-metal pool which has not yet solidified. In castings, this is accomplished by means of a riser, the use of which is shown in Figure 7-41.

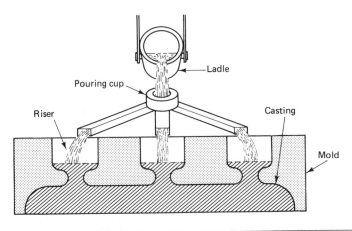

Figure 7-41 Use of risers to supply molten metal to region of solidification. (From C. W. Briggs, *The Metallurgy of Steel Castings*, McGraw-Hill Book Company, New York, 1946.)

Example 7-12

Calculate the total contraction in volume which occurs when 10,000 in.³ of steel cools from 3150°F to its melting point at 2760°F.

Solution

$$T = 3150°F - 2760°F = 390°F$$

From Figure 7-26 we obtain the volume contraction in the liquid: 0.009 %/°F.

$$\text{total contraction} = 10,000 \text{ in.}^3 (390°F)(0.009\%/°F)$$

$$\Delta V = 351 \text{ in.}^3 \quad \textit{Ans.}$$

A similar shrinkage cavity can occur in a solidifying ingot. Since the shape of the shrinkage cavity depends on how the metal around it solidifies and since the shape of most ingots is either upright, rectangular, or cylindrical, it is reasonable to expect centerline shrinkage in such ingots. This shrinkage is called *pipe* and is illustrated in Figure 7-42.

Piping is ordinarily an economic concern in metals' production, affecting production yields. The ingot is forged or rolled in order to consolidate the pipe. After rolling, any residual pipe is cut away and discarded. However, if the piped region is excessive, this defect could eventually produce a defective forged or rolled product.

Piping can be minimized by pouring ingot molds with their larger end up, and by adequate "hot-topping" procedures. A schematic example of poured ingots without hot topping is shown in Figure 7-42. The hot topping process consists of applying an exothermic material to the top surface of the liquid metal in the ingot, immediately following pouring. The ensuing exothermic reaction provides heat which retards

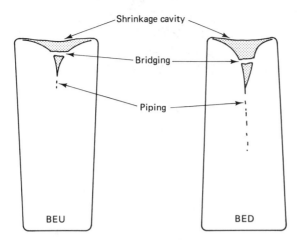

Figure 7-42 Schematic illustration showing pipe formation with two ingot forms: BEU, big end up; BED, big end down.

premature freezing-over in the top-central portions of the ingot. Thus this remaining liquid is available to feed the shrinkage that accompanies solidification.

Gas holes and porosity. These defects are, as their name implies, voids or holes in the cast structure. They occur as a result of the decreased solubility of gases as a function of the temperature of the liquid. As the temperature decreases, the gas solubility decreases and the gas is rejected at the solidification front. Figure 7-43 illustrates the decrease in solubility of hydrogen in copper as a function of temperature.

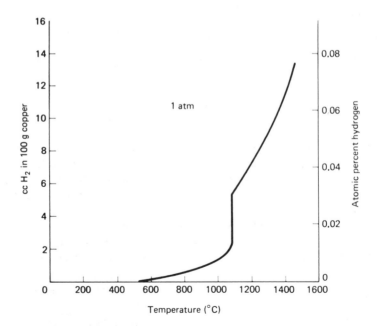

Figure 7-43 Decrease in hydrogen solubility in copper as a function of temperature. (From M. Hansen, *Constitution of Binary Alloys*, 2nd ed., McGraw-Hill Book Company, New York, 1958, p. 587.)

The evolved gas can either escape from a free surface or remain to form gas bubbles. Whether the defect is called porosity or gas holes is arbitrary and depends on the relative number and size of the voids. Large numbers of small voids are referred to as *porosity*, while a few large voids are called *gas holes*. Figure 7-44 shows gas holes in a steel casting which occurred as a result of excess gas in the melt.

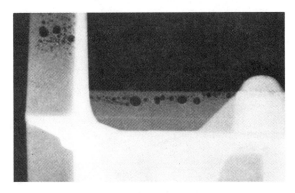

Figure 7-44 Radiograph showing gas holes in a steel casting.

Precipitation Hardening Reaction

In several important classes of engineering alloys, including aluminum, copper, nickel, and magnesium, the effect of decreasing solid solubility is utilized to produce a stronger material than can be achieved with solid-solution strengthening alone. This process is commonly referred to as *precipitation* or *age hardening*, and involves thermal treatment of the material to develop the final properties.

The actual method of precipitation hardening in these alloys involves a composition which can be heated into a single-phase region, as shown in Figure 7-45. The example used here is an aluminum alloy containing 3.5% copper, heated to approximately 550°C. This portion of the process is called *solution treatment* and serves to dissolve the copper in the aluminum, producing a homogeneous κ phase. Then the solid solution is rapidly cooled to room temperature, for instance by quenching in water. This produces a *supersaturated* solid solution. The κ phase now contains considerably more copper than it would under equilibrium conditions, and this component is uniformly distributed throughout κ.

The solution-treated alloy may *spontaneously* begin to increase in hardness and strength after a time at room temperature. This is known as *aging* and involves the precipitation of submicroscopic particles[9] of the θ phase, in this case $CuAl_2$, throughout the matrix (κ phase). The aging process is illustrated schematically in Figure 7-45. Since the precipitation reaction is diffusion controlled, the aging process can be accelerated by heating to an elevated temperature for an appropriate period of time. Such acceleration of the precipitation reaction is referred to as *artificial aging*. The aging temperature, however, must not exceed the solvus temperature for the alloy or dissolution of the precipitate will occur, with attendant softening. Furthermore, *overaging*, which consists of heating for too long a period, results in coalescence of the very fine precipitate particles. This process is also accompanied by a loss of strength and

[9] Figures 13–19 shows the precipitate particles in a nickel-based high-temperature alloy.

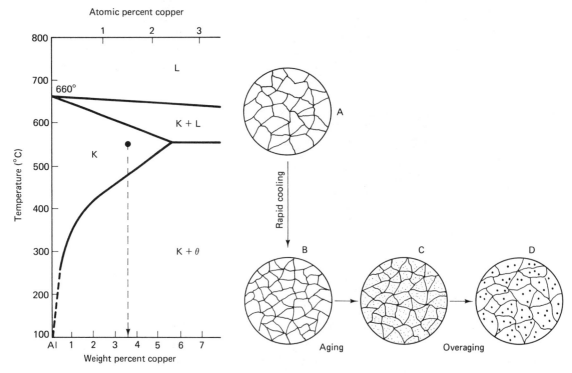

Figure 7-45 Effect of heat treatment on the structures produced in a precipitation-hardening alloy: A single phase structure (κ); B supersaturated solid solution (κ'); C fine submicroscopic precipitate (θ); D overaged structure with visible precipitate (θ).

hardness. For a discussion of the actual mechanism whereby precipitate particles strengthen a matrix, the student is referred back to Chapter 6.

STUDY PROBLEMS

7.1. Define what is meant by a phase in an alloy system. How can we distinguish between phases in a solid solution? How many phases coexist in the material shown in Figure 7-5?

7.2. In alloy systems what do we mean by a microstructural constituent? How many constituents are recognizable in Figure 7-17(a)? Can you identify the phases in this microstructure?

7.3. **(a)** Sketch the relationship between free energy and temperature for an alloy system which forms a solid solution at T_M. Identify the portion of your diagram in a manner similar to Figure 7-6 (i.e., stable phases, solid lines; unstable phases, dashed lines).
(b) What are the possible phases in this system?
(c) Where are these phases in equilibrium?

7.4. How many degrees of freedom are there in single-phase systems consisting of a pure metal? Interpret your result.

7.5. Determine the degrees of freedom for a mixture in the liquid region of Figure 7-8. What are the variables that represent these degrees of freedom?
Answer: $F = 2$

7.6. For the copper–nickel system illustrated in Figure 7-12, how many degrees of freedom are there for two-phase equilibrium? How do you interpret your answer?

7.7. (a) In a three-component alloy system, what are the conditions for three-phase equilibrium? Interpret your answer in terms of the system's variables.
(b) How does this result differ from three-phase equilibrium in a two-component system?

7.8. Given the following information regarding a binary system consisting of components A and B, plot the equilibrium phase diagram. Identify the various regions of your plot with the appropriate phases.

	Temperature (°C)	
Composition	Liquidus	Solidus
100% A	800	800
80% A–20% B	1125	925
60% A–40% B	1400	1120
40% A–60% B	1610	1350
20% A–80% B	1800	1625
100% B	1900	1900

7.9. An alloy of the nickel–copper system (Figure 7-12) contains 60% Cu.
(a) Determine the amounts (percentage) of the phases in equilibrium at 1250°C.
(b) What is the composition of the solid in equilibrium with the liquid at this temperature?
(c) Why is the composition in part (b) different from the nominal composition (C_0) we started with?
Answer: (a) % $\alpha = 37.5$, % liquid $= 62.5$

7.10. Describe what is meant by a eutectic reaction. Is the eutectic the highest freezing point in a binary system which displays such a reaction? Is this reaction irreversible?

7.11. Given the following information concerning a binary system of components A and B, sketch (to scale) the equilibrium phase diagram. Identify the phases that exist in this system. Also identify the microstructural constituents which you would expect to see in this material under the microscope. (Use of a straightedge is permissible for this exercise.)

- Melting temperature of component A = 1600°C
- Melting temperature of component B = 1400°C
- Eutectic reaction occurs at 800°C and 40% B
- Maximum solubility of B in α = 20%
- Solubility of B in $\alpha \simeq 0$ at room temperature
- Maximum solubility of A in β = 10%
- Solubility of A in $\beta \simeq 0$ at room temperature

7.12. General-purpose plumbing solder is composed of 50% Pb–50% Sn (Figure 7-18). If this composition is cooled to room temperature very slowly, a near-equilibrium structure will result. Determine the percentage of microstructural constituents that will be present under these conditions.
Answer: % α = 27.7, % eutectic = 72.3

7.13. An alloy of 10% Sn and 90% Pb (see Figure 7-18) is cooled to room temperature under near-equilibrium conditions. Calculate the percentages of the phases that result in this material.

7.14. Consider the alloy in problem 7.13 at 200°C.
(a) What phases exist at this elevated temperature?
(b) Where did the β come from in the room-temperature material above?

7.15. Sterling silver consists of 92.5% Ag and 7.5% Cu (see Figure 7-20).
(a) Calculate the percentages of the phases that would be formed by cooling this alloy to room temperature under near-equilibrium conditions.
(b) What is the maximum solubility of copper in α solid solution?
(c) How much silver can be dissolved in the β solid solution?

7.16. A brazing alloy contains 72% Ag and 28% Cu ("silver solder").
(a) Determine the percentages of the phases that will be formed if this material is slowly cooled to room temperature.
(b) Why is this particular composition used for a joining process such as brazing?
Answer: (a) % α = 76.9, % β = 23.1

7.17. An alloy consisting of 60% copper–40% silver (Figure 7-20) is cooled under equilibrium conditions.
(a) When does this material begin to solidify?
(b) When is the alloy completely solid?
(c) What are the percentages of microstructural constituents in this alloy after it has just solidified?

7.18. Microscopic examination of a Cu–Ag alloy (see Figure 7-20) reveals a structure containing approximately 70% eutectic and 30% β solid solution. Estimate the nominal composition of this alloy.
Answer: 47.3% Cu

7.19. Silver alloys used for jewelry purposes often contain 18% copper. In this particular alloy, 82% Ag–18% Cu, what are the respective atomic percents?

7.20. **(a)** Determine the composition of the hypothetical alloy designated as point Z in Figure 7-26(b).
(b) In this same ternary system, locate the alloy consisting of 55% A–35% B–10% C.

7.21. A certain Fe–Cr–Ni alloy (stainless steel) has a composition of 20% chromium, 12% nickel, and 68% iron.
(a) If this material is heated to 1400°C (see Figure 7-27), what phases are present?
(b) If the nickel concentration is increased to 20% and the chromium content decreased to 12%, what happens to the alloy at this temperature?
(c) Conversely, if we increase the chromium content to 25% with a corresponding decrease in nickel to 7%, what effect does this have on the steel?

7.22. What are the important differences between solidification in pure metals and in alloys?

7.23. During the solidification of an alloy the composition of the solute is measured in both the liquid and the solid (in equilibrium with it) at different temperatures. At 800°C the percent solute in the liquid is 6.5% while the solute concentration in the solid is 14%. What will the composition of the solid be at 500°C if the percent solute in the liquid is

4.2%? (Assume that K is independent of temperature.) What type of solidification mechanism is operating at the solid–liquid interface in this case?

7.24. An aluminum welding operation uses an aluminum filler wire which contains 6% silicon. During solidification of the weld, the composition of the weld puddle is chemically analyzed at time t_1 corresponding to temperature T_1, and at time t_2 corresponding to a lower temperature T_2. At T_1 the concentration of silicon in the liquid is 8.3%, while the solid formed at the same temperature contains 3.7% Si.

    (a) What type of solidification mechanism is operating at the solid–liquid interface?

    (b) At temperature (T_2) the composition of the liquid is 10.7%. What is the silicon content of the corresponding solid?

7.25. The nominal composition of an aluminum–magnesium alloy is 10% Mg. If dendritic solidification occurs according to the behavior displayed in Figure 7-36 and $K_0 = 0.5$:

    (a) What is the composition of the first solid to form?

    (b) What is the composition of the solid that forms at $X/X_c = 0.25$, where X_c is a characteristic distance $= DK_0R$.

    (c) What is the composition of the solid during steady-state growth?

    (d) What is the composition of the last portion of liquid to freeze?

7.26. A thin stream of liquid metal alloy is broken up into fine droplets by directing jets of nitrogen gas against it. This process is called "atomizing" and produces spherical particles (metal powder).

    (a) If this metal is cooled from 1650°C (3000°F) to 550°C (1022°F) in 0.10 sec, can you consider this a rapid solidification process?

    (b) If this cooling rate is not sufficient to produce an amorphous condition, what benefits can result from this process?

7.27. The contraction that accompanies cooling and solidification of steel is shown in Figure 7-40.

    (a) Which stage of contraction—liquid, solidification, or solid—has the greatest rate of shrinkage?

    (b) What is the percentage change in specific volume for a steel cooled from 1600°F (871°C) to room temperature?

    (c) What is responsible for the increase in specific volume in the neighborhood of 1100°F (593°C)?

7.28. What is the principal requisite for precipitation hardening in an alloy system? What are the steps involved in aging an alloy?

7.29. True or false? The following alloys can be precipitation hardened:

    (a) 90% Al–10% Si   **(b)** 40% FeO–60% MgO

    (c) 96% Al–4% Cu   **(d)** 20% Mg–80% Al

    (e) 80% Cu–20% Zn   **(f)** 90% Al–10% Mg

    (g) 30% Cu–70% Ni   **(h)** 0.5% C–99.5% Fe

    (i) 10% Sn–90% Pb   **(j)** 95% Al–5% Ni

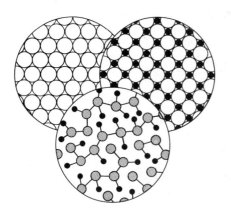

CHAPTER EIGHT:
Mechanical Testing
and Selection
of Engineering
Materials

The mechanical properties of a material may be defined as those properties of a material associated with its elastic and nonelastic reactions when force is applied. These properties include among others the following parameters: tensile strength, yield strength, hardness, toughness, and ductility. A knowledge of the mechanical properties is valuable for various reasons, among these, to determine if the material meets specifications, to establish design parameters, and to evaluate the effects of processing variables.

This chapter will acquaint the student with some of the more common tests and terms used in the mechanical tests of engineering materials. It should be kept in mind that these are tests of general applicability and do not include all the specialized tests available in the study and evaluation of materials. The topics covered in this chapter deal with the practical aspects of conducting mechanical property tests and with the type of information that can be determined from such testing. The topics are based on the fundamental principles regarding mechanical behavior discussed in Chapter 6.

Other mechanical tests of a more specific nature will be discussed in the appropriate chapters: for instance, fracture toughness testing and fatigue testing in Chapter 9.

The mechanical properties of a material—yield strength, ultimate tensile strength, elastic limit, ductility, and impact toughness—are usually determined by laboratory testing of a specimen of the material which is in a specific geometric form. These tests are used widely in engineering to provide basic design information and to evaluate materials relative to acceptance specifications.

TENSION TESTING

The tension test is conducted by applying an increasing load (at a constant rate of strain) to a specimen of specific geometry on an apparatus similar to that shown schematically in Figure 8-1.

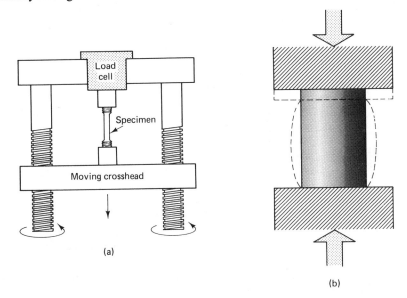

Figure 8-1 Schematic illustration of tensile testing apparatus. Typical tension and compression specimens are shown. (From H. W. Hayden, W. G. Moffatt, and J. Wulff, *The Structure and Properties of Materials*, John Wiley and Sons, New York, 1965, p. 2, p. 10.)

The specimens upon which the test is based may be of various geometrical shapes with different cross sections (i.e., cylindrical, rectangular, etc.), as typified in Figure 8-2. The primary requirement is that the cross section must be precisely known and uniform over the test length. A specific gauge length is inscribed on each specimen to establish the original reference length which will be used in calculations to determine ductility.

When a force is applied to a material, it deforms. When the force is applied in such a manner that the specimen is being pulled apart, this deformation results in an elongation. When the force pushes against the specimen, it results in compression. In either case, the resultant deformation may be expressed as strain (deformation per unit length) and the load converted to a stress—concepts that were described in Chapter 6.

As the specimen is increasingly stressed, first, elastic deformation occurs in response to Hooke's law, then plastic deformation occurs. This deformation is measured by a device called an *extensometer* attached to the tensile specimen. The device can be mechanically or electronically actuated, but in either case, it sensitively measures the increase in length relative to the original gauge length for each

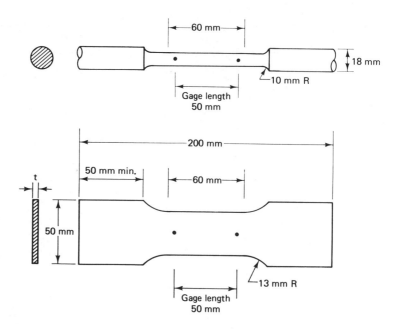

Figure 8-2 Geometry of typical tensile specimens.

load increment. The load–elongation data are then converted into stress–strain data as follows: the engineering stress (σ), as derived in Chapter 6 [see equation (6-8)], is obtained by dividing the load (P) by the original cross-sectional area (A_0), as follows:

$$\sigma = \frac{P}{A_0}$$

The average longitudinal strain (engineering strain), which was also derived in Chapter 6 [equation (6-1)], is obtained by determining the elongation of the original gauge length (Δl) and dividing by the gauge length (l_0) as follows:

$$e = \frac{\Delta l}{l_0}$$

Example 8-1

(a) Calculate the engineering stress (σ) for a tubular specimen which has an external diameter of 12 mm and an internal diameter of 10 mm at a load of 408 kg (4000 N).
(b) Assuming a gauge length of 50 mm, calculate the engineering strain (e) when the length between the gauge marks is (a) 57 mm, and (b) 65 mm.

Solution   (a) Engineering stress:

$$\sigma = \frac{P}{A} = \frac{P}{\dfrac{\pi(D_0^2 - D_i^2)}{4}}$$

$$= \frac{4000 \ \text{N}}{\dfrac{\pi(0.012^2 - 0.010^2) \ \text{m}^2}{4}}$$

$$= \frac{4000 \text{ N}}{3.45 \times 10^{-5} \text{ m}^2}$$

$$= 1.16 \times 10^8 \text{ N/m}^2 = 116 \text{ MPa} \qquad Ans.$$

(b) Engineering strain:

$$e = \frac{\Delta l}{l_0} = \frac{57 \text{ mm} - 50 \text{ mm}}{50 \text{ mm}}$$

$$= \frac{7 \text{ mm}}{50 \text{ mm}} = 0.14 \text{ mm/mm} \qquad Ans.$$

Similarly, when the length of the elongated specimen = 65 mm, elongation (Δl) = 15 mm and e = 0.30 mm/mm.

Note that the strain units are identical and may be canceled. Therefore, strain is frequently expressed as a dimensionless quantity.

Stress–Strain Curve

The data obtained in the tensile test are then plotted as stress–strain curves. Figure 8-3 illustrates some stress–strain curves that can be obtained with typical engineering materials. The shape of the stress–strain curve will depend on the material being

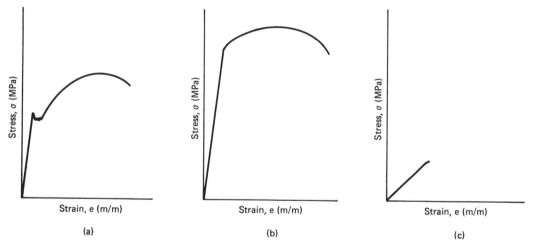

Figure 8-3 Strain curves for: (a) low-carbon steel, SAE 1025; (b) alloy steel, SAE 4340; (c) marble, $CaCO_3$.

tested, its process history, and the temperature at which the test is being conducted. When stressed in the elastic region, the specimen will return to its original dimensions if the load is removed. In metals and ceramics a plot of the elastic region is linear; that is, stress is proportional to strain and conforms to Hooke's law as follows:

$$\sigma = Ee$$

where E is a proportionality constant termed the elastic modulus or the Young's modulus, as described in Chapter 6.

In many other systems, elastomers for example, stress is not proportional to strain and the strain achieved at each load increment in the elastic region may vary. In rubber, for example, one gets large increases in strain with initial loads, but smaller increments in strain at higher loads.

With additional stresses, the material exceeds its elastic limit and plastic deformation occurs, with the specimen elongating and decreasing in diameter correspondingly. Initially, this response is uniform over the length of the specimen, but at some point in the structure instability occurs and localized plastic deformation is then concentrated in this region. The specimen diameter then constricts locally (i.e., "necks" down), and fracture of the specimen begins.

The stress–strain curve for a material can yield a great deal of valuable information about a material and its suitability for various applications. A typical stress–strain curve for a metallic system is presented in Figure 8-4. Let us examine this curve together with the definition of the various material parameters that can be determined directly from the plot.

1. *Proportional limit:* The proportional limit is the highest value of stress at which the stress–strain relationship is linear (i.e., proportional to strain). This occurs at point *A* in Figure 8-4.

2. *Elastic limit:* The highest stress imposed on the material such that there is no permanent deformation remaining when the load is removed (point *B*, Figure 8-4).

3. *Yield strength:* The yield strength (σ_{YS}) corresponds to the stress required to produce a small specific amount of plastic deformation. This is determined by graphically establishing an offset at some specified value of strain (e.g., at 0.2% strain, parallel to the straight-line portion of the stress–strain curve). The intersection of this offset line with the stress–strain curve gives the yield strength (point *C*, Figure 8-4). The yield-strength value should always be preceded by the offset at which it was established (i.e., 0.2% yield strength). The use of the offset method avoids the very practical difficulty of establishing exactly where the initiation of plastic behavior begins.

4. *Ultimate tensile strength.* The ultimate tensile strength (UTS) is a measure of the maximum load that a material can withstand under conditions of uniaxial loading. It is determined by taking the maximum load obtained during the test and dividing by the original cross-sectional area (point *D*, Figure 8-4).

$$\text{UTS, } \sigma_{TS} = \frac{P_{max}}{A_0} \qquad (8\text{-}1)$$

The ultimate tensile strength is not often utilized in designing structures or equipment, since at this stress value the component has already undergone substantial plastic deformation. Moreover, after the ultimate tensile strength has been exceeded the material may continue to plastically deform at stresses *lower* than this maximum value, as shown in Figure 8-4.

5. *Modulus of elasticity* (Young's modulus). The modulus of elasticity (*E*) defined in Chapter 6 can be determined graphically from the slope of the initial straight-line portion of the stress–strain curve.

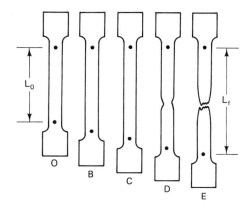

Increasing length of ga·ıge section

Decreasing cross-sectional area

(a)

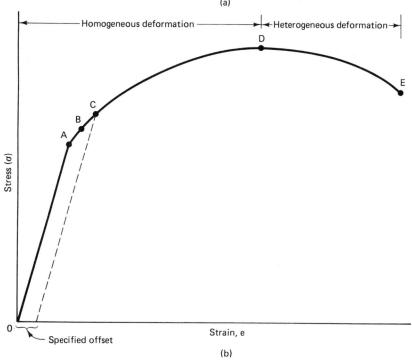

(b)

Figure 8-4 (a) Appearance of tensile specimen at various stages in the tension test; (b) typical stress–strain curve for a ductile metal: A, proportional limit; B, elastic limit; C, yield strength; D, ultimate tensile strength; E, fracture strength.

From Hooke's law [equation (6-13)],

$$\sigma = Ee$$

Solving for the modulus, we have

$$E = \frac{\sigma}{e}$$

or

$$E = \frac{\Delta\sigma}{\Delta e} \qquad \text{(in the linear region of the stress–strain curve)}$$

The modulus is a measure of the *stiffness* of the material; the higher the modulus, the lower is the elastic strain resulting from the application of a given stress. There are many structures, such as aircraft wings, in which the design is controlled not only by the strength of the material, but by its stiffness as well.

Example 8-2

A load of 1000 lb (4454 N) is suspended from each of two identically sized wires, 0.25 in. (6.4 mm) in diameter. One wire is steel, the other is aluminum. Determine the axial engineering strain produced in the two wires.

Solution The engineering strain (e) is related to σ by Hooke's law as shown above. From Table 6-1, the modulus of elasticity is given for the two metals as follows:

$$E_{\text{steel}} = 30 \times 10^6 \text{ psi (207,000 MPa)}$$

$$E_{\text{aluminum}} = 10.5 \times 10^6 \text{ psi (72,000 MPa)}$$

For both materials,

$$\sigma = \frac{P}{A_0} = \frac{1,000 \text{ lb}}{0.05 \text{ in.}^2} = 20,000 \text{ lb/in.}^2$$

The stress is independent of the type of metal!
    Now we can compute the strain as follows:

$$\text{Steel:} \quad e = \frac{\sigma}{E} = \frac{20,000 \text{ psi}}{30 \times 10^6 \text{ psi}} = 6.7 \times 10^{-4} \text{ in./in.} \qquad \textit{Ans.}$$

$$\text{Aluminum:} \quad e = \frac{20,000 \text{ psi}}{10.5 \times 10^6 \text{ psi}} = 19 \times 10^{-4} \text{ in./in.} \qquad \textit{Ans.}$$

This result shows that the strain in the aluminum wire is approximately three times greater than that of the steel wire. Since by observation it can be seen that steel has an elastic modulus three times greater than the aluminum ($E_s \simeq 3E_{\text{Al}}$), such a finding should not be too surprising.

Ductility

Ductility is a measure of the material's ability to deform plastically under the conditions of the test. Two measurements of ductility are derived from the tensile test and both are widely used: reduction in area (R.A.) and elongation. Measurements are obtained by placing the fractured test specimen back together after the test and making the appropriate measurements of final length and final diameter.

Reduction in area (R.A.) is the change in the cross-sectional area of the test bar expressed as a percentage and is computed as follows:

$$\% \text{ R.A.} = \frac{A_0 - A_f}{A_0} \times 100 \qquad (8\text{-}2)$$

where A_0 = original cross-sectional area

A_f = final cross-sectional area at fracture

Similarly, elongation is the increase that occurs in the gauge length of the specimen.

$$\% \text{ elongation} = \frac{L_f - L_0}{L_0} \times 100 \tag{8-3}$$

where L_0 = original gauge length

L_f = final length between gauge marks

Resilience

The ability of a material to absorb energy when elastically deformed and return it when unloaded is called resilience. This is commonly measured by the *modulus of resilience* (U_r), which is the strain energy per unit volume required to stress the material from a condition of zero stress to its yield strength, σ_{YS}.

Mathematically, the strain energy per unit volume for uniaxial tension is the area under the stress–strain curve. From the definition of resilience above it follows that

$$U_r = \tfrac{1}{2}\sigma e \tag{8-4}$$

in the linear region of the stress–strain relationship.[1] If the material is stressed to its yield strength (σ_{YS}), the expression becomes

$$U_r = \tfrac{1}{2}\sigma_{YS} e_{YS}$$

This relationship is shown graphically in Figure 8-5. However, from Hooke's law,

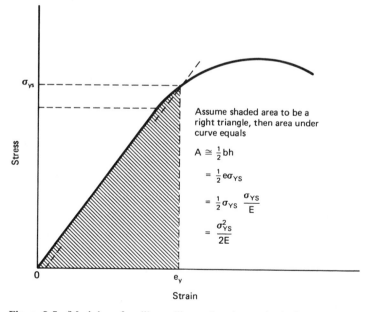

Figure 8-5 Modulus of resilience illustrating the method of calculation.

[1] Note that the area under the linear region of the stress–strain curve is a triangle, and the area of a triangle equals $\tfrac{1}{2}$(base)(height).

$e = \sigma/E$. Therefore, we can state that

$$U_r = \tfrac{1}{2}\sigma_{YS}\frac{\sigma_{YS}}{E}$$

$$= \frac{\sigma_{YS}^2}{2E} \qquad (8\text{-}5)$$

Equation (8-5) indicates that the resilience is maximized in materials with a high yield strength and a low modulus of elasticity. High resilience would be desirable in such applications as mechanical springs, for example. The values in Table 8-1 show the contrast between modulus of resilience and modulus of elasticity for several materials.

TABLE 8-1 TYPICAL PROPERTIES FOR VARIOUS ENGINEERING MATERIALS

Material	Elastic MPa	Modulus (E) 10^6 psi	Yield MPa	Strength 10^3 psi	Modulus of Resilience (U_r) 10^3 J/m^3	in-lb/in^3
Metals:						
low carbon steel	207,000	30	310	45	233	34
high carbon steel	207,000	30	966	140	2254	327
grey cast iron	131,000	19	207	30	163	24
aluminum, 1100 (hard)	69,000	10	152	22	165	24
aluminum 7075-T6	71,760	10.4	504	73	1766	256
magnesium, ZK60A-T5	44,850	6.5	276	40	848	123
titanium-6Al-4V	110,400	16	1035	150	4847	703
brass,[1] 70Cu-30Zn	110,400	16	276	40	345	50
stainless steel,[2] 302	193,200	28	966	140	2413	350
stainless steel, 17-7PH	193,200	28	1497	217	5797	841
Ceramics:[3]						
Al$_2$O$_3$	365,700	53	172	25	41	6
ZrO$_2$	144,900	21	55	8	10	2
pyroceram (glass-ceramic)	64,170	9.3	69	10	37	5
B$_4$C	289,800	42	152	22		6
WC	517,500	75	345	50		17
Polymers:						
ABS (high impact)	1,725	0.25	34	5	345	50
acrylic	2,760	0.40	55	8	552	80
epoxy	3,105	0.45	69	10	766	111
nylon 6/6	2,967	0.43	69	10	802	116
SB rubber	2	0.003	2	0.30	1034	150

1. 1/4 hard condition
2. 30% cold work
3. U_r determined from ultimate tensile strength values

Elevated-Temperature Effects

The tensile strength and ductility of most engineering materials are heavily influenced by elevated temperatures. Generally, as the test temperature is increased, tensile strength decreases while ductility increases. The elevated-temperature tensile behavior of a high-strength steel is shown in Figure 8-6. In particular, note the precipitous

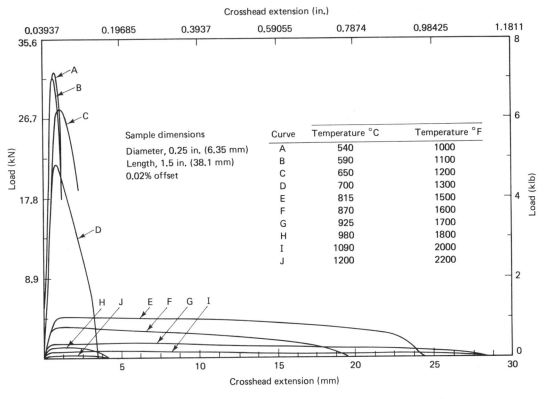

Figure 8-6 Elevated temperature tensile test of high-strength low alloy steel.

decrease in tensile strength between curves D and E in Figure 8-6. Correspondingly, the recrystallization temperature for this material is in the vicinity of 730°C.

Elevated-temperature tensile testing is useful in comparing the *short-time* response of materials to tensile loading or stress. In this regard, the test yields comparative data with a minimum of time and expense. However, we must emphasize that tensile testing at elevated temperatures is quite inadequate for design applications involving *sustained* times at high temperatures. Many engineering materials exhibit time-dependent deformation under relatively low applied stresses during elevated temperature service. This type of deformation is called *creep* and is very important with regard to the design and utilization of "High-Temperature Alloys."

COMPRESSION TESTING

Many common engineering materials, such as concrete, brick, and certain ceramics, are often weak in tension as a result of the presence of submicroscopic cracks and flaws. Tensile stresses tend to propagate those cracks which are oriented perpendicular to the axis of tension. The tensile strengths these materials exhibit are low and can vary considerably from sample to sample depending on the distribution of flaws;

however, they can nevertheless be quite strong in compression. These materials are therefore used chiefly in compression, where their strengths are much higher. Figure 8-7 permits a comparison of the compressive and tensile strengths of concrete. It can

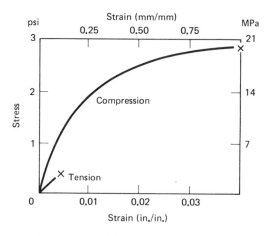

Figure 8-7   Stress–strain curves for concrete tested in tension and compression.

be seen that a much higher value of strength can be reached in compression than in tension.

The compression test is conducted in a manner similar to the tension test except that the forces act to push the ends of the specimen together, as shown in Figure 8-1. The specimen shape is most usually cylindrical, with its diameter large enough to prevent buckling. A stress–strain curve can then be drawn as shown in Figure 8-7. It should be noted that the compression test is rarely used for ductile materials; consequently, the curve usually does not extend much beyond the elastic region. The curve can be used to determine the modulus and compressive strength in the same fashion as described previously for the tension test.

We should point out that although the manually calculated and plotted stress–strain curves illustrated in Figures 8-3, 8-4, and 8-7 are completely valid in their representation of tensile and compressive behavior, testing of this nature is usually conducted on an automated basis. This is especially true in laboratories where many specimens are handled, such as in *quality control* situations. In the automated test, the apparatus may even be programmed to conduct the test, and the data are plotted (load versus strain) automatically or displayed on a digital readout. Such mechanization serves to speed up testing and interpretation of the results while maintaining experimental accuracy. This is very important from the standpoint of economics and reliability in any industry or research facility that must routinely perform tension or compression tests on engineering materials.

TORSION TESTING

In many applications, such as axles, coil springs, and drive shafts, an engineering material must have good resistance to stresses induced by twisting (torsion). The strains resulting from such torsional stresses can be determined by means of the torsion test.

The torsion test resembles the tension test in that a load–deformation curve is also developed. In the torsion test, a solid or hollow cylindrical specimen is twisted and the resultant deformation is measured as the angle through which the bar is twisted. The torsion curve is then plotted as the twisting moment versus the angle of twist (Figure 8-8) and is similar to the curve obtained in the tension test. The twisting

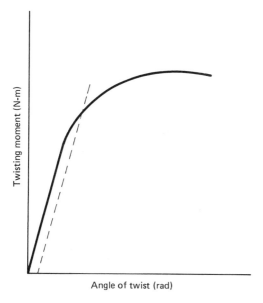

Figure 8-8 Torsion curve for a ductile material.

moment is the force applied to the shaft over the twisting moment distance that is applied. For a solid cylindrical bar, the shear stress (τ) can be calculated as follows:

$$\tau = \frac{16T}{D^3} \tag{8-6}$$

where T = twisting moment (torque) (N-m)
D = bar diameter (m)

Correspondingly, the shear strain (γ) is determined as follows:

$$\gamma = \tan \theta = \frac{r\theta}{l} \tag{8-7}$$

where θ = angle of twist (radians)
l = bar length (m)
r = radius of bar (m)

The torque (T) is related to the polar moment of inertia (J) of the cross section by the following relationship:

$$T = \frac{\tau J}{r} \tag{8-8}$$

where J is the polar moment of inertia in m^4. The polar moment of inertia is dependent on the shape and dimensions of the cross section. For a solid cylinder,

$$J = \frac{\pi d^4}{32}$$

Using this relationship and the shear modulus (G), which can be determined from the slope of the curve shown in Figure 8-8 and is expressed by $G = \tau/r$, then

$$G = \frac{Tr/J}{r\theta/l} = \frac{Tl}{J\theta}$$

$$\theta = \frac{Tl}{GJ} \tag{8-9}$$

In practice, these relationships can be used to determine the strain response of circular bars subjected to torsion so that one can calculate the degree of twist resulting from various stresses.

Example 8-3

Compute the diameter of a solid steel shaft necessary to carry a torque of 45,000 N-m if the twist is not to exceed 1° in 1.5 m. Assume that the shear modulus $G = 82.8$ GPa and the polar moment of inertia J for a solid cylindrical shaft $= \pi d^4/32$.

Solution

$$\theta = \frac{Tl}{GJ}$$

$$\frac{1°}{57.3} = \frac{(45,000 \text{ N-m})(1.5 \text{ m})}{(82.8 \text{ GPa})(\pi d^4/32)}$$

Solving for $d^4 = 4.76$ m^4, we have

$$d = (4.76 \times 10^{-4})^{1/4} = 0.147 \text{ m} \qquad Ans.$$

When a bar is twisted, tensile and compressive stresses are also developed. These stresses are always accompanied by shear stresses, as shown in Figure 8-9. The maximum shear stress occurs at a 45° angle to the principal stresses and has a value equal to half the difference between the minimum and maximum principal stresses.

$$\tau_{max} = \frac{\sigma_1 - \sigma_3}{2} \tag{8-10}$$

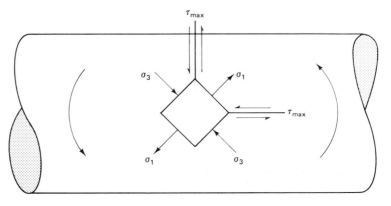

Figure 8-9 Orientation of stresses resulting from torsion.

This becomes especially significant in real systems where anisotropy is often encountered, since shear stresses from torque may occur in directions in which the material is substantially weaker than in the longitudinal direction. Consequently, failure may occur prematurely. The topic of anisotropic mechanical properties occurring in real materials is discussed later in this chapter.

TOUGHNESS

Toughness can be loosely defined as the ability of a material to resist fracture while under stress. However, the resistance of a material to fracture can vary broadly, depending on the strain rate. Unfortunately, the term "toughness" has been used to describe material behavior both at low strain rates (tension testing) and at high strain rates (impact). Herein we will attempt to describe the behavior under both loading conditions.

Low Strain Rates—Tensile Testing

In some applications, such as military armor, for example, the ability of a material to absorb energy while being plastically deformed is a useful attribute. This property, which is sometimes referred to as toughness, is the ability of a material to absorb energy up to the point of fracture. Therefore, metals that possess high yield strength together with good ductility exhibit high toughness. The energy absorbed by the specimen can be approximated by the area under the stress–strain curve.

The units can be determined by multiplying the stress times the strain as follows, which gives the work per unit volume that can be done on the material without causing it to fracture:

$$\left(\frac{\text{pounds}}{\text{in.}^2}\right)\left(\frac{\text{in.}}{\text{in.}}\right) = \frac{\text{in.-lb}}{\text{in.}^3} \quad \text{or} \quad \left(\frac{\text{newtons}}{\text{m}^2}\right)\left(\frac{\text{m}}{\text{m}}\right) = \frac{\text{N-m}}{\text{m}^3} = \frac{\text{J}}{\text{m}^3}$$

Figure 8-10 displays stress–strain curves for a high-carbon spring steel and a structural steel, which have been superimposed. This graph shows clearly that even though the structural steel has a lower yield strength, it has higher toughness, as measured by the area under the curve.

High Strain Rates—Impact Testing

The energy absorption that occurs for a material under impact loading is related to the size and geometrical configuration of the test specimen, the rate of impact loading, and the test temperature. These variables must be carefully controlled and for these reasons, standardized specimens and procedures are used, the most common being the Charpy V-notch or the Izod, which are notched bar tests.

The Charpy and Izod test specimens have precise configurations, as shown in Figure 8-11, with the dimensions and surface finish of the notch being very critical. Both tests employ a pendulum apparatus of the type schematized in Figure 8-12. The tests differ in that in the Charpy test the specimen is loaded as a simple beam with

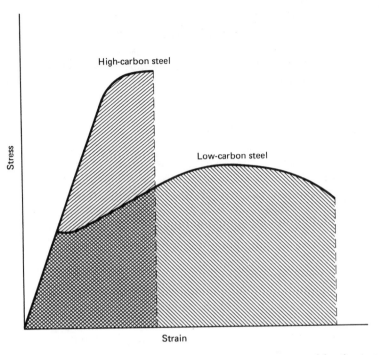

Figure 8-10 Comparison of toughness for two alloys as measured by the area under the stress–strain curve.

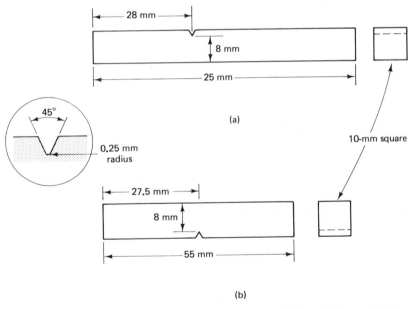

Figure 8-11 Configuration and dimensions of the Izod (a) and Charpy (b) V-notched impact specimens.

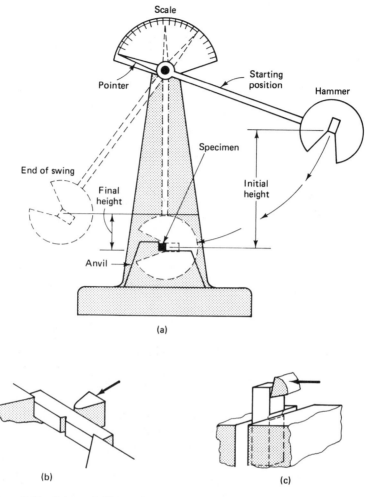

Figure 8-12 Schematic illustration showing the test apparatus (a), and specimen configuration for (b) Charpy V-notch and (c) Izod.

the anvil of the pendulum hammer striking on the side opposite the notch, whereas in the Izod test, the specimen is loaded vertically as a cantilever beam and the anvil strikes the notched face. In both tests, the specimen temperature should be carefully controlled. The kinetic energy of the pendulum at the point of impact is known because the starting point and the mass of the pendulum are known. The energy absorbed by the specimen is lost by the pendulum and the follow-through swing is reduced in proportion to the amount of energy absorbed.

Impact transition temperature. Unfortunately, many materials, including polymers and bcc metals, which exhibit high toughness at one temperature may display brittle behavior when tested at a lower temperature. In metals, this phenomenon is called the *ductile-to-brittle transition*. Figure 8-13 illustrates this behavior as

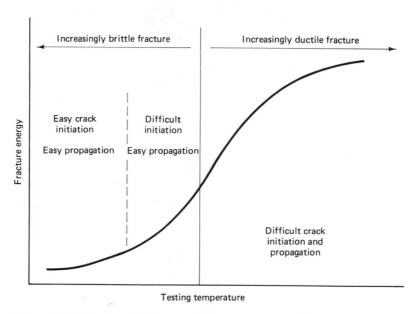

Figure 8-13 Ductile-to-brittle transition as revealed by Charpy V-notch testing.

it occurs with the Charpy V-notch impact test. As can be seen, there is a decrease from the toughness obtained at higher temperatures through a transition to a lower range of toughness values at lower testing temperatures. This loss in toughness is also characterized by a change in the appearance of the fracture surface as shown in Figure 8-14, with the percentage of brittle-appearing fracture increasing as the test temperature decreases. As the testing temperature decreases, the toughness decreases and the percentage of lateral expansion (or increase in the width of the test bar) also decreases.

Each of these observed phenomena may be the criterion used to establish the "transition temperature," the test temperature at which the behavior changes from ductile to brittle. However, the transition temperature is not a fixed value and can be arbitrarily set on any one of several bases. It is frequently set on an energy basis: for example, 50% of the spread between the upper and lower shelf values, or some acceptable energy value, such as 15 ft-lb (20.4 J). It may also be based on a fracture appearance transition temperature (FATT), depending on which value is the most conservative for the alloy in question. To determine the FATT, the fracture appearance is evaluated in terms of percent fibrosity or simply what percent of the fracture surface appears fibrous (ductile). A completely brittle fracture (i.e., the lower shelf energy) would exhibit 0% fibrosity, whereas a completely ductile fracture (i.e., the upper shelf energy shows 100% fibrosity. The most conservative transition temperature criterion is to select a temperature that corresponds to the temperature at which the fracture becomes 100% fibrous. However, this is not always practical and one of the above-mentioned arbitrary standards may be appropriate.

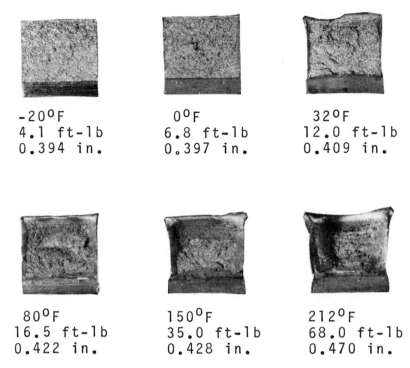

−20°F
4.1 ft-lb
0.394 in.

0°F
6.8 ft-lb
0.397 in.

32°F
12.0 ft-lb
0.409 in.

80°F
16.5 ft-lb
0.422 in.

150°F
35.0 ft-lb
0.428 in.

212°F
68.0 ft-lb
0.470 in.

Figure 8-14 Fracture appearance of Charpy V-notch specimens tested at various temperatures. Impact energy and lateral expansion is given for each specimen.

Regardless of the method chosen to establish the transition temperature, it can be seen that it is extremely critical that the transition temperature be below the expected service temperature; otherwise, catastrophic brittle failure can result if lower temperatures are encountered. Figure 8-15 demonstrates the behavior of two alloys with differing transition temperatures based on a criterion of 15 ft-lb (20 J). An examination of this diagram discloses that alloy A is the proper choice for the design based on the minimum service temperature. The use of alloy B would create under certain low-temperature service conditions an undesirable situation in which the structure would have severely impaired impact resistance.

Impact testing of nonmetallics. Impact testing of the Charpy and Izod types is applicable not only to metals but also to plastics and ceramics under certain circumstances. The test procedures are essentially similar, but reduced capacity machines are utilized since the impact energies encountered are lower.

An additional change is that the thickness of the test specimen is not fixed as with metals testing, but can be varied. This gives rise to a change in reporting the test results in that the energy units required to fracture are expressed per unit thickness of the test bar. For example, if a Charpy test specimen 0.250 in. thick requires 2 in.-lb

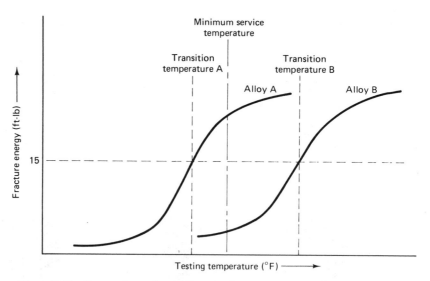

Figure 8-15 Comparison of two alloys having a transition temperature above and below the service temperature.

of energy to fracture, the results would be expressed as

$$\frac{2 \text{ in.-lb}}{0.250 \text{ in.}} = \frac{8 \text{ in.-lb}}{\text{in.}}$$

HARDNESS TESTS

Most hardness tests are a means of determining a material's resistance to indentation or penetration. There are some notable exceptions, such as Mohs' scratch test used in geology to measure the relative hardness of various minerals, and the scleroscope, which measures the dynamic hardness in terms of the height of rebound of the indenter. However, these systems are not commonly employed in materials engineering studies of an ordinary nature. As stated, most systems used in the evaluation of materials are based on the material's resistance to indentation while under load. The tests are divided into macro tests, i.e., those in which the indentation is visible to the naked eye, and micro tests, those in which a microscope is used to view the indentation.

As will be seen in the following discussion, the choice of a test method depends primarily on the size of the specimen available and the ultimate purpose of the testing. In multiphase materials and in materials that are subject to chemical segregation, each phase retains its individual characteristics, such as hardness. To determine the hardness of an individual phase, relatively small impressions must be used, giving rise to a need for micro hardness tests. For hardness testing on large castings and forgings, such pinpoint measurements cannot be made easily or reliably; therefore, larger systems are typically employed. In general, the larger the indentation, the greater the degree of averaging, since more material is included within the volume of the impression.

Macrohardness Tests

Brinell hardness.   The Brinell hardness test, in use since 1900, is used primarily to determine the bulk hardness of heavy sections, such as castings or forgings. Of all the indentation methods it necessitates the least surface preparation, requiring only that the surface be relatively smooth and free of dirt and scale.

The Brinell test is conducted by impressing a 10-mm diameter steel ball under a load of 3000 kg into the surface for a standard time, usually 30 sec. For nonferrous metals, the load is reduced to 500 kg, and for very hard metals, a tungsten ball is used. The average diameter of the resulting impression is read, from which the Brinell hardness number (BHN) can be determined using the formula

$$\text{BHN} = \frac{P}{(\pi D/2)(D - \sqrt{D^2 - d^2})} \tag{8-11}$$

where P = applied load (kg)
D = diameter of ball (mm)
d = diameter of impression (mm)

This formula simply represents the load (P) divided by the surface area of an impression of diameter d. In actual practice, no calculations are necessary; since the load is constant, the BHN values corresponding to various impression diameters are simply read from a table.

Attempts have been made over the years to correlate the BHN with the tensile strength of various materials and several approximations have been developed. For example, the following formula is frequently used for steel:

$$\text{UTS} \simeq 500 \times \text{BHN}$$

and for nonferrous material:

$$\text{UTS} \simeq 300 \times \text{BHN}$$

Rockwell hardness.   This test, which is conducted on a standardized machine, is one of the most widely used tests in the world. Variations on the basic test permit the testing of a wide variety of materials. The test utilizes the depth of penetration under constant load as a determinant of hardness. A minor load of 10 kg is applied to seat the specimen, then the major load is applied and the increment in depth of the indentation recorded on a dial gauge. This incremental penetration is illustrated in Figure 8-16. Various indenters may be employed: for example, a diamond cone (brale), a $\frac{1}{16}$-in. steel ball, or a $\frac{1}{8}$-in. steel ball, with major loads of 60, 100, and 150 kg. This capability gives the Rockwell test great versatility in testing a wide variety of materials. However, since the Rockwell hardness is dependent on both the load and the indenter, it is necessary to specify the combination used. This is done by prefixing the hardness number with a letter which indicates the combination of load and indenter used. Table 8-2(a) shows which combinations may be employed in ordinary Rockwell hardness testing.

The Rockwell test is also suitable for the measurement of hardness in special

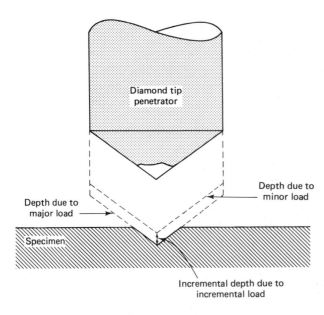

Figure 8-16 Schematic illustration of the Rockwell hardness test. The incremental depth of penetration due to an increment of load forms the linear relationship which is the basis of Rockwell hardness testing.

applications such as very thin materials through the use of a superficial hardness tester. This device employs various indenters and lighter loads than the conventional Rockwell tester. The combinations of load and indenter are shown in Table 8-2(b).

Microhardness Tests

There are many applications for hardness measurements which require the hardness testing of very small regions such as the determination of hardness of an individual phase in a microstructure or of the variation of hardness as a function of distance from a weld. There are two systems which are currently widely employed for the determination of microhardness: Vickers and Knoop.

Vickers hardness. The Vickers hardness test uses a square-base diamond pyramid, as shown in Figure 8-17. The Vickers hardness number (VHN) is determined by microscopically measuring the diagonals of the impression resulting from predetermined loads. Because of the precision required by the test and the minuteness of the impression, this test can be conducted only on carefully polished and prepared specimens. The Vickers test has received wide acceptance in research work on materials because it provides a continuous range of hardness ranging from 5 to 1500 DPH.

Knoop hardness. The Knoop hardness test utilizes a diamond indenter of the type shown in Figure 8-17 which produces an elongated diamond-shaped indentation at loads as low as 25 g. The length of the long diagonal is read (in mm) using a microscope attached to the microhardness tester. As with other microhardness tests, the low loads and the required precision of the measurements means that the specimen's surface must be carefully prepared.

TABLE 8-2 LOAD AND PENETRATOR CONFIGURATIONS FOR VARIOUS ROCKWELL HARDNESS SCALES

Scale symbol	Penetrator	Load (kg)	Dial figures
(a) Normal			
B	$\frac{1}{16}$-in. ball	100	Red
C	Brale	150	Black
A	Brale	60	Black
D	Brale	100	Black
E	$\frac{1}{8}$-in. ball	100	Red
F	$\frac{1}{16}$-in. ball	60	Red
G	$\frac{1}{16}$-in. ball	150	Red
H	$\frac{1}{8}$-in. ball	60	Red
K	$\frac{1}{8}$-in. ball	150	Red
(b) Superficial			
15N	N brale	15	
30N	N brale	30	
45N	N brale	45	
15T	$\frac{1}{16}$-in. ball	15	
30T	$\frac{1}{16}$-in. ball	30	
45T	$\frac{1}{16}$-in. ball	45	
15W	$\frac{1}{8}$-in. ball	15	
30W	$\frac{1}{8}$-in. ball	30	
45W	$\frac{1}{8}$-in. ball	45	

Hardness Testing of Nonmetallics

Some of the tests and instruments used to determine hardness in metals can also be applied to nonmetallic engineering materials. Such materials of commerce include wood, plastics, rubber, glass, minerals, and abrasives.

Several methods of hardness testing may be utilized for a specific nonmetallic application. Therefore, we shall comment briefly on the indentation tests discussed previously with respect to certain important classes of nonmetallics. For instance, the hardness of various plastics is routinely determined by all the methods illustrated in Figure 8-17, depending on the individual circumstances. These hardness data are valuable in correlating such material capabilities as punching, machining, buffing, and wear resistance of the plastics. Also, the hardness of minerals and glasses are evaluated quantitatively by the Knoop method. Moreover, indentation tests such as Rockwell E and H are used on a limited basis to grade and evaluate the hardness of abrasive grinding materials and wheels (i.e., silicon carbide, aluminum oxide).

Hardness Conversion

Since it is frequently desirable to know what hardness values would be equivalent to using other systems, conversion tables such as the one shown in Table 8-3 have been devised. These tables permit ready conversion from one system to another; however,

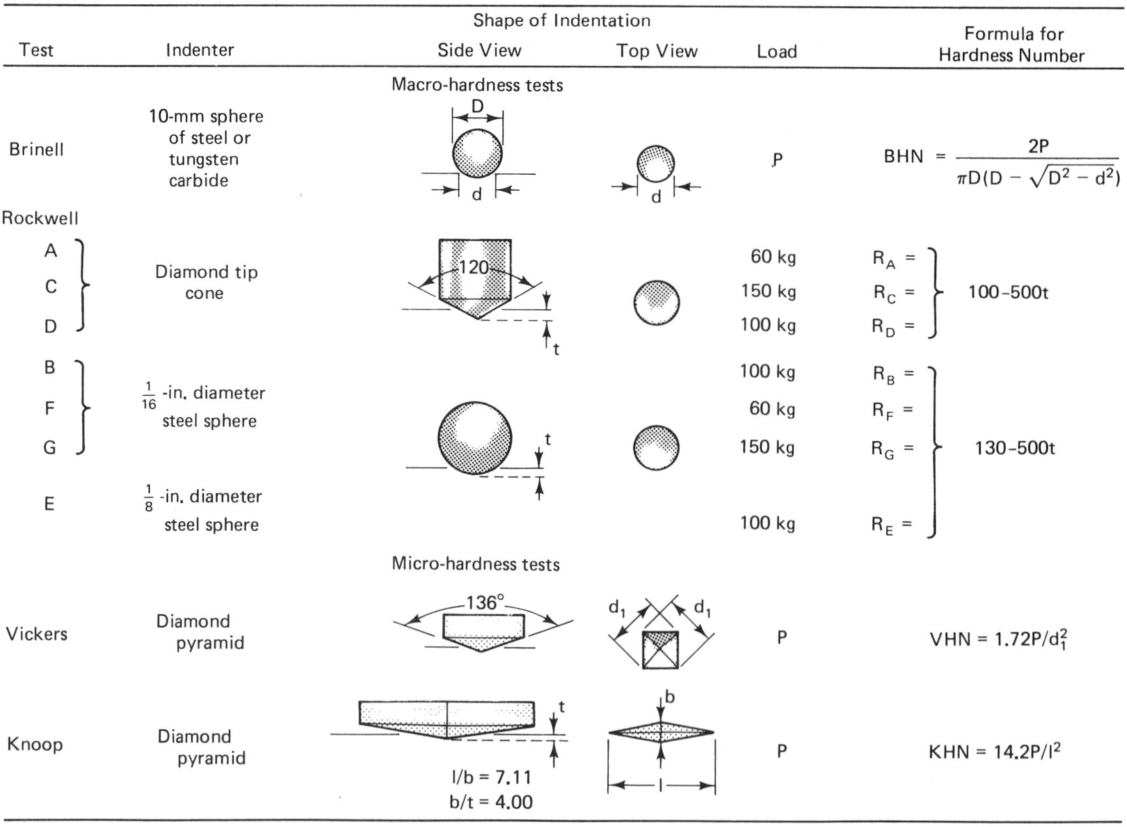

Figure 8-17 table:

Test	Indenter	Shape of Indentation Side View	Top View	Load	Formula for Hardness Number
		Macro-hardness tests			
Brinell	10-mm sphere of steel or tungsten carbide			P	$BHN = \dfrac{2P}{\pi D(D - \sqrt{D^2 - d^2})}$
Rockwell					
A	Diamond tip cone			60 kg	$R_A =$
C				150 kg	$R_C =$ 100–500t
D				100 kg	$R_D =$
B	$\frac{1}{16}$-in. diameter steel sphere			100 kg	$R_B =$
F				60 kg	$R_F =$
G				150 kg	$R_G =$ 130–500t
E	$\frac{1}{8}$-in. diameter steel sphere			100 kg	$R_E =$
		Micro-hardness tests			
Vickers	Diamond pyramid	136°	d_1 d_1	P	$VHN = 1.72P/d_1^2$
Knoop	Diamond pyramid	$l/b = 7.11$ $b/t = 4.00$	b	P	$KHN = 14.2P/l^2$

Figure 8-17 Comparison of various hardness methods. (Adapted from H. W. Hayden, W. G. Moffat, and J. Wulff, *The Structure and Properties of Materials*, Vol. 3: *Mechanical Behavior*, John Wiley & Sons, Inc., New York, 1965.)

it is important to realize that these are empirical relationships and may vary from material to material.

FACTORS RELATING TO MATERIALS SELECTION AND MECHANICAL PROPERTY UTILIZATION

Anisotropy

In many applications and design analyses, engineering materials, especially metals and ceramics, are treated as isotropic and continuous with respect to their mechanical behavior. Essentially, this means that they exhibit identical properties in all directions. In reality, however, engineering materials ordinarily contain dissolved and combined gases, nonmetallic inclusions, and variations in chemical composition referred to as segregation. These imperfections, particularly segregation and inclusions, together with the shaping process (if one is employed), such as rolling or forging, combine to produce variability in mechanical properties. Moreover, this property variation is dependent on the direction of mechanical working. Variable behavior such as this is termed *anisotropic*. Mechanical properties which are especially prone to anisotropy effects are ductility, toughness, and fatigue life. Resistance to corrosion, stress corrosion, and environmental attack in certain materials are also seriously affected by anisotropy.

Mechanical fibering. Anisotropy in engineering materials may take several recognizable forms. One of these is termed "fibering" in metals, named after the appearance of the longitudinal fibers in wood. Fibering consists of nonmetallic inclusions and chemical segregation of alloying elements aligned in the direction of mechanical working. As these constituents of the metal elongate during the shaping operation, planes and regions of weakness are created in the material. Depending on their concentration and distribution, the resulting fibered product may look and behave like a composite material. The structure of a press-forged, low-alloy steel, shown in Figure 8-18(a) for three mutually orthogonal planes, illustrates this appearance. "Banding" is a term which has been applied to worked products that exhibit a severely fibered structure. An example of microstructural banding in rolled, low-carbon steel plate is shown in Figure 8-18(b).

When mechanically worked metallic alloys exhibiting anisotropy are employed in engineering applications, particular consideration should be given to the direction of loading. Strength and ductility are generally better than average in the direction of fibering (working direction). But these properties may be drastically reduced in the direction normal to the fibering.

Crystallographic texturing. The deformation of metal during forming and shaping processes may also cause alignment of the crystal structure. Such a condition is called *texturing*, and involves the orientation of crystallographic planes in a preferred direction with respect to the direction of mechanical working. Texturing or preferred orientation is related to the slip planes and directions in a particular crystal structure. The tendency is for these planes and directions to lie in the direction of working. For example, in a cold-drawn wire, a definite crystallographic direction $[uvw]$ is parallel to the axis of the wire and the texture is symmetrical around this axis. Such a texture is illustrated in Figure 8-19 for bcc metals, which have preferred $\langle 110 \rangle$ direction parallel to the long axis of the wire.

Although texturing results in anisotropy, and we have implied that this condition is deleterious, there are exceptions. For instance, as you will see in Chapter 11, the $\langle 100 \rangle$ texture developed in 3% silicon–iron sheet improves the efficiency of electrical transformers, because iron is most easily magnetized in the $\langle 100 \rangle$ direction.

Fracture Modes

Engineering materials can be broadly classified into two groups depending on their mechanical behavior: (1) materials that behave in a ductile or plastic manner, and (2) those which behave in a brittle or glass-like fashion. Actually, the principal difference between ductile or brittle behavior is the ability of a material to undergo plastic deformation. For example, ductile materials such as aluminum and copper typically exhibit considerable plastic deformation prior to failing. Conversely, brittle materials

TABLE 8-3 HARDNESS CONVERSION FOR SEVERAL TYPES OF HARDNESS MEASUREMENTS

Rockwell C-scale hardness No.	Diamond pyramid hardness No.	Brinell hardness No., 10-mm ball, 3000-kg load			Rockwell hardness No.			Rockwell superficial hardness No., superficial brale penetrator			Shore scleroscope hardness No.	Tensile strength (approx), 100 psi	Rockwell C-scale hardness No.
		Standard ball	Hultgren ball	Tungsten carbide ball	A-scale, 60-kg load, brale penetrator	B-scale, 100-kg load, 1/16-in. diam ball	D-scale, 100-kg load, brale penetrator	15-N scale, 15-kg load	30-N scale, 30-kg load	45-N scale, 45-kg load			
68	940	...	...	...	85.6	...	76.9	93.2	84.4	75.4	97	...	68
67	900	...	...	...	85.0	...	76.1	92.9	83.6	74.2	95	...	67
66	865	...	...	...	84.5	...	75.4	92.5	82.8	73.3	92	...	66
65	832	...	...	739	83.9	...	74.5	92.2	81.9	72.0	91	...	65
64	800	...	...	722	83.4	...	73.8	91.8	81.1	71.0	88	...	64
63	772	...	...	705	82.8	...	73.0	91.4	80.1	69.9	87	...	63
62	746	...	...	688	82.3	...	72.2	91.1	79.3	68.8	85	...	62
61	720	...	...	670	81.8	...	71.5	90.7	78.4	67.7	83	...	61
60	697	...	613	654	81.2	...	70.7	90.2	77.5	66.6	81	...	60
59	674	...	599	634	80.7	...	69.9	89.8	76.6	65.5	80	326	59
58	653	...	587	615	80.1	...	69.2	89.3	75.7	64.3	78	315	58
57	633	...	575	595	79.6	...	68.5	88.9	74.8	63.2	76	305	57
56	613	...	561	577	79.0	...	67.7	88.3	73.9	62.0	75	295	56
55	595	...	546	560	78.5	...	66.9	87.9	73.0	60.9	74	287	55
54	577	...	534	543	78.0	...	66.1	87.4	72.0	59.8	72	278	54
53	560	...	519	525	77.4	...	65.4	86.9	71.2	58.6	71	269	53
52	544	500	508	512	76.8	...	64.6	86.4	70.2	57.4	69	262	52
51	528	487	494	496	76.3	...	63.8	85.9	69.4	56.1	68	253	51
50	513	475	481	481	75.9	...	63.1	85.5	68.5	55.0	67	245	50
49	498	464	469	469	75.2	...	62.1	85.0	67.6	53.8	66	239	49
48	484	451	455	455	74.7	...	61.4	84.5	66.7	52.5	64	232	48
47	471	442	443	443	74.1	...	60.8	83.9	65.8	51.4	63	225	47
46	458	432	432	432	73.6	...	60.0	83.5	64.8	50.3	62	219	46
45	446	421	421	421	73.1	...	59.2	83.0	64.0	49.0	60	212	45
44	434	409	409	409	72.5	...	58.5	82.5	63.1	47.8	58	206	44
43	423	400	400	400	72.0	...	57.7	82.0	62.2	46.7	57	201	43
42	412	390	390	390	71.5	...	56.9	81.5	61.3	45.5	56	196	42
41	402	381	381	381	70.9	...	56.2	80.9	60.4	44.3	55	191	41

40	392	371	371	371	70.4	...	55.4	80.4	59.5	43.1	54	186	40
39	382	362	362	362	69.9	...	54.6	79.9	58.6	41.9	52	181	39
38	372	353	353	353	69.4	...	53.8	79.4	57.7	40.8	51	176	38
37	363	344	344	344	68.9	...	53.1	78.8	56.8	39.6	50	172	37
36	354	336	336	336	68.4	(109.0)	52.3	78.3	55.9	38.4	49	168	36
35	345	327	327	327	67.9	(108.5)	51.5	77.7	55.0	37.2	48	163	35
34	336	319	319	319	67.4	(108.0)	50.8	77.2	54.2	36.1	47	159	34
33	327	311	311	311	66.8	(107.5)	50.0	76.6	53.3	34.9	46	154	33
32	318	301	301	301	66.3	(107.0)	49.2	76.1	52.1	33.7	44	150	32
31	310	294	294	294	65.8	(106.0)	48.4	75.6	51.3	32.5	43	146	31
30	302	286	286	286	65.3	(105.5)	47.7	75.0	50.4	31.3	42	142	30
29	294	279	279	279	64.7	(104.5)	47.0	74.5	49.5	30.1	41	138	29
28	286	271	271	271	64.3	(104.0)	46.1	73.9	48.6	28.9	41	134	28
27	279	264	264	264	63.8	(103.0)	45.2	73.3	47.7	27.8	40	131	27
26	272	258	258	258	63.3	(102.5)	44.6	72.8	46.8	26.7	38	127	26
25	266	253	253	253	62.8	(101.5)	43.8	72.2	45.9	25.5	38	124	25
24	260	247	247	247	62.4	(101.0)	43.1	71.6	45.0	24.3	37	121	24
23	254	243	243	243	62.0	100.0	42.1	71.0	44.0	23.1	36	118	23
22	248	237	237	237	61.5	99.0	41.6	70.5	43.2	22.0	35	115	22
21	243	231	231	231	61.0	98.5	40.9	69.9	42.3	20.7	35	113	21
20	238	226	226	226	60.5	97.8	40.1	69.4	41.5	19.6	34	110	20
(18)	230	219	219	219	...	96.7	...	...	...	...	33	106	(18)
(16)	222	212	212	212	...	95.5	...	...	...	...	32	102	(16)
(14)	213	203	203	203	...	93.9	...	...	...	...	31	98	(14)
(12)	204	194	194	194	...	92.3	...	...	...	...	29	94	(12)
(10)	196	187	187	187	...	90.7	...	...	...	...	28	90	(10)
(8)	188	179	179	179	...	89.5	...	...	...	...	27	87	(8)
(6)	180	171	171	171	...	87.1	...	...	...	...	26	84	(6)
(4)	173	165	165	165	...	85.5	...	...	...	...	25	80	(4)
(2)	166	158	158	158	...	83.5	...	...	...	...	24	77	(2)
(0)	160	152	152	152	...	81.7	...	...	...	...	24	75	(0)

(a) The values in bold face type correspond to the values in the joint SAE-ASM-ASTM hardness conversions as printed in ASTM E48, Table 2 Values in parentheses are beyond normal range and are given for information only.
Source: Metals Handbook, Vol. 1, "Properties and Selection of Metals", 8th Ed., 1961, ASM, Metals Park, Ohio.

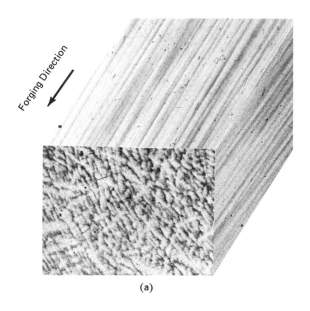

(a)

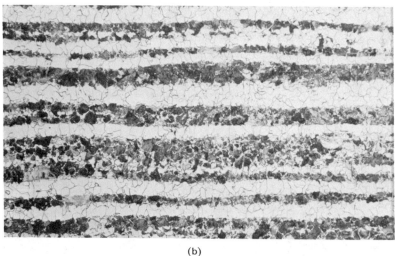

(b)

Figure 8-18 (a) Microstructure of a press-forged low-alloy steel (30×); (b) microstructural banding in rolled low-carbon steel plate, 225×. (Courtesy of F. A. Heiser.)

such as glass, tungsten carbide, and cast iron show very little evidence of plastic deformation in fracturing.

Ductile fracture. Ductile failures usually occur when the material in a component is overstressed. Because of this, ductile fractures are relatively high-energy fractures; they tend to absorb energy during their progression. Therefore, to

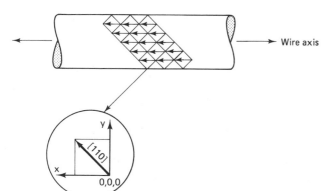

Figure 8-19 Schematic of crystallographic texturing in cold-drawn wire. Note that [110] is oriented parallel to axis of wire.

initiate and continue a ductile failure, energy must be continually supplied, usually in the form of an applied load. This type of failure is characterized by stable crack propagation, which means that if the load causing the fracture is removed, crack propagation ceases.

In addition to gross evidence of deformation such as bending or twisting, ductile failures also exhibit evidence of plastic deformation on their fracture surfaces. On a macroscopic scale, the fracture exhibits a rough appearance and shear lips (shear planes oriented approximately 45° to the fracture surface) can usually be found near the edges of the fracture. Examples of ductile fracture are shown in Figure 8-20. On a microscopic level, the fracture surface exhibits a dimpled appearance, which results from the coalescence of microvoids produced by localized deformation during the *transgranular* failure process (microvoid coalescence). These fracture features can be observed via electron microscopy techniques, either by replication of the fracture surface and examination in the transmission electron microscope (TEM), or by direct examination of the fracture in the scanning electron microscope (SEM). For instance, a ductile fracture mode as revealed by SEM and TEM is shown in Figure 8-21.

Brittle fracture. Brittle failures occur suddenly with little or no external signs of the impending fracture. This type of fracture can take place at stresses lower than the yield strength. Brittle fractures are frequently associated with cracks or other flaws in the material, and in contrast to ductile behavior, they are characterized by low energy absorption and a lack of plastic deformation in their gross appearance. On a macroscopic scale, the fractures tend to be rather flat, featureless, and frequently contain reflective facets. Also, they do not exhibit shear lips or other evidence of plasticity. Several examples of brittle fracture are shown in Figure 8-22. On a microscopic scale, this type of fracture in crystalline materials exhibits the morphology associated with transgranular cleavage (separation along crystallographic planes), quasi-cleavage (similar to cleavage because the features are relatively flat), or intergranular fracture (along the grain boundaries). Examples of these features are shown in Figure 8-23.

Brittle fractures are promoted by the following factors: low temperatures, high strain rates (rapidly applied loads), and triaxial stress states (associated with cracks or flaws).

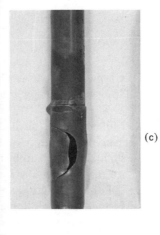

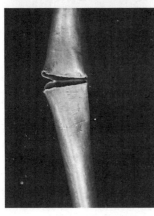

(a)

(b)

(c)

Figure 8-20 Examples of ductile fracture. (a) Plastic fork showing extensive "necking." (b) Bending failure in a 1 inch diameter aluminum tube. (c) Rupture of a 3/4 inch diameter copper tube due to water freeze-up. Note in particular the permanent deformation associated with these fractures. (Courtesy of J. A. Atchinson and J. R. Senick)

    The theory behind brittle fracture is based largely on the work of Griffith, who did much of the initial quantitative measurements of fracture strength. His theory considered the energy required to fracture a material containing a penny-shaped crack oriented perpendicular to the applied stress. He developed an energy balance with three main components:

1. Strain energy in a system without a flaw
2. Surface energy required by the creation of fracture surfaces
3. Energy released by a moving crack

    As a crack moves, strain energy is released, but surface energy is consumed. If there is a balance, the crack is stable and does not propagate. If there is an imbalance, the crack is unstable and grows. However, the Griffith concept reflects only the elastic energy required for brittle fracture, with no consideration of the energy absorbed in plastic deformation at the crack tip. The attention to this latter factor by others set the stage for the development of fracture mechanics. The fracture mechanics approach includes the effect of a crack on the actual stresses that exist in the vicinity of the crack tip, in addition to the nominally applied stress in the material. This intensification of stress due to a crack or flaw may be expressed in terms of a *stress intensity factor* (K). When the stress intensity factor reaches a critical value (K_c), fracture

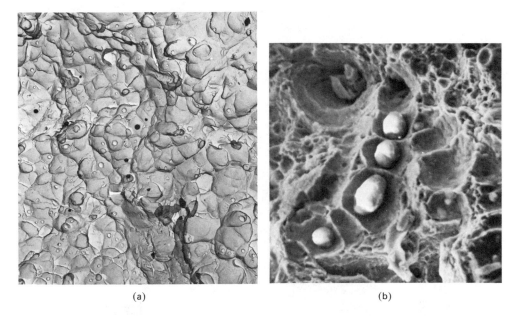

(a)   (b)

Figure 8-21 Ductile fracture as examined microscopically in the electron microscope. (a) Transmission replica showing equi-axed dimples (2000×). (b) Scanning electron micrograph (SEM) showing nonmetallic inclusions associated with the dimples (voids) at 1000×.

(a)

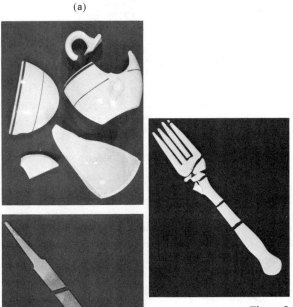

(b)

(c)

Figure 8-22 Fracture appearance of several materials that have experienced brittle fracture. (a) Ceramic cup. (b) Metal file. (c) Plastic fork. In all cases note the absence of permanent deformation and the resultant fragmentation. (Courtesy of J. R. Senick and J. A. Atchinson)

(a)

(b)

Figure 8-23 Transmission electron microscope fractograph in a high strength steel illustrating (a) cleavage, and (b) intergranular fracture (2600×). (Courtesy of L. J. McNamara.)

ensues. The engineering aspects of fracture mechanics and stress intensity are discussed in detail in Chapter 9.

Stress Concentration

The mechanical concepts that we have discussed in this chapter consider the stresses produced by applied forces to be uniform and continuous. In other words, the material is both homogeneous and free of defects or discontinuities. Alas, this is rarely the case.

Both mechanical (geometrical) and microstructural discontinuities commonly exist in engineering materials. In the first case, these discontinuities may take the form of notches, sharp radii, changes in cross section, holes, and so on. In the case of microstructural imperfections, nonmetallics, pores, cavities, and chemical segregation can act in the manner of a discontinuity with effects similar to those of their mechanical counterparts.

A geometrical discontinuity in a stress-carrying member or component can produce both nonuniform stress distribution and locally high stresses in its vicinity. Thus we say that a stress concentration occurs at the discontinuity; it acts as a *stress raiser*. Generally, the more abrupt a discontinuity or change, the greater its stress-concentrating effect.

Stress concentration is generally expressed as the ratio of the maximum stress (σ_{max}) to the nominal stress (σ_{nom}), where the nominal stress is based on load/area. The nominal stress may be based on the entire cross-sectional area (ignoring the dimensions of the discontinuity), or it may be based on the net area which takes into consideration the reduced cross section due to a stress concentrator such as a hole. The latter approach is more conservative, and as the size of a discontinuity increases, this aspect certainly becomes more realistic. The stress concentration factor (K_t) is expressed as follows:

$$K_t = \frac{\sigma_{max}}{\sigma_{nom}} \qquad (8\text{-}12)$$

For example, consider the circular hole in an infinitely wide plate as illustrated schematically in Figure 8-24. An analysis of this condition has shown that the maxi-

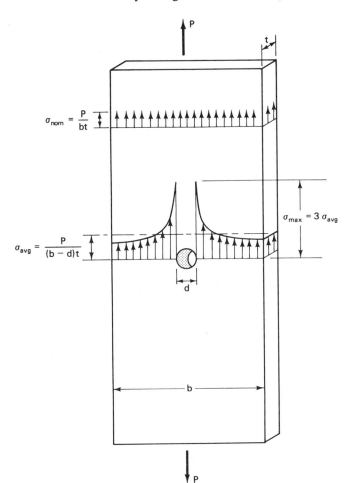

Figure 8-24 Stress concentration associated with a circular hole in a wide plate.

255

mum stress exists at the edges of the hole (as indicated by the arrow) and that σ_{max} = $3\sigma_{avg}$. In other words, the stress-concentrating effects of this rather smooth geometrical discontinuity result in a threefold increase in stress at the hole over the average stress (P/A) in the plate. Some of the more familiar and easily calculated stress concentrators are shown in Figure 8-25.

Example 8-4

A metal bar of the shape shown in Figure 8-25(b) is loaded in tension to 20,000 lb (89,000 N). If w = 2 in. (0.05 m) and the bar is $\frac{1}{2}$ in. (0.012 m) thick, determine the maximum stress developed in the bar for the following conditions:

$$b = \tfrac{1}{2} \text{ in. (0.012 m)}$$

$$r = \tfrac{1}{8} \text{ in. (0.003 m)}$$

Express the result in both English and SI units.

Solution   The nominal stress (based on the net area) in the bar is obtained from equation (6-9) as follows:

$$\sigma_{nom} = \frac{P}{A} = \frac{P}{(w - 2b)(t)} = \frac{20,000 \text{ lb}}{[2 - 2(\tfrac{1}{2})](\tfrac{1}{2}) \text{ in.}^2}$$

$$= \frac{20,000 \text{ lb}}{1/2 \text{ in.}^2} = 40,000 \text{ lb/in.}^2$$

Based on the relationships in Figure 8-25(b), the stress concentration factor (K_t) is determined as follows:

$$\frac{r}{h} = \frac{\tfrac{1}{8}}{w - 2b} = \frac{\tfrac{1}{8}}{2 - 1} = \frac{1}{8} = 0.125$$

$$\frac{b}{r} = \frac{\tfrac{1}{2}}{\tfrac{1}{8}} = 4$$

Utilizing these two factors, K_t is found (graphically) to be approximately 2.6. Therefore, from equation (8-12) we obtain

$$\sigma_{max} = K_t \sigma_{nom}$$

$$= (2.6)(40,000 \text{ psi})$$

$$= 104,000 \text{ psi (717.6 MPa)} \qquad Ans.$$

Comment: Even if this bar was initially designed to support 2.5 times the nominal stress (based on the entire cross section), the member would be seriously jeopardized by the stress concentrators that we have just analyzed.

Safety Factors

In the section "Plastic Deformation and Annealing" in Chapter 6, we introduced the topic of mechanical property variation and discussed briefly some of the factors that contribute to this variability. From a design standpoint, the problem of the inherent variability of engineering materials is compounded because uncertainties can exist regarding the direction and magnitudes of applied loads in service. This means that allowances are made in utilizing stress calculations in engineering structures and

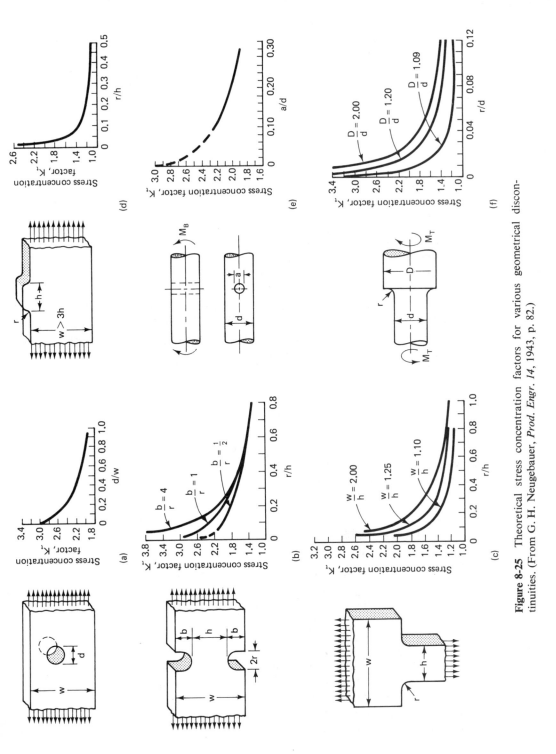

Figure 8-25 Theoretical stress concentration factors for various geometrical discontinuities. (From G. H. Neugebauer, *Prod. Engr. 14*, 1943, p. 82.)

components. A product design usually allows some latitude for contingency, which might result in abnormal loads.

This allowance is provided for by the use of a computational device called the *safety factor*. The allowable (or working) stress (σ_w) is simply taken as a fraction of the yield strength (σ_{YS}) for ductile materials and of the ultimate tensile strength (σ_{UTS}) for brittle materials. Thus N, the safety factor, is

$$\text{Ductile materials:} \quad N = \frac{\sigma_{YS}}{\sigma_w} \qquad (8\text{-}13)$$

$$\text{Brittle materials:} \quad N = \frac{\sigma_{UTS}}{\sigma_w} \qquad (8\text{-}14)$$

However, one is still left with the problem of selecting an appropriate value for N. If N is unnecessarily large, we are probably using more material or higher-strength material than is needed for the application. In other words, the component is overdesigned. This may be the conservative approach from a safety position, but it also may be economically unsound.

Interestingly, the importance of safety factors and material variability were recognized over 100 years ago. For example, Washington A. Roebling incorporated safety factors into the construction of the Brooklyn Bridge, circa 1870. In this particular instance, Mr. Roebling "overdesigned" the steel suspension cables in an effort to compensate for variability in strength and for substandard-quality steel which he suspected would be supplied by some unscrupulous contractors.

Overall, the value of N selected depends on previous experience, economics, engineering codes and regulations, and careful consideration of the consequences of a premature failure. Factors of safety based on yield strength are often taken between 1.5 and 4.0. For more reliable materials or familiar design and service conditions, the lower end of this range is appropriate. For untried materials and design or uncertain loading conditions, the higher factors are safer.

STUDY PROBLEMS

8.1. Calculate the engineering stress in the axial direction in a 60° cross section of tubing if the outer diameter of the tubing is 3.00 cm, the inner diameter is 1.5 cm, and the axial load on the tubing is 150 kg.
Answer: 16.6 MPa or 2413 psi

8.2. A bronze alloy has a 0.2% yield strength of 330 MPa and an elastic modulus of 111 $\times 10^3$ MPa.
(a) What is the stress required to elongate a 1.5-cm bar of this alloy by 0.2 cm?
(b) What is the cross-sectional area required to support a load of 28 $\times 10^3$ N without suffering any plastic deformation?

8.3. Listed below are the load–elongation values obtained in a tensile test for an engineering material. Original diameter = 0.357 in.; diameter at maximum load = 0.312 in. Plot the engineering stress–strain curve, and determine graphically the elastic modulus, the elastic limit, and the yield strength at 0.2% offset. Calculate the ultimate tensile strength and the ductility parameters (i.e., the percent elongation and the percent reduction in area).

Reading	Load (lb)	Length (in.)
1	0	2.0000
2	2,500	2.0017
3	5,500	2.0038
4	6,000	2.0043
5	6,500	2.0048
6	7,250	2.0052
7	6,950	2.0065
8	7,500	2.0080
9	9,050	2.0120
10	11,000	2.0200
11	13,500	2.057
12	20,000	2.472
13	17,000 (fracture)	2.840

8.4. A cast iron specimen with a diameter of 1.6 in. and a length of 6 in. has been tested in compression and the following data obtained. Plot the stress–strain curve and determine the modulus of elasticity.

Load (lb)	Displacement (in.)
4.0×10^3	6×10^{-4}
6.0×10^3	7.5×10^{-4}
8.0×10^3	12.6×10^{-4}
16.0×10^3	26.4×10^{-4}
24.0×10^3	42.0×10^{-4}
32.0×10^3	54.0×10^{-4}
40.0×10^3	69.0×10^{-4}
56.0×10^3	102.0×10^{-4}
64.0×10^3	120.0×10^{-4}
80.0×10^3	174.0×10^{-4}

8.5. A steel bar 39 in. long and 0.505 in. in diameter is subjected to a tensile load of 10,000 lb. Calculate the axial strain, the increase in length, and the change in diameter if the modulus of elasticity is 30×10^6 psi and Poisson's ratio is 0.26.

8.6. A $\frac{1}{2}$-in.-diameter steel rod fractures under a load of 30,000 lb (133 kN). Its final diameter is 0.351 in.
   (a) What is the true fracture strength?
   (b) What is the nominal fracture strength?
   (c) What is the true strain at fracture?
   Answer: (a) 3.1×10^5 psi (b) 1.5×10^5 psi (c) 0.706

8.7. Assume that a steel cable 1000 ft long expands 12.1×10^{-6} in./in./°C when heated or contracts by the same amount when cooled. What is the stress in the cable if it is pinned at both ends so that contraction cannot occur, after it has been cooled from 40°C to room temperature (25°C)?

8.8. A string of steel drill pipe is suspended in an oil well being drilled. The cross-sectional area of the pipe is 4 in.^{2} The strain produced in the pipe due to its own weight is 0.00083. If the density of steel is 0.285 lb/in.^{3}, calculate the depth of the well.

8.9. If the yield strength of the alloy in problem 8.6 is 135,000 lb/in.^{2} and a 100% safety factor is required, how deep can the well be drilled using the current specifications for the drill pipe?
Answer: 19,736 ft

8.10. Compare the requirements for a tensile test bar and a Charpy V-notch impact test. Which specimen has the most rigid requirements? List several of them.

8.11. A component for a drilling rig is planned for use in Alaska, where service temperatures can reach $-40°C$. The component can be manufactured from three steels A, B, and C, which have a transition temperature of $-20°C$, $-40°C$, and $-60°C$, respectively. Which of the steels is most suitable on the basis of transition temperature? Why?

8.12. Is there any basis for relating the Brinell hardness to the tensile strength of a material? What is it?

8.13. The hardness of a steel component was tested with a Brinell machine, creating an impression 3.05 mm in diameter.
(a) What is the Brinell hardness number?
(b) What is the approximate tensile strength?
Answer: (a) BHN = 400 (b) 200,000 psi

8.14. In problem 8.13, if the component was an aluminum alloy, what would the hardness be? What would be the tensile strength?

8.15. The following components must be checked to verify that the hardness levels are in accordance with specifications: large casting for a lathe bed, razor blade, steel ball-bearing race, forging (not machined) section from a heart-pacemaker lead. Which hardness method would you use in each case, and why?

8.16. Define an anisotropic material, an isotropic material, and discuss the effects of anisotropy on the selection of an engineering material.

8.17. **(a)** What is the basis for the development of "fibering" in mechanically worked metallic systems?
(b) What is the effect of fibering on mechanical properties?

8.18. Describe the characteristics of a brittle fracture and a ductile fracture on the basis of macroscopic and microscopic features that might be present.

8.19. In the calculation of safety factors for brittle materials, why is the tensile strength used rather than the yield strength?

8.20. A component having the general configuration shown in Figure 8-25(a) is 30 mm wide, 12.5 mm thick, and has a hole 9 mm in diameter through the center. The component has an axial load of 70,000 N. If the application requires a safety factor of 2, is the component safe to use if its yield strength is 1100 MPa?

8.21. Referring to problem 8.20, calculate the maximum localized stress in the component, using the cross-sectional area rather than the net sectional area, and compare the values obtained. Which is more conservative?

8.22. In problem 8.8, assume that the drill pipe has the following dimensions: outer diameter = 3.02 in., inner diameter = 2.00 in. How deep can the well be drilled if the pipe contains a hole $\frac{3}{8}$ in. in diameter and a safety factor of 2 must be maintained? Use the net section stress in your calculations.
Answer: 6750 ft

8.23. A cylindrical shaft must fit inside a bushing as illustrated below. The maximum diameter of the shaft therefore is 25.4 mm. If the tensile force on the shaft is occasionally as high as 222.5 kN and the appropriate codes specify that a safety factor of 1.75 must be applied to this design, can high-strength aluminum alloy (yield strength = 497 MPa) be used? If not, what other alloys might be acceptable from the strength standpoint?

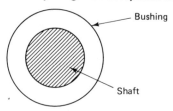

8.24. To secure a microwave receiver/transmitter antenna, high-tension "guy" wires must be produced which will individually withstand a load of 5000 lb without permanent deformation. To ensure that the system operates in the elastic region, a safety factor of 2.0 is utilized. What is the minimum-diameter steel wire (yield strength = 140,000 psi) that can be applied to this structure?

Answer: 0.302 in.

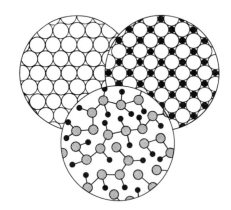

CHAPTER NINE:
Fracture Toughness and Fatigue of Engineering Materials

FRACTURE TOUGHNESS

During our discussion of mechanical property utilization in Chapter 8, we introduced the subject of fracture in engineering materials. At that point, certain general categories of fracture, such as ductile and brittle, were discussed briefly. Thus it is important and appropriate now to address some of the mechanical and material parameters that influence brittle fracture behavior.

Basically, *fracture toughness* can be qualitatively defined as the resistance that a material offers to rapid (brittle) cracking under conditions of constant or rising load. Admittedly, this is an oversimplified view because, as you will soon realize, many factors affect this particular property of materials. Actually, the concept of fracture toughness is an extension of our previous discussions of such areas as deformation, fracture, impact toughness, and stress concentration.

Thus our first objective in this chapter is to present the fundamental concepts of fracture toughness development. From this basic treatment you will see that the propensity toward sudden fracture is related to the presence of cracks or other flaws, the size of these flaws, the stress placed on a component, and certain material properties.

Our second objective will be eventually to combine information regarding the fracture toughness of a material with its *fatigue* behavior to produce analytical solutions for damage tolerance, fail-safe designs, and predictions of the useful or safe life for many components and structures. At this point, fatigue may be simplistically defined as the condition whereby a material cracks progressively as a result of repeated loading. Such an approach to materials' utilization and design cannot be overemphasized because of the potentially catastrophic consequences associated with crack growth and subsequent fracture. In order to impress the student with the serious implications of this type failure, just such a situation is pictured in Figure 9-1. This particular incident involved the sudden fracture of a tank barge while it was moored in New York harbor. As the picture clearly shows, the barge virtually broke in two!

Figure 9-1 Dockside view of the fractured Tank Barge I.O.S. 3301, which failed in New York Harbor, Jan. 10, 1972. (Courtesy of S. T. Rolfe, Univ. of Kansas)

Background

As we mentioned briefly in Chapter 8, during the early part of this century, Griffith formulated a quantitative theory for fracture in brittle solids. He noted in his studies of fracture in glasses that as the length of an individual glass rod decreased due to a successive number of tensile failures, the observed strength correspondingly increased in the remaining portions. This observation was astutely attributed to the progressive elimination of surface flaws which initiated the previous failures. In other words, one after another the "weak links in the chain" were eliminated during his tests. The eventual result was the first analytical expression for determining the load-carrying capacity of a material that contained a crack.

Griffith energy criteria. Essentially, Griffith's analysis is based on an energy balance. When a stress is applied to a material, strain energy or potential energy (U) is produced due to the deformation. This process relates back to the bonding forces between atoms described in Chapter 2 (see Figure 2-3). If no flaws (cracks) were present, this condition would persist until the ultimate tensile strength was exceeded and the material would fail. However, flaws such as cracks, pores, cavities, and inclusions (minute as they may be) are usually present in most commercial engineering materials. In this case, when the strain energy increases to a certain level, the flaws or cracks may extend or enlarge. The change in energy of the system due to crack extension depends on two terms: (1) the decrease in strain energy due to extension of the crack, and (2) the increase in surface energy[1] (W) required to create the new crack surfaces. Analytically, we can express this energy balance as follows:

$$\Delta U = \Delta W \tag{9-1}$$

where ΔU is the change in potential energy or the decrease in strain energy due to cracking, and ΔW is the corresponding increase in surface energy due to the new

[1]In an ideal brittle solid (i.e., no plastic deformation) the released energy can be offset only by the surface energy absorbed. However, in solids that exhibit plastic deformation, the deformation energy is considerably larger than the surface energy term.

Fracture Toughness

portions of the crack. The threshold or condition for cracking can then be stated as

$$\frac{dU}{d\mathbf{a}} = \frac{dW}{d\mathbf{a}} \qquad (9\text{-}2)$$

where $d\mathbf{a}$ represents an increment of crack extension. Based on an elliptical crack, as illustrated in Figure 9-2, Griffith determined the change in potential energy to be

$$\frac{dU}{d\mathbf{a}} = \frac{2\pi\sigma^2\mathbf{a}}{E} \qquad (9\text{-}3)$$

where σ = applied stress
$\quad \mathbf{a}$ = one-half crack length
$\quad E$ = elastic modulus

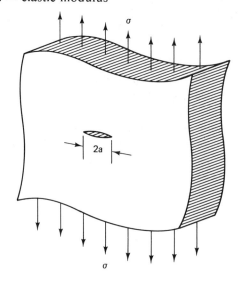

Figure 9-2 Elliptical crack in a large plate subjected to tension.

The attendant change in surface energy is

$$\frac{dW}{d\mathbf{a}} = 4\gamma_s \qquad (9\text{-}4)$$

where γ_s is the specific surface energy for a given material. Substituting these terms in equation (9-2), the following expression is obtained:

$$4\gamma_s = \frac{2\pi\sigma^2\mathbf{a}}{E}$$

This equation can be simplified slightly and restated in terms of the energy necessary to produce the new crack surface area as follows:

$$2\gamma_s = \frac{\pi\sigma^2\mathbf{a}}{E} \qquad (9\text{-}5)$$

This term, $2\gamma_s$ (the 2 represents the two mating surfaces of the crack), is also known as G, the energy release rate. The critical value of G is called the *critical energy release* *rate*² (G_c), and cracking occurs when this value is exceeded. Griffith's cracking criteria can then be expressed as

$$G_c = \frac{\pi\sigma^2 \mathbf{a}}{E}$$

This condition for fracture may be more meaningful if we express it in terms of the applied stress necessary to cause fracture:

$$\sigma_c = \sqrt{\frac{2E\gamma_s}{\pi \mathbf{a}}} \tag{9-6}$$

where σ_c is the critical value of applied stress for crack extension. Experimentally, the critical value of energy release rate (G_c) can be determined by measuring the stress required to fracture a plate containing an elliptical crack (**2a**), as shown in Figure 9-2.

Example 9-1

A large plate of soda–lime glass containing a crack, as shown in Figure 9-2, is subjected to tensile loading. If the surface energy (γ_s) for this material is 0.3 J/m² and $E = 69,000$ MPa, determine the fracture stress (σ_c) if the flaw is 0.1 mm (0.0001 m) long.

Solution   Utilizing the Griffith criteria expressed in equation (9-6) and remembering that $2\mathbf{a} = 1 \times 10^{-4}$ m, we obtain the following:

$$\sigma_c = \sqrt{\frac{2\gamma_s E}{\pi \mathbf{a}}}$$

$$= \sqrt{\frac{2(0.3 \text{ J/m}^2)(69,000 \text{ MPa})}{\pi(5 \times 10^{-5} \text{ m})}}$$

Since J = N-m, we can rearrange our units to yield

$$\sigma_c = \sqrt{\frac{(2)(0.3 \text{ Pa-m})(69,000 \times 10^6 \text{ Pa})}{\pi(5 \times 10^{-5} \text{ m})}}$$

$$= \sqrt{2.64 \times 10^{14} \text{ Pa}^2}$$

$$= 1.62 \times 10^7 \text{ Pa} = 16.2 \text{ MPa} = (2,349 \text{ psi}) \qquad \textit{Ans.}$$

Crack tip stresses.   Basically, the stresses operating on cracks that exist in engineering materials can be categorized into three modes. These modes, denoted as opening (I), sliding (II), and tearing (III), are illustrated in Figure 9-3. In actual experience, the opening mode (I) is most frequently encountered, and therefore this mode is generally considered for purposes of example and discussion. The ensuing treatment of fracture toughness in this chapter will reflect this convention. For instance, the stresses on an element in the vicinity of a hypothetical crack tip as depicted

---

²In reality, the critical strain energy release rate (G_c) is equal to twice an effective surface energy (γ_{eff}), where γ_{eff} is predominately the plastic energy absorption associated with the growth of a crack or flaw. Only a small portion of γ_{eff} is due to the surface energy of the crack surfaces.

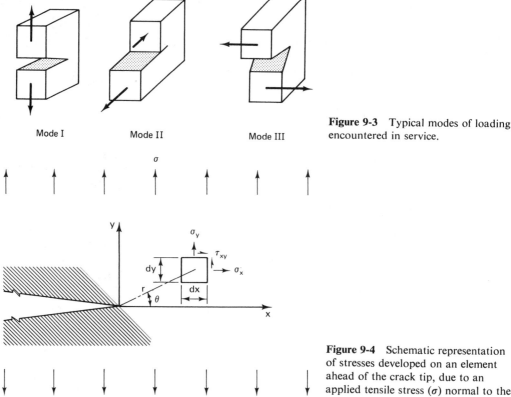

Mode I   Mode II   Mode III

Figure 9-3 Typical modes of loading encountered in service.

Figure 9-4 Schematic representation of stresses developed on an element ahead of the crack tip, due to an applied tensile stress (σ) normal to the crack plane.

in Figure 9-4 are given as follows for mode I loading:[3]

$$\sigma_x = \sigma\sqrt{\frac{\mathbf{a}}{2r}} \cos\frac{\theta}{2}\left(1 - \sin\frac{\theta}{2}\sin\frac{3\theta}{2}\right)$$

$$\sigma_y = \sigma\sqrt{\frac{\mathbf{a}}{2r}} \cos\frac{\theta}{2}\left(1 + \sin\frac{\theta}{2}\sin\frac{3\theta}{2}\right)$$

$$\tau_{xy} = \sigma\sqrt{\frac{\mathbf{a}}{2r}} \sin\frac{\theta}{2}\cos\frac{\theta}{2}\cos\frac{3\theta}{2}$$ (9-7)

$$\sigma_z = 0 \quad \text{(for plane stress)}$$

$$\sigma_z = \nu(\sigma_x + \sigma_y) \quad \text{(for plane strain)}$$

These expressions show that the nominal applied stress (σ) is affected quantitatively by the presence of a crack ($\mathbf{a}$), and that the stress varies directly with the square root of the crack length and inversely with the square root of the distance (r) from the crack tip. For large values of r, σ_y tends to approach the nominal stress (P/A). How-

---

[3]See the next paragraph for an explanation of plane stress and plane strain. ν is Poisson's ratio [see equation (6-7)].

ever, as r gets very small, the normal stress in the y direction increases sharply and controls the stress and the fracture state near the crack tip. This situation is schematized in Figure 9-5.

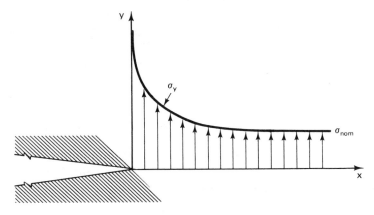

Figure 9-5 Schematic illustration of tensile stress magnitude, with respect to distance from the crack tip.

Finally, the concepts of plane strain and plane stress should be explained. In terms of the stresses and strains produced in a material under load, these terms mean that one of the *principal* strains or stresses is zero. These conditions are given as follows:

Plane Strain		Plane Stress	
$\sigma_x \neq 0$	$e_x \neq 0$	$\sigma_x \neq 0$	$e_x \neq 0$
$\sigma_y \neq 0$	$e_y \neq 0$	$\sigma_y \neq 0$	$e_y \neq 0$
$\sigma_z \neq 0$	$e_z = 0$	$\sigma_z = 0$	$e_z \neq 0$

Because of the physical constraints imposed, plane strain conditions tend to apply in the interior of a specimen or material, whereas plane stress conditions apply at the surface because stresses normal to a free surface are zero.

Stress intensity factor. Therefore, we see that certain stresses in the vicinity of a crack tip can be substantially greater than the nominal or engineering stress produced by P/A. This increase in stress has been defined as the *stress intensity factor* (K).[4] The stress intensity factor is a function of the applied stress (σ) and the crack size $(\mathbf{a})$. This relationship is expressed as follows:

$$K = f(\sigma, \mathbf{a}) \tag{9-8}$$

[4]K, the stress intensity factor, should not be confused with K_t, the stress concentration factor, which is the ratio of $\sigma_{max}/\sigma_{min}$ in a notched specimen (see Equation 8-12).

The stress intensity factor is also influenced by the geometrical configuration of the cracked component and the cracking mode (I, II, or III), as mentioned previously. Many functions for expressing crack tip stress intensity have been developed and applied to various specimens or component geometries. Some of the analytical expressions for stress intensity are given in the next section.

In a manner similar to Griffith's energy concept, fracture occurs when the stress intensity factor reaches a critical value (K_c). For example, if we consider the stress intensity factor for a large plate as depicted in Figure 9-2, the critical stress intensity is as follows:

$$K_c = \sigma_c \sqrt{\pi a_c} \, f(a/W) \tag{9-9}$$

where $f(a/W)$ is a correction factor which depends on the geometry of the component [i.e., the width (W) and the flaw size (a)]. For $a/W \rightarrow 0$, $f(a/W) \rightarrow 1.0$.

The critical value of stress intensity (K_{Ic}) is commonly known as the plane strain (thick plate) *fracture toughness* for a material. Fracture toughness is a material property; thus K_{Ic} may be established for a given material and this parameter in turn

TABLE 9-1 TYPICAL PLANE STRAIN FRACTURE TOUGHNESS VALUES FOR CERTAIN ALLOYS

Material	K_{Ic}		Yield strength	
	MPa-m$^{1/2}$	ksi-in.$^{1/2}$	MPa	ksi
Aluminum				
2014-T651	24	22	455	66
2024-T3	44	40	345	50
2024-T851	26	24	455	66
7075-T651	24	22	495	72
7178-T651	23	21	570	83
7178-T7651	33	30	490	71
7475-T7651	47	43	462	67
Titanium				
Ti–6 Al–4 V	115	105	910	132
Ti–6 Al–4 V	55	50	1035	150
Steel				
4340	99	90	860	125
4340	60	55	1515	220
4335 + V	72	66	1340	194
4335 + V	132	120	1035	150
17-7 pH stainless	77	70	1435	208
15-7 Mo stainless	50	45	1415	205
H-11 tool	38	35	1790	260
350 marage	55	50	1550	225
350 marage	38	35	2240	325
52100 ball bearing	14	13	2070	300

Source: R. W. Hertzberg, *Deformation and Fracture Mechanics of Engineering Materials*, John Wiley & Sons, Inc., New York, 1976, p. 286.

applied to various components or designs which utilize that material. For example, the fracture stress (σ_c) of a material containing a crack of known size can be predicted. Or vice versa, the largest crack or flaw that can be tolerated in a given material can be determined for a given applied stress level. These concepts are the basis for the application of fracture mechanics to engineering materials design problems, and what makes this "relatively new" branch of mechanics such a powerful design tool. We can now quantitatively address structures and components that contain flaws, many of which are inherent because of fabrication and processing (as described in Chapter 7), and relate this information to the safe or allowable working stresses mentioned in Chapter 8.

Fracture toughness data are now fairly widespread in the literature and usually not too difficult or expensive to establish for new engineering materials. Typical values of plane strain fracture toughness (K_{Ic}) for certain engineering alloys of interest are presented in Table 9-1.[5] In addition, fracture toughness data for selected ceramics and polymers are given in Table 9-2. Remember, K_{Ic} is the *critical* value of stress intensity for mode I loading, as described in Figure 9-3.

TABLE 9-2 FRACTURE TOUGHNESS DATA FOR SELECTED NONMETALLIC ENGINEERING MATERIALS

Material	K_{Ic}	
	MPa $\sqrt{m}$	ksi $\sqrt{in.}$
Mortar	0.14–1.4	0.13–1.3
Concrete	0.25–1.57	0.23–1.43
Al_2O_3	3.3–5.8	3–5.3
SiC	3.7	3.4
Si_3N_4	4.6–5.7	4.2–5.2
Soda–lime silicate glass	0.8–0.9	0.7–0.8
Electrical porcelain ceramics	1.1–1.37	1.03–1.25
WC–3% Co	11.6	10.6
WC–15% Co	18.1–19.9	16.5–18
Indiana limestone	1.09	0.99
PMMA	0.9–1.92	0.8–1.75
PS	0.9–1.2	0.8–1.1
Polycarbonate	3.02–3.6	2.75–3.3

Source: R. W. Hertzberg, *Deformation and Fracture Mechanics of Engineering Materials*, John Wiley & Sons, Inc., New York, 1976, p. 370.

[5]Although certain properties are well established for many engineering materials, the student is cautioned against using "typical" data published in the literature for anything but sample calculations. To apply mechanical property values confidently to a specific situation, one must possess accurate data or actually measure the property in question.

Example 9-2

The design stress on a certain structural component is 690 MPa (100,000 psi) in tension. For this condition, determine the critical crack size (a_c) for (a) a 4335 + V alloy steel and (b) a titanium alloy (Ti–6 Al–4 V), both at a yield strength level of 1035 MPa (150,000 psi).

Solution Typical fracture toughness (K_{Ic}) data are given in Table 9-1 for the alloys and strength levels involved. K_{Ic} equals 132 MPa-m$^{1/2}$ (120 ksi-in.$^{1/2}$) for the steel and 60 MPa-m$^{1/2}$ (55 ksi-in.$^{1/2}$) for the titanium alloy.

The critical crack size can be determined from equation (9-9) [assuming a value of unity for $f(a/W)$ for the purpose of this example] as follows:

$$K_{Ic} = \sigma\sqrt{\pi a}\, f(a/W)$$
$$= \sigma\sqrt{\pi a}\,(1)$$

Solving for **a**, we have

$$\sqrt{a} = \frac{K_{Ic}}{\sigma\sqrt{\pi}}$$

$$a = \frac{K_{Ic}^2}{\sigma^2\pi}$$

(a) The critical crack size for the steel is

$$a_{\text{Steel}} = \frac{(132 \text{ MPa-m}^{1/2})^2}{(690 \text{ MPa})^2\pi}$$

$$= \frac{17,424 \text{ MPa}^2\text{-m}}{4.76 \times 10^5 \text{ MPa}^2(\pi)}$$

$$= 1.2 \times 10^{-2} \text{ m}$$

$$= 12 \text{ mm } (0.47 \text{ in.}) \qquad \textit{Ans.}$$

(b) The critical crack size for the titanium is

$$a_{\text{Ti}} = \frac{(60 \text{ MPa-m}^{1/2})^2}{(690 \text{ MPa})^2\pi}$$

$$= \frac{3600 \text{ MPa}^2\text{-m}}{4.76 \times 10^5 \text{ MPa}^2(\pi)}$$

$$= 2.4 \times 10^{-3} \text{ m}$$

$$= 2.4 \text{ mm } (0.09 \text{ in.}) \qquad \textit{Ans.}$$

Clearly, the titanium alloy is more crack sensitive in this case ($a_{\text{steel}} = 5a_{\text{Ti}}$).

Comment: Note that the critical crack size (a_c) in this example is one-half ($a/2$) the crack length of an internal flaw, as illustrated in Figure 9-2. Furthermore, the stress intensity factor for an edge crack under these conditions may be expressed as

$$K_{Ic} = 1.12\sigma\sqrt{\pi a}\, f(a/W)$$

Fracture Toughness Testing

Although many types of test methods have been developed for evaluating fracture toughness in engineering materials, the dimensions of the available material or component frequently determine the specimen geometry and size. Specimen configurations

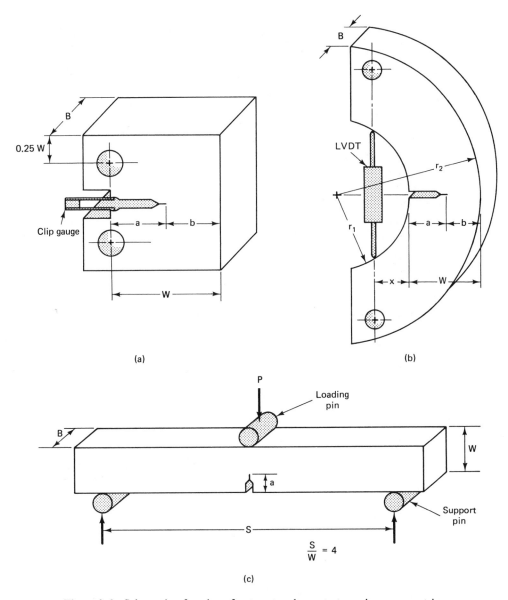

(a)

(b)

(c)

Figure 9-6 Schematic of various fracture toughness test specimen geometries: (a) modified compact tension specimen with clip gauge in position to measure displacement; (b) arc-shaped specimen with a linear voltage displacement transducer (LVDT) in place; (c) three-point bend specimen.

of several standard fracture toughness bars are shown in Figure 9-6. The exact details of these specimens and their test criteria are explained for the interested reader in ASTM Standard E399 (*Plane Strain Fracture Toughness Tests for Metallic Materials*). Normally, these standards are adhered to in manufacturing fracture toughness test specimens. However, if thin sheet material is involved, it would be appropriate to test a sample prepared from the "bulk" sheet even though it may not meet the exact criteria of the standard test method. In such a case, the fracture toughness value obtained may be *invalid* according to ASTM Standards, but it still furnishes an indication of the fracture toughness relative to this particular geometry or application. Therefore, in this instance (and similar instances) the value may be used as a guideline.

Similarly, it is important to orient the test specimen taken from a component or a large section of material (i.e., plate, billet, etc.), so that the specimen orientation corresponds to the actual loading conditions that will be imposed on the material in service. In other words, the design load and orientation are duplicated in the test specimen. Only then will there be a realistic correspondence between the laboratory results and actual service conditions. Such a practice takes *anisotropy* into consideration. This is illustrated for rectangular and cylindrical shapes in Figure 9-7, where the crack planes are oriented in three different directions with respect to the working direction (direction of fibering). You may wonder just how important this particular aspect of fracture toughness testing is. In answer to this rhetorical question, the student is referred back to Figures 8-18 and 8-19 for a better appreciation of the effects that microstructural orientation may have on fracture toughness.

K computation.   As mentioned previously, the stress intensity factor (K) depends on a geometrical function related to the shape of the test bar, in addition to stress level (σ) and crack length (**a**). This dependence can be expressed mathematically as follows:

$$K = \sigma W^{1/2} f(a/W) \qquad (9\text{-}10)$$

where W is the width of the specimen. The expressions for K are shown in Figure 9-8 for several important specimen configurations. Referring to those test specimen examples, we see that the term $f(a/W)$ is presented graphically for values of a/W. Also note that the applied stress (σ) is replaced by (P/BW), where P is the applied load and BW is the area of the specimen.

The actual polynomial expressions for the specimens under consideration are given as follows:

Bend specimen:

$$f(a/W) = \frac{3(a/W)^{1/2}[1.99 - (a/W)(1 - a/W)(2.15 - 3.93a/W + 2.7a^2/W^2)]}{2(1 + 2a/W)(1 - a/W)^{3/2}}$$

Compact tension specimen:

$$f(a/W) = \frac{(2 + a/W)(0.886 + 4.64a/W - 13.32a^2/W^2 + 14.72a^3/W^3 - 5.6a^4/W^4)}{(1 - a/W)^{3/2}} \qquad (9\text{-}11)$$

Arc-shaped specimen:

$$f(a/W) = \frac{(a/W)^{1/2}}{(1 - a/W)^{3/2}}[3.74 - 6.30a/W + 6.32(a/W)^2 - 2.43(a/W)^3]$$

Plastic-zone considerations.   Another factor that must be considered in fracture toughness testing is the extent of the *plastic zone* associated with the crack tip. Just ahead of the tip of a crack under load, a region of the material experiences plastic deformation when the localized stresses described by equation (9-7) exceed the

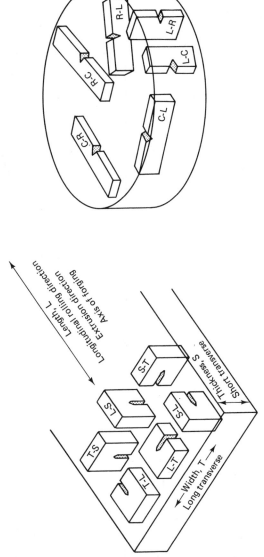

Figure 9-7 Standarized notation for test specimen orientation with respect to the "working" direction of plates and bars. (From ASTM Standarized E-399, *Plane Strain Fracture Toughness Testing of Metallic Materials*, American Society for Testing Materials, Philadelphia, Pa.. 1981, p. 72.) (Copyright, ASTM, 1916 Race Street, Philadelphia, Pa., 19103. Reprinted with permission.)

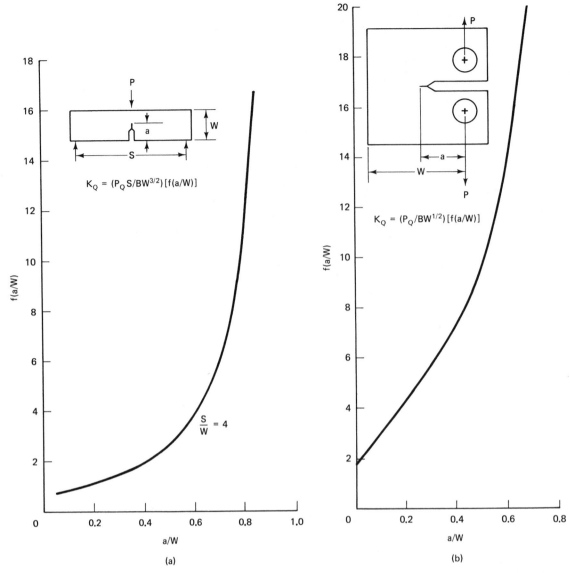

$$K_Q = (P_Q S / BW^{3/2})[f(a/W)]$$

$$\frac{S}{W} = 4$$

$$K_Q = (P_Q / BW^{1/2})[f(a/W)]$$

(a)

(b)

Figure 9-8 Graphical presentation of $f(a/W)$ for several standard fracture toughness specimens: (a) three-point bend specimen; (b) compact tension specimen; (c) arc-shaped specimen.

yield strength of the material. This condition is illustrated in Figure 9-9. The extent of such deformation is important because fracture toughness by definition is a measure of cracking resistance—not resistance to plastic deformation. Unfortunately, fracture toughness is affected by plastic deformation. As the size of the plastic zone increases, toughness generally increases, but the validity and significance of the fracture toughness *test* decreases.

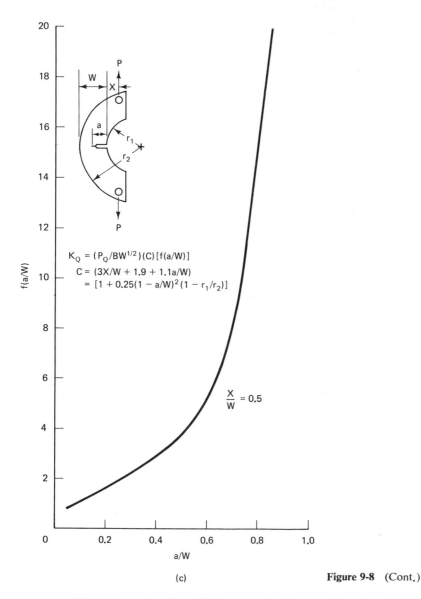

$$K_Q = (P_Q/BW^{1/2})(C)[f(a/W)]$$
$$C = (3X/W + 1.9 + 1.1a/W)$$
$$= [1 + 0.25(1 - a/W)^2(1 - r_1/r_2)]$$

$$\frac{X}{W} = 0.5$$

a/W

(c)

Figure 9-8 (Cont.)

Yield Strength. In addition to plastic zone size, the yield strength (YS) of a material influences the test results. As the YS decreases, the plastic zone size generally increases, since plastic deformation is easier. Conversely, as K_{Ic} decreases, the size of the plastic zone decreases, because the material exhibits a tendency to crack rather than deform plastically. The size of the plastic zone (r_p) is proportional to the quantity $(K_{Ic}/YS)^2$, and this factor is useful in determining the specimen size necessary to meet the requirements of ASTM Standard E399. Based on this standard test method, the following conditions or criteria are necessary for a valid measurement

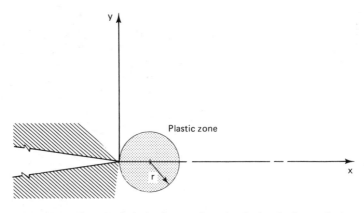

Figure 9-9 Hypothetical plastic zone (envelope) ahead of a crack tip.

of fracture toughness:

$$\text{crack length } (a) \geq 2.5(K_{Ic}/\text{YS})^2$$

$$\text{specimen thickness } (B) \geq 2.5(K_{Ic}/\text{YS})^2 \qquad (9\text{-}12)$$

$$\text{ligament of specimen } (W - a) \geq 2.5(K_{Ic}/\text{YS})^2$$

Specimen Thickness. A final note on specimen thickness. The variation of fracture toughness with specimen thickness is depicted schematically in Figure 9-10. If the specimen is thin, the material exhibits plane stress behavior and high apparent fracture toughness. As the thickness is increased, however, plane strain conditions

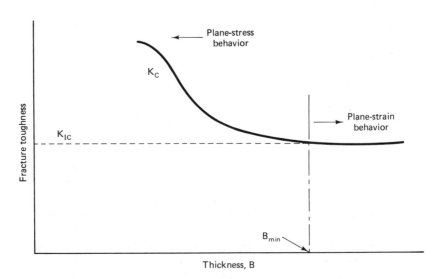

Figure 9-10 Effect of specimen thickness (*B*) on fracture toughness behavior. (From S. T. Rolfe and J. M. Barsom, *Fracture and Fatigue Control in Structures— Applications of Fracture Mechanics*, Prentice-Hall, Inc., Englewood Cliffs, N.J., 1977, p. 100.)

prevail and toughness decreases to a certain level beyond which no further change occurs. This is the material property, K_{Ic}, and is conservatively taken as the accepted value of *plane strain fracture toughness* because it does not decrease with further increases in thickness.

Test Procedures

Although other mechanical property specimens also utilize carefully machined notches (i.e., Charpy V-notch and notched tensiles), fracture toughness specimens employ a *fatigue crack* in addition to the machined notch. This type of crack, as you will discover, represents a very sharp flaw, because of its extremely small dimensions. The radius at the tip of a fatigue crack is considerably smaller than any radius that could be produced by conventional machining methods. This *sharp* flaw also eliminates crack tip deformation and stresses that could be introduced by machining the notch. The initial crack size or flaw (a_0) includes the notch and the fatigue crack as shown in Figure 9-11.

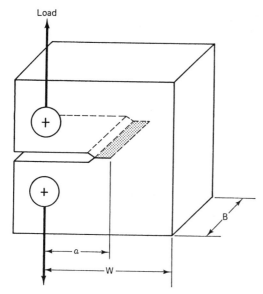

Figure 9-11 Schematic of a "precracked" toughness test specimen (shaded portion represents fatigue crack).

The "precracked" specimen is placed in an appropriate testing machine or fixture. In the case of a compact tension specimen, a tensile testing machine or similar apparatus may be used, as shown in Figure 8-1. Indeed, the actual test is conducted in much the same manner as a conventional tensile test. The applied load is measured and plotted versus displacement of the crack faces. Sensitive measuring instruments such as a linear voltage displacement transducer (LVDT) are used to detect this displacement. Some typical results of this test are shown schematically in Figure 9-12 for several types of fracture behavior.

Examination of these curves reveals that in the initial portions of the test, the material behaves elastically and the load–displacement curve is a straight line. But at

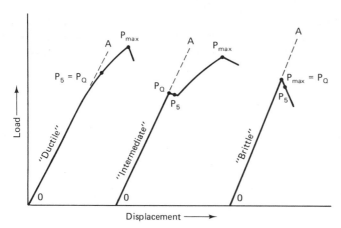

Figure 9-12 Hypothetical load–displacement curves from fracture toughness tests of different materials.

some critical point on this curve, the load–displacement behavior changes, depending on the material. For instance, in a brittle material, the load drops sharply and the bar fractures abruptly. This type of response is illustrated by type III. Less brittle materials may exhibit the behavior shown by type II. Finally, for more ductile materials the load–displacement curve may behave like the material depicted by type I. In each case, the load at the intersection of the 5% secant is used as illustrated in Figure 9-13.

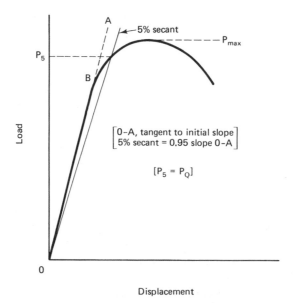

Figure 9-13 The 5% secant load value for establishing the critical load (P_Q).

This load (P_5) is taken to be the provisional critical load (P_Q) for fracture toughness calculations. Again, for the purposes of test validity, the ratio P_{max}/P_5 is calculated. If $P_{max}/P_5 \geq 1.10$, the test is considered invalid according to ASTM Standard E399, because such a result indicates that crack tip blunting and deformation have occurred rather than crack growth.

Interpretation of results. Once the specimen is broken, the fracture surfaces are examined and the average crack depth (**a**) is carefully measured. The mating fracture surfaces in a compact tension specimen are shown in Figure 9-14.

Figure 9-14 Mating halves of fractured compact tension specimen ($B = 1$ in.). This view shows the depth and shape of fatigue precrack (light gray region denoted with arrow).

This figure shows a view of the machined notch and the subsequent cracked portions. The depth of the precrack is measured in three locations, at $\frac{1}{4}$, $\frac{1}{2}$, and $\frac{3}{4}$ of the thickness (B), as shown in Figure 9-15. The average value of **a** is used in the fracture toughness expression [equation (9-10)]. However, if the difference between any two measurements of **a** exceeds 5% of the average value (indicating a skewed crack front), the test is considered *invalid* according to ASTM Standard E399.

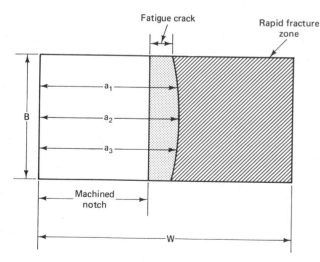

Figure 9-15 Schematic of fracture surface showing various important regions and dimensions.

Determination of K_{Ic}. Using the average value of **a**, the critical load (P_Q), and the appropriate K calculation, the critical value of stress intensity is computed. At this stage of the analysis, however, the value of K_{Ic} obtained is not assumed to be a valid measurement of fracture toughness. In fact, it is labeled K_Q (provisional) until it has been tested against the criteria in equation (9-12). The various specimen dimensions (**a**, *B*) are recalculated using the value of K_Q obtained in these expressions. If the dimensions are adequate, then $K_Q = K_{Ic}$, a valid, acceptable measurement of *plane strain fracture toughness*. On the other hand, if these dimensions turn out to be inadequate, $K_Q \neq K_{Ic}$ and further testing with larger dimensions (based on K_Q) is necessary to establish valid fracture toughness results.

Example 9-3

A compact tension specimen, as illustrated in Figure 9-11, was machined from a newly developed alloy. Given the following information, determine the fracture toughness estimate (K_{IQ}) for this alloy.

$$W = 50.8 \text{ mm}$$

$$B = 25.4 \text{ mm}$$

$$\mathbf{a} = 30.8 \text{ mm}$$

$$P_5 = 58 \text{ kN}$$

Solution Utilizing the *K* calibration function given in Figure 9-8(b), we obtain the following: In this instance, **a**/$W = 0.6$; therefore, $f(\mathbf{a}/W) = 14$. Our expression for fracture toughness then is

$$K_{IQ} = \frac{(P_5)f(\mathbf{a}/W)}{(B)(W)^{1/2}}$$

$$= \frac{(58 \times 10^3 \text{ N})(14)}{(0.0254 \text{ m})(0.051 \text{ m})^{1/2}} \left(\frac{\text{m}^{1/2}}{\text{m}^{1/2}}\right)$$

$$= 1.4 \times 10^8 \text{ (N/m}^2)(\text{m}^{1/2})$$

Recall that $\text{Pa} = \text{N/m}^2$. Therefore,

$$K_{IQ} = 140 \text{ MPa-m}^{1/2} \text{ or } (127 \text{ ksi-in.}^{1/2}) \qquad Ans.$$

Comment: At this point K_{IQ} is an estimate of the plane strain fracture toughness (K_{Ic}). It is not a valid K_{Ic} until the criteria of ASTM Standard E399 have been met.

Example 9-4

The alloy tested in Example 9-3 has a yield strength of 1380 MPa (200 ksi). Determine if the fracture toughness results can be considered a valid K_{Ic}.

Solution Subjecting these results to the criteria specified in equation (9-12), we obtain the following:

1. *Required crack length ($\mathbf{a}_R$):*

$$\mathbf{a}_R \geq 2.5(K_Q/\text{YS})^2$$

$$\geq 2.5(140/1380)^2$$

$$\geq 0.0257 \text{ m} \geq 25.7 \text{ mm}$$

$\mathbf{a}_{\text{act}} = 30.8 \text{ mm} > \mathbf{a}_R$. Thus the crack length (**a**) is sufficient.

2. *Required specimen thickness* (B_R):

$$B_R \geq 2.5(K_Q/\text{YS})^2 \geq 25.7 \text{ mm}$$

$$B_{\text{act}} = 25.4 \text{ mm}$$

$B_{\text{act}} \simeq B_{\text{req}}$; the specimen thickness is just barely sufficient.

3. *Required specimen ligament* $(W - a)$:

$$W - a \geq 2.5(K_Q/\text{YS})^2 \geq 25.7 \text{ mm}$$

$$(W - a)_{\text{act}} = 20 \text{ mm}$$

$(W - a)_{\text{act}} < (W - a)_{\text{req}}$. Therefore, the ligament is insufficient!

Conclusion: $K_Q \neq K_{\text{Ic}}$. *Ans.*

Comment: Based on this validity analysis, the fracture toughness value K_Q is not a valid K_{Ic}. However, this does not mean that K_Q is useless as an estimate of fracture toughness. This value may be used for comparative purposes in selection of materials. Furthermore, if there is sufficient material for testing, a larger specimen should be prepared, taking into consideration the *required* dimensions calculated in this example.

Fracture Toughness and Design

Fracture toughness, like other material properties, is influenced by a number of factors. Basically, these factors include *strain rate, temperature,* and *yield strength,* as alluded to previously. Let us briefly examine these factors individually. In general, as the strain rate increases, K_{Ic} decreases. Therefore, rapidly applied loads tend to lower fracture toughness. As temperature increases, K_{Ic} increases correspondingly. Finally, as the yield strength (or ultimate tensile strength) increases, K_{Ic} decreases. These effects are demonstrated for high-strength low-alloy steel (HSLA) in Figures 9-16 and 9-17.

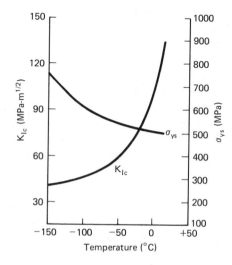

Figure 9-16 Effect of temperature on K_{Ic} and yield strength in high-strength low-alloy steel. (From L. Engel and H. Klingele, *An Atlas of Metal Damage*, Prentice-Hall, Inc., Englewood Cliffs, N.J., 1981, p. 64.)

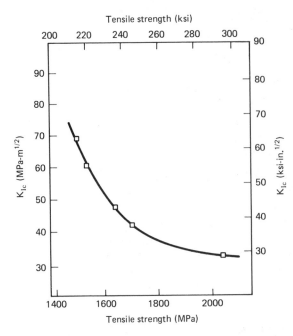

Figure 9-17 Influence of tensile strength on K_{Ic} in high-strength low-alloy steel. (From A. J. Birkle, R. P. Wei, and G. E. Pellissier, *Trans. ASM 59*, 1966, p. 981.)

Unfortunately, these factors influencing fracture toughness are not readily separable (just as yield strength and plastic zone size are related), since as strain rate increases, YS also increases. Furthermore, as temperature increases, YS decreases. Although this may sound slightly confusing, there is a synergistic interaction between these factors, and although the trends mentioned are usually observed, the behavior of specific materials can vary.

Influence of microstructure.   On several previous occasions we have mentioned the effects of microstructure on mechanical properties. For instance, in Chapter 7, we discussed certain types of microstructural defects which contribute to property variation and degradation. In Chapter 8, the role of microstructural banding or fibering and its influence on mechanical property anisotropy was examined (see Figure 8-18). Fracture toughness is also sensitive to microstructure and to microstructural orientation (fibering). For example, the toughness of steels that contain a fine distribution of transformation products is considerably better than the same steels containing a coarse microstructure. Similarly, the fracture stress (σ_f) in many polycrystalline engineering materials shows a strong dependence on grain size according to a Petch–Hall type of relationship [see equation (6-20)]. Just such a relationship is demonstrated in Figure 9-18 for low-carbon steel. These particular data clearly show a substantial increase in the stress necessary to cause fracture as the grain size (average diameter) decreases.

Fracture toughness in certain nonferrous materials such as titanium and aluminum also depends on microstructural features, including the size, shape, and distribution of the phases or constituents. For example, in titanium the β phase alloys (bcc) exhibit higher toughness levels compared to microstructures containing α and

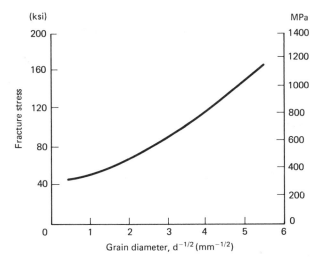

Figure 9-18 Relationship between fracture stress and grain size for low carbon steel (at $-196°C$). (Adapted from J. R. Low, Jr., *Relation of Properties to Microstructure*, American Society for Metals, Metals Park, Ohio, 1954, p. 163.)

β (where α is a hcp phase). Similarly, the toughness of *age hardenable alloys* (see Figure 7-45), for example, aluminum, magnesium, nickel, and certain stainless steels, generally increases with decreasing precipitate particle size. These microstructural considerations are extremely important when such materials are used in structural and often critical applications: for instance, titanium in air frame structures or surgical implants, and nickel-based alloys in jet engine turbines.

Microstructural Orientation. This term refers to the alignment of certain microstructural constituents, such as nonmetallic inclusions, second phases, and chemical segregation, in a preferred direction. This direction is usually parallel to the long axis of sheet, plate, or bar when the material has been mechanically worked (e.g., rolled, forged, drawn, extruded). We refer to this condition as fibering (or in severe cases, as banding), because the microstructure can take on the appearance of a fibrous product, much like the structure of wood parallel to its grain (see Figure 16-33).

Preferred orientation of microstructural constituents can seriously affect the fracture toughness of engineering materials, as shown by the data in Table 9-3. This information demonstrates that K_{Ic} in different aluminum alloys manufactured as plate is greatest in the L-T orientation and poorest in the S-L orientation (these designations were explained in Figure 9-7). Differences in fracture toughness on the order of 20 to 30% between these particular orientations are not uncommon, and are directly attributable to the difference in microstructural arrangement on these mutually orthogonal planes, in the *same* bar or plate.

The design aspects. Let us briefly examine some of the important aspects of fracture toughness. Admittedly, there are certain drawbacks to the widespread use and application of fracture toughness. First, even though K expressions are available for many specimen configurations, they may not be available for a specific application being considered. In this instance, the value of K_{Ic} obtained experimentally can only be used as a guideline. Second, as we previously cautioned, there may be insufficient material to provide a valid size specimen. Therefore, only comparative data can be

Fracture Toughness

TABLE 9-3 EFFECT OF ORIENTATION ON PLANE STRAIN FRACTURE
TOUGHNESS VALUES OF SEVERAL HIGH-STRENGTH ALUMINUM ALLOYS

Product	Alloy	Temper	Yield strength (longitudinal) (ksi)	Plain strain fracture toughness, K_{Ic} (ksi-in.$^{1/2}$)		
				L-T	T-L	S-L
Plate	2014	T651	64	22	20	17
	2024	T351	47	33	30	24
	2024	T851	66	22	21	16
	2124	T851	64	29	23	22
	2219	T851	63	35	33	—
	7050	T73651	66	32	27	26
	7075	T651	73	26	23	18
	7075	T7651	68	27	22	18
	7075	T7351	63	29	26	19
	7475	T651	72	39	34	29
	7475	T7651	67	43	35	28
	7475	T7351	62	48	38	32

obtained from these tests. Third, for very ductile materials, fracture toughness testing may just be inappropriate. Tensile and Charpy impact testing may provide sufficient design information for such materials and they are less expensive.

On the other hand, there are also many advantages to utilizing fracture toughness. Its acceptability is more widespread than other tests for a brittle material. This is due principally to the background development in basic mechanics and because the inherent problems of machined notch tolerances and reproducibility are circumvented by fatigue precracking. Finally, and most important, the relationship between K_{Ic}, the applied stress (σ), and a flaw size (**a**) governs the conditions for fracture in a component or structure. For instance, if a designer selects a material for a particular application, then accurately determines its fracture toughness in the laboratory, the flaw size that will produce failure can be reliably predicted for the anticipated design stresses. How realistic is this approach to design analysis? Well, presently there are many nondestructive test techniques (NDT) commercially available for both surface and internal examination of engineering materials. These NDT methods can be used to "screen out" defects or flaws that approach the critical size, thus ensuring against a brittle, catastrophic failure.

Finally, we have not considered the effects of fatigue in our fundamental discussions of fracture toughness. Our elementary analyses therefore deal only with single or constant-load applications, as discussed in the introduction to this chapter. Certainly, this view is not completely realistic.

For instance, a *cyclically* applied load may vary continuously from zero to a value in tension, then pass through zero to a value in compression, and so on. When such a loading cycle is repeated over and over, we call this condition *fatigue*. Although the subject of fatigue will be studied in the next section, it is not premature to point out that the insidious nature of this type of loading can facilitate the initiation and growth of cracks at stress levels *well* below the yield strength or the allowable design (working) stresses. During the fatigue process, a subcritical crack or flaw can grow progressively through a component or structure, unsuspected and undetected! Often these flaws are the result of manufacturing processes or materials defects, as we saw in Chapter 7, and can affect performance in a critical manner. The interaction between fracture toughness and fatigue is discussed later in the section "Fatigue Crack Propagation."

FATIGUE

As we mentioned previously, *fatigue* is the condition whereby a material cracks or fails as a result of repeated (*cyclic*) stresses. From an engineering standpoint, fatigue should be rigorously defined as the progressive localized permanent structural change that takes place in a material subjected to repeated or fluctuating strains. Generally, these strains occur at stresses below the ultimate tensile strength of the material, and often occur at stresses below the yield strength. This type of failure is characterized by three stages: (1) crack initiation, (2) crack propagation, and (3) fast fracture. Although these stages are usually treated as separate portions, a certain amount of overlap occurs in actual fatigue failures.

The subject of fatigue is of particular importance to engineers and designers due to the *insidious* nature of this failure mechanism and the potential catastrophic results that it may cause. To be more specific, fatigue cracking can initiate and progress at stress levels considerably *lower* than the yield stress or the allowable design stress (working stress) that we discussed in Chapter 8. If the fatigue process goes undetected, and in many cases it does, the fatigue crack may eventually reach a critical size (a_c), such as we defined in the section "Fracture Toughness." When the critical crack size is exceeded, rapid fracture of the component ensues, almost instantaneously. Unfortunately, as a result of the fatigue process, a structure very often fails suddenly and catastrophically, with little or no warning. The serious consequences that can be associated with fatigue must therefore never be overlooked.

Fatigue can affect virtually all engineering materials that are subjected to cyclic stresses. Cyclic stresses include repeated applied stresses from external forces, and thermal stresses as a result of alternating heating and cooling. Such circumstances apply to a broad spectrum of engineering materials: metals and alloys, ceramics, polymers, and composites, in a wide variety of applications, including mechanical, electrical, architectural, chemical, and biomedical applications.

The process of fatigue in an engineering material is affected by a number of factors, which can be classified into the following general categories: *mechanical factors*, *microstructural factors*, and *environmental factors*. Therefore, in addition to studying the fundamental aspects of fatigue, in this section we will also examine the role of these factors in fatigue behavior. Hopefully, a better understanding of the conditions that influence fatigue will enable the student to recognize the pitfalls and potential threat that this failure mechanism presents to many engineering applications.

Cyclic Stresses

We mentioned previously that fatigue occurs as a result of *cyclic stresses*; but what are cyclic stresses, and how are they produced? Fundamentally, this type of stress may arise from various external loading situations, particularly those which are fluctuating or repetitive in nature. Also, cyclic stresses can be produced by repeated or alternating changes in the temperature of a material. The latter stresses are commonly referred to as *thermal stresses* and arise from the expansion and contraction that accompany heating and cooling.

There are numerous ways in which fluctuating stresses may actually be produced in both the laboratory and actual service situations. Therefore, it will be helpful to examine several that illustrate the basic character of fatigue stresses. However, we must caution the student that in actual fatigue failures (the *real world*), variations of these modes can be much more complex and often occur in combinations. Let us examine two important categories of constant-amplitude cyclic stresses: *reversed* and *repeated*.

Reversed stress cycles. This type of loading or stressing is represented schematically (sinusoidally) in Figure 9-19 for two hypothetical reversed stress situations; (1) a continuously changing stress function [Figure 9-19(a)], and (2) a varying stress which remains constant for a portion of the cycle [Figure 9-19(b)]. In both cases, however, the reversed stress cycle consists of stresses alternating from tension to compression. Also, in this example, the maximum tensile stress equals the maximum compressive stress. Certain symbols and terminology used in fatigue testing are indicated in this figure. For instance, *one cycle* of the curve is denoted. What do we actually mean by *fatigue cycle* (stress cycle)? This is a single segment of the stress function which is periodically repeated. In our example, then, one cycle consists of zero–tension–zero–compression–zero. Another common fatigue term is the stress range (σ_r). The stress range refers to the difference between the maximum and minimum applied stress. In the case of Figure 9-19, $\sigma_r = 2\sigma_a$, where σ_a is the stress amplitude. The manifestations of this type of cycling will be discussed in the section "Fatigue Life."

Repeated stress cycles. In contrast to the reversed type of cycle, *repeated stress cycles* consist of alternating stresses in one mode of loading, such as tensile. This type of stress cycle can also be represented by a sinusoidal function, as illustrated in Figure 9-20. In our example of repeated stress cycles, the stress amplitude (σ_a) alternates uniformly about the mean stress (σ_m). We have illustrated a continuously changing stress pattern in Figure 9-20(a), and a situation where stress remains constant for a portion of the cycle in Figure 9-20(b).

The reversed and repeating types of cyclic stressing are commonly used in laboratory testing, since they are not too difficult to reproduce in a standard test specimen. However, we must again remind the student that actual fatigue load–time sequences in service are very likely to be more complex than the simple situations we have demonstrated here. The complexity of cyclic stresses is vividly emphasized by the

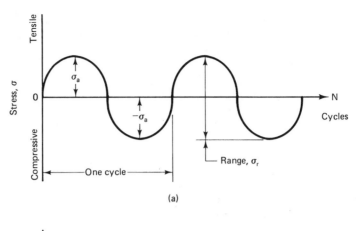

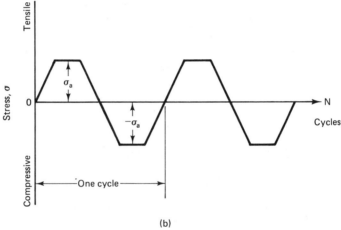

Figure 9-19 Schematic representation of reversed stress fatigue cycles: (a) continuously varying stress; (b) reversed stress cycle containing a portion constant with time.

load–time histories shown in Figure 9-21 for such varied applications as automobiles, aircraft, pipelines, and nuclear reactors.

Fatigue Life

The *fatigue life* of a component or a material is defined as the total number of stress cycles to cause failure (N_f). As we indicated earlier, this life can be separated into three stages, consisting of *crack initiation* (N_i), *crack propagation* (N_p), and *rapid fracture*. The fatigue life in terms of number of cycles can therefore be stated as

$$N_f = N_i + N_p \tag{9-13}$$

where N_i = number of cycles required to initiate a *discernible* crack; influenced primarily by such factors as stress level, mechanical stress concentrators,

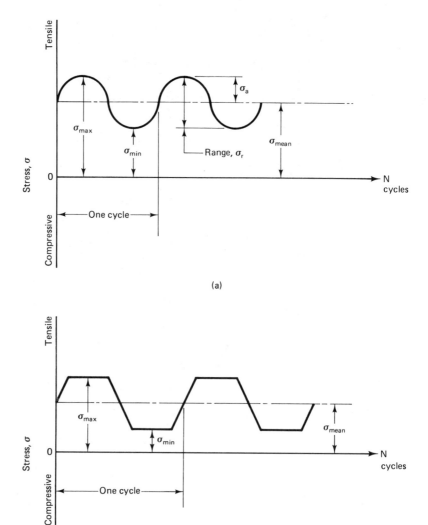

Figure 9-20 Examples of repeated stress cycles.

second-phase particles, crystalline imperfections, and environmental conditions

N_p = number of cycles required to propagate or grow the crack in a stable manner to a critical size, whereupon fast fracture ensues; affected primarily by stress level, microstructural orientation, and environment

You will note that the rapid fracture stage is not included in N_f since it is the terminal phase in the fatigue failure of a material, occurs very quickly, and contributes very little to the total life.

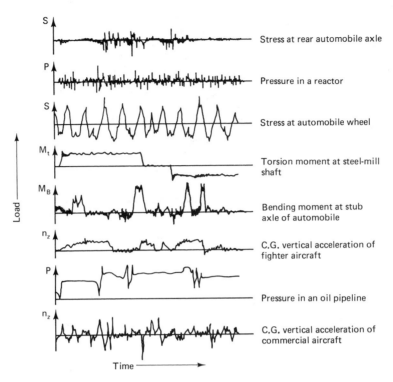

Figure 9-21 Typical examples of load-time histories for various engineering applications.

Initially, fatigue cracks form due to a *slip* mechanism. This portion of the fatigue crack is crystallographically oriented along the slip plane and is often referred to as *stage I* crack growth. The mechanism of fatigue crack initiation is discussed in more detail later in this chapter. Eventually, the crack plane becomes *macroscopically* evident and is perpendicular to the direction of maximum tensile stress. This portion is called *stage II* crack growth or the *propagation stage*. Fatigue crack propagation is also discussed separately, following the section on initiation. These two stages of fatigue crack growth are shown schematically in Figure 9-22 for a polycrystalline material subjected to an alternating tensile stress in the axial direction. Note that we have illustrated the cracking as *transgranular* (through the grains) in both stages. This is the common mode of failure, unless the grain boundaries are extremely susceptible to cracking.

The relative number of cycles involved in stage I (initiation) and stage II (propagation) depends primarily on the stress level. At low stress levels many cycles are required to develop and initiate a crack. As the stress level increases, the crack initiation phase (N_i) decreases. At very low stress levels, therefore, most of the fatigue life is utilized in crack initiation. Conversely, at relatively high stress levels, cracks form early in the fatigue life. Based on this principle, fatigue failure is usually divided into two categories:

Fatigue

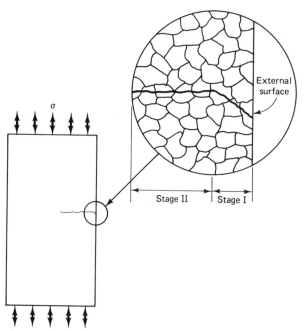

Figure 9-22 Hypothetical fatigue crack showing initiation (stage I) and propagation (stage II) portions on the side of the specimen.

1. *High-cycle fatigue:* associated with low stresses (generally involves greater than about 10^5 cycles)

2. *Low-cycle fatigue:* associated with high stresses (generally involves less than about 10^5 cycles)

These categories simply indicate the relative number of stress cycles during the fatigue life of a component or a material. Although their separation is not distinct, generally the low-cycle region results from stress levels that are often high enough to produce significant *plastic strains* in the component under test or in service.

Stress-life (S–N) behavior. The fatigue life of a material is an important concept which is frequently studied in the laboratory so that this information can be applied to designs and components in actual service.

A common method of analyzing fatigue life employs constant-amplitude cyclic loading. Two types of tests used for this purpose are illustrated in Figure 9-23. For example, the rotating bending apparatus pictured in Figure 9-23(a) will result in the reversed stress cycles shown in Figure 9-19(a). In this particular test, the specimen undergoes bending over its entire length. Treating this as a simple beam deflecting downward, we see that the upper portion experiences compression while the lower portion simultaneously sustains tension, in the longitudinal direction. However, since this "beam" is continuously rotating, the material in the bar experiences an alternating stress (i.e., compression–zero–tension–zero–compression, etc.).

Another type of fatigue test customarily used to assess fatigue life is shown in Figure 9-23(c). This technique can produce the types of cyclic stresses illustrated in

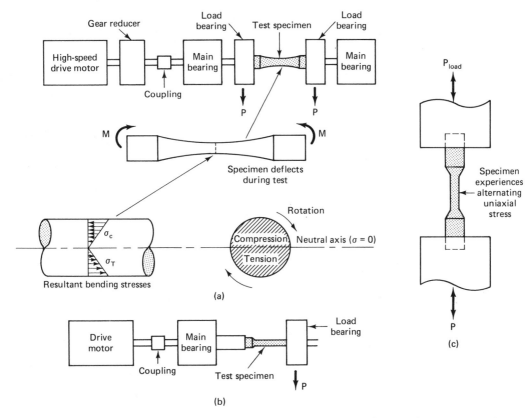

Figure 9-23 Typical fatigue life test setups: (a) rotating bending showing stress state in specimen; (b) cantilever rotating bending; (c) alternating uniaxial tension or compression.

both Figure 9-19 and 9-20. Note that in this configuration the cycle's fatigue load is applied axially (as in the tensile test).

We should reemphasize that in both of the fatigue test methods above, the maximum load does not change (constant-amplitude loading), even though changes may occur within the material (strain hardening), or a crack may initiate and grow. Other test methods may employ a varying peak load, if such a condition represents the actual service conditions one is attempting to reproduce in the laboratory.

S–N curves. Data from the fatigue life tests is usually displayed in the form of $S–N$ curves, where the alternating stress (S) is plotted against the number of cycles to failure (N). Although such curves are a useful approach to analyzing the overall fatigue life of a material, we should point out that they do not differentiate between the initiation and propagation stages of fatigue. These curves simply indicate the *total* number of cycles to failure at a given stress level. Typical $S–N$ curves for certain engineering alloys are illustrated in Figure 9-24.

Examination of these "typical" $S–N$ curves reveals that as the level of the alternating stress decreases, the number of cycles to failure increases. Ferrous alloys (notably the low-strength steels) and some titanium alloys tend to exhibit a stress limit below which failure does not occur by fatigue. This limit is referred to as the *fatigue limit* or the *endurance limit*, and is defined as the limiting stress below which the material will withstand an indefinitely large number of stress cycles without fracture. However, since infinite life tests are impossible and impractical to even attempt, the endurance limit for many materials is frequently specified as that stress which will not produce failure in 10^7 cycles.

Most nonferrous metals and alloys, such as aluminum, copper, and magnesium, do not exhibit a specific fatigue limit. Rather, they display a continuous sloping curve for all stress levels, as indicated in Figure 9-24.

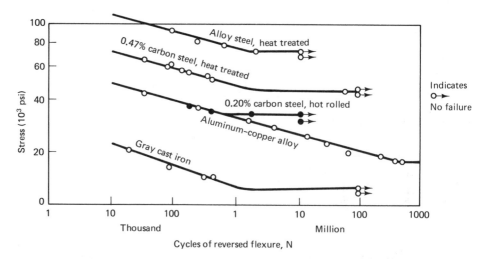

Figure 9-24 *S–N* behavior in selected engineering materials (log-log plot). (From *Metals Handbook*, 1948 ed., American Society for Metals, Metals Park, Ohio, 1948, p. 118.)

Example 9-5

An alloy substitution has been suggested in order to reduce the weight of the axle shaft assembly shown in Figure 9-25. One criterion specified by the designers is that the shaft must withstand 321,800 km (200,000 mi) service at a given stress level. (a) In laboratory testing, therefore, what is the *minimum* number of cycles that the fatigue life test should be conducted for that particular stress level? (b) How long a period of time

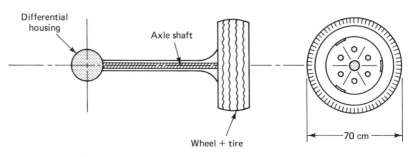

Figure 9-25 Example of shaft assembly subjected to rotational bending.

will this take if the rotating-bending apparatus operates at a frequency of 8.3 Hz (500 rpm).

Solution (a) The revolution of the wheel on the road corresponds to one revolution of the shaft, which represents one cycle in the fatigue test. The number of kilometers per revolution (same as the circumference of the tire) is

$$c = \pi d$$

$$= \pi(0.7 \text{ m})$$

$$= 2.2 \text{ m} = 0.0022 \text{ km}$$

So the minimum number of test cycles must be

$$N = \frac{321,800 \text{ km}}{0.0022 \text{ km/rev}}$$

$$= 1.46 \times 10^8 \text{ cycles} \qquad Ans.$$

(b) At an operating frequency of 8.3 Hz (500 rpm), the test time would be

$$t = \frac{1.46 \times 10^8 \text{ cycles}}{3 \times 10^4 \text{ cycles/hr}}$$

$$= 4866 \text{ hr or 203 days} \qquad Ans.$$

Comment: This example serves to illustrate the lengthy time periods (and attendant expense) involved in fatigue life testing.

Statistical aspects. Unfortunately, fatigue life data generally exhibit large amounts of scatter. Therefore, the examples shown in Figure 9-24 actually represent the "best-fit" curves, constructed through median values. Thus approximately 50% of the data (the *material response*) will actually fall *below* the S–N curve! Occasionally, the lower boundary of the test data is used to represent the S–N behavior since this tends to be conservative. Nevertheless, even this practice is not completely safe.

To establish even more conservative data for design purposes and material selection, statistical methods are often utilized. One technique consists of establishing an average curve through the experimental data, and performing a statistical analysis to compute the standard deviation (*s*). The use of the standard deviation helps determine the degree of *risk* or the percentage of test points that will fall within a certain range. The standard deviations for a large number of "normally" distributed points are as follows:

Deviation	Percent test points within deviation	Percent test points below minimum deviation
$\pm 1s$	68.2	15.9
$\pm 2s$	95.4	2.3

This means that if boundary curves corresponding to $\pm 2s$ are drawn above and below the average curve, such boundaries will include 95.4% of all test data. Consequently, in a large sample, only 2.3% of the specimens would fall below the $-2s$ curve. Since in the majority of laboratory testing the sample size is small, the $-2s$ curve usually

Fatigue

falls below all the data points. Therefore, it is a more reliable estimate of the stress levels at which there is a low probability of failure than a curve drawn arbitrarily through the lowest data points. This statistical approach is illustrated in Figure 9-26.

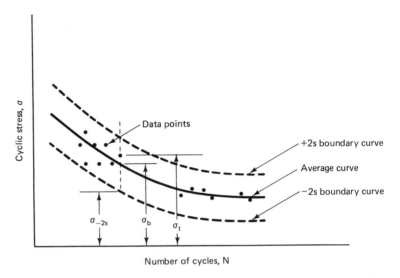

Figure 9-26 Typical fatigue life (*S–N*) curve showing statistical boundaries (±2*s*).

Example 9-6

A fatigue life experiment resulted in an *S–N* curve, with the stress data given below, for a particular value of *N* cycles. Statistically determine the value of stress (σ_{-2s}) below which no more than 2.3% of the specimens in a large, "normal" sample will fall.

Specimen	Stress (MPa)	Specimen	Stress (MPa)
1	242	6	262
2	272	7	259
3	276	8	248
4	231	9	245
5	287	10	262

Solution   For this sample we computed

$$\text{average stress } (\bar{\sigma}) = 258 \text{ MPa}$$

$$\text{standard deviation } s = \pm 17$$

Therefore, the "lower bound" $= \bar{\sigma} - 2s$ and

$$\sigma_{-2s} = 258 - (2)(17)$$

$$= 224 \text{ MPa} \quad Ans.$$

How does this stress value compare with the experimental data?

Factors Affecting Fatigue Life

In the section "Fatigue" we alluded to three main categories of factors which have a strong influence on fatigue behavior in engineering materials. These factors are *mechanical, microstructural,* and *environmental.* Let us now examine the principal effects in each category.

Mechanical factors. This group of factors deals with the geometrical and mechanical considerations of fatigue life. In particular, *mean stress, stress concentration,* and *surface effects* will be discussed, since they can have an appreciable influence on fatigue life.

Mean Stress. The stress amplitude (σ_a) exerts the major influence on fatigue life, as the S–N tests indicate. However, the mean stress (σ_m) also strongly affects fatigue life. These stresses, which were depicted in Figures 9-19 and 9-20, are expressed quantitatively as follows:

$$\sigma_a = \frac{\sigma_{max} - \sigma_{min}}{2} \tag{9-14}$$

$$\sigma_m = \frac{\sigma_{max} + \sigma_{min}}{2} \tag{9-15}$$

Basically, the effect of increasing the mean stress (going from compressive to tensile) is a reduction in the fatigue life, as illustrated in Figure 9-27(a).

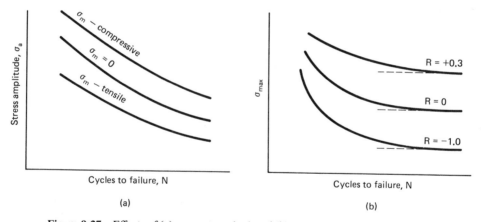

Figure 9-27 Effects of (a) mean stress (σ_m) and (b) stress ratio (R) on fatigue life.

Another term used to describe the relationship between σ_{max} and σ_{min} is denoted as R (stress ratio), where

$$R = \frac{\sigma_{min}}{\sigma_{max}} \tag{9-16}$$

This ratio (R) may vary from −1 to +1. For example, if the stress is completely reversed (Figure 9-19), that is, $\sigma_{min} = -\sigma_{max}$, then R = −1. As the value of R approaches +1, the stress range (σ_r) approaches zero and the loading becomes monotonic (constant). The effect of R becoming more positive is an increase in the fatigue life, as shown in Figure 9-27(b).

Stress Concentration. The overall effect of *stress concentrators* is to reduce fatigue life. As discussed in Chapter 8, stress raisers can consist of keyways, fillets, grooves, holes, or threaded regions. Virtually any notch or discontinuity can produce stress-concentrating effects, although generally the sharper the geometric change (the smaller radius at the notch root), the greater its effect.

The influence of notches are evaluated by comparing notched versus unnotched *S–N* data for the same material. The *S–N* behavior for notched and unnotched specimens of a low-alloy (3% Ni–Cr) steel is presented in Figure 9-28. From these curves,

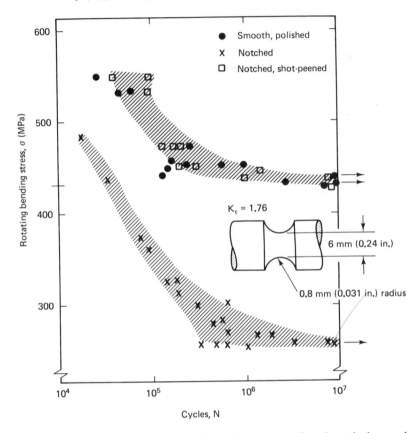

Figure 9-28 *S–N* behavior for smooth, notched unpeened, and notched peened specimens of low-alloy steel (3% Ni–Cr). (From W. J. Harris, *Metallic Fatigue*, Pergamon Press Ltd., London, 1961, p. 48.)

a notch sensitivity factor (q) can be developed as follows:

$$q = \frac{K_f - 1}{K_t - 1} \qquad (9\text{-}17)$$

where q = notch sensitivity factor

$K_f = \dfrac{\text{fatigue limit (unnotched)}}{\text{fatigue limit (notched)}}$

K_t = geometric stress concentration factor (for notch)

Relatively high values of q indicate a notch-sensitive material, while low values imply a material that is less sensitive to notches. The relative notch sensitivity increases with increasing tensile strength and the severity (acuity) of the notch root.

Example 9-7

Determine the notch sensitivity factor (q) for the alloy in Figure 9-28. K_t is given as 1.76.

Solution   From Figure 9-28 we can interpolate the following data:

$$\text{Fatigue limit (unnotched)} \simeq 430 \text{ MPa}$$

$$\text{Fatigue limit (notched)} \simeq 260 \text{ MPa}$$

Therefore,

$$K_f = \frac{430}{260} = 1.65$$

Substituting into equation (9-6), we have

$$q = \frac{1.65 - 1}{1.76 - 1} = \frac{0.65}{0.76}$$

$$= 0.86 \quad Ans.$$

Surface Effect.   In most fatigue tests and service applications, the maximum stresses tend to occur at the surface of a material. Therefore, in addition to stress concentrators, fatigue life is especially sensitive to surface conditions. Just what are these surface conditions? Well, the principal factors to consider are *surface finish* (roughness) *surface properties*, and *residual stresses*.

1. *Surface finish:* Machining marks essentially are small notches or grooves in the surface of a component produced by the cutting action of a tool. As the surface finish becomes more coarse, the depth of such notches increase. Correspondingly, fatigue life as measured by total cycles to failure decreases. For example, consider the effects of surface finish and the type of finishing operation presented in Table 9-4. These data clearly demonstrate that the type of finishing operation and the resultant surface finish have a considerable influence on fatigue life.

2. *Surface properties:* The properties of many engineering materials are altered by fabrication and processing. If such changes are produced exclusively at the mate-

TABLE 9-4 EFFECT OF SURFACE FINISH ON FATIGUE LIFE[a]

Operation	Surface finish		Fatigue life (cycles)
	μ m	μ in.	
Lathe turned	2.67	105	24,000
Hand polished (partly)	0.15	6	91,000
Hand polished	0.13	5	137,000
Ground	0.18	7	217,000
Ground and polished	0.05	2	234,000

[a]Material SAE 3130 steel; reversed stress = 655.5 MPa (95 ksi).

rial's surface, its fatigue life may be affected significantly. These effects may be classified into those which *decrease* fatigue life and those which *increase* it. Electrochemical deposition (electroplating) of one metal on the surface of another, for example chromium plating on steels for improved appearance or wear resistance, may decrease fatigue life due to the tensile stresses and chemical effects associated with the plated layer. If this deposit cracks as shown in Figure 9-29, in effect the crack initiation stage of fatigue may be significantly reduced, with a corresponding decrease in fatigue life. Basically, any process that assists or encourages cracking at the surface of a material contributes to a decrease in fatigue life.

Conversely, some processes that affect surface properties can have a beneficial

(a)

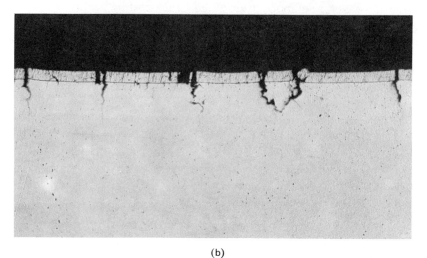

(b)

Figure 9-29 Heat checking (crazing) in chrome plated alloy steel: (a) surface of material exhibiting crazed cracking pattern (4×). (b) cross section normal to surface showing extension of cracks into the base metal (15×). (Courtesy of J. Renslow and S. Tauscher)

effect on fatigue life by retarding crack initiation in the surface layers. These processes include selective surface alloying and surface modification by laser techniques (see our discussion of rapid solidification processes in Chapter 7).

3. *Residual stresses:* Residual stresses due to strain hardening are frequently left in a material which has undergone plastic deformation during processing. If such stresses are tensile in nature, they can promote fatigue cracking and *decrease* fatigue life, depending on their magnitude and orientation. Therefore, surface residual tensile stresses should be avoided in any material or component that will be subjected to fatigue conditions.

On the other hand, compressive residual stresses at the surface of a material offset the effect of applied tensile stresses. Since in many applications, the maximum tensile stresses occur at the surface, the overall fatigue life is usually enhanced by residual compressive stresses. Such a condition can be developed when nonuniform plastic deformation[6] is induced in a material. For example, consider the situation illustrated in Figure 9-30. The elastic stress distribution produced by bending is shown in Figure 9-30(a). If a residual compressive stress (σ_R) is developed at the surface as

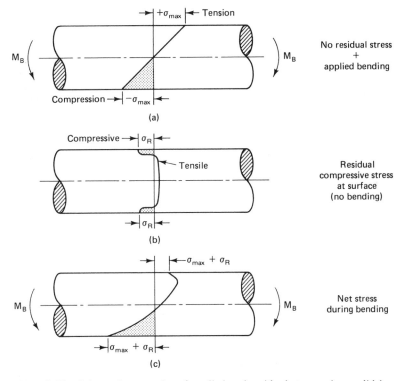

Figure 9-30 Schematic examples of applied and residual stresses in a solid bar subjected to bending. (From G. E. Dieter, *Mechanical Metallurgy*, McGraw-Hill Book Company, New York, 1976, p. 431.)

[6]A common practice for developing residual compressive stresses in the surface of a material is *shot peening*, which consists of directing a high-velocity stream of steel shot (pellets) against the surface of the workpiece.

Fatigue

shown in part (b), the final stress distribution is the algebraic sum of the applied bending stress and the residual stress. This result is shown in Figure 9-30(c). Note the change in stress distribution between parts (c) and (a); the maximum tensile stress at the surface is reduced by an amount equal to σ_R.

Although any technique that induces residual compressive stresses is generally beneficial to fatigue life, one note of caution should be mentioned. Residual compressive stresses in the surface of a part must necessarily be balanced by an equal amount of residual tensile stress elsewhere in the component in order for equilibrium conditions to exist. For example, the stress distribution below the surface of a shot-peened component is depicted in Figure 9-31. Therefore, under certain conditions, *subsurface crack initiation* may occur, if the corresponding tensile stresses are too high.

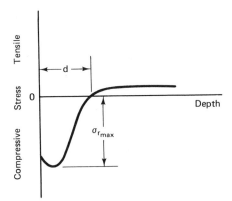

Figure 9-31 Residual stress distribution below a shot-peened surface. Residual compressive stresses (σ_r) extend to a depth (d).

In summarizing these surface aspects, it is worthwhile to reexamine the *S–N* behavior displayed in Figure 9-28. This example shows data for notched and unnotched specimens of the same steel, *plus* the influence of shot peening. Most noteworthy is that the decrease in fatigue strength due to notches was completely restored when the notched bars were shot peened. This certainly demonstrates the beneficial effect of residual compressive stresses.

Microstructural factors. Fatigue in engineering materials is heavily influenced by structural design and by the mechanical factors we have examined previously. Thus it is not surprising that major improvements in fatigue life are attainable through proper design and appropriate consideration of the mechanical factors involved in a particular fatigue situation. Increases in fatigue life through changes to the microstructure are perhaps not as consequential, but can still play an important role in the fatigue behavior of engineering materials. Therefore, it will be beneficial for the student to review the effects of such factors as *grain size*, *second-phase particles*, and *microstructural orientation*.

Grain Size. The fatigue life of many engineering materials has shown an increase with decreasing grain size. In this context, grain size denotes the average grain diameter (Figure 4-16). This effect of grain size is observed principally in high-cycle fatigue application ($N_f > 10^5$ cycles).

Small grain size appears to inhibit the initiation stages (stage I) of fatigue because the crack is growing on a crystallographic scale at this point. Since grain boundaries are, in essence, crystalline defects and small grain size implies more grain boundary region, this feature tends to impede stage I crack growth.

Second-Phase Particles. This group of microstructural constituents can be separated into *nonmetallic inclusions*, which we discussed briefly in Chapter 7 and *precipitate particles*, mentioned in Chapter 6 in the section "Precipitation Hardening." The latter second-phase particles are purposely developed in certain alloys to increase their strength. For example, very fine particles of $CuAl_2$ are precipitated in high-strength aluminum alloys (see Chapters 6 and 13) to develop their strength properties.

Second phase particles have a strong influence on fatigue crack initiation (stage I), since they act as stress concentrators, thus assisting initiation. In the case of a propagating crack (stage II), their effect can be mixed. Depending on certain circumstances these particles can either assist or inhibit crack propagation. For example, if their interface with the matrix is weak, separation between particles and matrix in the region ahead of the crack tip can actually *accelerate* crack growth. On the other hand, if the particles are strongly bonded to the matrix, fracture of the particle itself or detour of the growing crack around the particle (delamination) can result in slower crack propagation.

Microstructural Orientation. As we described in Chapter 7, the alignment of nonmetallics and chemical segregation by mechanical working operations (rolling, forging, etc.) produces *fibering* or *banding* of microstructural constituents in engineering alloys. Fatigue life is sensitive to microstructural orientation; it is generally lower in a specimen transverse to the direction of fibering. By this we mean that the *specimen* is oriented transverse to the direction of mechanical working, which implies that the *crack plane* is parallel to the working direction (e.g. specimen C-L or R-L illustrated in Figure 9-7). As the applied stress level and yield strength (YS) increase, this effect becomes more pronounced. Such anisotropic behavior has been described by a very general relationship that shows "transverse" fatigue life on the order of approximately 0.6 to 0.7 of the longitudinal life. However, we must be extremely careful in using such relationships, because specific cases may deviate significantly from this general trend.

Environmental factors. In addition to mechanical and microstructural factors, there are several other *environmental* factors that affect fatigue life. In essence, they may be considered special cases of fatigue and include *thermal effects, contact fatigue,* and *corrosive effects.*

Thermal Effects. The effect of temperature on fatigue behavior can be considered from two different viewpoints: fatigue at a constant temperature (elevated or low), and fatigue due to *thermal cycling* (also known as thermal fatigue).

1. *Constant temperature:* The effect of temperature on fatigue is often consistent with its effect on tensile strength; that is, as temperature decreases, strength

increases. Thus, as temperature decreases, the fatigue strength of unnotched specimens increases. However, most engineering materials become more notch sensitive as temperature decreases (e.g., their critical crack size decreases). Consequently, the fatigue strength of notched specimens decreases.

2. *Thermal cycling:* As we indicated in the introduction, there is a fatigue problem associated with components subject to cyclic thermal conditions. Such parts may or may not experience significant mechanical stresses in addition to thermal stresses and include the following applications: internal combustion engines, turbine components, heat treatment furnaces and fixtures, electrical and electronic systems, and nuclear reactors.

When the temperature of a material is varied significantly, *thermal* stresses that can cause fatigue may be developed, even in the absence of additional mechanical stress. The cyclic stresses (σ) associated with thermal cycling may be estimated from the following expression, in conjunction with Hooke's Law (equation 6-13):

$$\Delta l = l_0 \, \alpha \Delta T \qquad (9\text{-}18)$$

Rearranging and substituting the appropriate terms, we can express the "thermal"

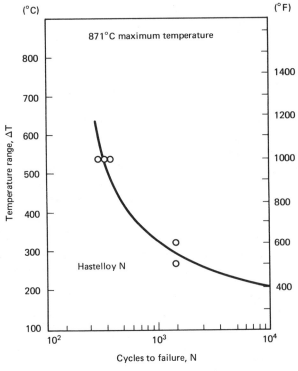

Figure 9-32 Effects of thermal fatigue in a high-temperature (nickel-based) alloy. (From A. E. Carden, *Trans. ASME 87*, Series D237, 1965.)

stress as follows:[7]

$$\sigma = \alpha E \, \Delta T \qquad (9\text{-}19)$$

where α = coefficient of thermal expansion
$\quad E$ = modulus of elasticity
$\quad \Delta T$ = thermal gradient

As the thermal gradient is increased, the resultant stresses are also increased. Furthermore, it is not necessary to subject an entire component to a large temperature gradient in order to produce thermal fatigue. If one face is held at a fixed temperature and another face is cyclically heated, the thermal gradient within the component can cause fatigue cracking. This type of cracking is often referred to as *heat checking* or incipient cracking and routinely occurs in such components as turbine buckets, furnace parts, and torch nozzles. The effects of thermal fatigue are illustrated in Figure 9-32 for a high-temperature, nickel-based alloy (Hastelloy N). Inspection of this figure shows that as the thermal gradient, and thus the thermal stresses, is increased, fatigue life decreases.

Example 9-8

Compare the stresses produced by a thermal gradient of 200°C in identical bars of (a) a high-strength aluminum alloy (2024-T6) and (b) a high-strength steel (4340) if these materials are constrained from expanding or contracting by their design.

Solution Utilizing equation (9-19) and the physical property data given in Appendix 1, the following results are obtained.

(a) 2024-T6 Al:

$$\sigma = \alpha_{Al} E_{Al} \, \Delta T$$
$$= (22 \times 10^{-6}/°\text{C})(70{,}000 \text{ MPa})(200°\text{C})$$
$$= 308 \text{ MPa } (44{,}660 \text{ psi}) \qquad \textit{Ans.}$$

(b) 4340 steel:

$$\sigma = \alpha_{steel} E_{steel} \, \Delta T$$
$$= (11 \times 10^{-6}/°\text{C})(205{,}000 \text{ MPa})(200°\text{C})$$
$$= 451 \text{ MPa } (65{,}395 \text{ psi}) \qquad \textit{Ans.}$$

Comment: Although the stress developed in the 4340 steel is certainly greater than that in the aluminum, the ratio of the stress to the YS of the two materials is:

$$\text{2024 Al:} \quad \frac{\sigma}{\sigma_{YS}} = \frac{308 \text{ MPa}}{331 \text{ MPa}} = 0.93$$

$$\text{4340 Steel:} \quad \frac{\sigma}{\sigma_{YS}} = \frac{451 \text{ MPa}}{1034 \text{ MPa}} = 0.44$$

Thus we see that the conditions above produce a stress in the aluminum alloy on the order of 93% of its yield strength. But in the steel the stress is only about 44% of its yield strength. This is a very important factor in determining fatigue behavior.

---

[7]This expression may be used for relatively simple shapes (e.g., round bars) but would necessitate a more rigorous analysis for geometrically complex shapes.

Fatigue

Contact Fatigue. Failure due to *contact* fatigue occurs when surfaces contact repeatedly. The stresses developed are more complex than those in standard bending or axial fatigue. In many cases the contact area tends to be small and the resultant stresses are quite large. Contact fatigue generally produces three types of damage: *surface pitting*, *subsurface pitting*, and *spalling*.

Corrosion Fatigue. Corrosion fatigue may be considered a special case of general fatigue, with certain modifying effects resulting from the environment. But the combination of a *corrosive environment* and *cyclic stressing* can be much more detrimental than either factor acting separately. The time or cycles necessary for fatigue crack initiation can be markedly reduced by corrosion reactions that create *pits* (as you will see in Chapter 17) or other surface damage. In addition, the fatigue crack propagation rate can be significantly increased by a corrosive environment. The relative fatigue life behavior is typically illustrated by the relationship shown in Figure 9-33. The physical and mechanical aspects of corrosion fatigue are discussed in more detail in Chapter 17.

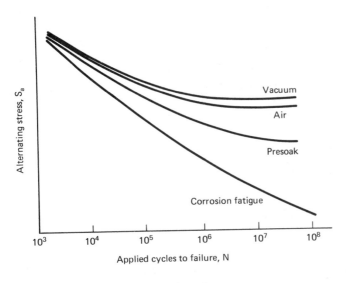

Figure 9-33 Comparison of fatigue life behavior of typical engineering alloys under various environmental conditions. Presoaked condition implies soaking specimens in corrosive liquid prior to testing in air. (From H. O. Fuchs and R. I. Stephens, *Metal Fatigue in Engineering*, John Wiley & Sons, Inc., New York, 1980, p. 220.)

Fatigue Crack Initiation

The initiation of a fatigue crack in an engineering material may take place at microstructural imperfections such as second-phase particles, shrinkage cavities, and gas pores. These flaws were described in Chapter 7 in the section "Solidification Defects." In this sense, the existing flaws serve to assist crack initiation by acting as stress concentrators or as preexisting cracks. Furthermore, when a crystalline engineering material is subjected to cyclic loading, very localized regions of *slip* may develop in its surface. Recall from Chapters 4 and 6, that slip occurs on an atomic scale and consists of the movement or displacement of atoms on adjacent (parallel) planes in the crystal.

Slip mechanism. The slip mechanism of fatigue crack initiation is explained on the basis of a series of *intrusions* and *extrusions* which develop at the surface during stress cycling. This process is shown schematically in Figure 9-34. When slip takes

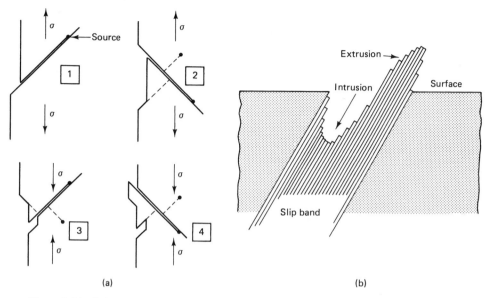

Figure 9-34 Schematic example of mechanisms leading to a fatigue crack initiation: (a) extrusion and intrusion in slip region due to cyclic stress; (b) slip band intrusions and extrusions prior to crack initiation (note the stress concentrator formed in the surface by this process). (From H. O. Fuchs and R. I. Stephens, *Metal Fatigue in Engineering*, John Wiley & Sons, Inc., New York, 1980, p. 28.)

place, a step forms at the free surface, due to the relative displacement of atoms along the slip plane. During an alternate or reverse stress cycle the slip that occurs could be the exact negative of the original displacement, overriding any deformation effects and restoring the crystal lattice to its original condition. But such recovery usually does not happen and some residual deformation results as shown in Figure 9-34(a).

Eventually, this deformation, accentuated by cyclic stressing, accumulates until a discernible crack forms (stage I), as illustrated schematically in Figure 9-34(b). Unfortunately, crack growth rates in this stage are on the order of angstroms/cycle, making it an extremely difficult process to observe and study.

Fatigue Crack Propagation

The subject of fatigue crack propagation (stage II crack growth) has received considerable interest in recent years. There are two basic reasons for this interest; (1) most commercial engineering materials and structures contain *inherent flaws*, thus allowing initiation (stage I) to occur relatively early in the fatigue life; and (2) the rate at which fatigue cracks grow is a function of certain *material properties, crack length*, and *applied stress level*. These factors can be expressed in fracture mechanics terms, as we will demonstrate shortly.

The advantage of combining fatigue data with fracture mechanics is, of course, that reliable predictions of fatigue life and thus, safe life, can be made for many materials and applications. However, before proceeding to these quantitative relationships, let us examine the micromechanisms by which fatigue cracks propagate in stage II.

Crack propagation mechanism. The *micro* mechanisms of local fracture in stage II crack growth are somewhat easier to study than stage I because the crack propagation rates can reach values on the order of 25 μm (10^{-3} in.) per cycle. This facilitates observation of the cracks during laboratory testing and also allows their fracture features (fracture surfaces) to be intimately examined by electron microscopy after the test.

Recall that we have frequently referred to fatigue as a *progressive* failure mechanism. In stage II, the fatigue crack propagates (advances) incrementally with each stress cycle. The morphology of the cracking has been explained on the basis of a *plastic blunting process*. By "plastic blunting process" we mean that the advancing tip of the fatigue crack becomes blunted during the tensile portion of the stress cycle, followed by resharpening of the tip during the compression or relaxation portion. This type of process is illustrated schematically in Figure 9-35. In this model the fatigue crack goes through a stress cycle and advances one "increment." Notice the mating notches or grooves that are produced on the surfaces of the fatigue crack. These features, called fatigue *striations*, are tell-tale markings on a fatigue fracture surface which signify the incremental advancement of the crack during a loading cycle. Striations on an actual fatigue fracture surface are shown in Figure 9-36. In the following section we describe fatigue striations in more detail.

Crack growth rate. As we have indicated previously, in stage II the fatigue crack may grow from barely discernible size to critical length, whereupon rapid fracture of the material ensues. In the laboratory, stage II crack growth can be measured by a number of techniques, including simply monitoring crack progression on the side of a specimen with a calibrated traveling telescope, and nondestructive test (NDT) techniques such as ultrasonic inspection.[8] Regardless of what method is actually used, the progression of the fatigue crack (crack length) is recorded together with the corresponding number of stress cycles. These data typically produce a curve, as shown in Figure 9-37. Examining this figure, we see two curves of crack length (**a**) versus the number of stress cycles (N). Both tests start at the same initial crack depth or notch (a_0), but because stress level has a major effect on fatigue crack propagation, curve 2 (σ_2) is growing considerably faster than curve 1 (σ_1). In other words, for the same number of cycles (N) the crack is longer in case 2. Furthermore, this relationship

---

[8] Ultrasonic inspection basically consists of directing ultrasound waves through an opaque material and observing "reflections" from cracks or other flaws.

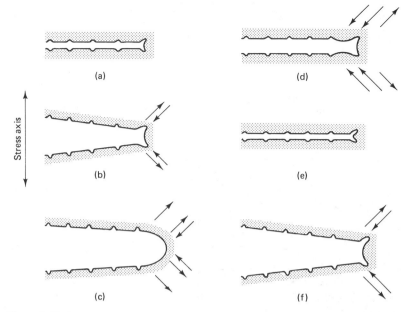

Figure 9-35 Schematic of plastic blunting process during fatigue crack propagation (stage II): (a) zero load; (b) small tensile load; (c) maximum tensile load; (d) small compressive load; (e) maximum compressive load; (f) small tensile load. The double arrowheads in parts (c) and (d) signify the greater width of slip bands at the crack in these stages of the process. (From C. Laird, "Fatigue Crack Propagation," *ASTM STP 415*, 1967, p. 136.) (Copyright, ASTM, 1916 Race Street, Philadelphia, Pa., 19103. Reprinted with permission.)

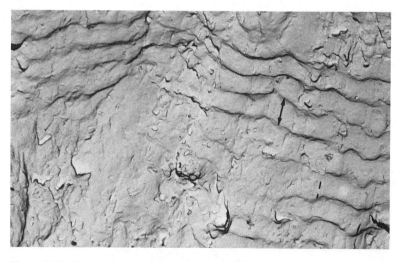

Figure 9-36 Transmission electron micrograph of fatigue fracture showing striations (4300×). Arrow denotes an increment of crack growth (Δa) during one stress cycle (ΔN).

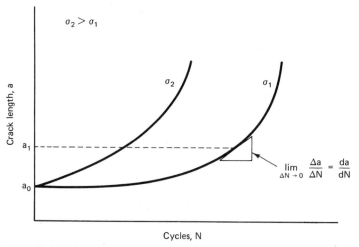

Figure 9-37 Plot of hypothetical fatigue data (**a** versus N) for two stress levels. Slope of the **a** versus N curve is the rate of crack growth (da/dn).

is obviously not linear. As crack length increases, the number of cycles required to propagate the crack a specific distance decreases. Indeed, the crack growth *rate* is increasing. What do we mean by "crack growth rate"? This is the *slope* (da/dN) of the **a** versus N curve at any point (e.g., a_1, as indicated in Figure 9-37). In other words, as the crack length **a** increases, the change in **a** gets larger with each successive stress cycle (N).

Therefore, two very important factors become evident from this analysis; (1) the applied stress level has a pronounced effect on fatigue crack propagation (stage II growth); that is, increasing the stress level shortens life in stage II; and (2) crack growth rate da/dN increases with increasing crack length (**a**), for nearly all specimen geometries.

Stress intensity relationship. In fracture mechanics terms, the crack growth rate (da/dN) during a major portion of the propagation stage has been successfully related to the applied stress level (σ), the crack length (**a**), and certain properties of a particular material as follows:

$$\frac{d\mathbf{a}}{dN} = C(\Delta K)^m \tag{9-20}$$

where ΔK = stress intensity factor range ($K_{max} - K_{min}$)[9]
 C, m = "constants" depending on material variables, environment, temperature, frequency, etc.

This analytical expression relating crack growth rate to the stress intensity factor is especially useful in evaluating the fatigue behavior of engineering material under laboratory conditions. For instance, a specimen of known K calibration (see fracture

[9]See equation (9-8).

toughness specimens) is prepared from the material under consideration, and tested under the appropriate fatigue conditions (stress levels, orientation, frequency, etc.). The crack growth rate and stress intensity data from such a test may be plotted as illustrated hypothetically in Figure 9-38. This curve depicts a somewhat sigmoidal relationship with three distinct regions (I, II, III). In region I, crack growth rates are very low. Conversely, region III shows the crack accelerating and approaching the rapid fracture stage. In region II, however, the log of crack growth rate may vary linearly with respect to the log of stress intensity parameter (K). Such a linear relationship corroborates the power-law dependence of equation (9-20) and is expressed in the

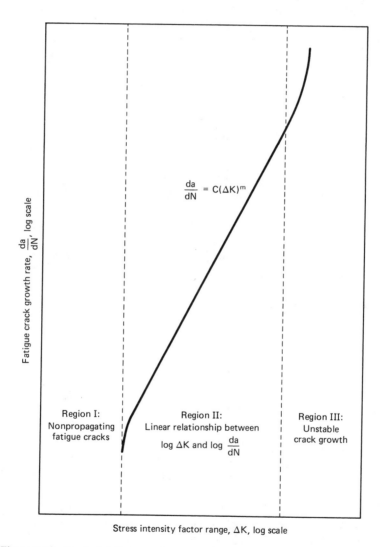

Figure 9-38 Typical fatigue crack growth behavior for engineering materials depicting the three primary regions of fatigue. (From W. G. Clark, Jr., *Met. Engr. Quart. 14*, August 1974, p. 17.)

Fatigue

following manner:

$$\log \frac{d\mathbf{a}}{dN} = \log [C(\Delta K)^m]$$

which can be simplified to

$$\log \frac{d\mathbf{a}}{dN} = m \log \Delta K + \log C \qquad (9\text{-}21)$$

As the student will no doubt recognize, the latter expression is the equation for a straight-line relationship between $\log (d\mathbf{a}/dN)$ and $\log \Delta K$, with slope m and intercept $\log C$. Fatigue crack growth data are frequently displayed in this manner, excluding the upper (III) and lower (I) portions, which are nonlinear. Generally, the linear portion representing stage II growth is the most important part of the curve, since this is the portion of fatigue where the crack is growing steadily, in other words, the "stable" portion of the fatigue process. One can see by inspection of Figure 9-38, that as the transition from stage II (linear portion) occurs, crack growth rate accelerates and failure is imminent ($\Delta K \longrightarrow K_{\mathrm{Ic}}$).

Fatigue crack propagation rates as a function of stress intensity are shown in Figure 9-39 for several engineering materials with different crystal structures. Not only do these data verify the exponential relationship given by equation (9-20) but they also demonstrate that the exponent (m) is approximately equal to 4 for many materials.

Safe-Life Prediction. Based on the relationship between crack growth rate and stress intensity (Eq. 9-20), and plane strain fracture toughness (K_{Ic}), we can develop an analytical expression for the fatigue life or safe life of a component. Laboratory data for the material under consideration are combined with quality control and service information as illustrated in Figure 9-40.

The calculation of fatigue life (N_f) involves integrating equation (9-20) between the initial flaw size ($\mathbf{a_0}$), which may be measured by nondestructive examination (NDE), and the critical flaw size ($\mathbf{a_c}$) as determined by plane strain fracture toughness testing. This integration is performed as follows:

$$N_f = \int_0^{N_f} dN = \int_{a_0}^{a_c} \frac{d\mathbf{a}}{C(\Delta K)^m} \qquad (9\text{-}22)$$

$$N = \frac{1}{C} \int_{a_0}^{a_c} \frac{d\mathbf{a}}{(\Delta K)^m} \qquad (9\text{-}23)$$

where $K = \sigma W^{1/2} f(\mathbf{a}/W)$ from equation (9-10)

$C = $ constant from equation (9-20) in the appropriate English or SI units

A glance back at the polynomial expressions in equation (9-11) will convince most students that the integral above is not a simple function with respect to the stress intensity factor, which contains **a**. Such integrals may be solved using numerical analysis or computer methods. However, in a number of instances the polynomial function $f(\mathbf{a}/W)$ does not vary significantly for small values of $\mathbf{a}/W$ and therefore this factor can be treated as a constant. The stress intensity factor for a shallow edge crack in many types of specimens is expressed as follows:

$$K_{\mathrm{I}} = 1.12\sigma(\pi \mathbf{a})^{1/2} f(\mathbf{a}/W) \qquad (9\text{-}24)$$

If the pertinent portion of the fatigue life is spent in this region (region II, Figure

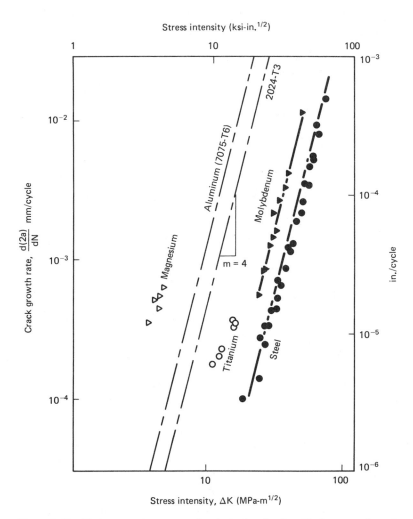

Figure 9-39 Fatigue crack propagation behavior (region II) for several engineering materials on a log-log scale. Data fit the power relationship given in Eq. 9-20. (From P. C. Paris, "Fatigue—An Interdisciplinary Approach," *Proc. 10th Sagamore Conf.*, Syracuse University Press, Syracuse, N.Y., 1964, p. 107.)

9-38), then the number of cycles to failure can be estimated as follows:

$$N_f = \frac{2}{(m-2)C(1.12)^m(\sigma)^m Y^m(\pi)^{m/2}} \left[\frac{1}{a_0^{(m-2)/2}} - \frac{1}{a_c^{(m-2)/2}} \right] \quad \text{for } m \neq 2$$

where σ = applied stress

  for bend specimens: $\sigma = 6M/BW^2$, where $M = PS/4$
  for compact tension specimens: $\sigma = P/BW$

m, C = constants established by fatigue crack growth testing
a_0 = initial flaw size: a_c = critical flaw size
$Y = f(a/W)$

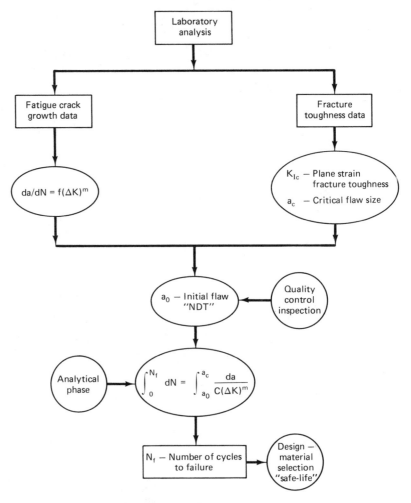

Figure 9-40 Flow diagram depicting analytical relationship between fatigue and fracture toughness in the design of safe-life components and structures.

However, we must emphasize that to *precisely* calculate the number of cycles to failure, the crack growth rate expression [equation (9-20)] must be known over the entire range of data, from a_0 to a_c.

Example 9-9

A high-strength aluminum alloy bar is subjected to cyclic bending as illustrated in Figure 9-41. The maximum moment (M) is 200 N-m. During final quality control inspection, surface flaws as deep as 0.64 mm (0.025 in.) were detected. If this material displays a relationship between fatigue crack growth rate and stress intensity as described by equation (9-20) and $K = \sigma a^{1/2} f(a/W)$, determine the number of cycles (N_f) required for the initial flaw to grow to critical dimensions, when the critical crack size is 2.54 mm (0.100 in.)

The following information regarding this material has been developed by

---

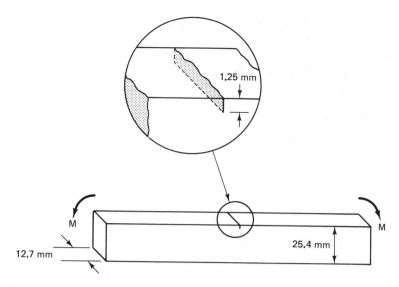

Figure 9-41 Beam type component subjected to cyclic bending. Surface flaw along top edge is shown inset.

laboratory testing:

> crack growth rate exponent $(m) = 3$
>
> C [in equation (9-20)] $= 2 \times 10^{-10}$ (in appropriate SI units)
>
> $f(a/W) \simeq 1$ for the range of a/W involved

Solution   From equation (9-24),

$$N_f = \frac{1}{C} \int_{a_0}^{a_c} \frac{da}{(\Delta K)^3}$$

where $K^3 = \sigma^3 a^{3/2} f(a/W)^3$.

$$N_f = \frac{1}{C \sigma^3 f(a/W)^3} \int_{a_0}^{a_c} \frac{da}{a^{3/2}}$$

Evaluating the constants of integration, we have

$$\sigma^3 = \left(\frac{6M}{BW^2}\right)^3 = \left[\frac{(6)(200 \text{ N-m})}{(0.0127 \text{ m})(0.0254 \text{ m})^2}\right]^3 = \left(\frac{1200 \text{ N}}{8.2 \times 10^{-6} \text{ m}^2}\right)^3$$

$$= (1.5 \times 10^8 \text{ N/m}^2)^3 = (150 \text{ MPa})^3$$

$$= 3.4 \times 10^6 \text{ MPa}^3$$

$$N_f = \frac{1}{(2 \times 10^{-10})(3.4 \times 10^6)(1)^3} \left[\frac{-2}{a^{1.2}}\right]_{a_0}^{a_c}$$

$$= \frac{-2}{6.8 \times 10^{-4}} \left(\frac{1}{\sqrt{a_c}} - \frac{1}{\sqrt{a_0}}\right)$$

$$= -2941 \left(\frac{1}{\sqrt{0.00254 \text{ m}}} - \frac{1}{\sqrt{0.00064 \text{ m}}}\right)$$

$$= -2941 \,(19.8 - 39.5)$$

$$= -2941(-19.7)$$

$$= 57{,}938 \text{ cycles} \qquad Ans.$$

> *Comment:* This analysis presumes that the initial flaw (a_0) accounts for the initiation of the fatigue crack and that crack growth commences with the first loading cycle. Such a presumption is, of course, a conservative approach and N_f may be considered an estimate of the safe "life" of this bar.

Factors affecting fatigue crack propagation. In addition to the major influence exerted by stress level and crack length, the fatigue crack propagation rate (da/dN) can be affected by material variables which influence C and m in equation (9-20). For instance, increases in the following properties or parameters tend to *decrease* the value of C: elastic modulus, yield strength, K_{Ic}. The exponent m (slope of log da/dN versus log Δk curve) is correspondingly *affected* by *orientation*, *material toughness*, and *yield strength*.

The effects of microstructural orientation, that is, the orientation of the crack plane with respect to the direction of mechanical working, are shown in Figure 9-42 for rolled steel plate. For a given value of stress intensity (Δk), the crack growth rate is greater for a direction parallel to the rolling direction (TR) than for directions normal to it. Although the difference pictured here may not appear to be significant, the student is reminded that these slopes (m) represent *exponential* values in equation (9-20), and can have a profound influence on crack growth rate.

As far as material properties are concerned, the following generalizations can be made; (1) as toughness (K_{Ic} or Charpy impact) increases, m decreases, indicating that material which is tough resists fatigue crack propagation; and (2) as yield strength increases (material is harder), m usually increases; fatigue cracks tend to propagate at a faster rate.

Appearances of Fatigue

Fatigue failures characteristically exhibit fracture features which identify such aspects as the cyclic nature of their progression, the direction of crack growth, their initiation sites, and other important information that may be related to the mechanical, microstructural, and environmental conditions involved in the failure process. In the case of analyzing a service failure, this type of information or evidence may be instrumental in determining the cause or causes of the failure. Furthermore, accurate failure analysis information may be directed back to the design and production stages of manufacturing, thereby improving the quality and reliability of a component (see Figure 1-7).

Macroscopic. Fatigue failures generally display very little permanent (plastic) deformation on a *macro* scale. However, as the toughness of a material increases, the tolerable crack size increases, as does the amount of *shear lip* associated with the final fracture. Macroscopically, fatigue failures often exhibit *beach marks*, which represent the cyclic progression of the fatigue crack. Such fracture features are shown in Figure 9-43.

The location and shape of the beach marks vary with the geometry of the component and the loading conditions. As we indicated in the section on crack initiation,

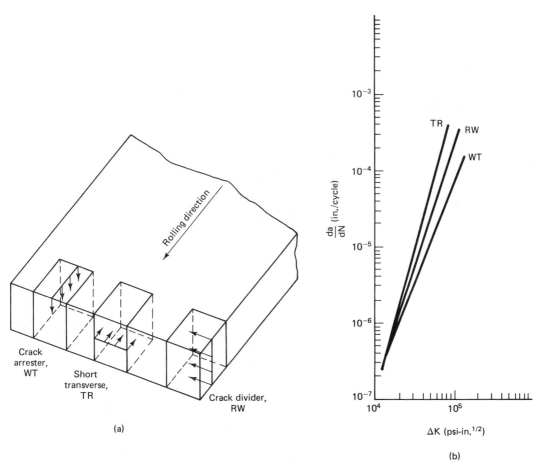

Figure 9-42 Effect of test specimen orientation on crack growth rate: (a) orientation of crack plane (test specimen) with respect to rolling direction; (b) corresponding fatigue crack growth rates. (Courtesy of F. A. Heiser.)

fatigue failures usually start at stress concentrators, such as notches, fillets, keyways, and sharp radii.

Since fatigue cracks propagate normal to the principal tensile axis, this characteristic can often be used to determine probable crack origins and to identify areas of maximum stress in a component.

Fatigue failures can often be positively identified by macroscopically examining the fracture surfaces. This is sometimes accomplished by a simple visual examination; otherwise, a magnifying device such as a magnifying glass or a stereomicroscope is utilized. The "stereo" is capable of magnifications typically from about 2 to 50X. The location and typical fatigue fracture appearance for smooth and notched components under two types of loading conditions are illustrated in Figure 9-44. However, beach marks and other evidence of fatigue are not always resolved at these magnifications, and it may be necessary to use *electron microscopy*.

Fatigue

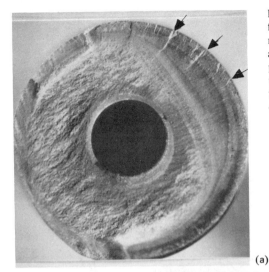

Figure 9-43 Macroscopic appearance of fatigue. (a) Fatigue fracture appearance in a rock drill bit. Arrows denote multiple origins and also indicate the direction of crack progression. "Beach marks" are located in the upper right hand quadrant. (b) Fatigue fracture in polystyrene component. Arrow denotes origin.

(a)

(b)

Microscopic. In the microscopic examination of fatigue fractures, the light microscope is of marginal value because of its limited depth of field. The intimate examination of fracture surfaces requires instrumentation which is capable of resolving fracture features that are characteristically very fine, yet detailed when viewed on a micro scale. The electron microscope[10] has proved to be an extremely valuable tool for resolving the microfracture features of most engineering materials. Although the details of electron microscopy are beyond the scope of this text, it will be sufficient at this point simply to point out that the fracture features that identify fatigue were promoted by the use of the electron microscope.

In describing the mechanism of fatigue crack propagation we alluded to *striations* as the tell-tale markings of fatigue. A striation represents the advancement of the fatigue crack during one stress cycle. These fracture features are often observed

---

[10]Basically, the electron microscope is an instrument that employs a beam of electrons to produce an image of a fracture surface, in contrast to the optical microscope, which uses visible light radiation.

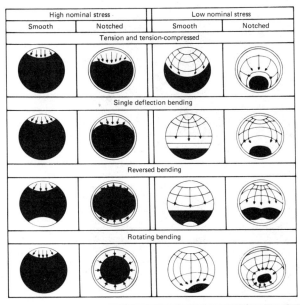

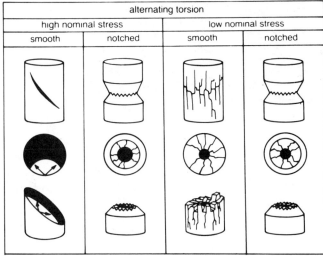

Figure 9-44 Schematic representation of fatigue fracture surface appearance under various conditions of stress and stress concentration.

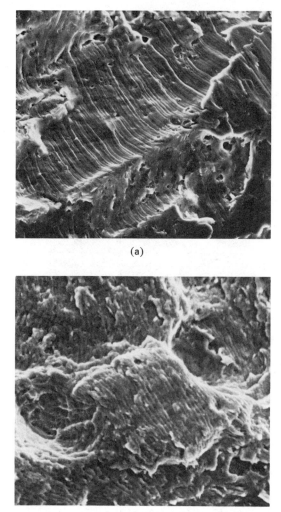

(a)

(b)

Figure 9-45 Examples of fatigue striations: (a) high-strength aluminum alloy (440×); (b) high-strength steel (1100×).

by means of the electron microscope, as illustrated in Figure 9-45. The size and spacing of fatigue striations depends on the type of material and its properties, the stress level and stress state (plane stress or plane strain), and environmental interactions such as a corrosive atmosphere.

STUDY PROBLEMS

9.1. The surface energy for MgO is 1000 ergs/cm². Based on the data in Table 15-8, what is the stress that will cause fracture for a 0.5-mm-long flaw? Express your answer in both English and SI units.

9.2. An alumina (Al_2O_3) component fractures in a brittle manner. Examination of the fracture surface reveals a preexisting crack approximately 1.2 mm long. If the surface energy of this material is 900 dyn/cm, determine the stress that produced this failure (see Table 15-8 for $E_{Al_2O_3}$).
Answer: 18.7 MPa or 2,712 psi

9.3. Titanium carbide (TiC) will be used in an application where the design stress is 2500 psi. If the surface energy for this material is 4.7×10^{-5} in.-lb/in.^{2} and $E = 45$ million psi, what is the maximum-size flaw that can be tolerated without causing failure?

9.4. What is the critical energy release rate?

9.5. Determine the normal stresses σ_x and σ_y in the vicinity of a crack tip, as shown in Figure 9-4, for an applied stress (σ) of 138 MPa (20,000 psi) under the following conditions:

Condition	θ (deg)	r (mm)	a (mm)
1	30	0.1	5
2	60	0.1	5
3	30	1.0	5
4	60	1.0	5

9.6. If a structural material containing a flaw 0.250 in. long is subjected to an applied stress (σ) of 40,000 psi as illustrated in Figure 9-4, calculate the normal stress σ_y produced on the crack plane ($\theta = 0$) at $r = 0.050$ in., $r = 1$ in., and $r = 5$ in. What can you conclude from your answers?

9.7. The critical stress intensity (K_c) for a certain material is 40 ksi-$\sqrt{\text{in.}}$ If a structure made from this material contains a crack 0.10 in. long, as illustrated in Figure 9-2, what is the applied stress (σ) that will result in fracture? Assume that $f(a/W)$ is 1 in this case.
Answer: 71,000 psi or 493 MPa

9.8. Determine the flaw size that will produce fracture in a material, as shown in Figure 9-2, under the following conditions: $K_c = 100$ MPa$\sqrt{\text{m}}$, $\sigma = 690$ MPa, and $f(a/W) = 3.6$.
Answer: 0.51 mm

9.9. Your quality control department has the NDT capability to detect and identify internal flaws 1 mm in length or greater in a certain component. In service, this part will be subjected to a tensile stress of 55,000 psi (380 MPa). Determine whether or not you can safely use either of the following alloys for the component: aluminum alloy 7178-T651 (YS = 83 ksi), and titanium alloy Ti–6 Al–4 V (YS = 132 ksi). Assume that $f(a/W)$ for this application is equal to 2.

9.10. An elevated-temperature engine application involves an applied tensile stress of 35,000 psi. The use of alumina (Al_2O_3) in this situation would seem to be ideal from the standpoint of high-temperature strength and stability. However, we can only detect internal flaws in this part larger than 0.5 mm, by conventional nondestructive examination techniques. If brittle fracture is a concern and $f(a/W)$ is 2.1 for this application, is it safe to use Al_2O_3?

9.11. Titanium (Ti–6 Al–4 V) at the 132,000-psi (910-MPa) yield strength level is being considered for a new prothesis device to be surgically implanted. As you will learn in

Chapter 13, titanium is attractive for such applications because of its strength, low density, and chemical stability at ambient temperatures. If this particular prothesis will probably encounter tensile stress on the order of 75,000 psi (518 MPa), what size internal flaws may be tolerated without causing fracture? Assume that the applied stresses or loads are not cyclic and that $f(a/W) = 1.5$.

9.12. If internal flaws and cracks as small as 1 mm can be reliably detected in the prothesis discussed in problem 9.11, can we safely substitute a less expensive material, such as 7178-T651 aluminum alloy (treated to 83 ksi yield strength)?

9.13. Polymethyl methacrylate (PMMA) is used to manufacture a lightweight component which occasionally experiences minor tensile loading in service. If this part contains "edge" flaws on the order of 0.5 mm deep, estimate the stress that must not be exceeded if we wish to avoid a brittle fracture. [Assume that $f(a/W) = 2.0$.]
Answer: 10.3 MPa or 1,496 psi

9.14. (a) Determine the value of $f(a/W)$ for an edge-cracked specimen [as illustrated in Figure 9-8(b)] if the bar is 50.8 mm wide and $a = 25.4$ mm.
(b) What is the stress intensity factor (K_I) for this specimen at a load of 44.5 kN if the bar is 25.4 mm thick?

9.15. What is the stress intensity factor K_I for a bar 2 cm wide by 1 cm thick containing a notch 0.5 cm deep if a bending moment is produced by a load of $P = 2225$ N over a span of $S = 16$ cm? [See Figure 9-8(a).]

9.16. A bend specimen is made from an experimental ceramic material and tested in three-point bending. The bar has the following dimensions:

$$B = 1 \text{ in.}$$
$$W = 1.5 \text{ in.}$$
$$S = 6 \text{ in.}$$
$$a = 0.5 \text{ in.}$$

If the load–displacement behavior is similar to the "brittle" material in Figure 9-12 and $P_5 = 540$ lb, what is the fracture toughness (K_{Ic}) of this material?
Answer: 2.82 ksi $\sqrt{\text{in.}}$

9.17. (a) A compact tension specimen is made from 2024 aluminum alloy, according to Figure 9-6(a). The dimensions of the specimen are as follows:

$$W = 2 \text{ in.}$$
$$B = 1 \text{ in.}$$
$$a = 0.75 \text{ in.}$$

Utilizing the data in Figure 9-8(b), determine the plane strain fracture toughness (K_{IQ}) for this material if $P_5 = 5550$ lb.
(b) If the yield strength of the alloy in part (a) is 60 ksi, is the value of K_{IQ} that you obtained a valid K_{Ic}?

9.18. A fracture toughness test has been performed on a center-cracked specimen with the following dimensions: $W = 75$ mm, $B = 12.5$ mm, $a = 20$ mm. Plot the load–displacement curve for the data listed below, and determine K_{IQ} for this material (see Figure 9-13). For center-cracked specimens, $K = \sigma\sqrt{a}\, f(a/W)$, and for this situation $f(a/W) = 2.15$.

Load (kN)	Displacement (mm)
0	0
15	0.25
30	0.50
47	0.75
62	1.00
79	1.25
96	1.50
106	1.75
118	2.00
122	2.25
119	2.50
107	2.75
90	3.00

9.19. Test the specimen in problem 9.18 for a valid K_{Ic} determination if the yield strength of the material is 725 MPa.

9.20. In wrought products such as rolled plate or forged bar, fracture toughness, like other structure-sensitive properties, depends on orientation. Identify the specimens in Figure 9-7 for both plate and round bar which typically should exhibit the highest toughness and the lowest toughness.

9.21. Why is fatigue cracking a potentially serious problem in engineering materials?

9.22. (a) Explain the difference between reversed stress cycles and repeated stress cycles.
(b) Which case results in the greater mean stress?

9.23. Name and briefly discuss a current or recent example of fatigue failure which has received attention in the news media or engineering literature. Does your example point out the serious nature of fatigue failures?

9.24. Determine the approximate endurance limit for the alloys shown in Figure 9-24.

9.25. Rotating-bending fatigue tests are conducted on an alloy steel with the following results. Plot the S–N curve for this material on four-cycle semilog paper using the average number of cycles to failure for each stress level.

Applied stress (ksi)	Cycles to failure, N
140	556
	800
	695
120	1503
	1340
	1476
100	2650
	2900
	3142
80	5×10^5
	1×10^6
	7.5×10^5
60	10^7
	10^7
	10^7

(a) Determine the endurance limit for this material.

(b) Determine the average number of cycles to failure at an applied stress of 130 ksi in this material.

(c) Is it safe to use this alloy for 1 million cycles at a stress level of 70 ksi?

9.26. A fatigue life (S–N) test is conducted on a high-strength aluminum alloy. The results of this test are given below. Plot the mean fatigue life curve on five-cycle semilog paper and establish the approximate −2s boundary for these data.

(a) If a certain application for this material must withstand 100,000 cycles, what is the maximum stress that can be applied without producing failure?

(b) At a stress level of 300 MPa, what is the safe life, that is, number of cycles below which failure does not occur?

Applied stress (MPa)	Cycles to failure, N
345	8,142
	6,500
	7,850
	6,935
276	46,200
	51,200
	41,950
	38,356
207	294,300
	250,500
	275,450
	229,630
138	1×10^6
	1.01×10^6
	0.99×10^6
	1.008×10^6
69	5×10^7
	4.9×10^7
	4.8×10^7
	5.2×10^7

9.27. Determine the stress ratio (R) for the following fatigue test conditions.

(a) $\sigma_{max} = 65,000$ psi, $\sigma_{min} = 20,000$ psi

(b) $\sigma_{max} = 40,000$ psi, $\sigma_{min} = 1000$ psi

(c) $\sigma_{max} = 200$ MPa, $\sigma_{min} = 34.5$ MPa

(d) $\sigma_{max} = 172$ MPa, $\sigma_{min} = 172$ MPa

9.28. Notched bar fatigue life tests are conducted on a 0.47% carbon steel and the fatigue limit for these tests is found to be 21,000 psi. The unnotched S–N behavior for this material is shown in Figure 9-24. If the stress concentration factor for the notched bar is 1.5, what is the notch sensitivity factor (q) for this material?
Answer: 2.29

9.29. A heat exchange system fabricated from a titanium alloy is subjected to cyclic changes in temperature from approximately 27 to 350°C. Calculate the stresses produced by this thermal gradient if the material is constrained from expanding or contracting. Assume that the modulus of elasticity and coefficient of thermal expansion are approximately equivalent to that of pure titanium.

9.30. Certain high-temperature jet engine parts are made from Hastelloy N (see Figure 9-32). Suppose that these parts experienced maximum temperatures of 871°C. What would their fatigue life be if they were subjected to cyclic heating and cooling with a temperature gradient of 600°C?

9.31. Crack growth tests have been conducted on a high-strength aluminum alloy and a high-strength steel. Plot the test data and compare the fatigue behavior of these two materials.

	N (10^3 cycles)	
a (in.)	Steel	Aluminum
0.520	98.0	52.5
0.560	104.1	56.0
0.600	109.7	59.1
0.640	115.0	61.7
0.720	123.8	65.6
0.800	130.3	68.9
0.900	135.9	71.5
1.000	139.0	73.2
1.100	140.1	73.9
1.200	140.6	74.4
1.300	140.9	74.6

9.32. The following fatigue data have been obtained from a uniaxial tension–zero–tension test performed on specimens of the same alloy steel. Specimen 1 was tested at $P = 2000$ lb, while specimen 2 was tested at $P = 3000$ lb. Plot the **a** versus N data and comment on your results.

	N (10^3 cycles)	
Crack length, **a** (in.)	Specimen 1	Specimen 2
0.520	73.3	11.0
0.560	79.3	32.6
0.600	86.2	50.7
0.640	91.5	66.5
0.700	98.6	81.7
0.760	104.4	93.0
0.800	107.7	98.7
0.900	113.3	109.2
1.000	116.8	115.6
1.100	118.3	118.9
1.200	118.9	120.2
1.300	—	120.6

9.33. The region II results of a fatigue crack growth test are given below. Plot the crack growth rate versus stress intensity parameter (ΔK) on log paper (3 × 2 cycles). What is the crack growth rate exponent for these data?

da/dN (in./cycle)	ΔK [a] (psi-in.)
6.8×10^{-6}	8,400
7.4×10^{-6}	8,991
1.0×10^{-5}	10,123
1.5×10^{-5}	11,000
2.0×10^{-5}	12,128
3.8×10^{-5}	13,452
9.1×10^{-5}	14,568
2.0×10^{-4}	16,278
3.3×10^{-4}	18,326

[a] $\Delta K = K_{max} - K_{min}$. ΔK typically taken as K_{max} where $K_{min} \longrightarrow 0$.

9.34. Crack growth rate studies are conducted on two polymers, polymethyl methacrylate (PMMA) and nylon 66, with the following results. Plot da/dN versus ΔK for these data on log paper and determine the crack growth rate exponent for each material. Which material displays better fatigue crack propagation behavior?

da/dN (in./cycle)	ΔK (psi-in.$^{1/2}$)
PMMA	
3×10^{-6}	500
6×10^{-6}	600
1×10^{-5}	700
5×10^{-5}	900
1×10^{-4}	1000
2×10^{-4}	1100
3×10^{-4}	1200
8×10^{-4}	1300
Nylon 66	
5×10^{-6}	2200
8×10^{-6}	2500
1×10^{-5}	3000
2×10^{-5}	3500
5×10^{-5}	4000
9×10^{-5}	4800
2×10^{-4}	5000

9.35. A constant-amplitude fatigue crack growth test is conducted on "Thoralloy" with a modified compact tension specimen. $P_{max} = 13.4$ kN and $P_{min} \simeq 0$. This specimen [see Figure 9-6(a)] has dimensions $W = 50.8$ mm, $B = 25.4$ mm. Assuming that the following data obey the relationship expressed by equation (9-20), plot da/dN versus ΔK, and determine the values of m and C. The stress intensity parameter for this specimen is given in Figure 9-8(b).

Crack length, **a** (mm)	N (10^3 cycles)
20.0	70.0
21.0	75.6
22.0	80.8
23.0	85.6
24.0	89.8
25.5	93.7
27.0	97.3
29.0	100.1
31.0	101.3
34.1	102.1
37.9	102.5

9.36. Perform a linear regression analysis on log da/dN and log K for the following crack growth data. How well do these data fit a straight-line relationship? Write an equation for crack growth rate as a function of stress intensity for this material.

da/dN (in./cycle)	K (psi-in.$^{1/2}$)
2.4×10^{-6}	14,000
5.7×10^{-6}	18,000
8.0×10^{-6}	22,000
1.5×10^{-5}	28,000
2.0×10^{-5}	31,000
3.2×10^{-5}	35,000
5.5×10^{-5}	41,000
9.0×10^{-5}	52,000
1.0×10^{-4}	57,000

9.37. Utilizing the crack growth data that you developed in problem 9.35, perform a linear regression on the log da/dN and corresponding log K values. What is the equation of the "best-fit" straight line which represents these data? How does the slope (m) compare with the value you obtained in problem 9.35?

9.38. A bridge structural member experiences a cyclically applied tensile load. This constant-amplitude loading produces a $\sigma_{max} = 55,000$ psi and $\sigma_{min} = 0$. The stress intensity parameter for this situation may be approximated by $K = \sigma \sqrt{a} \, f(a/W)$. Estimate the number of loading cycles this member can withstand before failing under the following conditions:

$$\mathbf{a}_0 = 0.0625 \text{ in.}$$
$$\mathbf{a}_c = 1.00 \text{ in.}$$
$$m = 3.5$$
$$C = 10^{-14}/\text{cycle (ksi)}^{3.5} \text{ (in.)}^{0.75}$$
$$f(\mathbf{a}/W) = 5.0$$

Answer: 2.7×10^6 cycles

9.39. The critical crack size in a titanium alloy component is determined to be 2.4 mm (see Example 9-2). If fatigue crack growth in this alloy can be described by $da/dN = C(K)^4$, where $C = 6 \times 10^{-12}$ cycle^{-1} MPa^{-4} m, and $K = \sigma\sqrt{\pi a}\, f(a/W)$, determine the "safe life" of this part under the following constant-amplitude cyclic loading conditions:

$$\sigma_{max} = 276 \text{ MPa} \qquad f(a/W) = 2.0$$
$$\sigma_{min} = 0 \qquad\qquad a_0 = 0.254 \text{ mm}$$

9.40. The crack growth rate behavior of a carbon steel plate is shown in Figure 9-42.
   (a) At a stress intensity level of 60,000 psi-in.$^{1/2}$, how much farther will a fatigue crack advance in the TR orientation than in the WT orientation, for each loading cycle?
   (b) Explain the reason for this behavior.

9.41. You are requested to perform a failure analysis on a fractured component which possibly experienced fatigue loading conditions. How would you attempt to verify that fatigue cracking actually occurred? If fatigue crack growth did indeed occur, how would you analyze the direction in which the crack propagated? Also, how would you estimate the crack growth rate at various locations, based on fracture surface evidence?

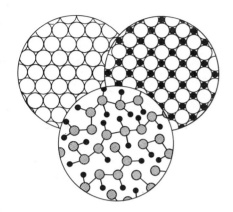

CHAPTER TEN:
Electrical Properties of
Engineering Materials

At this point in our study of engineering materials, it should come as no surprise that electrical and electronic properties originate from the basic nature of the atomic structure, particularly electronic configuration, and the interaction between atoms joined together in both crystalline and noncrystalline (amorphous) structures. Many of these relationships were presented in Chapters 2 through 4.

In the case of metals and metallic alloys, the delocalized or "free" electrons associated with the metallic bond are not firmly attached to a particular atom or atoms. Rather, they tend to wander or drift throughout the lattice structure. The movement of these electrons is random unless an external electric force (field) is imposed on the material. When an electric field or *voltage* is applied, the conduction electrons respond by moving in the direction of the field, and we say that an *electric current* flows.

Although electric currents may be produced in materials other than metals, emphasis is usually placed on metals because they are the most efficient and practical class of conductors. Again, referring to Chapter 2, the fundamental reasons for this behavior involves the electronic configuration of the metallic structure.

On the other hand, in ionic materials such as ceramics, the electrons are more tightly bound to the atoms and it is easier for electrical conduction to occur by the movement of the ions themselves. Transport of ions in the solid state was briefly discussed in Chapter 5 in the section "Diffusion in Nonmetallic Solids." In the case of covalently bonded materials such as polymers, electrical conduction may occur by the movement of molecules. However, the movement of charge carriers in both ceramics and polymers is usually very low in the solid state. Therefore, such materials are typically considered *insulators*, under ordinary conditions of temperature and voltage.

We may define still another class of electrical materials which exhibit properties intermediate to the conductors and the insulators. This class is referred to as *semi-*

conductors. Basically, semiconductors contain electrons which ordinarily are tightly held to their atoms, but can become mobile (conductive) when they are excited by an externally applied energy source such as heat or electromagnetic radiation. For example, vanadium sesquioxide (V_2O_5) is a semiconductor material above $-123°C$ ($-189°F$), but behaves as a good insulator just below this temperature.

Furthermore, some semiconductor materials are electrically sensitive to light. When these materials are subjected to illumination (or the lack of it), their electrical conductivity may change significantly. These materials are called *photoconductors*, and due to their light-sensitive characteristics, find many applications in the areas of controls and detectors. For example, some smoke detectors may utilize the photoelectric cell in their circuitry for the detection of smoke particles. If smoke enters the unit, thereby reducing the light falling on the photoelectric device, the current in the circuit changes and an alarm is triggered.

Semiconductors are principally responsible for the miniaturization and proliferation of many familiar solid-state devices, including radios, televisions, calculators, and watches. In addition, the capabilities of computers and communication equipment have been tremendously enhanced due to the integrated circuitry, which utilizes semiconductive materials and their unique fabrication techniques.

Even more dramatic changes in electrical behavior are exemplified by certain materials called *superconductors*. For instance, some alloys of niobium (Nb) exhibit superconductivity when they are cooled to temperatures approaching absolute zero ($0°K$). In these unique alloys all electrical resistance vanishes if their temperature drops below $23°K$. Even though the term "superconductivity" connotes an electrical property, this topic will be discussed in Chapter 11, because the most practical applications for this unprecedented behavior presently exist in magnetics.

It should be obvious by now that the various classes of engineering materials demonstrate considerable variability in their electrical and electronic properties. Moreover, these differences are also strongly affected within specific materials by such factors as *temperature* and *composition*. Thus in this chapter we explain the important relationships between material structure and electrical properties.

ELECTRICAL RESISTIVITY

Basic Relationships

Electrical resistivity is a measure of the resistance a particular material presents to the passage of charge carriers or current, under the influence of an electric field. The resistivity (ρ) of a conductor with a uniform cross-sectional area (A) can be expressed as follows:

$$\rho = \frac{RA}{L} \tag{10-1}$$

In this relationship, R, the resistance to current flow, is measured in ohms, and L is the length of the conductor. Resistivity is typically expressed in units of Ω-m, or $\mu\Omega$-cm. However, in some engineering applications, the length of the conductor may be expressed in feet and its cross-sectional area in circular mils (a circular mil is the area of a circle with diameter equal to 0.001 in.). In this case, the units for resistivity would be ohm-circular mils per foot.

The resistance (R) in equation (10-1) is obtained from *Ohm's law* as follows:

$$I = \frac{V}{R} \qquad (10\text{-}2)$$

where I is the current measured in amperes, passing through a conductor, under the influence of an electric field (V) measured in volts. In electrical engineering terms, the electrical power (P) is related to the resistance of a conductor by the following expression, which is known as Joule's law of heating:

$$P = \frac{V^2}{R} \qquad (10\text{-}3)$$

Since $V = IR$ [from equation (10-2)] we may also express power as follows:

$$P = \frac{(IR)^2}{R}$$

or simply

$$P = I^2 R \qquad (10\text{-}4)$$

The room-temperature values of electrical resistivity for a number of important engineering materials are listed in Table 10-1. A brief inspection of this list shows that resistivity ranges from relatively low values for the metals and metallic alloys to extremely high values for certain ceramics, glasses, and polymers.

Example 10-1

A size 12-2 (two conductors) annealed copper wire with a diameter of 2.1 mm (80.8 mils) is used for a 110- to 115-V household circuit. (a) What is the resistance per meter length of this wire? (b) If a heating device with a hot value (constant) resistance of 5 Ω, connected to 10 m (two-conductor) of this wire, operates for 8 hours, how much electrical energy is used?

Solution (a) From Table 10-1, the resistivity of copper at room temperature is given as 1.7 $\mu\Omega$-cm. Rearranging equation (10-1) and solving for resistance,

$$R = \frac{\rho L}{A} = \frac{(1.7 \times 10^{-8}\ \Omega\text{-m}(1\ \text{m})}{(\pi/4)(2.1 \times 10^{-3}\ \text{m})^2}$$

$$= 4.9 \times 10^{-3}\Omega/\text{m} \qquad Ans.$$

(b) The resistance of the circuit is determined as follows:

$$R_{\text{wire}} = (4.9 \times 10^{-3}\ \Omega/\text{m})(20\ \text{m}) = 9.8 \times 10^{-2}\ \Omega$$

$$R_{\text{total}} = R_{\text{wire}} + R_{\text{device}} = 0.098\ \Omega + 5.0\ \Omega$$

$$= 5.098\ \Omega$$

From equation (10-3) the power used is

$$P = \frac{V^2}{R} = \frac{(110\ \text{V})^2}{5.098\ \Omega}$$

$$= 2373\ \text{W} = 2.37\ \text{kW}$$

Therefore, the electrical energy (power $\times$ time) is

$$U = (P)(t)$$

$$= (2.37\ \text{kW})(8\ \text{hr})$$

$$= 19.0\ \text{kW-hr or } (68.4\ \text{MJ}) \qquad Ans.$$

TABLE 10-1 ELECTRICAL RESISTIVITY OF SELECTED MATERIALS
AT ROOM TEMPERATURE

Material	Resistivity $(\mu\Omega\text{-cm})$	Temperature coefficient of resistivity, (α) $(1/°C)$
Pure metals		
Aluminum	2.6	0.0039
Copper	1.7	0.004
Gold	2.4	0.0034
Iron	9.7	0.0065
Silver	1.6	0.0038
Alloys		
87 Cu–13 Mn (Manganin)	48	0.00001
57 Cu–43 Ni (Constantan)	49	0.00001
96 Fe–4 Si	59	0.0008
18-8 stainless steel	73	0.00094
62 Fe–21 Ni–12 Al–5 Co (Alnico 1)	75	0.002
55 Fe–37.5 Cr–7.5 Al (high-resistance alloy)	166	0.001
80 Ni–20 Cr (Nichrome V)	108	0.0001
92.5 Ag–7.5 Cu (sterling silver)	2.0	0.004
85 Ag–15 Cd (contact alloy)	5.0	0.004
Ceramics		
SiC (dense)	10^7	
Boron carbide	5×10^5	
Ge	4×10^7	
Fe_3O_4	10^4	
Al_2O_3	$10^{17}-10^{20}$	
Porcelain	$10^{18}-10^{20}$	
TiO_2	$10^{18}-10^{24}$	
Glass		
Fused silica	10^{18}	
Soda–lime, general purpose	$10^{12}-10^{13}$	
Borosilicate	$10^{14}-10^{17}$	
Aluminosilicate	10^{17}	
Polymers		
Urea-formaldehyde	10^{18}	
Vulcanized rubber	10^{20}	
Polystyrene	10^{24}	
Polyethylene	10^{19}	
PMMA	10^{22}	
PVC	10^{16}	
Nylon	10^{20}	

Source: Data compiled from *Handbook of Tables for Applied Engineering Science*, 2nd ed., CRC Press, Inc., Boca Raton, Fla., 1973; and *Handbook of Materials Science*, Vol. 1: *General Properties*, CRC Press, Inc., Boca Raton, Fla., 1974.

As you no doubt have surmised by this point, the electrical conductivity of a material is related to its resistivity. Indeed, as resistivity increases we should intuitively expect the conductivity to decrease. A relationship between these two properties may be established by considering the electric current density or flux in a conductor. The current density (j) is defined as the charge transported through a unit area (I/A) under the influence of an electric field (E) which has a strength (V/L). Substituting in equation (10-1), we see that

$$\rho = \frac{A}{L}\frac{V}{I}$$

where $V/I = R$ from Ohm's law. Rearranging we have

$$\rho = \frac{A}{I}\frac{V}{L}$$

where A/I is the reciprocal current density and V/L is the strength of the field. Therefore,

$$\rho = \frac{E}{j} \tag{10-5}$$

Solving this expression for the current density (j), or really the movement of charge carriers through a conductor,

$$j = \frac{1}{\rho}E \tag{10-6}$$

Thus we see that charge transport is inversely proportional to the electrical resistivity. Equation (10-6) can be restated as follows:

$$j = \sigma E \tag{10-7}$$

where $1/\rho$ is the electrical conductivity σ and has units of siemens[1] per meter.

Factors Affecting Resistivity

The electrical resistivity of materials (and therefore their conductivity) can be heavily influenced by several factors, including temperature, pressure, and composition, as in the case of solid solutions. Pressure is important in polymers and electrolytes (conduction liquids) but is insignificant in metals and ceramics under ordinary circumstances. Therefore, we will not be concerned with pressure at this time.

Temperature. Above temperatures in the vicinity of absolute zero ($-273°C$), the resistivity of most metals and alloys increases with increasing temperature. Recall our example of the effects of temperature on the vibrational amplitude of an atom or ion in Chapter 4 (see Figure 4-5). As temperature increases, the amplitude of the atomic vibrations in the crystal lattice increases. Consequently, the probability of collision between conduction electrons and atoms or ions increases, resulting in greater resistance to the flow of current (i.e., greater resistivity).

The relationship between ρ and temperature for several commercially important materials is shown in Figure 10-1. From the standpoint of resistivity, it is apparent

---

[1] $1/\rho$ has units of $1/\Omega$-m or mho/m, and conductivity is expressed as $1/\Omega$-m. Another convention of units for σ is called the siemen (S), where S $= 1/\Omega$. Therefore, we may also express σ as S/m.

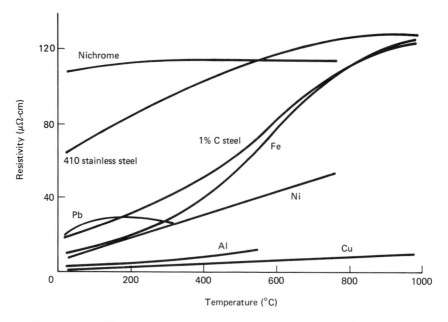

Figure 10-1 Effect of temperature on resistivity for certain engineering materials.

why copper and aluminum are used for conductors and Nichrome is used for heating elements. Furthermore, the change in ρ with increased temperature is very slight compared with the steels, for example.

Conversely, the resistivity of some common polymers decreases as temperature increases. However, one should bear in mind that these materials are generally *insulators* and their resistivity is extraordinarily high to begin with. The reason for this condition (high resistivity) lies in the fact that they are covalently bonded and therefore lack the appropriate delocalized, valence electrons for conduction.

Chemical composition. In general, the greater the degree of chemical purity in a metal, the lower its resistivity. As we witnessed in Table 10-1, the "pure" metals exhibited the lowest values of resistivity and therefore tend to produce the most efficient conductors. Solid solution alloying, which benefits certain properties such as strength, toughness, and corrosion resistance, increases the resistivity of metals. This increase occurs because the solute atoms introduce strains (distortion) and valency effects in the lattice which impede the mobility of the charge carriers. The increase in resistivity due to alloying is illustrated in Figure 10-2 for a hypothetical binary system that exhibits complete solid solubility. A number of commercially important alloys display this type of behavior, e.g., Copper-nickel system (see Figure 13-18).

The preceding example considers the resistivity of a single-phase, homogeneous material. But as we should now realize, many commercial alloys are multiphase and usually are inhomogeneous. For instance, consider the two-phase material illustrated in Figure 10-3. The resistivity of this material depends on the volume fractions (v) of the phases and their respective resistivities, and may be expressed as follows:

$$\rho_T = \rho_\alpha v_\alpha + \rho_\beta v_\beta \tag{10-8}$$

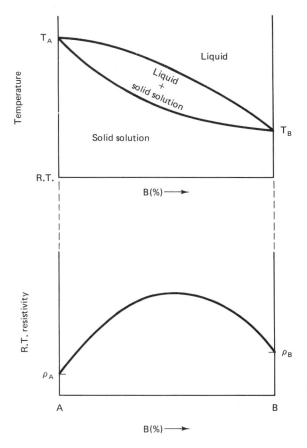

Figure 10-2 Schematic example of the influence of alloying on resistivity for a binary system exhibiting complete solid solubility.

where ρ_T is the "composite" resistivity of the alloy. In multiphase systems when the densities of the phases are similar, mass fraction can be used in the foregoing *law of mixtures* relationship [equation (10-8)], and the resistivity determined as a function of composition.

Applications Relating to Resistivity

In addition to the inverse relationship with conductivity [equation (10-6)], resistivity is very important in applications where specific resistive characteristics are required. The properties we are referring to include uniform resistivity over the length of a conductor, and low *temperature coefficient* of resistivity (α). The latter property refers to the ratio of ρ at a particular temperature to room temperature resistivity (ρ_0), and is expressed as follows:

$$\rho = \rho_0 + \alpha\,\Delta T\rho_0 \qquad (10\text{-}9)$$

A high temperature coefficient means that a small temperature change can produce a large change in resistivity. Correspondingly, the resistance of a circuit or resistive device can be significantly affected. Values of some temperature coefficients, in units of °C^{-1}, are given in Table 10-1. For most pure metals, α is in the neighborhood of 0.004/°C, and for alloys, this coefficient is usually lower.

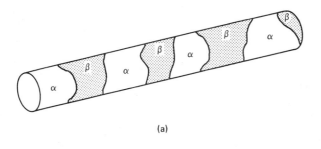

(a)

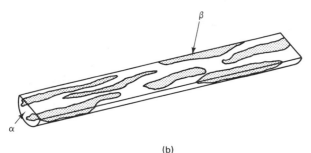

(b)

Figure 10-3 Schematic of a chemically inhomogeneous material containing two nonuniformly distributed phases (α and β): (a) transverse view; (b) longitudinal view.

Example 10-2

Compare the resistivity change experienced (a) by copper and (b) by Constantan when temperature is increased from 27°C to 150°C. (c) What effect does this change have on the resistance of a conductor or a device utilizing these materials?

Solution Based on the data in Table 10-1 and the relationship given in equation (10-9), we find the following.

(a) For copper:

$$\rho_0 = 1.7 \ \mu\Omega\text{-cm} \qquad \alpha_{Cu} = 0.004/°C$$

$$\rho_{Cu} = \rho_0(1 + \alpha \ \Delta T) = 1.7 \ \mu\Omega\text{-cm} \ [1 + 0.004(150 - 27)]$$

$$= 1.7 \ \mu\Omega\text{-cm} \ [1 + (0.004/°C)(150 - 27°C)]$$

$$= 1.7(1 + 0.492) \ \mu\Omega\text{-cm}$$

$$= 2.54 \ \mu\Omega\text{-cm} \qquad Ans.$$

(b) For constantan:

$$\rho_0 \doteq 49 \ \mu\Omega\text{-cm} \qquad \alpha_C = 0.00001/°C$$

$$\rho_C = 49 \ \mu\Omega\text{-cm}[1 + (10^{-5}/°C)(123°C)]$$

$$= 49(1 + 0.00123) \ \mu\Omega\text{-cm}$$

$$= 49.06 \ \mu\Omega\text{-cm} \qquad Ans.$$

(c) In the case of copper, this resistivity change amounts to approximately 50%, while in constantan the change amounts to about 0.1%. Thus the resistance in the copper system increases by 50%, but the constantan system displays essentially no increase in resistance! Such insensitivity to temperature changes in the vicinity of room temperature is an important characteristic for certain applications (e.g., electric resistance strain gauge devices). *Ans.*

Some specific applications for "resistive materials" may help reinforce the concepts that we have just discussed. Although other physical properties, such as thermal expansivity, melting temperature, and corrosion or oxidation resistance, play an important role in the design and performance of precision resistive applications, we will restrict our discussion to the electrical characteristics at this time.

Resistance thermometer.   This is an instrument in which a change in electrical resistance indicates a corresponding change in temperature. Such devices are used in certain thermal or temperature measuring experiments, and the most common metals employed are nickel and copper. However, platinum wire is occasionally used for very exacting temperature measurements, such as the melting points of pure metals.

Resistance strain gauge.   This device, shown in Figure 10-4, is adhesively bonded to the surface of a material in order to measure the strain in that surface.

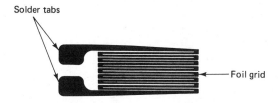

Figure 10-4   Schematic of an electrical resistance strain gauge.

Deformation or changes in strain in the surface under the gauge are sensitively detected due to corresponding changes in resistance of the gauge material, which is formed from small-diameter wire or etched from thin foil sheets. Constantan (see Table 10-1) is commonly employed in strain gauges and precision resistors, because of its extremely low temperature coefficient of resistivity (α) at room temperature (0.00001/°C). Nichrome V and platinum–tungsten alloys are used for elevated temperature tests (approximately 650 to 1000°C).

Heating element.   Resistive heating elements for a wide range of applications, such as kitchen ranges, ovens, furnaces, and radiant and convective heaters, are commonly produced from Nichrome. In high-temperature furnaces, however, platinum is sometimes used for the elements because of its resistance to oxidation (although it is considerably more expensive).

CONDUCTIVITY

Metallic Conduction

The movement of electrical charge (current) by mobile charge carriers is probably the most basic and most important electrical property of electrical engineering materials. In 1900, the German physicist P. K. Drude suggested that current passing through a conductor was due to the movement of charged particles (electrons) under the influence of an electric field. But as we have just discussed, these charged particles encounter resistance to their passage through the crystal lattice of real materials. Present-day scientific information has shown this resistance to be comprised of interactions between the conduction electrons and other particles, dislocations, vacancies, impurity atoms, and other crystalline imperfections that we examined in Chapter 4. Essentially, these interactions cause the electrons to be scattered, thereby decreasing their net movement in the direction of the applied field.

As we discussed earlier in this chapter, conductivity (σ) is the reciprocal of resistivity ($1/\rho$) and is expressed in S/m. The electrical conductivities of some commercially important materials are given in Table 10-2. A quick inspection of these data reveals (probably to no one's surprise) that pure metals such as silver, copper, gold, and aluminum are the best conductors of electricity. But here is an interesting puzzle. Ordinarily, one might think that metals with greater numbers of valence electrons would be better conductors, since more electrons *seem* available for charge transport. Yet if we compare iron (which has two valence electrons) to silver (which has one valence electron) in Table 10-2, we see that the conductivity of silver is about six times that of iron! The answer to this enigma lies in the electron band theory of conductivity, which we will discuss very shortly.

TABLE 10-2 ELECTRICAL CONDUCTIVITY FOR SELECTED MATERIALS AT ROOM TEMPERATURE

Material	Electrical conductivity, σ (S/m)*
Silver, commercial purity	6.3×10^7
Copper, OFHC	5.85×10^7
Copper + 2% Be	2.0×10^7
Gold	4.25×10^7
Aluminum, commercial high-purity	3.5×10^7
Aluminum + 1% Mn	2.31×10^7
Brass, yellow	1.56×10^7
Tungsten, commercial	1.82×10^7
Ingot, iron, commercial	1.07×10^7
1010 steel	0.7×10^7
Nickel, commercial	1.03×10^7
Stainless steel, type 301	0.14×10^7
Graphite	$(10)^5$ (av)
Silicon	5×10^{-4}
Window glass	$2–3 \times 10^{-5}$
Bakelite	$1–2 \times 10^{-11}$
Lucite	$10^{-17} - 10^{-14}$
Borosilicate glass	$10^{-10} - 10^{-15}$
Mica	$10^{-11} - 10^{-15}$
Polyethylene	$10^{-15} - 10^{-17}$

*A siemen (S) is the unit of electrical conductivity in the SI system. It is equivalent to a mho or $1/\Omega$.

Source: R. M. Rose, L. A. Shepard, and J. Wulff, *The Structure and Properties of Materials*, Vol. 4: *Electronic Properties*, John Wiley & Sons, Inc., New York, 1966, p. 75.

Further examination of the data presented in Table 10-2 shows that the conductivity of certain glasses and polymers is extremely low, classifying them as good insulators. These electrical insulating materials are discussed later in this chapter in the section on dielectrics.

In order to compare conductor materials readily, the grade of *standard annealed copper* is frequently used as a reference unit for comparison purposes. Hence, by definition, it has 100% conductivity. Other materials are then expressed as a percent of this standard or international annealed copper standard (% IACS). A comparison of some conductor materials by this standard is given in Table 10-3. Note that aluminum exhibits approximately 65% the conductivity of copper. This characteristic,

TABLE 10-3 ELECTRICAL CONDUCTIVITY COMPARISON OF SELECTED MATERIALS

Material	Percent IACS[a]
Aluminum (99.99%)	65
5052 alloy	35
6101 (T6)	56
Copper (pure)	103
70 Cu–30 Zn (cartridge brass)	28
Leaded bronze	42
Beryllium–copper	22–30
87 Cu–13 Mn (Manganin)	4
57 Cu–43 Ni (Constantan)	3
Gold	75
Iron (99.99%)	18
55 Fe–37.5 Cr–7.5 Al	1
Alnico	3–4
1% C steel	8
1010 steel	14
4% Si steel	3
Stainless steel	2–4
Molybdenum	34
Molybdenum disilicide ($MoSi_2$)	4
Nickel (99.8%)	23
80 Ni–20 Cr (Nichrome V)	2
Platinum	16
95 Pt–5 Ir	9
95 Pt–5 Ru	6
Silicon carbide (SiC)	1–2
Silver (fine)	106
92.5 Ag–7.5 Cu (sterling)	85
97 Ag–3 Pt	50
85 Ag–15 Cd	35
Tantalum	14
Tungsten	30

[a]International Annealed Copper Standard.
Source: Metals Handbook, 8th ed., Vol. 1: *Selection and Properties of Metals*, American Society for Metals, Metals Park, Ohio, 1961, p. 56.

together with other desirable properties, such as relatively low density and good wire drawing capabilities, has led to the development of an aluminum conductor as a feasible substitute for copper wire.

Example 10-3

An aluminum conductor is suggested as a replacement for more expensive copper wire. This substitute must be capable of carrying the same amount of current, however, as the original No. 10 (AWG) copper wire, which has a diameter of 2.6 mm (102 mils). Determine the diameter of aluminum conductor that will safely meet this requirement.

Solution In this problem the conductance must be equal in the two wires; therefore, the resistance must also be equal. By utilizing the data for ρ in Table 10-1, and the relationship expressed in equation (10-1), we can solve for the unknown area as follows:

$$R_{Al} = R_{Cu}$$

$$\frac{(\rho_{Al})(L_{Al})}{A_{Al}} = \frac{(\rho_{Cu})(L_{Cu})}{A_{Cu}}$$

For same length of conductor,

$$\frac{\rho_{Al}}{A_{Al}} = \frac{\rho_{Cu}}{A_{Cu}}$$

Solving for the area of aluminum wire, we have

$$A_{Al} = \frac{\rho_{Al}}{\rho_{Cu}} A_{Cu}$$

Substituting our known data gives us

$$A_{Al} = \frac{2.8 \ \mu\Omega\text{-cm}}{1.7 \ \mu\Omega\text{-cm}} \frac{\pi}{4} (2.6 \text{ mm})^2$$

$$= (1.65)(5.3 \text{ mm}^2)$$

$$= 8.76 \text{ mm}^2$$

$$\frac{\pi d_{Al}^2}{4} = 8.76 \text{ mm}^2$$

$$d_{Al}^2 = \frac{(4)(8.76 \text{ mm}^2)}{\pi}$$

$$d_{Al} = \sqrt{11.1 \text{ mm}^2}$$

$$= 3.34 \text{ mm } (131 \text{ mils}) \qquad Ans.$$

In terms of the American Wire Gauge (AWG) number, this diameter lies between No. 8 and No. 7.

It is worth mentioning that even when aluminum conductors are properly sized, potentially serious problems can develop in their usage. For example, if the aluminum wire is connected to copper terminals in switches and junctions, the aluminum can become oxidized. This process dramatically increases its resistance to current flow, thus causing significant resistance heating. Such heating has been responsible for many structural fires which started in proximity to electrical junctions.

Ionic Conduction

In the case of ionic materials, such as certain ceramics, charge movement occurs by *ion diffusion*, as mentioned in Chapter 5. Ionic conduction therefore depends on the presence of lattice defects, such as ion-pair vacancies (Schottky defect) and displaced ions (Frenkel defect). These point defects are illustrated in Figures 4-2 and 4-3.

From our study of diffusion in Chapter 5, we know that the concentration of crystalline defects increases significantly with temperature. Diffusion in ionic solids, and thus electrical conductivity, generally occurs more rapidly as temperature increases. The mobility of ions (μ), the diffusion coefficient (D), and the conductivity (σ) are related by the Nernst–Einstein equation [equation (5-16)]. Thus the conductivity in ionic materials may be expressed as

$$\sigma = \frac{nZ^2e^2}{kT}D_0e^{-Q/kT} \qquad (10\text{-}10)$$

where n = total number of ions/cm^3
e = charge on the ions
Z = valence of the ions
k = Boltzmann's constant
Q = activation energy for diffusion
D_0 = diffusion coefficient for ion in question
T = absolute temperature

For example, the electrical conductivity (cation conductivity) of sodium chloride (NaCl), shown as a function of temperature in Figure 10-5, increases roughly by a factor of 10^4 in the temperature range 300 to 700°C.

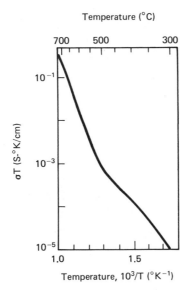

Figure 10-5 Cation conductivity in pure NaCl. (From Kirk and Pratt, *Proc. British Ceramic Soc.* 9, 1967, p. 215.)

Conductivity

Band Theory

As we discussed in Chapter 2, when atoms are joined together in a solid, their valence electron structures no longer behave in a classical orbital sense. Although the inner-shell electrons remain essentially unchanged, the valence shell in a material containing N number of atoms splits into N energy levels. Thus an s orbit, which classically can contain two electrons, becomes an s band that will hold $2N$ electrons. The p level becomes a p band with room for $6N$ electrons, the d level a d band with $10N$ electrons, and so on. This process is shown schematically in Figure 10-6. The

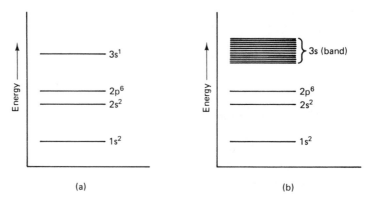

Figure 10-6 Schematic representation of electronic energy bands in sodium as atoms are joined in a metallic solid. (a) Classical representation of energy levels. (b) Band model: the $3s$ band can accommodate $2N$ electrons, where N is the number of atoms in the solid.

uppermost band associated with the atoms joined together in the solid is called the conduction band *if* it is partially filled, because the electrons in this band can be excited to higher (unoccupied) energy states. For example, consider the energy bands for conductors such as sodium (Na), magnesium (Mg), and aluminum (Al), as illustrated in Figure 10-7. In the case of sodium, which contains one valence electron, the $3s$ band is half full (remember that the $3s$ band can contain twice as many electrons as atoms present in the solid). Therefore, there are many unoccupied higher-energy states available in the band to which the electrons may rise in response to an applied field. Although magnesium contains two valence electrons, the $3s$ and $3p$ bands overlap and some electrons "spill over" into the second band. The unoccupied states in these bands serve as higher-energy states for the electrons during conduction. Finally, in aluminum, which contains three valence electrons, the first band is full, but the second overlapping band is only half full. Once again, unoccupied energy states are readily available to the electrons and the material is a conductor of electric charge.

Remember our enigma involving the difference in conductivity between iron (divalent) and silver (monovalent) from the section "Metallic Conduction"? A simplified explanation for the greater conductivity of silver is based on the energy band theory. In the monovalent silver the $5s$ band is only half filled and electrons in the upper part of the band can be easily excited to the conduction state, since there are many open (unoccupied) higher-level energy states readily accessible. In the divalent

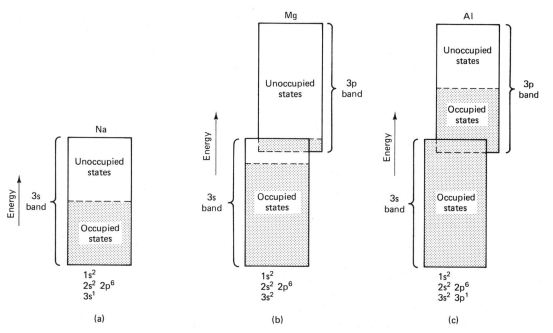

Figure 10-7 Schematic of energy bands for three conductors together with their respective classical electric configuration: (a) sodium (monovalent) displays half-filled 3s band; (b) in magnesium (divalent) the 3p band overlaps the 3s band and contains some electrons; (c) in aluminum (trivalent) the 3s band is full but the 3p band is only half-filled.

iron, there is overlap of the adjacent bands; that is, s and d, and complex energy levels develop, reducing the open higher-energy levels available for conduction. Therefore, even though iron has more valence electrons, it is much easier for the valence electrons in silver to become mobile charge carriers.

Differences in conductivity between metals that exhibit relatively lower conductivity, such as nickel, iron, titanium, and so on, can be attributed to the complex energy levels produced in the regions of overlap between adjacent bands. In general, the electronic behavior of conductors may be summarized by saying that their valence bands are not completely filled and the valence electrons can be readily excited to unoccupied higher energy levels, where they become conduction electrons.

Now we will consider the case where an energy band is completely full and separated from the next higher band, which is empty. In other words, the energy bands do not overlap. Under these circumstances none of the electrons in the full energy band may respond to an electric field, unless they can cross the *energy gap* between the bands. Just such a condition occurs in silicon (Si), as illustrated in Figure 10-8, where the two 3s and two 3p valence electrons form a hybrid group of four electrons which *behaves* like a filled valence band. Indeed, for all practical purposes, silicon is not a conductor at room temperature (see Table 10-2), but its energy gap is sufficiently small (1.06 eV) at room temperature and some electrons are able to be excited across the energy gap into the conduction band. Materials that display these characteristics are called *semiconductors*.

However, if the energy gap between a filled band and an empty band is large

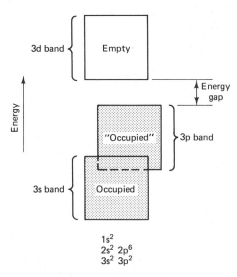

3d band — Empty

Energy

Energy gap

"Occupied" — 3p band

3s band — Occupied

$1s^2$
$2s^2\ 2p^6$
$3s^2\ 3p^2$

Figure 10-8 Energy bands for silicon (tetravalent). An energy gap exists between the second band (which behaves like a filled band) and the empty third band.

enough to prevent electrons from crossing it, the material will not conduct electrical charge. Such materials are *insulators*.

The values of energy gaps in some commercially important materials are presented in Table 10-4. An energy gap of about 4 eV is taken as an arbitrary separation between semiconductors and insulators.

TABLE 10-4 ENERGY GAP VALUES AND CONDUCTIVITY FOR SELECTED GROUP IV MATERIALS

Material	eV	Energy gap, $J\ (\times\ 10^{-19})$	Conductivity, σ (S/m)	Electrical classification
C (diamond)	5.3	8.48	10^{-16}	Insulator
Si	1.1	1.76	5×10^{-4}	Semiconductor
Ge	0.7	1.12	2	Semiconductor
Sn (gray)	0.1	0.16	10^6	Conductor

SEMICONDUCTORS

By now the student surely recognizes that semiconductors play an extremely important role in our daily lives. The unique electrical characteristics of these materials are responsible for revolutionary changes and developments in electronic components and circuitry. The list of equipment and devices that utilize semiconductors is most impressive and extensive. Therefore, we must be content for the moment to mention just some of the more important fields of application, such as communications, computing and data processing, photography, home entertainment, medicine and surgery, timekeeping and measurements, and controls.

To utilize semiconductor materials efficiently, we must understand how they function, what special properties they exhibit, and what factors (chemical, microstructural, environmental) influence their performance. The following section explores the basic concepts of semiconductors and their applications.

Theory

We have introduced the basic theory of semiconductors in the section "Band Theory." There we saw that materials with a full valence band separated from the conduction band by an energy gap could conduct electric charge only if some of the electrons crossed the energy gap. This transition of electrons to a higher energy state, shown schematically in Figure 10-9, occurs in semiconductor materials because the energy gap is relatively small, usually less than 2 eV. Electric fields (voltage), electro-

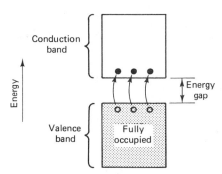

Figure 10-9 Schematic portrayal of a semiconductor material. Certain valence electrons obtain the energy necessary to "jump" the gap into the conduction band. Electron "holes" are left in the valence band.

magnetic radiation, heat, and magnetic fields are potential energy sources for assisting electrons to jump this gap. If the gap is much more than this, very high electric fields are required to move the electrons to the conduction band, and the material is an insulator. However, if the field is sufficiently high, even insulators may conduct electric current. This process is referred to as *dielectric* breakdown and is discussed in the section "Insulators."

Intrinsic semiconductors. In materials where just the valence band and the conduction band are involved in charge transport, the semiconductor is called *intrinsic*. As we explained previously, if the energy gap is narrow enough, some of the valence band electrons can be elevated to the conduction band. Each electron that jumps the gap leaves behind an electron hole which in turn behaves like a positive charge (see Figure 10-9). Electrical conduction is promoted if this process occurs at a sufficient rate.

To produce high-quality intrinsic semiconductors, very pure, near-perfect, crystals are required. Typical intrinsic semiconductors can be made from silicon, germanium, gray tin, selenium, and tellurium. In addition, compounds that show this type of semiconduction include Cu_2O, ZnO, Fe_2O_3, and $PbTe$. Impurity atoms in these materials are usually controlled to 1 ppm or less. Unfortunately, this level is not usually detectable by ordinary chemical analysis techniques and must be measured by electrical conductivity. The conductivity in this case is expressed as

$$\sigma = q(\mu_p p + \mu_n n) \tag{10-11}$$

where q = charge on an electron (0.16×10^{-18}C)

μ_p = mobility of holes (m²/V-sec)

μ_n = mobility of electrons (m²/V-sec)

p = number of holes per cm³

n = number of electrons per cm³

For the intrinsic semiconductor, the number of holes (p) is equal to the number of electrons (n).

Example 10-4

If the conductivity of "pure" silicon is determined to be 5×10^{-4} S/m, estimate the concentration of conduction electrons per cm³ in this material.

Solution The conductivity is related to the charge carrier concentration by equation (10-11). If we assume that the number of conduction electrons (n) equals the number of holes in the valence band (p) in this intrinsic semiconductor, we can state

$$\sigma = qn(\mu_p + \mu_n)$$

Solving for n (the total charge carriers), we have[2]

$$n = \frac{\sigma}{q(\mu_p + \mu_n)}$$

We substitute for q and μ (Table 10-5):

$$n = \frac{5 \times 10^{-4}}{(0.16 \times 10^{-18} \text{ C})(0.19 \text{ m}^2/\text{V-sec})} \frac{1}{\Omega\text{-m}}$$

$$= \frac{5 \times 10^{-4}(\text{m}^3)(\text{V-sec})}{(0.16 \times 10^{-18})(0.19)(10^6 \text{ cm}^3)(\text{A-sec})(\text{m}^2)} \frac{1}{\Omega\text{-m}}$$

$$= 1.64 \times 10^{10} \text{ carriers/cm}^3$$

In other words, 8.2 billion *electrons* per cm³ are available for conduction.   *Ans.*

Extrinsic semiconductors. Conversely, extrinsic semiconductors owe their conduction characteristics to the controlled presence of impurity atoms, or in some instances, to an excess of one of the elements in a compound. For example, silicon and germanium are tetravalent and form covalent bonds. If a substitutional impurity atom containing five valence electrons is introduced to their lattice, only four of the five valence electrons participate in the bond. This situation is illustrated for an arsenic impurity atom in germanium in Figure 10-10(a). The fifth valence electron in arsenic can be readily excited to the conduction band, as shown in Figure 10-10(b). Such an impurity is called a *donor*, because it produces conduction electrons without leaving holes in the valence band. In this case, the conduction electrons are the majority charge carriers while the holes are the minority carriers and we have an *n*-type extrinsic semiconductor. The addition of group V elements such as phosphorus, arsenic, or antimony to silicon or germanium produces *n*-type semiconductors.

On the other hand, consider what the effect of introducing a trivalent impurity to the tetravalent lattice might be? Right! A hole exists in the valence state of the impurity atom. This condition is illustrated in Figure 10-11. If the hole moves away

---

[2]From equation (10-2), Ω = V/A. Also, C = A-sec.

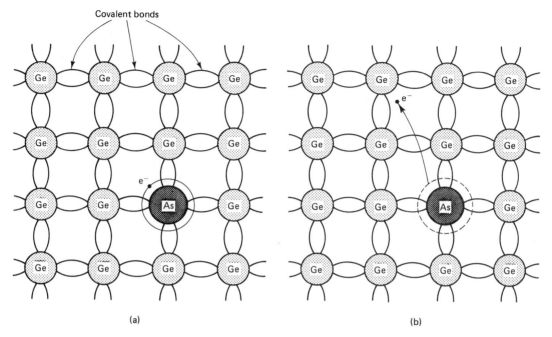

Figure 10-10 *N*-type extrinsic semiconductor—germanium lattice (tetravalent) containing an arsenic impurity (pentavalent): (a) fifth valence electron associated with the donor atom (A_s); (b) electron jumps to conduction band—no hole is left in valence band.

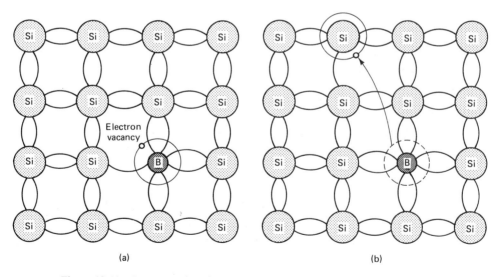

Figure 10-11 *P*-type semiconductor—silicon lattice (tetravalent) containing a boron impurity atom (trivalent): (a) one bonding state is empty and the "hole" is associated with the boron (acceptor) atom; (b) hole leaves acceptor atom (B), and boron accepts a valence electron from the silicon lattice.

Semiconductors

from the impurity, the bonding state is filled by accepting a valence electron from a tetravalent atom. In this case, the impurity is termed an *acceptor* and a net negative ion is produced. When trivalent impurity atoms (group III atoms such as boron, aluminum, etc.) are substituted in "pure" silicon or germanium, the majority charge carriers are holes. Thus the material is called a *p*-type semiconductor. Impurity levels in extrinsic semiconductors are typically about 1 to 100 ppm (0.0001 to 0.01 w/o). The electrical properties of the group IV semiconductor materials, and compounds consisting of group III and IV elements are given in Table 10-5.

TABLE 10-5 SEMICONDUCTIVE PROPERTIES OF SELECTED GROUP III AND IV MATERIALS (20°C)

Material	Energy gap (eV)	Electron mobility, μ_n (m^2/V-sec)	Hole mobility, μ_p (m^2/V-sec)
Carbon (diamond)	5.3	0.18	0.12
Silicon	1.1	0.14	0.05
Germanium	0.7	0.39	0.19
Tin (gray)	0.1	0.20	0.10
Compounds:			
AlP	3.0	—	—
AlAs	2.3	—	—
GaP	2.2	0.04	0.002
AlSb	1.5	0.14	0.02
GaAs	1.3	0.85	0.45
InP	1.3	0.60	0.02
GaSb	0.7	0.50	0.08
InAs	0.3	2.30	0.01
InSb	0.2	8.00	0.07

Example 10-5

Pure silicon is "doped" with 0.5×10^{22} aluminum atoms/m^3 (0.1 ppm). How will this addition affect the conductivity of Si?

Solution The conductivity of pure Si is given in Table 10-2 as 5×10^{-4} S/m. The addition of trivalent atoms to Si produces holes which are available for conduction of electric charge. Rearranging equation (10-11), we can determine the increase in conductivity from this addition as follows:

$$\sigma = pq\mu_p$$

$$= (0.5 \times 10^{22} \text{ atoms/}m^3)(0.16 \times 10^{-18} \text{ C})(0.05 \text{ } m^2/\text{V-sec.})$$

$$= (0.5 \times 10^{22})(0.16 \times 10^{-18})(0.05)\frac{\text{A-sec-}m^2}{m^3\text{-V-sec}}$$

$$= 40(\Omega\text{-m})^{-1} \text{ or } 40 \text{ S/m} \quad Ans.$$

Comment: The addition of just 0.1 ppm aluminum to pure silicon increases its conductivity by 80,000!

Temperature effects. As we saw in the preceding section, thermal energy provides a means for valence electrons in certain materials to cross the energy gap into the conduction band. Therefore, we can expect the concentration of charge carriers, both electrons and holes, to increase with increasing temperature. In fact, the density of charge carriers (n_i) in an intrinsic semiconductor depends principally on the energy gap (E_g) and temperature as follows:

$$n_i \propto e^{-E_g/2kT} \tag{10-12}$$

where the value $E_g/2$ is the average energy at the midpoint of the energy gap, and n_i is the number of conduction electrons per unit volume (note the number of electrons excited to the conduction band equals the number of holes left behind). Such temperature-dependent behavior is illustrated for two semiconductor materials in Figure 10-12.

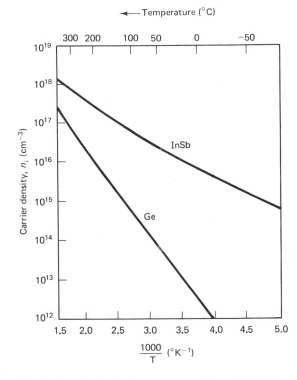

Figure 10-12 Temperature dependence of the intrinsic carrier density in two semiconductors.

Notice the similarity between equations (10-12) and (10-10), which relates conductivity to charge density and temperature. The conductivity in an intrinsic semiconductor can be expressed in a similar manner as follows:

$$\sigma = \sigma_0 e^{-E_g/2kT} \tag{10-13}$$

where σ_0 = proportionality constant (depends on charge mobility, etc.)
E_g = energy gap (eV or J)
k = Boltzmann's constant
T = absolute temperature (°K)

Now the student no doubt recognizes the form of this expression to be an Arrhenius relationship. Thus $-E_g/2k$ is the slope of the logarithmic plot σ versus $1/T$, as shown in Figure 10-13. Such an experimental plot yields values of the energy gap (E_g) as well as conductivity, at various temperatures of interest.

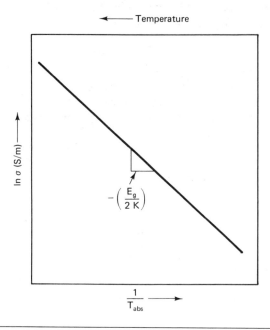

Figure 10-13 Schematic illustration of the relationship between conductivity and temperature in intrinsic semiconductors.

Example 10-6

Compare the conductivity of silicon at room temperature (20°C) and at a temperature of 50°C (122°F).

Solution From equation (10-13) we learned that conductivity (σ) is related to temperature by an Arrhenius-type expression. Therefore, we may state

$$T_1 = 20°C = 293°K \qquad \sigma_1 = \sigma_0 e^{-E_g/2kT_1}$$

$$T_2 = 50°C = 323°K \qquad \sigma_2 = \sigma_0 e^{-E_g/2kT_2}$$

Thus

$$\frac{\sigma_2}{\sigma_1} = \frac{e^{-E_g/2kT_2}}{e^{-E_g/2kT_1}}$$

$$\ln \frac{\sigma_2}{\sigma_1} = \frac{T_1}{T_2}$$

$$\ln \sigma_2 - \ln \sigma_1 = \frac{293°K}{323°K}$$

From Table 10-5, the conductivity of Si at 20°C is 5×10^{-4} S/m. Substituting for σ_1, we have

$$\ln \sigma_2 = \ln (5 \times 10^{-4}) + 0.907$$

$$= -7.6 + 0.907$$

$$= -6.69$$

$$\sigma_2 = e^{-6.69}$$
$$= 1.24 \times 10^{-3} \text{ S/m}$$

Comparing the two values of conductivity yields

$$\frac{\sigma_2}{\sigma_1} = \frac{1.24 \times 10^{-3}}{5 \times 10^{-4}}$$

$$= 2.48 \quad \textit{Ans.}$$

Comment: In other words, a 30°C increase in temperature increased the conductivity of silicon $2\frac{1}{2}$ times.

In extrinsic semiconductors, the concentration of charge carriers exhibits a somewhat different behavior with regard to temperature. The overall behavior is illustrated in a graphical manner in Figure 10-14. Let us consider separately the three

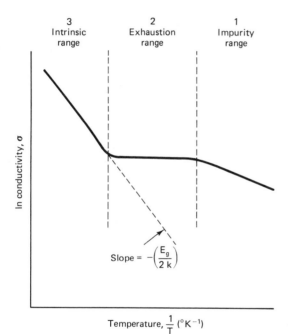

Figure 10-14 Conductivity as a function of temperature in an *n*-type (donor) extrinsic semiconductor.

regions of this figure. Initially (region 1), the charge carriers can be mobilized or excited at ordinary temperatures, because the required energy is relatively small. In this region (also known as the *impurity range*) charge density, and thus conductivity, increases gradually as temperature rises. As temperature continues to increase, eventually the donor levels become exhausted (or acceptor levels become saturated). In this region (2), the conductivity becomes relatively constant and this temperature interval is called the *exhaustion range*. Finally, at yet higher temperatures (region 3) sufficient thermal energy is available to produce intrinsic behavior and conductivity increases according to equation (10-12). The temperature at which intrinsic conduction develops is very important because many semiconductors are designed to operate in the exhaustion range where σ is relatively constant. The start of intrinsic conduction may be designated as the upper temperature limit in certain devices. For example, this upper limit for extrinsic conductivity in germanium-based materials is about 100°C, and about 200°C for silicon-based materials.

Recombination. In the impurity range and exhaustion range of extrinsic semiconductors (Figure 10-14), the density of the majority charge carriers is far greater than that of the minority carriers. Suppose that such a material is irradiated by photons of energy sufficient to raise electrons to the conduction band. An equal number of conduction electrons and holes are produced. When irradiation is stopped, the excess conduction electrons can return to the valence band and combine with the excess holes. This process is called *recombination* and results in the emission of photons (electromagnetic radiation). As you will see in the section on semiconductor applications, when this radiation is in the visible spectrum, the process of recombination is instrumental in some very familiar devices.

Semiconductor Applications and Devices

Since semiconductive materials are utilized in many important applications, and this field of materials in electrical engineering has an extremely bright future, we would like to introduce a few basic uses of semiconductors. The student will find that these applications and devices are based primarily on the concepts that we studied in the preceding section.

Rectifying junctions. Ideally, a rectifier is a device that conducts electrical current in one direction only. This is illustrated in Figure 10-15. By convention we

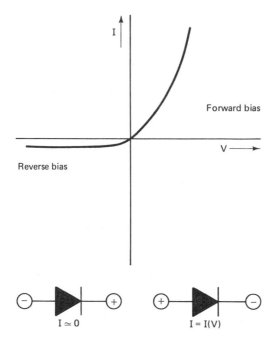

Figure 10-15 Electrical characteristics of a rectifier. Very little current flows for the reverse (negative) bias condition. Current readily flows for the forward (positive) bias condition.

depict current as the flow of positive charge. In this schematic rectifier, current will not flow when the potential at the right side of the connection (+) is greater than that at the left side (−), as pictured on the left. In electrical engineering terminology, this is called *reverse bias*. However, if the potential at the left-hand connection is higher than the right-hand connection, as shown to the right, current is conducted. This situation is called *forward bias*.

The boundary between *p*- and *n*-type extrinsic semiconductor material can be used as a rectifier. Remember that in *p*-type semiconductors the majority carriers are holes (+), whereas in *n*-type the majority carriers are electrons (−). Electrical current (positive charge) can flow from the *p*-side of the junction to the *n*-side, but not in the opposite direction. Therefore, *p-n* junctions can be used to rectify alternating current; that is, convert it to direct current (dc). A *p-n* junction is illustrated in Figure 10-16.

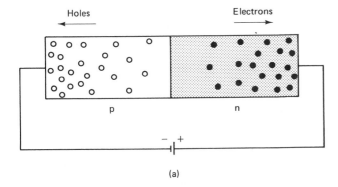

(a)

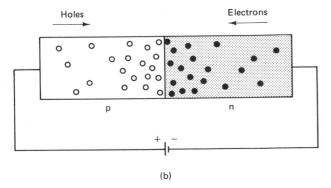

(b)

Figure 10-16 Schematic of a *p-n* rectifier: (a) reverse bias—very little conduction; (b) forward bias—recombination at junction causes large currents to flow.

Furthermore, the recombination events that take place in the vicinity of the *p-n* junction not only allow large currents to flow, but as we hinted in the preceding section, also may cause luminescence.[3]

When gallium phosphide (GaP), or gallium arsenide (GaAs) is used at room

[3]As opposed to incandescence (broad-band radiation due to thermal vibration of atoms), luminescence is the narrow-band radiation emitted as a result of electrons changing their energy states when the material is excited by an energy source that does not significantly change the temperature of the sample.

temperature, the photons emitted appear as red light. This junction device we have just described is the principle of the light-emitting diode (LED) used in the digital displays of calculators, clocks, and instrumentation.

In certain cases, *p-n* junction rectifiers are designed to *break down* and conduct current under a reverse bias. If the voltage across the diode becomes sufficiently high, valence band electrons may pass from the *p* material to the *n* material. This occurs at the *breakdown voltage* (V_B), as illustrated in Figure 10-17.

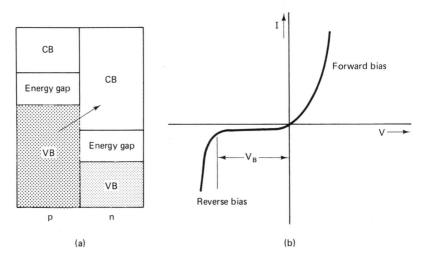

Figure 10-17 Electrical characteristics of a zener diode: (a) valence band of *p* material overlaps conduction band of *n* material; (b) diode breaks down at a reverse bias equal to V_B. (In practice, zener diodes always contain a series resistance added to prevent the current from increasing without limit until the device is destroyed.)

The breakdown voltage in these devices, called *zener diodes*, is controlled by the concentration of impurity atoms in the semiconductor material. This limiting type of behavior can be used in regulating circuit voltage.

Tunnel diodes. Another important application of *p-n* junctions are tunnel diodes, which have the unique characteristic of exhibiting *negative* resistance under certain conditions. In these devices (also known as Esaki diodes), both the *p* and the *n* materials are heavily loaded (doped) with acceptor and donor impurities. Also, the width of the junction region is made as thin as possible to present the least barrier to electron movement between valence bands. The tunnel diode is illustrated schematically in Figure 10-18. Normally, no current flows until a considerable voltage is applied. But at a small forward bias, valence electrons from the *n* material are able to cross into the empty states in the valence band of the *p* material. This process is called *tunneling*.

The electrical characteristics (current versus voltage) for a typical tunnel diode are shown in Figure 10-19. In a certain range of forward bias (V_1 to V_2), the current

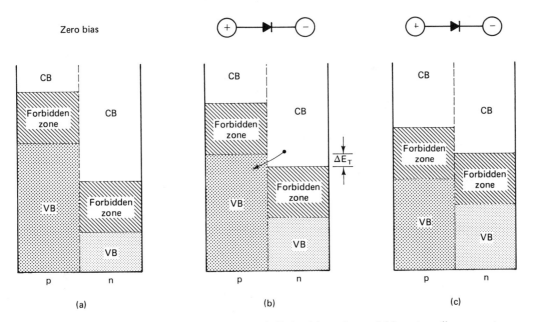

Zero bias

Figure 10-18 Energy bands for a tunnel diode: (a) no forward bias—tunneling current equals zero; (b) small forward bias produces tunneling in the energy region ΔE_T; (c) large forward bias—no tunneling permitted.

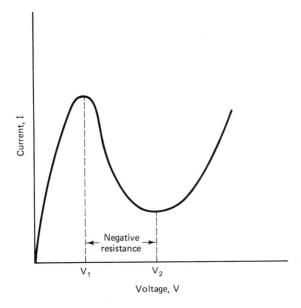

Figure 10-19 Electrical characteristics of a tunnel diode. Tunneling occurs in the bias range V_1 to V_2.

Semiconductors

decreases as voltage increases. Such behavior corresponds to a *negative* resistance. So when the diode is biased in this region, it may be considered an amplifier. Tunnel diodes, in conjunction with a capacitor and an inductance, are also used as high-frequency oscillators.

Transistors. The transistor is a semiconductor device developed in the late 1940s, which is utilized for amplification of electrical current. The term "transistor" has become practically a household word, since it finds applications in so many types of circuitry and instruments.

Basically, a transistor consists of two *p-n* junctions arranged as shown in Figure 10-20. The *p-n-p* type illustrated in (a) has two *p* regions separated by an *n* region.

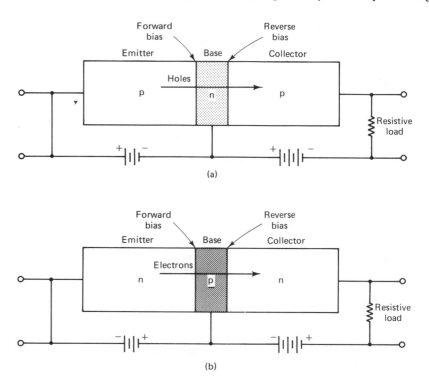

Figure 10-20 Schematic representation of a transistor: (a) transistor is *p-n-p* type and current consists of "hole" flow; (b) transistor is *n-p-n* type and charge carriers are electrons.

The *p-n* junction on the left is forward biased so that electrons move to the left and holes move to the right across the junction. The base is intentionally thin so that current (holes in this case) may readily pass into the collector through the *n-p* junction, which is reverse biased. A relatively small increase in the applied voltage across the emitter–base junction produces a large increase in hole flow into the base. Accordingly, a large increase in current occurs across the base–collector junction. This increase in collector current is manifested in a large voltage increase across the load resistor.

Therefore, a small change in voltage on the emitter results in a large voltage change in the collector circuit. The transistor acts as a voltage amplifier.

A similar explanation may be applied to the *n-p-n* type transistor shown in Figure 10-20(b). However, in this device the charge carriers are conduction electrons rather than holes. The electrons are injected into the base and collected at the *p-n* junction (base–collector).

Transistors, like other semiconductor devices, are temperature-sensitive components. High temperatures (and irradiation) can change their electrical characteristics markedly, due to the onset of intrinsic semiconduction (see Figure 10-14). For this reason, many applications of semiconductive transistors require temperature control, such as cooling by fans or dissipation of heat to a chassis via heat transfer compounds (silicone compounds) or metallic heat sinks.

Photoconductors. Electromagnetic radiation has been mentioned previously in this chapter as another source of energy for excitation of valence electrons to the conduction band. The condition for this process is $hv \geq \Delta E_g$, where h is Planck's constant, v is the frequency of the radiation, and ΔE_g is the energy gap. Light photons (visible radiation) generally have sufficient energy to excite electrons across the energy gap in such semiconductor materials as silicon and germanium, plus some compounds. The process of photoconduction may be represented by the example shown in Figure 10-9, where photons interact with the electrons in the valence band of a semiconductor material.

Increases in conductivity caused by photoconduction may be used to measure radiation intensity, in other words, to detect changes in light intensity (i.e., photocells used in detection units such as smoke alarms, light meters, burner controls, etc.). Moreover, if radiation produces electron–hole pairs in the vicinity of a reverse-biased *p-n* junction [Figure 10-16(a)], the current across this junction can increase sharply due to the increased concentration of *minority* carriers. If this process continues, a voltage will be generated, causing current to flow in a circuit connected to the junction. This type of device is known as a *solar battery* when the junction responds to visible (or near-visible) light. Solar photovoltaic systems composed of many individual cells (Figure 10-21) are capable of producing electrical power for direct use or for storage

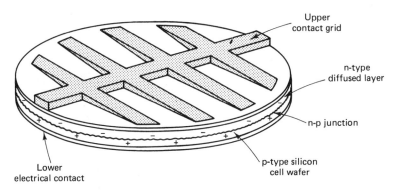

Figure 10-21 Silicon-type solar photovoltaic cell.

by batteries. Such power generation is particularly important when conventional electrical power sources are unavailable (e.g., spacecraft, satellites, remote facilities).

Thermistors. Earlier in this section we discussed the effects of temperature on semiconductors (see Figure 10-14). The resistivity of certain semiconductors changes with temperature. When these materials are utilized in temperature-measuring instruments, they are referred to as *thermistors*.

Although thermocouples are a more convenient method of measuring temperature changes, thermistors have sensitivities of millidegrees and therefore are useful in applications that require precision measurements or extremely rapid temperature sensing and compensation in electrical components.

INTEGRATED CIRCUITS

Integrated circuits, or microelectronic circuits as they are sometimes called, are basically a combination of transistors, resistors, capacitors, and diodes fabricated on a single-crystal wafer of silicon. The silicon wafer containing these miniature electronic components is referred to as a "chip." Originally, integrated circuits (ICs) contained about a dozen components per chip. Presently, these circuits may contain upward of 100,000 components, principally transistors, on chips measuring only a few millimeters on a side. Such microelectronic circuits are referred to as very large scale integrated circuits (VLSI). The electronic (digital) watch chip shown in Figure 10-22 illustrates our point. This chip, which controls the watch function and the *liquid-crystal display*[4] (LCD), measures approximately 2 mm by 4 mm.

The extremely high component densities in ICs are presently achieved using metal-oxide semiconductors (MOS). The sequential stages of the MOS technique are shown in Figure 10-23. The initial step consists of oxidizing the surface of an *n*-type (e.g., phosphorus doped) silicon wafer (a). Then, the surface is masked and selectively etched to open channels in the oxide so that *p*-type (e.g., boron) material may be diffused (vacancy mechanism from Chapter 5) or implanted, into the silicon (b). The existing region of oxide between the *p* regions (gate) is removed and a new, uniformly thick layer is formed over the entire surface (c). Once again, the oxide layer is etched to expose the *p* material, so that electrical contact can be made to these areas (d). This contact is accomplished by vapor depositing a layer of aluminum over the surface (e). Finally, the contact metal (Al) is selectively removed by etching, to electrically separate the appropriate regions of the transistor, as shown in (f). This last stage results in a completed component, in this case a *p*-type transistor. Of course, this represents only one component out of perhaps thousands on a single chip. To design and build such complex circuitry on these very small wafers, it has become necessary to use computer-aided design techniques such as interactive graphics displays and special programs in automated circuit design.

[4] Liquid crystals are fluids containing large molecules which exhibit certain crystalline properties. These materials respond to an electric field by becoming cloudy or opaque, and also by changing color. Such an effect is very useful for the display of images.

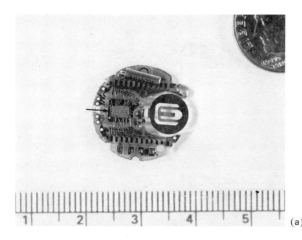

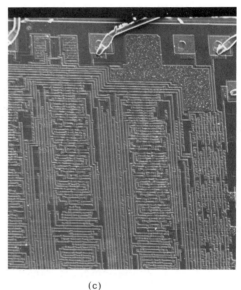

Figure 10-22 Integrated circuitry of a digital wrist watch: (a) overall view of watch components (arrow denotes "chip"); (b) enlarged view of chip (25 ×); (c) close-up upper right hand corner showing microcircuit components (60 ×).

(a)

(b)   (c)

MOS transistors differ from the bipolar transistors we discussed previously (Figure 10-20) in that only one type of charge carrier is active in a single component. A p-type MOS such as Figure 10-23 employs holes, whereas an n-MOS transistor utilizes electrons. The operation of these devices is illustrated schematically in Figure 10-24 for a typical n-type device. Note that the *source*, *gate*, and *drain* serve the same function as the emitter, base, and collector in the bipolar transistors previously discussed. A positive potential applied at the drain exerts an attractive force on the conduction electrons in the source, but they are restricted from passing through the p-type material. However, when a positive charge is applied to the gate, the resultant field attracts electrons to a thin layer under the gate and current flows as indicated by the arrow in Figure 10-24(a). Such a device is referred to as an *enhancement-mode* transistor.

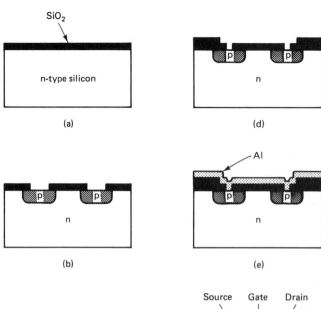

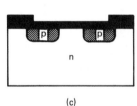

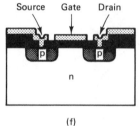

Figure 10-23 Production sequence of an MOS transistor: (a) oxide layer produced on a doped (*n*) silicon wafer; (b) *p*-type material diffused into wafer; (c) oxide layer produced on surface; (d) channel to *p*-region etched through oxide; (e) aluminum conductor deposited on surface; (f) aluminum contacts selectively separated.

Alternatively, in the *depletion-mode* transistor shown in Figure 10-24(b), the *n*-type silicon extends between the source and drain regions in a thin channel under the gate. Ordinarily, this device conducts current. But when a negative potential is applied to the gate, electrons are expelled from the channel and current ceases to flow. In this sense, the transistor acts as an electrical *switch*.

Other circuit components may also be fabricated in the IC by variations in the MOS technology and by the addition of discrete circuit elements. For instance, resistors, capacitors, and diodes can be integrated on the chips. As a matter of fact, many passive components such as resistors can be eliminated by the substitution of active components such as transistors. Indeed, a transistor can be regarded as a current-controlled or voltage-controlled *resistor*. In microelectronic circuits fabricated by MOS techniques, this practice may often be more economical than fabricating resistors or integrating discrete components into the circuit. A portion of a silicon chip containing several components is shown in Figure 10-25.

No doubt as IC technology continues to emerge, improvements in materials and processing will promote more "solid-state" applications. Very often, discoveries or developments in one area complement state-of-the-art technology in another field. Perhaps this is how advances in microelectronics will be realized in our immediate future.

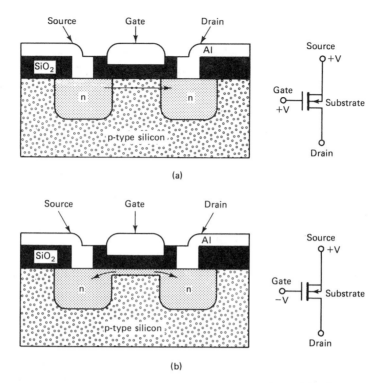

(a)

(b)

Figure 10-24 Schematic operation of *n*-type MOS transistor. (a) enhancement mode—positive potential at the gate produces current (electrons) from source to drain. (b) Depletion mode—thin continuous channel of *n*-silicon between source and drain. Transistor normally conducts, but current stops when a negative potential is applied to gate.

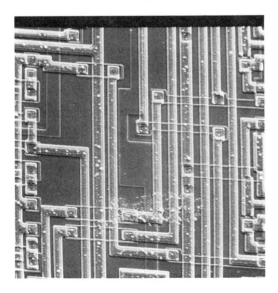

Figure 10-25 Portion of an integrated microelectronic circuit (chip) showing several semiconductor components (400×).

DIELECTRIC PROPERTIES

In our study of electron band theory we explained the reasons for differences in electrical conductivity between various materials. The electron band structure concept provided a lucid explanation of why some materials are excellent conductors of electrical current and others are very poor conductors—in other words, *insulators*.

When we place an insulating material between the plates of a capacitor to increase its capability to store charge, or when we consider the resistance of the insulating material to electric breakdown, the material is commonly referred to as a *dielectric*.

Dielectric materials are used to increase the capacitance of electrical capacitors and also as various insulators. The effectiveness and usefulness of these materials are graded by the following factors: *dielectric constant* (relative permittivity), *dielectric power loss*, and *dielectric strength*. We shall briefly examine the role of these factors in classifying dielectrics.

Dielectric Constant

The dielectric constant or relative permittivity (ϵ_r) of a material is defined as the ratio of the permittivity of the material (ϵ) to that of a vacuum (ϵ_0) and may be expressed as follows:

$$\epsilon_r = \frac{\epsilon}{\epsilon_0} \qquad (10\text{-}14)$$

This essentially means that if a material with a dielectric constant of ϵ_r was used in a capacitor, it would increase the capacity of that device by a factor of ϵ_r. Values of the dielectric constant for selected plastics, ceramics, and glasses are given in Table 10-6.

Example 10-7

A parallel-plate capacitor having a capacitance of 1 μF is to be fabricated from the following materials:

> *Plates (electrodes):* aluminum foil, 4 in. (10.2 cm) square and 1 mil (0.0025 cm) thick
>
> *Dielectric:* paper, 4.5 in. (11.4 cm) square, and 2 mils (0.0051 cm) thick
> $$\epsilon_r = 2.4$$

(a) How many plates (electrodes) and dielectric sheets are required, and (b) how thick will this capacitor be?

Solution (a) The total capacitance (C) in microfarads $(\mu$F$)$ of a simple parallel-plate capacitor can be determined from the following expression:

$$C = \frac{\epsilon_r A}{36\pi d \times 10^5}$$

where A = area of the plate
  d = distance between plates
  ϵ_r = dielectric constant

In our case, the area of a plate is

$$A = (10.2 \text{ cm})^2 = 104 \text{ cm}^2$$
$$d = 0.0051 \text{ cm}$$

Therefore,

$$C = \frac{(2.4)(104 \text{ cm}^2)}{36\pi(0.0051 \times 10^5) \text{ cm}}$$

$$= \frac{260}{5.77 \times 10^4} = 4.5 \times 10^{-3} \; \mu F$$

$$= 0.0045 \; \mu F \text{ "between two electrodes"}$$

So,

$$\frac{1 \; \mu F}{0.0045 \; \mu F} = 222 \text{ sections needed}$$

In other words, 223 electrodes and 222 sheets of paper will be needed.   *Ans.*

    (b) The capacitor thickness will be:

        Electrodes: (223)(0.0025 cm) = 0.558 cm

        Dielectric: (222)(0.0051 cm) = $\underline{1.132 \text{ cm}}$

                                1.690 cm   *Ans.*

TABLE 10-6 DIELECTRIC PROPERTIES OF SELECTED MATERIALS AT ROOM TEMPERATURE

Material	Dielectric constant (at 10^6 cycles)	Dielectric strength (kV/mm)	Loss factor
Air	1	3.0	—
Alumina	8.8	1.6–6.3	0.0002–0.01
Asbestos (paper)	3.0	—	—
Bakelite (paper base)	5–10	0.4–1.2	—
Calcium titanate	168	20–11.8	0.0001–0.02
Cambric (varnished)	4.2	32	—
Cellulose acetate	3.5–5.5	9.8–11.8	—
Fused silica	3.8	—	0.0002
Glass (soda–lime)	7.0–7.6	1.2–5.9	0.004–0.011
Mica	4.5–7.5	2.0–8.7	0.0015–0.002
Nylon	3.0–3.5	18.5	0.03
Phenol-formaldehyde resin (no filler)	4.5–5.0	11.8–15.8	0.015–0.03
Polyester resin	2.8–5.2	9.8–19.7	0.023–0.052
Polyethylene	2.3	18.1	0.0005
Polystyrene	2.4–2.6	19.7–27.6	0.0001–0.0004
Porcelain	5.5–7.0	1.6–15.8	0.003–0.02
Rubber (hard)	2.8	18.5	0.06
Shellac	2.9–3.7	7.9–23.6	—
Silicone molding compound (glass-filled)	3.7	7.3	0.0017
Steatite	5.5–7.5	7.9–15.8	0.0002–0.004
Styrene (shock resistant)	2.4–3.8	11.8–23.6	0.0004–0.02
Titanates (Ba, Sr, Mg, Pb)	15–12,000	2.0–11.8	0.0001–0.02
Titanium dioxide	14–110	3.9–8.3	0.0002–0.005
Urea-formaldehyde	6.4–6.9	11.8–15.8	0.028–0.032
Vinyl chloride (unfilled)	3.5–4.5	31.5–39.4	0.09–0.10
Wood	2.6	—	—

In general, temperature does not significantly affect the dielectric constant of ionic materials at low temperatures. Recall, however, that the conductivity of ionic materials depends chiefly on the mobility of ions and that this increases rapidly with temperature [see equation (10-9)]. For example, consider the combined effects of frequency and temperature on the relative permittivity of soda-lime glass as shown in Figure 10-26. At temperatures below about 60°C, ϵ_r is constant. Above this tempera-

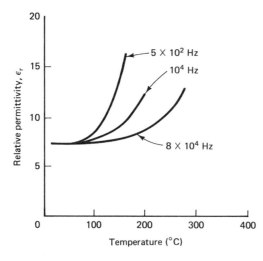

Figure 10-26 Effect of frequency and temperature on the permittivity of soda–lime silica glass. (From R. M. Rose, L. A. Shepard, and J. W. Wulff, *The Structure and Properties of Materials*, Vol. 4, John Wiley & Sons, Inc., New York, 1966, p. 261.)

ture, the permittivity increases significantly as temperature rises, and this effect is more pronounced the lower the frequency. Similar behavior has also been observed in some ceramics (e.g., Al_2O_3).

Insulators. Many of the common polymers (discussed in Chapter 14), are generally associated with insulating behavior. These "plastics" include polyethylene, polystyrene, and PVC; rubber should also be included as a familiar example of an insulator. Although the electrical conductivity of polymers depends on the type of chain-like structure and its attendant bonds, in the common polymers all the valence electrons are utilized in the covalent bond. Thus these valence electrons are not available for conduction unless they can be excited across a large energy band gap (see Table 10-4).

The electrical conductivity of the common polymers is usually very low (sometimes referred to as *leakage current*) and behaves oppositely from the metallic conductors. For instance, as temperatures increase, the conductivity of polymers increases, in a manner similar to the ionic (ceramic) materials. The temperature dependence of several amorphous polymers exhibit the familiar Arrhenius type of relationship, as shown in Figure 10-27. The abrupt change in slope noted on these curves occurs at the known *glass transition temperature* for the particular materials. This temperature (T_g) is the point below which no further atomic rearrangement occurs in the structure upon cooling.

Such behavior suggests that electrical conductivity in common polymers occurs by an ionic mechanism, as in certain ceramic materials. Therefore, the factors (i.e.,

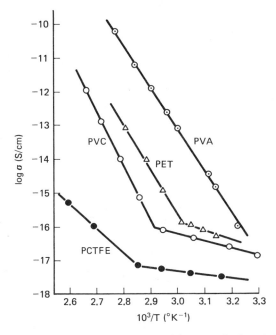

Figure 10-27 Temperature dependence of electrical conductivity in polymeric materials. PVC, polyvinyl chloride; PVAc, polyvinyl acetate; PCTFE, polychlorotrifluoroethylene; PET, polyethylene terephthalate. (From S. Saito et al., *J. Polymer Sci. 6A-2*, 1965, p. 1297.)

temperature, pressure, composition, etc.) that influence conductivity in ionic structures will also affect conduction in polymers.

Dielectric Power Loss

In capacitors that contain a dielectric material, the applied current leads the voltage by $90° - \phi$, where the angle ϕ is a measure of the dielectric power loss. The *loss factor* is the product of ϵ_r times the power factor (tan ϕ) and represents energy lost as heat. The usefulness of a dielectric is indicated by the value of the loss factor. Low values of this factor are sought, because power losses are undesirable and heating may contribute to breakdown. The power loss factor of selected dielectrics is also included in Table 10-6.

Dielectric Strength

The ability of a material to resist dielectric breakdown is an indication of its *dielectric strength*. More accurately, this factor may be defined as the maximum voltage gradient that a dielectric can withstand before breaking-down or failing. The units of dielectric strength are typically expressed as V/mm or V/mil. For example, the dielectric strength of air is about 3000 V/mm, whereas rubber and varnished cambric[5] exhibit dielectric strengths of approximately 16,000 V/mm and 32,000 V/mm, respectively. Values of the dielectric constant for selected materials are given in Table 10-6.

The process of dielectric breakdown involves electrons being able to jump the energy gap to the conduction band. These electrons may result from impurities

[5]Cambric is a fabric tightly woven from linen or cotton.

(donors), surface contamination, structural imperfections in the dielectric, and so on. In any event, once the breakdown process initiates, electrical current increases rapidly and the material fails. The failure may be interpreted as a loss of dielectric strength (material conducts current) or the material may physically fail by melting.

Applications for dielectric materials depend on the three factors we have just discussed plus the operating temperature they will experience. Generally, fabrics, paper, and polymers are utilized for service temperatures below about 90°C. High-temperature applications require dielectrics such as mica, glass, porcelain, and other ceramic materials.

Ferroelectric Behavior

Electric dipoles or the asymmetric distribution of electrical charge exists naturally in some materials and can be induced in others. Some molecules, notably the water molecule H_2O, exhibit a permanent dipole. This means that the centers of positive and negative charges are not coincident (as they are in electrically neutral molecules). The formation of an electric dipole is shown in Figure 10-28(a).

The spontaneous alignment of electric dipoles in a dielectric is referred to as

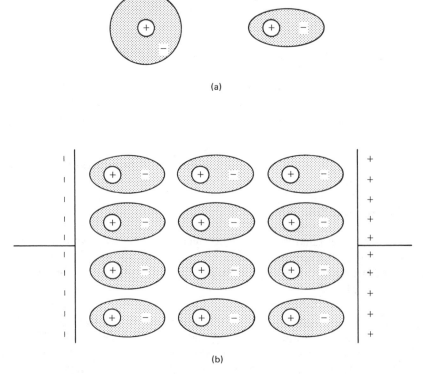

(a)

(b)

Figure 10-28 Schematic of an electric dipole: (a) dipole formed by an applied electric field; (b) alignment of dipoles in an electric field.

ferroelectricity. This ferroelectric behavior results from the fact that the local applied field (E) is increased by these parallel arrays of dipoles (polarization). Such a condition is illustrated in Figure 10-28(b). Barium titanate ($BaTiO_3$) is a ferroelectric ceramic material which changes its ionic structure slightly at approximately 115°C. Below this temperature (called the ferroelectric Curie point), the anions and the cations shift slightly in opposite directions, resulting in an electric dipole, as illustrated in Figure 10-29.

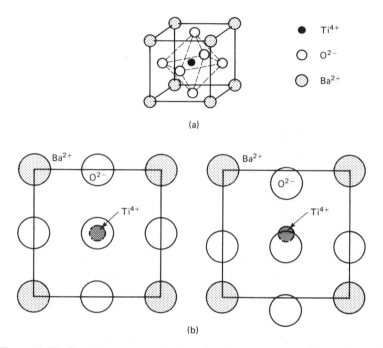

Figure 10-29 Structure of $BaTiO_3$: (a) unit cell arrangement; (b) top view of dipole formation below the Curie point.

Although barium titanate has been the most widely studied ferroelectric material, a number of other materials also display this behavior. Among these ferroelectrics are lead titanate ($PbTiO_3$), lead meta-niobiate ($PbNbO_3$), lead zirconate titanate $Pb(ZrTi)O_3$, Rochelle salt ($KNaC_4H_4O_6$), potassium dihydrogen phosphate (KH_2PO_4), tungsten oxide (WO_2), and other ceramic structures (e.g., perovskites and ilmenties). These particular materials are discussed in more detail in Chapter 15.

Piezoelectric crystals. The ferroelectric materials we have been discussing do not ordinarily exhibit a center of symmetry, in other words, a point about which the ions are symmetric. In these materials, a potential (electrical) difference is produced when the crystal is mechanically deformed (strained). Alternatively, the application of a voltage to such crystals produces dimensional changes or distortion. The change in dimensions results from the alignment of the electric dipoles in response to the applied field (Figure 10-28). Materials that exhibit this type of behavior are called *piezoelectric*. The efficiency of piezoelectric crystals is measured in terms of a *coupling coefficient*.

This factor indicates the fraction of applied mechanical energy that is converted into electrical energy (or vice versa). A list of some commercially important piezoelectric materials and certain properties they display is given in Table 10-7.

TABLE 10-7 PROPERTIES OF SOME PIEZOELECTRIC MATERIALS

Property	Material					
				Lead zirconate titanate		Lead meta-niobate
	Quartz	Lithium sulfate	Barium titanate	PZT-4	PZT-5	
Density (g/cm³)	2.65	2.06	5.6	7.6	7.7	5.8
Maximum operating temperature (°C)	550	75	70–90	250	290	500
Dielectric constant, ϵ_r	4.5	10.3	1700	1300	1700	225
Coupling coefficient (thickness mode)	0.1	0.35	0.48	0.64	0.68	0.42
Curie temperature (°C)	575	—	115	320	365	550

Source: Handbook of Tables for Applied Engineering Science, 2nd ed., CRC Press, Inc., Boca Raton, Fla., 1973, p. 799.

Piezoelectric crystals are utilized in a variety of transducer applications. Basically, a transducer is any device that transforms energy from one form to another. In this case, we are considering the conversion of electrical energy to mechanical, and vice versa. Transducers are an integral part of many instruments and systems, including sonar devices, microphones, phonograph pickup cartridges, accelerometers, strain gauges, and ultrasonic detectors. For example, in typical ultrasonic nondestructive examination (NDE) methods, the crystals listed in Table 10-7 are cut to produce the desired frequency (typically on the order of 1 to 15 MHz). These ultrasonic vibrations (ultrasound waves) are used to internally examine opaque materials, such as metals, plastics, and ceramics, for flaws. Furthermore, ultrasonic detection techniques have been extended to certain medical applications where X-ray radiology is sometimes not feasible because of possible health hazards.

STUDY PROBLEMS

10.1. A heart pacemaker lead consists of a nickel–cobalt alloy in a silver matrix. This conductor is encased in a polyurethane insulating sheath. If the resistivity of this composite wire is $\simeq 25\ \mu\Omega$-cm, what is the resistance for a segment 375 mm long by 1 mm in diameter?
Answer: $1.2 \times 10^5\ \mu\Omega$ or $0.12\ \Omega$

10.2. If the lead in problem 10.1 was made from copper, what would the resistance be?

10.3. Referring to problems 10.1 and 10.2, which conductor will dissipate less electrical energy? Discuss reasons for the use of the nickel–cobalt plus silver composite wire instead of pure copper or pure silver.

10.4. What is the resistance of 100 ft of a size 14-2 annealed copper conductor (wire diameter = 64.1 mils)?

Answer: $1.24 \times 10^{-1}\ \Omega$

10.5. An electric baseboard heater is wired with 15 m of size 10-2 annealed copper wire (2.59 mm diameter). If this heater has a resistance of 46 Ω and is used an average of 2.5 hr per day, how much electrical energy is consumed when the circuit operates on 220 V? Is the resistance contribution of the wire significant in this case?

10.6. Compare the current density (j) for silver and vulcanized rubber if the strength of an applied electric field is 400 V/m.

10.7. An 96% iron (Fe)–4% silicon (Si) alloy is used for transformer and generator laminations because of its magnetic properties. During service the temperature of a particular transformer increases from 75°F to 250°F.

(a) What is the change in resistivity for this condition?

(b) What effect does this change have on the performance of the transformer?

10.8. A strain gauge similar to the one depicted in Figure 10-4 is made with metallic foil having cross-sectional dimensions 0.50 mm × 0.25 mm. The grid contains 14 foil ribbons 10 mm in length.

(a) Determine the resistance of the gauge (omitting the tab portions) if the foil material is *constantan*.

(b) What would be the change in resistance if the gauge is used in an application where the temperature increases from 75°F (24°C) to 250°F (121°C) during the test?

(c) What effect do you think the result in part (b) would have on the attendant strain measurements?

Answer: (b) $\Delta R = +0.5\ m\Omega$

10.9. A coiled heating element consisting of a 0.064-in.-diameter wire 300 in. long is made from *high-resistance alloy* (see Table 10-1).

(a) What is the power requirement of this device in a 120-V circuit at room temperature?

(b) What is the power requirement if the temperature of the system increases from 75°F to 350°F during operation of the heating element?

10.10. In addition to the temperature coefficient of resistivity, other properties must be considered in many resistive applications. Name and discuss some of these properties.

10.11. In a certain critical aerospace application, it is suggested that a 1.0-mm-diameter copper conductor be replaced by gold because of the latter's better corrosion resistance.

(a) Since gold is considerably more expensive than copper, what is the minimum-diameter wire you can substitute that will carry the same current?

(b) Based on current market values, estimate how much more costly the gold substitute is if the total length of wire is 10 m.

Answer: (a) 1.19 mm

10.12. In problem 10.1 we examined a special "composite" wire conductor for a heart pacemaker application. If the maximum-diameter wire that can be transvenously implanted is 1.30 mm and the wires must carry the same current, can this conductor be made from 18-8 stainless steel? (Assume that the leads are the same length.)

10.13. Sketch a possible energy band configuration for lithium (Li). Do you expect Li to be a conductor or an insulator?

10.14. Explain why beryllium (Be), which contains two electrons in the $1s$ shell and two electrons in the $2s$ shell (see Table 2-2) is a conductor. Sketch the probable energy band configuration for Be.

10.15. (a) Explain the basic difference between intrinsic and extrinsic semiconductors. (b) What is the magnitude of the energy gap in a typical semiconductor? Express your answer in joules. (c) A YAG (yttrium-aluminum-garnet) LASER produces electromagnetic radiation in the near infra-red spectrum with a wavelength (λ) of 1.064 μm. Since the wavelength and frequency (ν) of electromagnetic radiation are related to the velocity of light (c) as follows: $c = \lambda \nu$, determine if a photon of the radiation has sufficient energy to cause an electron to jump the energy gap in gallium arsenide (GaAs).

10.16. (a) Determine the percentage of the conductivity of silicon which is due to conduction electrons. (b) What percentage is due to electron holes in the valence band?
Answer: (a) 74% (b) 26%

10.17. At 20°C, the semiconductive compound InAs exhibits an intrinsic conductivity of 10^4 $(\Omega\text{-m})^{-1}$. How many electrons have jumped the energy gap?
Answer: 2.7×10^{22} per m^3

10.18. What is the total number of charge carriers per unit volume in problem 10.17?

10.19. What is meant by a *donor*-type atom with regard to semiconductors? What does the term *acceptor* mean in this same regard? List some donors and acceptors for silicon and the periodic table group in which they reside.

10.20. The conductivity of germanium is 2 S/m. How will the addition of 1×10^{16} phosphorus atoms/cm^3 affect this value?

10.21. Silicon is used for a certain semiconductor chip application. If the conductivity of one portion of this semiconductor must be 0.10 S/m, and a *p*-type material is required, how much boron needs to be added to the Si?
Answer: 1.24×10^{19} atoms/m^3

10.22. (a) What is the concentration of boron in the solid solution produced in problem 10.21?
(b) How would such a concentration be measured?

10.23. (a) The conductivity of germanium is 2 S/m at 20°C. What is the conductivity of this material at 75°C.
(b) What is the resistivity of Ge at 75°C?

10.24. At 300°C, the intrinsic conductivity of silicon is equal to the room-temperature extrinsic conductivity of a certain Al-doped silicon semiconductor. How many Al atoms are required to produce this effect?

10.25. An experimental semiconductor material exhibits a conductivity of 1×10^{-2} S/m at 20°C and 5 S/m at 100°C. What is the size of its energy gap?
Answer: 1.53 eV

10.26. A certain underground cable application requires 1000 ft of No. 4 AWG single conductor (diameter = 204.3 mils). The conductor is insulated with a silicone rubber compound $\frac{1}{8}$ in. thick. Measurements show the capacitance of the cable to be 0.100 μF. What is the ϵ_r of the silicone rubber compound? [*Hint:* $C = 0.039\epsilon_r/\log(r_2/r_1)$ in μF/mi for coaxial cylinders.]

10.27. Design a parallel-plate capacitor with a total capacitance of 10 μF using aluminum foil (1 mil thick) and calcium titanate (CaTiO$_3$) 3 mils thick if the capacitor must fit within a rectangular area 2.5 in. by 3.75 in. How much volume will this device occupy?

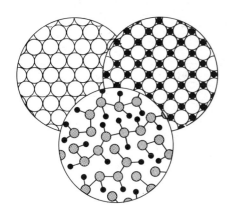

CHAPTER ELEVEN:
Magnetic Properties of
Engineering Materials

MAGNETIC BEHAVIOR

The magnetic properties of engineering materials is a subject that presently includes both metals and ceramics. Magnetic behavior has traditionally been associated with metals, that is, iron, nickel, cobalt, and their alloys, such as Alnico (Al–Ni–Co–Fe). These materials are widely used as permanent magnets in transformers, motors, induction coils, controls, and so on. However, certain ceramic materials also exhibit magnetic behavior. Indeed, *lodestones* utilized by early sailors to assist navigation consisted of pieces of magnetite (Fe_3O_4). Recently, considerable attention has been focused on the properties of *ceramic magnets*. This group of oxides, also known as "ferrites," includes $PbFe_{12}O_{19}$, $BaFe_{12}O_{19}$, and $NiFe_2O_4$. Currently, ferrites are being developed for application to high-frequency devices and computer memory units.

Any discussion of magnetic characteristics and behavior in engineering materials should properly begin with a review of the physics of magnetism. A magnetic field is analogous to an electrical field and can be produced either by current-carrying electrical conductors or by permanent magnets.

In the case of an electrically induced magnetic field, a coil of wire in space carrying a current will produce a magnetic field in the center of the coil. The field strength, termed H, measured in the coil is the product of the number of turns per meter of coil length and the current, so that

$$H = Ni \tag{11-1}$$

where N = number of turns per meter
i = current (A)

Within the coil, lines of force or flux exist as shown in Figure 11-1. The flux

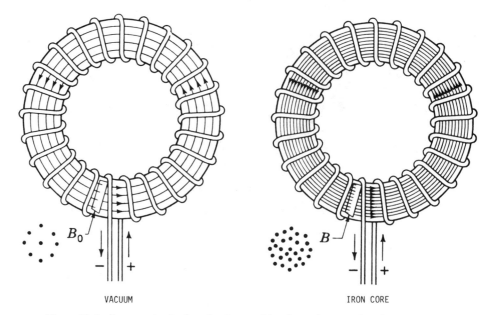

VACUUM   IRON CORE

Figure 11-1 Increase in the flux density resulting from the use of an iron core.

density (B) within the coil can be determined by the expression

$$B = \mu_0 H \tag{11-2}$$

where H = field intensity (A-turns/m)
B = flux density (Wb/m²)

These units are expressed in rationalized MKS units. Since there is frequently some confusion concerning magnetic units, the equivalent CGS and English units are presented in Table 11-1, and Table 11-2 gives factors for converting from one system to another. The term μ_0 represents the permeability of the vacuum within the coil = $4\pi \times 10^{-7}$ Wb/A-m. If the space within the coil contains a material other than a vacuum, the flux density will be altered. For example, if an iron core is used, an increase in the lines of flux results. The increase in flux density due to the insertion of a ferro-

TABLE 11-1 MAGNETIC UNITS IN VARIOUS SYSTEMS

Quantity	CGS	Rationalized MKS	English
Flux, ϕ	maxwell	weber	line
Flux density, B	gauss	weber/square meter	line/square inch
Field intensity, H	oersted	ampere-turn/meter	ampere-turn/inch
Length	centimeter	meter	inch
Equation (11-1)	$H = 0.4\pi NI$	$H = NI$	$H = NI$
Permeability of vacuum, μ_0	1	1.257×10^{-6}	3.20

TABLE 11-2 CONVERSION FACTORS FOR MAGNETIC UNITS

Multiply number of:	By:	To obtain number of:
	Magnetic flux, ϕ	
kilolines (kilomaxwells)	1000	maxwells (lines)
kilolines (kilomaxwells)	10^{-5}	webers
maxwells (lines)	10^{-8}	webers
webers	10^5	kilolines (kilomaxwells)
webers	10^8	maxwells (lines)
	Magnetic flux density, B	
gauss	6.452	lines per square inch
gauss	10^{-4}	webers per square meter
lines per square inch	0.1550	gauss
lines per square inch	1.550×10^{-5}	webers per square meter
webers per square meter	10^4	gauss
webers per square meter	6.452×10^4	lines per square inch
	Magnetic field intensity, H	
ampere-turns per inch	39.37	ampere-turns per meter
ampere-turns per inch	0.495	oersteds
ampere-turns per meter	0.0254	ampere-turns per inch
ampere-turns per meter	0.01257	oersteds
oersteds	2.021	ampere-turns per inch
oersteds	79.58	ampere-turns per meter

magnetic material can be expressed by the following:

$$B = \mu_0 H + \mu_0 M \tag{11-3}$$

where M = magnetization and the term
$\mu_0 M$ represents the increase in field strength

Alternatively, the flux density can be expressed as

$$B = \mu H \tag{11-4}$$

where μ is the permeability of the material.

Magnetic Permeability

The magnetic permeability is defined as the ratio of B/H as shown in equation (11-4). This term is a useful and distinguishing feature in characterizing the magnetic behavior of materials. In actual systems, however, the relative permeability μ_r is commonly used. This is simply the ratio of the real permeability of the material to the permeability of a vacuum:

$$\mu_r = \frac{\mu}{\mu_0}$$

Diamagnetic behavior. Diamagnetic materials usually exhibit a linear relationship of B versus H, with μ_r being slightly less than 1. Figure 11-2(a) illustrates the appearance of the magnetization curve for a diamagnetic material. Diamagnetic metals include copper, silver, gold, and bismuth.

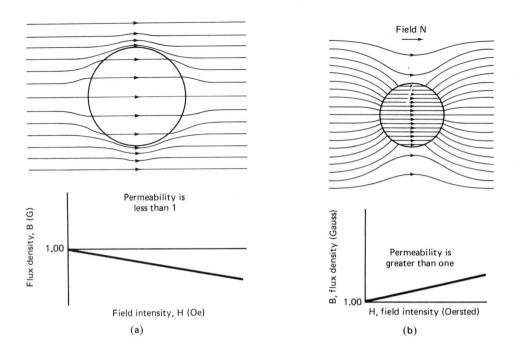

Permeability is
less than 1

1.00

Flux density, B (G)

Field intensity, H (Oe)

(a)

Field N →

B, flux density (Gauss)

Permeability is
greater than one

1.00

H, field intensity (Oersted)

(b)

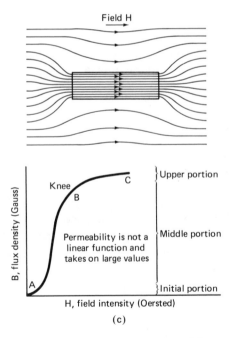

Field H →

B, flux density (Gauss)

Knee

C — Upper portion

B

Permeability is not a
linear function and
takes on large values

Middle portion

A

Initial portion

H, field intensity (Oersted)

(c)

Figure 11-2 Flux density (B) versus magnetic field strength (H) curves for (a) diamagnetic, (b) paramagnetic, and (c) ferromagnetic materials. (From J. K. Stanley, *Metallurgy and Magnetism*, American Society for Metals, Metals Park, Ohio, 1949.)

Paramagnetic behavior. When a material displays paramagnetic behavior, the relationship is also approximately linear, with μ_r being slightly greater than 1, as illustrated in Figure 11-2(b). Most metals are paramagnetic: e.g., lithium, sodium, potassium, calcium, strontium, magnesium, molybdenum, and tantalum.

Ferromagnetic behavior. For a ferromagnetic material μ_r is generally much larger than 1 and exhibits a nonlinear relationship between B and H, as illustrated in Figure 11-2(c). The initial permeability is low, then rises sharply and levels off at high field intensities. An important feature of ferromagnetic materials is that they retain their magnetism after the magnetizing field has been removed. This is the reason that iron, cobalt, nickel, and gadolinium are used for permanent magnets.

B versus H. The magnetization curve (i.e., B versus H curve) for ferromagnetic materials provides a useful tool for evaluating the behavior of ferromagnetic materials. The curve depicting the magnetic response is not reversible. The initial curve for virgin material resembles that shown in Figure 11-2(c); however, when the applied field (H) is reduced to zero, from its initial peak, a residual flux density (B_r) called *remanence* remains. To remove this remanence, an applied field in the opposite direction must be applied. The magnitude of the applied reverse field is termed H_c, the coercive force. If the field intensity is increased to its maximum in this direction, then reversed again, a plot such as the one presented in Figure 11-3 results. The area enclosed within the plot is referred to as the hysteresis loop and represents the energy loss per cycle.

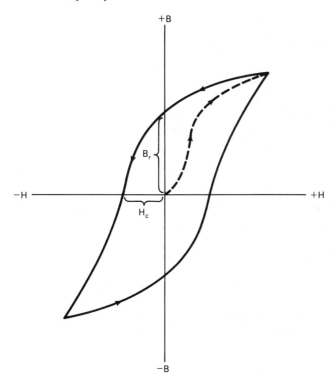

Figure 11-3 Typical hysteresis curve for a ferromagnetic material: B_r, remanence; H_c, coercive force.

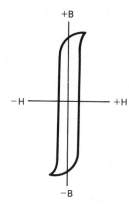

Figure 11-4 Hysteresis curve for a *soft* magnetic material. Typical magnetic core material: small coercive force (H_c) required to eliminate large residual magnetism (B_r).

The characteristics of various ferromagnetic materials with regard to their hysteresis behavior is sometimes used to classify them. Materials exhibiting low hysteresis and low coercive force are labeled *soft*. Their behavior is illustrated by the curve presented in Figure 11-4. Such materials find their greater usage in power applications, such as transformer cores. It is desirable in these applications to keep the hysteresis loop to a minimum, since each cycle represents a power loss. On the other hand, for certain other applications, such as permanent magnets, it is desirable to have a high coercive force to maintain the field strength. These materials are referred to as *hard* and their behavior is represented by Figure 11-5.

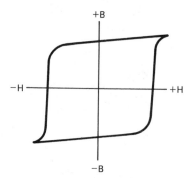

Figure 11-5 Hysteresis curve for a *hard* magnetic material. Typical of a class of permanent magnet material: large coercive force (H_c) required to eliminate large residual magnetism (B_r).

Magnetic materials with square hysteresis loops are used as memory storage devices in computers. Since they can exist in two distinctly different magnetic states, they are ideally suited for the binary (on–off) codes used in computers.

Theory of Ferromagnetism

What forms the basis for ferromagnetic behavior? On the basis of physical studies, it is believed that each electron exhibits a magnetic moment, and when placed in a magnetic field, aligns itself with the applied field, the direction of alignment being dependent on the spin direction. Since an electron can spin in either of two directions, opposite alignments are possible. Each spinning electron has a magnetic moment associated with it which can be calculated from the expression:

$$\frac{eh}{4\pi mc} = \text{one Bohr magneton} \qquad (11\text{-}5)$$

where e = charge on the electron
h = Planck's constant
m = mass of the electron
c = velocity of light

This quantity, termed the Bohr magneton, equals 0.927×10^{-20} erg/G (9.27×10^{-24} A-m^2). Since any quantum state can be filled by two electrons of opposite spin, in a completely filled shell the magnetic moment, due to spin, is zero. However, each electron not balanced by one spinning in the opposite direction, contributes a net magnetic moment of one magneton.

An examination of the literature reveals that of all the elements, only the metals iron, cobalt, nickel, gadolinium, and disprosium exhibit ferromagnetic behavior. These transition elements have unfilled shells of electrons, in particular, the third shell. The net magnetic moment due to spin in a partly filled electron shell is not always apparent. However, *Hund's rule* states that the spins of electrons in a shell add together in such a manner that they contribute the maximum magnetic moment. If one examines the electronic structure of the third shell with respect to the direction of spin, one will see that there are 10 states, 5 with spin up and 5 with spin down, as shown in Figure 11-6. The maximum magnetic moment occurs when the state contains five electrons, all spinning in the same direction. The addition of the sixth electron reduces the net magnetic moment, since it cancels the spin of the first electron. On this basis, one would expect that the metal manganese would have the highest magnetic moment. In actuality, this does not occur, because the individual manganese atoms align themselves in such a way that the magnetic contribution of one atom is canceled by the next. This unfavorable *exchange* energy interaction between atoms accounts for the fact that metallic manganese does not exhibit ferromagnetism. In manganese, the unpaired electrons between adjacent atoms are arranged so that the magnetic moments tend to be neutralized. Therefore, ferromagnetism is related not only to the electronic configuration of individual atoms, but also the interaction of each atom to its neighbors in the lattice.

Element	Number of electrons	Electronic structure 3d shell	Magnetic moment
Mn	25	↑ ↑ ↑ ↑ ↑	5
Fe	26	↑↓ ↑ ↑ ↑ ↑	4
Co	27	↑↓ ↑↓ ↑ ↑ ↑	3
Ni	28	↑↓ ↑↓ ↑↓ ↑ ↑	2
Cu	29	↑↓ ↑↓ ↑↓ ↑↓ ↑↓	0

Figure 11-6 Electronic configuration for third electron shell.

Magnetic Behavior

Domain Theory

Since quantum theory has shown that certain of the transitional elements display ferromagnetic behavior, how is it possible for these elements, such as iron, to exist in the unmagnetized state? Weiss (1907) hypothesized that in a ferromagnetic material there existed microscopic regions called *domains* (10^{-6} to 10^{-5} cm³) within which the magnetic moments of all the spinning electrons are parallel to each other (i.e., magnetized to saturation). In unmagnetized iron, the domains are arranged so that the net magnetic moment of the specimen is zero, as shown in Figure 11-7(a). The boundary between two domains oriented in different directions is termed a *Bloch wall* and its structure is illustrated in Figure 11-7(b). The domain wall has a thickness of approximately 300 lattice parameters.

As a magnetic field is applied, the domains that are favorably oriented with respect to the field grow at the expense of adjacent domains which are not as favorably oriented (i.e., those in which the angle of orientation is large relative to the direction of the applied field). At higher field values, entire domains may rotate into parallelism with the applied field. These *Barkhausen* jumps provide evidence that the magne-

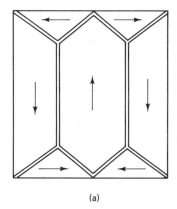

(a)

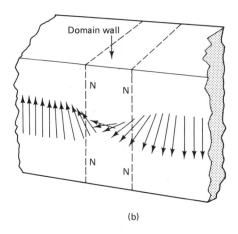

(b)

Figure 11-7 (a) Two-dimensional representation illustrating domain orientation; (b) schematic presentation of Bloch wall between two domains.

tization proceeds by a series of discontinuous steps as the domains rotate through large angles. Finally, at the highest field strengths, essentially all the domains are oriented favorably and the specimen is magnetically saturated. The ease of movement of the domain walls affects the magnetic characteristics of the material. If the domain walls are easy to move, the coercive force is low and the material is *soft*. If domain walls are hard to move, the coercive force is large and a magnetically *hard* material results. Domain walls may be physically pinned by structural imperfections, such as inclusions, precipitated particles, or voids. Therefore, the presence of these imperfections has a significant effect on the magnetic behavior of the material and must be controlled.

Antiferromagnetic behavior. In some materials such as manganese, the exchange energy balance has the magnetic moment of adjacent atoms aligned in the opposite direction. The net result is that the spins nullify one another, as shown in Figure 11-8. An interesting observation which can be made with such materials is that the permeability tends to increase with increasing temperature, up to a point. This increase in permeability is the result of the increased mobility of the individual atoms and the resulting decrease in the order of the system, the order of the system being the very foundation of the antiferromagnetic behavior.

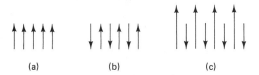

Figure 11-8 Illustration of magnetic moment alignments for (a) ferromagnetic, (b) antiferromagnetic, and (c) ferrimagnetic materials.

Magnetostriction

It is generally characteristic of ferromagnetic materials that they exhibit a change in dimension when a magnetic field is applied. For example, iron expands in the direction of magnetization and contracts laterally at small values of H, whereas nickel and cobalt act oppositely, as shown in Figure 11-9. This elastic deformation phenomenon is termed magnetostriction and is quantified by the coefficient of magnetostriction (λ), where

$$\lambda = \frac{l_f - l_0}{l_0} = \frac{\Delta l}{l_0} \tag{11-6}$$

where l_0 = original length
l_f = final length

For the same reason that the elastic constants of a solid are not isotropic, magnetostriction will also be anisotropic and a function of the crystal orientation and the orientation of the applied field. The *process is reversible* and the application of a stress will decrease the magnetization of nickel, whereas the same stress will increase the magnetization of an alloy such as Permalloy. Magnetostriction has found uses in alternating-current applications for transducers where electrical energy is converted

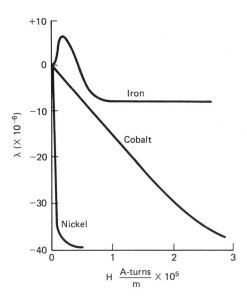

Figure 11-9 Coefficient of magnetostriction as a function of the applied magnetic field intensity for iron, nickel, and cobalt.

into mechanical energy and vice versa. As for example, the magnetic cartridge of an audio turntable which converts vibrations contained in the record grooves into electrical energy.

Effect of Temperature on Magnetic Behavior

The permeability of ferromagnetic material decreases as the temperature rises and the relative permeability falls to practically unity at a temperature called the *Curie temperature* (T_θ), which is different for each material. This behavior is illustrated for iron in Figure 11-10. Below the Curie temperature (780°C), iron is ferromagnetic;

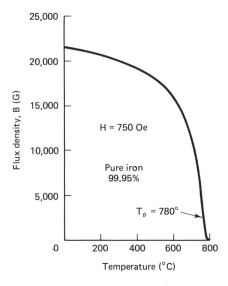

Figure 11-10 Effect of temperature on the magnetization of iron.

however, above this temperature iron behaves paramagnetically. This behavior is believed to be the result of the increased disorder introduced into the structure at higher temperatures.

Anisotropy

It should not be surprising that magnetic behavior, as with other properties, can display differences depending on crystal orientation. Studies conducted on single ferromagnetic crystals demonstrate that there are crystallographic directions in which *easy* magnetization occurs and others in which magnetization is more difficult. A portion of the *B* versus *H* curve for bcc iron is presented in Figure 11-11. As can be seen, the [100] direction is easy and the [111] direction is the most difficult (i.e., *hard*). It should be noted that the values of B at saturation are the same, but at low and intermediate magnetic fields, the values are lower.

As you will see in the next section, this phenomenon is used in the development of improved materials for transformer cores in which the power losses are minimized by using grain-oriented sheet.

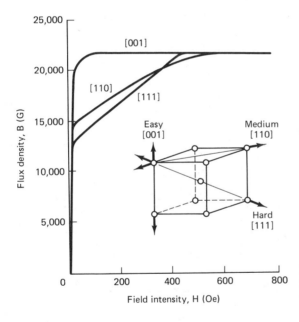

Figure 11-11 Magnetization curves for iron in various crystallographic directions.

TYPES OF MAGNETIC MATERIALS

By far the largest use of magnetic materials is in the field of electrical power handling and transmission equipment. Examples of such applications include transformer cores, motor armatures, and generator cores. These materials must be easy to magnetize to a

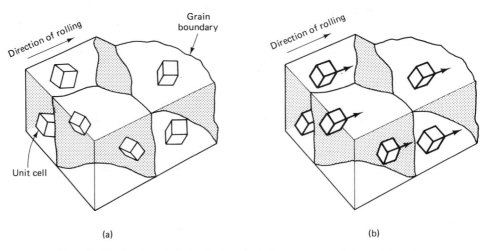

Figure 11-12 Grain orientation in polycrystalline silicon steel sheet: (a) random orientation; (b) preferred orientation after cold rolling and annealing. The small cubes indicate the orientation of individual grains.

high value of *B*, but must also be easy to demagnetize as the current is reversed. In these alternating-current applications, the electrical field is rapidly cycled many times per second. Since the area enclosed by the hysteresis curve represents an energy loss, it should be obvious that to minimize such losses, soft magnetic materials are the best engineering choice.

At the common power frequencies (50 to 60 Hz), the Fe–Si alloys are the most commonly used core materials. These steels, which range in silicon from 0.5 to 5%, were first introduced in 1907. Steels with silicon contents of 0.5 to 3.5% are usually employed for rotating power generation equipment, while those with higher silicon contents (up to 5%) are used for transformer cores and other stationary electrical equipment. The reason for this is because the higher silicon tends to make the steel more brittle and less able to withstand the stress associated with the rotating components. Silicon plays a dual role in these alloys; it increases the electrical resistance and thereby reduces "eddy current" losses (described below) and renders the carbon and oxygen, which are present as impurities, less harmful. If the processing of the Fe–Si sheet, in particular the rolling and annealing, is carefully controlled, an oriented crystal structure as shown in Figure 11-12 can be obtained. The direction of easy magnetization, then, is in the rolling direction. It is therefore easier to magnetize this preferred orientation sheet in the rolling direction than it is to so magnetize ordinary randomly oriented sheets. To take advantage of this orientation effect, the transformer cores should be stamped from the sheet in such a manner that the longest magnetic path is in the best magnetic direction.

As the magnetic field passes through the material, electrical currents, called *eddy currents*, are generated within the material. Eddy currents are induced by the fluctuating magnetic field and appear as shown in Figure 11-13. This eddy current generation results in another form of power loss as heat and is directly related to the frequency of the electrical current. These losses can be reduced by decreasing the current path through the use of thin plates or laminations (Figure 11-14) or by increasing the

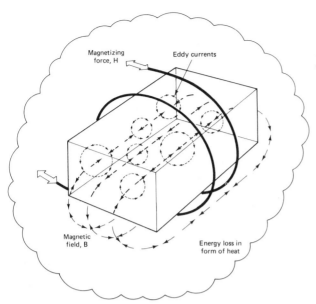

Figure 11-14 Reduction of eddy currents by laminating and insulating transformer core sections. (a) Unlaminated core: eddy currents cause high energy losses and heat generation. (b) Laminated core—uninsulated: eddy currents are reduced by laminating; energy losses and heat generation are reduced. (c) Laminated core—insulated: eddy currents limited within each lamination; energy losses and heat generation reduced still further.

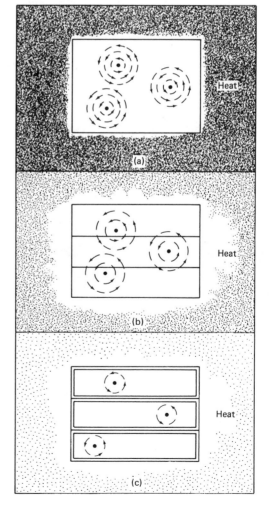

resistivity of the material. Consequently, at very high frequencies, materials with high resistivity, such as ceramic ferrites, are used to reduce power losses.

Due to recent advances in solidification technology, one particular group of materials, *amorphous metal alloys*, has become particularly attractive for transformer cores. For instance, amorphous Fe–B–Si alloys produced by rapid solidification (see RSP in Chapter 7) in the form of ribbon or strip are relatively easy to magnetize. Therefore, very little heat energy is generated in the process of magnetizing a transformer core in order to induce a voltage in the secondary windings. Even though transformer efficiency is extremely good in conventional core materials, certain "low-loss" core materials, such as amorphous Fe–B–Si, can further reduce the economic consequences of continuous (24 hr per day) energy dissipation, in the form of heat loss, from an energized transformer core.

Ceramic Magnetic Materials

It is important to recognize that not all magnetic materials are metallic in nature. A certain group of metallic oxides known as "ferrites" (not to be confused with α-iron, which is also called ferrite), which have a cell structure resembling that of the spinels (Chapter 15), exhibit *ferrimagnetic* behavior. A ferrimagnetic material is one in which the net magnetic moment for ions on adjacent crystal planes is unequal, resulting in a net macroscopic magnetic moment (see Figure 11-8c). These ferrites have the general formula XM_2O_4, where X is a divalent metal such as Fe, Mn, Ni, or Cu and M is a trivalent metal ion. The oxygen ions are arranged to form eight unit cells with an fcc structure as shown in Figure 11-15. You will discover in our discussion of oxide structures in Chapter 15 that the spinel structure has 32 potential octahedral vacancy sites and 64 tetrahedral vacancy sites; however, not all are occupied. In an *inverse* spinel structure, the eight divalent (X^{2+}) ions occupy octahedral sites, whereas in a normal spinel structure they would occupy tetrahedral sites. The 16 trivalent (M^{3+}) ions are divided equally between octahedral and tetrahedral sites. With ferrites as well as other magnetic materials, the magnetism is the result of the net magnetic moment of

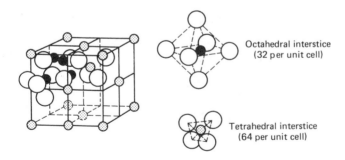

Octahedral interstice
(32 per unit cell)

Tetrahedral interstice
(64 per unit cell)

○ Oxygen

● Cation in octahedral site

◉ Cation in tetrahedral site

Figure 11-15 Structure of a spinel showing location of interstitials. (From A. R. von Hippel, *Dielectrics and Waves*, John Wiley & Sons, Inc., New York, 1954.)

the atoms comprising the structure. In the case of ferrites, the moment results from the divalent (X^{2+}) ions occupying the octahedral sites. These are all aligned in the same direction, whereas the (M^{3+}) trivalent ions are aligned oppositely and equally with respect to spin direction (and tend to balance magnetically) depending on whether they occupy octahedral or tetrahedral sites. By varying the composition of the divalent ionic species, or by substituting nonmagnetic species, the magnetic behavior of the resulting ferrites can be adjusted to suit specific applications.

Applications of spinel ferrites are usually separated into three broad categories according to the frequency of the magnetic field of the device in which they will be used: (1) low-frequency, high-permeability applications; (2) high-frequency, low-loss applications; and (3) microwave applications. Specifically, the "magnetic" tape utilized in tape recording devices consists of ferrite powder particles deposited on a plastic (acrylic) substrate. The ferrite in this case is most commonly γ-Fe_2O_3. Ferrites are also used as the beam deflectors in television receivers and for transformer cores.

Example 11-1

Calculate an approximate value of the magnetic moment per unit cell for (a) $[MgAl_2O_4]_8$ and (b) $[Fe_3O_4]_8$.

Solution (a) The $[MgAl_2O_4]_8$ is considered a normal spinel in which the trivalent aluminum ions occupy the octahedral sites and the divalent magnesium ions occupy the tetrahedral sites. Since there are no unpaired electrons, the number of Bohr magnetons for each of these ion species is zero and the magnetic moment for the unit cell is also zero.   *Ans.*

(b) The $[Fe_3O_4]_8$ may also be considered as $[Fe^{+2}Fe_2^{+3}O_4]_8$ an inverse spinel. In this case the trivalent Fe^{+3} ions cancel since half occupy tetrahedral sites and half occupy octahedral sites, and the spins cancel. The divalent Fe^{2+} ions contribute four Bohr magnetons, so that

$$M = \frac{4\ \text{BM}}{\text{ion}} \left(\frac{8\ \text{ions}}{\text{unit cell}} \right) = 32 \text{ Bohr magnetons/unit cell}   \textit{Ans.}$$

In addition to the spinel structure, the garnet crystal structure also exhibits very interesting magnetic properties. The rare earth garnets have the general formula $M_3Fe_2Fe_3O_{12}$, where M is a rare earth ion such as lanthanum, samarium, gadolinium, or yttrium. Just as in the spinels, a net magnetic moment originates from the antiparallel electron spins of the ions.

Thin films of magnetic garnets may be deposited on nonmagnetic substrates. This process produces a preferred magnetization normal to the plane of the film. When observed under polarized light in the microscope, small magnetic domains with spin up, separated by regions with spin down, appear as *bubbles*. This magnetic array can provide the binary input for a digital computer and is referred to as the *bubble memory*.

Permanent Magnets

The strength of a permanent magnet material is related to the product of the coercive force (H_c) and the residual flux density (B_r), since this is a measure of the available energy. Quite frequently in the evaluation of permanent-magnet materials this energy

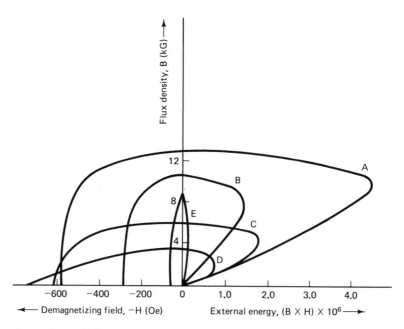

Figure 11-16 *B–H* curves for several permanent magnet materials. A, Alnico V; B, Vicalloy I; C, Cunife; D, Cunico; E, carbon steel.

product $(B \times H)$ is plotted against B, and the demagnetization force $(-H)$ is also plotted against B, in a combined curve. For metallic magnets the maximum energy product (BH) can be approximated by $\frac{1}{2}(H_c B_r)$. For ceramic permanent-magnet materials, the value of the maximum energy product is best approximated by the relation $B_r^2/4$. Exhibited in Figure 11-16 are these combination curves for a variety of magnet materials.

These alloys fall into two main product classifications:

1. High-carbon and alloy steels bearing Cr, Co, W, or Al that depend on the martensite transformation and carbide formation to develop a microstructure which retards domain growth and rotation.

2. Precipitation hardening materials such as the Cu–Ni–Co (Cunico), Cu–Ni–Fe (Cunife), Ag–Mn–Al (Simanal), and Al–Ni–Co (Alnico) alloys, whereby the precipitation produces a still finer substructure.

As can be seen from an examination of Table 11-3, the Alnico alloys are usually produced by casting, since they are hard and brittle, whereas the Cunife and Cunico alloys can be formed and machined. The properties of the Alnico group can be enhanced by controlled solidification and heat treatment in a magnetic field. This results in a microstructure which is comprised of oriented needle-like particles (Figure 11-17) whose dimensions approach those of a domain.

This technique of utilizing small particles of Alnico whose dimensions are less than that of a domain (> 1 μm) has been used to produce permanent magnets by pressing and sintering powders using nonmagnetic binder materials.

TABLE 11-3 MAGNETIC PROPERTIES AND COMPOSITION OF SEVERAL PERMANENT MAGNET ALLOYS

Grade	Average chemical composition	Average magnetic values			Density (g/cm³) ρ	Average physical properties		Form	Finish	Optimum heat treatment	
		Residual flux density (G)[a] B_r	Coercive force (O)[b] H_c	Maximum energy product $(10)^6$ (G-O) BH max.		Rockwell C hardness Annealed	Rockwell C hardness Hardened			Hardening temperature (°C)[c]	Time (min) at temperature (depends on mass)
Designations:											
3.5% Cr steel	3.5 Cr, 1 C, bal. Fe	10,300	60	0.302	7.8	15–25	60–65	Rolled or cast	Hot or cold form, machine or grind	850	5–15
3% Co steel	3.25 Co, 4 Cr, 1 C, bal. Fe	9,700	81	0.382	7.8	15–25	60–65	Rolled or cast	Hot or cold form, machine or grind	880	5–15
18.5% Co steel	18.5 Co, 3.75 Cr, 5 W, 0.75 C, bal. Fe	10,700	160	0.690	8.0	25–35	60–65	Hot rolled	Hot form, machine or grind	955	5–15
37% Co	38 Co, 3.8 Cr, 5 W, 0.75 C, bal. Fe	10,400	230	0.982	8.2	30–35	58–62	Cast	Machine or grind	940	10–15
						38–45	60–65	Hot rolled or cast	Hot form, machine or grind	925	5–15
Alnico 1	12 Al, 22 Ni, 5 Co, 0.35 Ti, bal. Fe	6,800	530	1.38	6.9	—	45	Cast	Grind	Furnished heat-treated	
Alnico 2	10 Al, 17 Ni, 12.5 Co, 6 Cu, 0.45 Ti, bal. Fe	7,500	550	1.65	7.1	—	45	Cast	Grind	Furnished heat-treated	
Alnico 2C	10 Al, 21 Ni, 12.5 Co, 6 Cu, 0.30 Ti, bal. Fe	7,100	580	1.60	7.0	—	45	Cast	Grind	Furnished heat-treated	
Alnico 3	12 Al, 26 Ni, bal. Fe	6,600	450	1.20	6.9	—	45	Cast	Grind	Furnished heat-treated	
Alnico 4	12 Al, 25 Ni, 5 Co, 0.40 Ti, bal. Fe	6,300	630	1.40	7.0	—	45	Cast	Grind	Furnished heat-treated	
Alnico 5[d]	8 Al, 14.5 Ni, 24 Co, 3 Cu, bal. Fe	12,700	640	5.25	7.3	—	50	Cast	Grind	Furnished heat-treated	
Alnico 5E[d]	8 Al, 15 Ni, 22.5 Co, 3.6 Cu, 0.45 Ti, bal. Fe	11,800	700	4.60	7.3	—	50	Cast	Grind	Furnished heat-treated	
Alnico 6[d]	8 Al, 15 Ni, 24 Co, 3.4 Cu, 1.25 Ti, bal. Fe	10,400	750	3.65	7.4	—	56	Cast	Grind	Furnished heat-treated	

[a] 10^4 G = tesla.
[b] 4π oersteds = 1 kA/m.
[c] For best results a hardening furnace with neutral atmosphere is recommended.
[d] Directional magnetic properties.

385

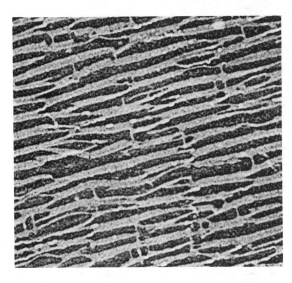

Figure 11-17 Substructure of Alnico alloy as it appears in the electron microscope at approx. 125,000×.

SUPERCONDUCTIVITY

Although superconductivity is in essence an electrical property, we have deferred the discussion of this topic until now because the practical applications and developments of this emerging technology currently pertain principally to magnetic applications.

A superconducting material is one that has no resistance to the flow of electricity when it is below its critical temperature (usually within a few degrees of absolute zero at -273°C) and its critical current density and critical magnetic field are not exceeded. This phenomenon was discovered by H. Kamerlingh Onnes in 1911 as a result of his experiments with mercury. Figure 11-18 shows the drop in resistivity that occurs with α-mercury at 4.15°K. Superconductivity will disappear if the critical temperature is exceeded or if a critical magnetic field (H_c) or a critical current density (I_c) is applied.

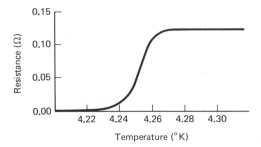

Figure 11-18 Resistance as a function of temperature for mercury.

The critical field is related to the temperature according to the expression

$$H_c = H_0\left(1 - \frac{T^2}{T_c^2}\right)$$

where H_0 = critical field at 0°K
T_c = critical temperature (°K)

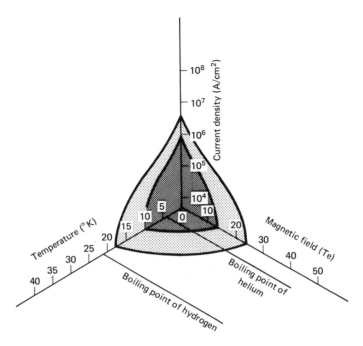

Figure 11-19 Critical temperature, current density, and magnetic field boundaries for Nb₃Sn (light region) and NbTi (darker region). (Courtesy of Intermagnetics General Corporation, Albany, N.Y.)

If one examines this relationship, one sees that the curves generated are approximately parabolic and represent the phase boundaries for superconductivity. Figure 11-19 illustrates the critical boundaries for two superconducting materials, niobium–titanium (Nb–Ti) and an intermetallic compound of niobium and tin, Nb₃Sn.

The magnetic response of superconductors is of two types, type I and type II. The difference between the two types is as follows: with a type I superconductor, the magnetic field is completely excluded from the body of the conductor, except for a thin region near the surface (i.e., it is diamagnetic until the critical field (H_c) is exceeded). This phenomenon, called the *Meissner effect*, is depicted in Figures 11-20 and 11-21(a).

With type II superconductors, the behavior is similar at low applied fields, but at higher fields the conductor is gradually penetrated. This behavior is illustrated in

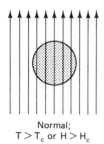

Normal;
$T > T_c$ or $H > H_c$

Superconducting;
$T < T_c$ or $H < H_c$

Figure 11-20 Meissner effect. When the specimen becomes superconducting, all of the magnetic field is expelled from it. Ideal type I superconductors behave in this manner, as do ideal type II superconductors in magnetic fields below H_{c_1}.

Superconductivity

387

Figure 11-21(b). At applied fields below the lower critical field (H_{c_1}) the field is excluded from the specimen; at fields above H_{c_1}, the applied field begins to penetrate the specimen increasingly until the upper critical field (H_{c_2}) is reached and the specimen exhibits normal behavior. The elements that exhibit superconductivity of types I and II are shown in Figure 11-22 together with their critical temperatures.

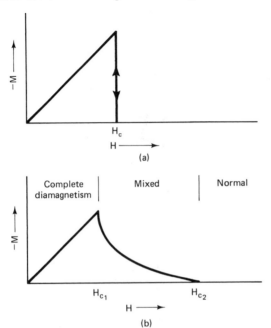

Figure 11-21 Penetration of superconductors by an applied magnetic field: (a) type I; (b) type II.

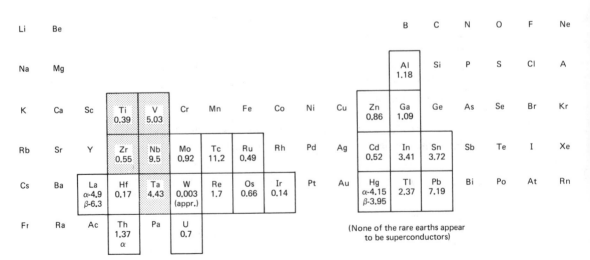

Figure 11-22 Periodic table indicating elements exhibiting superconductivity. Numerals indicate the absolute temperature ($^\circ$K) at which the elements become superconducting. Shading indicates type II superconductors.

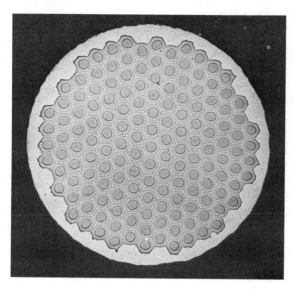

Figure 11-23 Microstructure of multifilamentary NbTi wire (125×). (Courtesy of T. A. Miller, Intermagnetics General Corporation.)

Superconducting Materials

Of the various elements and compounds that have exhibited superconductivity in the laboratory, only three have presently reached the production stage. These are the alloys of niobium–zirconium (Nb–Zr) and niobium–titanium (Nb–Ti) and the niobium–tin compound, Nb_3Sn.

The superconductive niobium–titanium has a lower critical current (I_c) and critical field (H_c) than does the Nb_3Sn compound, as shown in Figure 11-19. However, it is more ductile and is therefore more easily produced. The product is usually produced in the form of a composite multifilamentary wire, a typical cross section of which is shown in Figure 11-23. These wires usually contain copper as well as Nb–Ti for stability and protection of the system. Should it revert from the superconducting to the normal state, the copper can carry the current briefly until the superconducting condition can be stabilized or the system shut down.

Niobium–tin (Nb_3Sn) is an intermetallic compound with the highest overall critical current density of any high-field superconductor. Although more costly and less ductile than NbTi, Nb_3Sn is very important for applications requiring ultrahigh fields or operating temperatures above $4.2°K$. Nb_3Sn can also be produced as a composite tape using a diffusion process. The tape, which remains superconductive to $18°K$, is comprised of a Nb_3Sn strip metallurgically bonded to strips of copper. Using these tapes, superconducting magnets of up to 16 T* have been constructed. Although tape conductors remain the most efficient and reliable method of constructing magnets producing fields over 12 T, in some applications it is preferable to use multifilament wire. These applications include those where small radius bends in two directions are required, where very low residual magnetization is necessary, or where low alternating-current losses are desired. Such a multifilament wire is shown in Figure 11-24.

*Tesla (T) is the international unit (SI) for magnetic flux density. 1 T = 1 w/m².

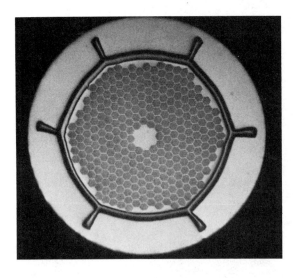

Figure 11-24 Microstructure of a Nb_3Sn multifilament wire, 21,390 Nb_3Sn filaments in a bronze matrix encased within a copper stabilizer and protected by a diffusion barrier. (Courtesy of T. A. Miller, Intermagnetics General Corporation.)

Applications of Superconductors

Applications of superconductive materials are limited primarily by two factors: the need to cool the superconductor to cryogenic temperatures, and their generally inadequate performance with alternating current. In an alternating field, magnetic hysteresis occurs. This occurrence creates localized regions of overheating which tend to drive the superconductor normal.

In spite of these limitations, there are numerous applications currently being investigated. The applications for superconducting materials fall into two primary categories, power transmission and superconducting magnets.

Since once a current is generated in a superconductor it continues to flow with essentially no loss, it is easy to see why superconducting power transmission is receiving great attention from the utility industry and the U.S. Department of Energy. The potential for superconductors to carry huge amounts of power is extremely important. Estimates have been made that four superconductive lines 15 to 20 in. in diameter could satisfy the power needs of New York City. Prototypes for such lines have already been constructed and are currently being evaluated. One design for the construction of the superconducting core is illustrated schematically in Figure 11-25.

Electromagnets using superconductive wire coils are of interest wherever large magnetic fields are required. The advantages of superconducting magnets can be illustrated by considering a high-field magnet of 50,000 G. A superconducting magnet of this size would weigh less than 1 kg, would require no steady-state power consumption, and would require only a small amount of power to start the current. Contrast this with a conventional magnet, which would weigh several tons, would require more than 50 kW of power, and would require 4000 liters of cooling water per minute to remove the heat generated. Even when the refrigeration system for the superconducting magnet is considered, there is a considerable savings in cost, energy, and space.

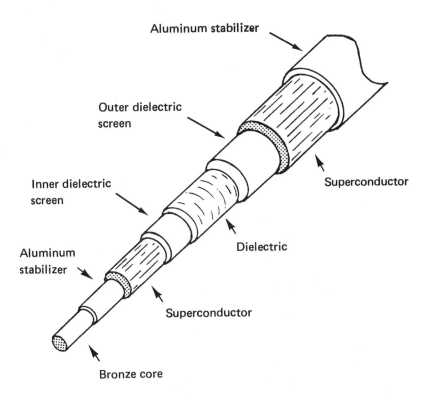

Figure 11-25 Prototype design for a superconducting power transmission line.

A number of intriguing systems based on these superconducting magnets are currently being evaluated. One of these is the superconducting electrical generator. Such generators are similar in concept to a conventional generator. Electrical conductors in a spinning motor create a revolving magnetic field which produces an electrical current in the conductors of the surrounding cylinder called a *stator*. A superconducting generator would use a rotor wound with superconducting wire. The rotor would be surrounded by a cryogenic stator. The benefits of such a generating system are smaller size and weight, higher efficiency, and lower operating costs.

Another application of superconducting magnets is in the development of linear motors which can be used to propel high-speed trains. The speed and safety of conventional trains are limited by their propulsion systems. The use of magnetically levitated and driven trains would overcome these problems. The principle of operation is as follows. When a superconducting magnet located beneath the train passes over a series of coils laid in the guideway, an electric current is generated in the ground coil, causing it to behave as an electromagnet whose polarity is the same as that of the superconducting magnet on board the train. This is illustrated in Figure 11-26. The repulsive forces between the ground coil and the vehicle produce levitation. Several train systems using magnetic levitation are presently in experimental operation in Europe and Japan.

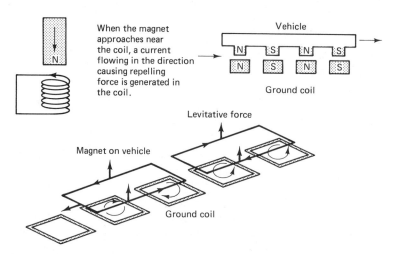

Figure 11-26 Schematic illustration of a magnetically levitated vehicle. (From *Compressed Air*, p. 9, Sept. 1980.)

Superconducting magnets are also being used to study magneto-hydrodynamic and thermonuclear power systems. In both these systems, very hot ionized gases (plasma) are contained in a magnetic field. Only superconductive magnets can generate the intense magnetic fields necessary to confine the plasma and keep it from contact with the walls of the vessel.

Finally, one very important current commercial use of superconducting magnets is in nuclear magnetic resonance (NMR) scanners. These devices are used in the medical field as a diagnostic tool. Essentially, the intense magnetic field generated by the superconductor stimulates the nuclei of atoms to emit radiation. In turn, this radiation is utilized to produce a cross-sectional image of a patient's body or internal organs. In fact, not only the shape and density of an organ is obtained, but this technique is also capable of *chemically analyzing* the tissue being scanned. The obvious advantage of such a process is early detection and accurate diagnosis of disease or abnormalities in human tissue.

It is clear from these few current applications that important applications for superconductive materials will continue to increase, and that superconductivity will likely play a major role in future energy-related technological developments.

STUDY PROBLEMS

11.1. A specimen of annealed iron has a permeability of 6.75×10^{-3} Wb/A-m when it is in a magnetic field of 152 A-turns/m. What is the flux density in the iron?

11.2. What is the permeability of a particular steel in which a magnetic field strength of 543 A-turns/m produces a flux density of 1.13 Wb/m²?
Answer: 2.08×10^{-3} W/A-m

11.3. A current of 5.5 A passes through a closely wound solenoid with 300 turns and a length of 39 cm. What is the field intensity at the center of the coil?

11.4. The iron core of a electromagnet with a field intensity of 150 A-turns/m has a flux density of 1.07 Wb/m². What is its relative permeability?

11.5. Define *paramagnetic*, *diamagnetic*, and *ferromagnetic* behavior. Describe an experiment to demonstrate such behavior.

11.6. The table below lists corresponding values of H and B for a specimen of annealed iron. Construct graphs showing the magnetization curve and permeability for this material.

Field intensity, H (A-turns/m)	Flux density, B (Wb/m²)
0	0
20	0.01
50	0.05
75	0.20
100	0.67
150	1.01
200	1.18
250	1.25
500	1.44
1000	1.58

11.7. Explain why hysteresis occurs in many magnetic materials. What is the significance of hysteresis in transformer applications?

11.8. Draw the typical hysteresis loops for a material that might be used for the following applications: permanent magnet, transformer core, computer memory core, and audio recording tape.

11.9. What are magnetic domains? How do they respond when an external applied magnetic field is removed? What measures could be taken to minimize the changes occurring after removal?

11.10. Two metal rods originally 40 cm long are subjected to a magnetic field of 629 Oe. One is cobalt and the other is pure iron. What is the final length of each of the bars?

11.11. Compare the magnetic characteristics of iron as related to crystallographic direction. Which direction would produce the least hysteresis? Which direction would be the most desirable for a transformer core?

11.12. How would you expect the power loss in alternating-current applications to vary with the following: the frequency of the current, the volume of the specimen, and the maximum flux density attained?

11.13. Nickel ferrite has eight $NiFe_2O_4$ units per cubic unit cell. Each side of the unit cell has a length of 0.838×10^{-9} m. Calculate the magnetization for the cell if all the unit cells are oriented with the same polarity.
Answer: 252×10^3 A/m

11.14. What would the remaining magnetization be in problem 11.13 if 37% of the unit cells undergo a complete polarity reversal when the external field is removed?

11.15. A new ceramic ferrite has been developed to replace $[NiFe_2O_4]_8$ in which 50% of the nickel atoms have been replaced by zinc atoms. Assume that the zinc ions displace some of the Fe^{3+} from their fourfold sites. What would the change in magnetic moment be?

11.16. Using the data in Table 11-3 for 3% Cr steel, 3% Co steel, and 18.5% Co steel, calculate the maximum energy product and compare the results to the actual values listed in the table.

11.17. If the coercive force for barium ferrite is 1.6×10^5 A/m and the residual flux density is 0.38 W/m^2, calculate the maximum energy product in CGS units.

11.18. Explain how the following permanent-magnet materials are manufactured into their final shape and how a structure conducive to strong residual magnetization is achieved: high-carbon steel, Cunico, and Alnico.

11.19. What is the difference in behavior between a type I and a type II superconductor? In what range of fields is each superconducting?

11.20. Niobium has a critical magnetic field of 20×10^4 A-turns/m at 0°K and a critical temperature of 10.5°K. What is its critical field at 5°K?
Answer: 15.5×10^4 A-turns/m

11.21. Vanadium exhibits a critical magnetic field of 9×10^4 A-turns/m at 0°K. Construct a plot of H_c versus temperature for vanadium.

11.22. With respect to problem 11.21, what is the general form of the curve obtained? What is its significance?

11.23. What are the relative advantages and disadvantages of Nb_3Sn over Nb–Ti?

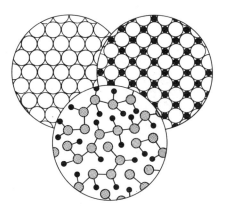

CHAPTER TWELVE:
Ferrous Engineering Alloys and Their Applications

In this chapter we examine the relationships between structure and properties, together with the applications of ferrous alloys. In the most basic sense, *steel* is an alloy of iron and carbon. By definition the carbon content of steel ranges from approximately 0.008 to 2.0%. Ferrous alloys that contain carbon contents greater than 2% are classified as *cast iron*. Although carbon is the alloying element that produces the most profound effects, other elements are generally added to iron, depending on the desired properties or the requirements that are imposed on a particular ferrous alloy.

It should come as no surprise that steel is a very important engineering material and correspondingly contributes the largest share of metals production in the industrialized world. Just how high is this contribution? Well, for example, in the United States, the average production of raw steel—this is steel melted for all purposes—has averaged approximately 121 million tons per year over the last two decades. Compared to aluminum, which is the principal nonferrous metal, this output is about 25 times greater than the tonnage of primary aluminum produced annually in this country. Our point is that in spite of the increased usage of nonferrous alloys, ceramics, plastics, and composites, steel is still the most widely used engineering material. Furthermore, the demand for steel and other ferrous alloys will, according to all projections, remain at or above current levels through the end of this century.

Logical questions by the student studying engineering materials are: Why are steels so important? Why are they in such great demand? The widespread use of steels is due to a number of factors, including their availability, versatility, properties, and economic advantage over certain comparable nonferrous alloys. Let us briefly examine these factors.

Iron is the fifth most common element in the earth's crust, and many iron ore deposits (taconite) are large and concentrated, making mining operations very feasible. In addition, ferrous scrap is readily recycled from within the steel plant itself and from obsolescent equipment or structures such as junk automobiles, worn-out railway cars and track, ship plate, old machinery, and structural (building) razings. These features are beneficial to both the availability and the economics of steel production.

The versatility of steels as an engineering material is demonstrated by the flow diagram shown in Figure 12-1. This chart depicts the steelmaking process from its

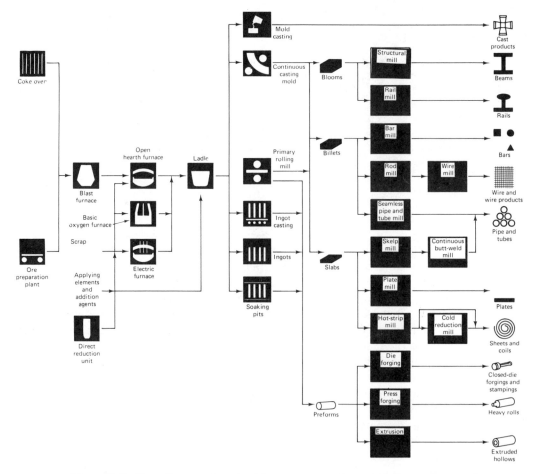

Figure 12-1 Flow diagram of steelmaking processes and conversion to various products. (Adapted from J. Szekely, *Met. Trans. B*, Vol. 11B, September 1980, p. 355.)

inception at the ore reduction stage, through various processing steps to both semifinished and finished products. Moreover, we should recognize that the end items pictured here represent a very small sample of the actual ferrous alloy products turned out by the mills and intermediate processors.

The mechanical and physical properties of steels may be accurately controlled by regulating the chemical composition of a particular alloy and the processing, such as mechanical working and/or heat treatment, that the material receives. For instance, small variations in composition and different heat treatments can result in steels suited for the following applications: (1) soft, ductile material for use in automobile fenders, appliance panels, roofing, and so on; (2) hard, strong materials for gears, bearings, shovels, blades, tools, and so on; and (3) tough materials utilized in shafts, pressure vessels, landing gear, and other applications where impact or shock resistance is required.

Greater changes in composition can also produce more pronounced effects in the properties of steels. Consider the addition of 18% chromium and 8% nickel to a low-carbon steel. The result is a material highly resistant to corrosion, which we commonly know as *stainless steel*. Stainless steels play such an important role in our lives that virtually no home, automobile, aircraft, or restaurant, is without them. Tool steels are yet another example where significant changes in properties are developed by adding such elements as cobalt, molybdenum, and tungsten to the basic steel composition. These materials find applications ranging from the home workshop as hammers, chisels, saw blades, and drills, to the metalworking industry in the form of forging dies, press hammers, cut-off blades, shears, and rolls.

Hopefully, the student should now be convinced that the subject of ferrous alloys is very important, regardless of one's engineering or scientific discipline. Furthermore, as we study the forthcoming topics in this chapter, many of the basic structure–property relationships discussed in earlier chapters will be reinforced and will help us to understand the behavior of ferrous alloys.

THE IRON–CARBON SYSTEM

Although steels and most ferrous alloys contain a certain amount of carbon in solid solution with iron, let us first consider the polymorphic behavior of pure iron. These characteristics are very important to any alloy of iron and may be illustrated by the schematic cooling curve shown in Figure 12-2. From this diagram we see that iron is body-centered cubic (bcc) between its melting point (1539°C) and 1400°C. In this region the material is called *delta* (δ) iron. Between 1400 and 910°C, the structure is face-centered cubic (fcc) and is called *gamma* (γ) iron. Below 910°C, the material is once again bcc and referred to as *alpha* (α) iron. In addition, pure iron undergoes a magnetic transformation at 780°C (Curie temperature), as indicated in Figure 12-2.

Fe–Fe₃C Phase Diagram

The iron–iron carbide equilibrium phase diagram, or the iron–carbon diagram as it is usually referred to, serves as the basic diagram for analyzing plain carbon and many low-alloy steels. It is very useful in studying both the equilibrium and nonequilibrium changes that iron–carbon alloys undergo during heating and cooling. Such usefulness stems from the fact that the iron–carbon diagram is relatively unaffected by the small concentrations of manganese, silicon, sulfur, and phosphorus ordinarily present in conventionally refined steels. The metastable iron–iron carbide equilibrium diagram is shown in Figure 12-3 for carbon contents up to 6.7%. This contains the portion of the diagram that is commercially useful, since the carbon content in cast irons is usually less than about 4.5%.

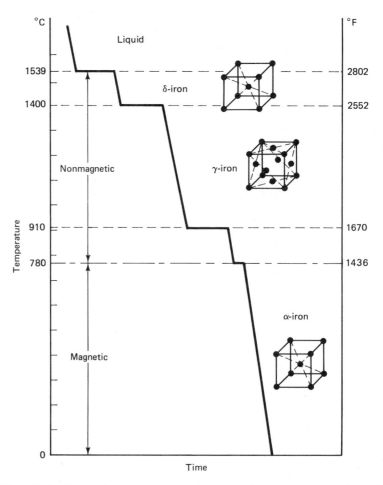

Figure 12-2 Schematic cooling curve for iron showing polymorphic and magnetic transformation.

As this diagram shows, irons (wrought iron) are ferrous materials that contain essentially no carbon. This material is restricted to concentrations of less than 0.008 % carbon, but intentionally contains a few percent slag (nonmetallic siliceous material), as shown in Figure 12-4. Wrought iron exhibits good resistance to corrosion and fatigue failure. Since the iron contains no alloying elements, it is relatively soft and ductile; therefore, wrought iron finds application in bolts, pipe, tubing, nails, and other parts which do not require high strength and hardness.

Steels, on the other hand, contain carbon additions up to 2 %. As you will soon see, 2 % carbon is a very important composition in this alloy system, because it is the maximum amount of carbon that is soluble in *steel* in the solid state. This category of ferrous alloys is further classified into hypoeutectoid (containing less than 0.8 % C) and hypereutectoid (containing more than 0.8 % C). The term *eutectoid* refers to a solid-state reaction which takes place at a *specific* composition and temperature, in a

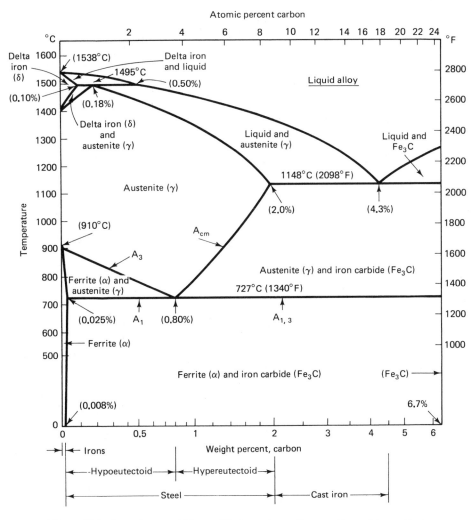

Figure 12-3 Iron–iron carbide equilibrium phase diagram (iron-rich portion). (Adapted from *Metal Progress Data Sheets*, No. 38, American Society for Metals, Metals Park, Ohio, 1954.)

manner similar to the eutectic reactions discussed in Chapter 7. We will analyze the eutectoid reaction in steel in considerable detail in the next section.

Compositions containing greater than 2% carbon are defined as *cast irons*. Note that these alloys have exceeded the maximum solubility of carbon in steel and contain a eutectic constituent which forms at 4.3% carbon and 1148°C. Such relatively high concentrations of carbon produce a material that is hard and/or brittle and cannot be formed or shaped by mechanical working. Therefore, these alloys are poured as castings to obtain the desired geometrical configurations, hence the term "cast iron."

Let us now consider some important phases and constituents that exist in steels. The student is referred to Figure 12-3 during our discussion of these regions.

Figure 12-4 Microstructure of wrought iron showing slag (siliceous) stringers in ferrite matrix. Slag constituent is dark gray (75×).

Ferrite (α). This phase consists of an interstitial solid solution of carbon in bcc iron. An interstitial type of solid solution rather than substitutional forms because the difference in radius between the two atomic species (approximately 60%) is far in excess of the 15% criteria, according to the Hume–Rothery rules discussed in Chapter 6. It is evident from Figure 12-3 that the solubility of carbon in α-iron is indeed very limited. The maximum solubility is 0.025% at 727°C (approximately 1 carbon atom per 1000 iron atoms) and decreases to about 0.008% at room temperature (approximately 1 carbon atom per 2500 iron atoms). This restricted solubility is due to the size of the interstitial spaces in the iron lattice compared with the size of a carbon atom. Such a comparison is made in Example 12-1, where we see that the carbon atom is approximately twice as large as the largest interstitial hole in the bcc iron lattice at room temperature. The largest interstitial holes in bcc are located at $\frac{1}{2}$, $\frac{1}{4}$, 0 sites and equivalent lattice positions. However, there is considerable X-ray diffraction evidence, indicating that carbon atoms in ferrite are also located at 0, 0, $\frac{1}{2}$ positions (the midpoints of the cell edges) rather than exclusively at the largest interstices. In bcc ferrite, the latter interstitial holes are only about 0.38 Å in diameter. Therefore, the iron cannot accommodate carbon interstitially without producing a significant lattice disturbance. You will soon see that this feature—low solid solubility of carbon in bcc iron—forms the basis for the hardening effects in steels, under both equilibrium and nonequilibrium conditions.

Example 12-1

Compare the size of the largest interstitial hole in bcc iron lattice with the diameter of a carbon atom at room temperature. This hole is illustrated in Figure 12-5, where R = radius of Fe and a is the lattice parameter.

Solution Determine the radius of the hole (r) from the triangular relationship shown as follows:

$$r + R = \sqrt{\left(\frac{a}{4}\right)^2 + \left(\frac{a}{2}\right)^2}$$

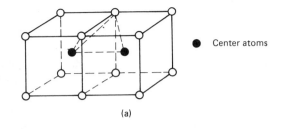

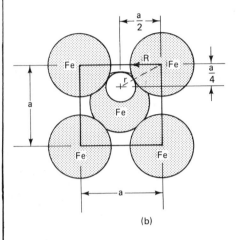

Center atoms

(a)

(b)

Figure 12-5 Schematic showing 4-f (tetrahedral) interstitial site in bcc structure: (a) interstice located between adjacent unit cells; (b) center of interstitial hole (with radius r) is located at lattice position $\frac{1}{2}$, 0, $\frac{3}{4}$.

But from Table 3-3, we know that $a_{bcc} = 4R/\sqrt{3}$.

$$r + R = \sqrt{\left(\frac{4R}{4\sqrt{3}}\right)^2 + \left(\frac{4R}{2\sqrt{3}}\right)^2}$$

$$= \sqrt{\frac{R^2}{3} + \frac{4R^2}{3}}$$

$$= \sqrt{\frac{5R^2}{3}} = \sqrt{\frac{5}{3}}\,R$$

$$r = \sqrt{\frac{5}{3}}\,R - R$$

$$= 1.29R - R$$

$$= 0.29R$$

From endpapers, $R_{Fe} = 1.24$ Å; thus

$$r = 0.29(1.24\ \text{Å}) = 0.36\ \text{Å}$$

$$\text{diameter of hole} = 2r = 0.72\ \text{Å}$$

$$\text{diameter of carbon atom} = 1.54\ \text{Å}$$

Therefore, the carbon atom is approximately twice as large as the largest interstitial hole in bcc iron at room temperature.   *Ans.*

Austenite (γ). In steels, this phase consists of an interstitial solid solution of carbon in fcc iron. Figure 12-3 shows that under equilibrium conditions, austenite can exist from a temperature of 727°C up to 1495°C and can dissolve up to 2% carbon at 1148°C. This amounts to approximately 80 times the weight percent of carbon that is soluble in α-iron. The difference in solubility arises from the crystallographic difference between the two polymorphs of iron. Simply put, there are fewer interstitial void sites in the close-packed gamma iron (fcc) lattice, but these holes are larger than those in the bcc form. Example 12-2 compares the size of a carbon atom and the largest interstitial hole in fcc austenite. This example shows that the largest interstitial hole in fcc iron is about 1½ times larger than that of bcc iron.

Example 12-2

Compare the size of the largest interstitial hole in the fcc iron lattice with the size of a carbon atom. The interstitial hole is illustrated in Figure 12-6. Assume that the iron atom has a radius (R) of approximately 1.27 Å at the temperature range (austenite) of interest.

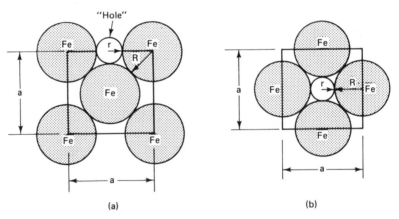

Figure 12-6 Octahedral void sites (6-f) in fcc structure: (a) interstitial site located in face of unit cell at $\frac{1}{2}, \frac{1}{2}, 0$; (b) void located in center of unit cell at $\frac{1}{2}, \frac{1}{2}, \frac{1}{2}$.

Solution We can determine the radius (r) of the hole in either view (plane) as follows:

$$a = 2R + 2r$$

From Table 3-3 we find that

$$a = \frac{4R}{\sqrt{2}} \quad \text{for fcc structure}$$

Thus

$$2R + 2r = \frac{4R}{\sqrt{2}}$$

$$2r = \frac{4R}{\sqrt{2}} - 2R$$

$$= 2.83R - 2R$$

$$= 0.83R$$

For $R = 1.27$ Å,

$$r = 0.41R$$

$$r = 0.41(1.27 \text{ Å})$$

$$= 0.52 \text{ Å}$$

Therefore, the diameter of the interstitial hole is $2r$ or 1.04 Å. The diameter of a carbon atom is 1.54 Å (see back cover). So in this case, the carbon atom is approximately $1\frac{1}{2}$ times larger than the interstitial hole.   *Ans.*

By definition, steels contain carbon contents up to 2%. Inspection of our "iron–carbon" diagram shows that these alloys can have their carbon content entirely dissolved when they are heated into the austenite region. This is an extremely important aspect in the transformation and strengthening behavior of steels, as we will see in later topics dealing with *heat treatment*. Furthermore, in the austenitic state, steel is nonmagnetic and relatively soft or ductile. Consequently, most shaping and forming operations, such as forging, rolling, and extrusion, are performed at temperatures sufficiently into this region. Typical hot-work processes are conducted on carbon and alloy steels in the neighborhood of 980 to 1200°C (1800 to 2200°F).

Delta (δ).   This phase consists of a solid solution of carbon in bcc iron. It has essentially the same crystal structure as α-iron, except that it exists above a temperature of 1400°C. The maximum solubility of carbon in δ-iron is 0.10%.

Although we will not study the delta region in detail, an interesting reaction occurs in the iron–carbon system at a temperature of 1495°C and composition of 0.18% carbon. This reaction consists of a solid phase plus a liquid phase, transforming to one solid phase. Such a transformation is called a *peritectic* reaction and may be expressed for steel as follows:

$$\text{delta } (\delta) + \text{liquid} \underset{\text{cooling}}{\overset{\text{heating}}{\rightleftarrows}} \text{austenite } (\gamma)$$

The delta (δ) region of the Fe–Fe$_3$C phase diagram is important in the analysis of welds where temperatures in the weld zone exceed the solidus. However, in other processes, such as heat treatment, temperatures are normally limited to the lower portion of the austenite region, to avoid excessive grain growth and localized melting, both of which are usually injurious to room-temperature mechanical properties. Furthermore, in service, ferrous alloys are generally utilized at temperatures well below the austenite range in order to avoid recrystallization and the softening which accompanies this transformation.

Iron carbide (Fe$_3$C).   In contrast to ferrite (α), which contains very little carbon, a second solid phase containing a relatively large amount of carbon (6.7%) forms in iron–carbon alloys. This phase, the compound iron carbide (Fe$_3$C), has an orthorhombic structure (see Figure 3-1) with 12 iron atoms and 4 carbon atoms per unit cell. The weight percent carbon in this intermetallic compound is considerably greater than ferrite, as demonstrated by Example 12-3.

As one might suspect, the increased carbon content of Fe$_3$C produces a very hard but brittle phase. However, when combined with ferrite, iron carbide beneficially increases the strength of ferrous alloys.

Example 12-3

Based on a unit cell of iron carbide (Fe_3C), determine the weight percent of carbon in this intermetallic compound.

Solution   The unit cell contains 12 iron atoms and 4 carbon atoms. From the atomic weights listed on the inside back cover we can determine the individual weight of iron (W_{Fe}) and carbon (W_C) per unit cell as follows:

$$\text{Carbon:} \quad W_C = \frac{(4 \text{ atoms})(12 \text{ amu/atom})}{6.02 \times 10^{23} \text{ amu/g}}$$

$$= 8.0 \times 10^{-23} \text{ g/unit cell}$$

$$\text{Iron:} \quad W_{Fe} = \frac{(12 \text{ atoms})(55.8 \text{ amu/atom})}{6.02 \times 10^{23} \text{ amu/g}}$$

$$= 111.2 \times 10^{-23} \text{ g/unit cell}$$

$$\text{Total weight} = W_C + W_{Fe}$$

$$= 8.0 \times 10^{-23} \text{ g} + 111.2 \times 10^{-23} \text{ g}$$

$$= 119.2 \times 10^{-23} \text{ g/unit cell}$$

$$\text{w/o C} = \frac{8.0 \times 10^{-23} \text{ g}}{119.2 \times 10^{-23} \text{ g}} \times 100 = 6.7\% \qquad \textit{Ans.}$$

Eutectic (ledeburite).   The iron–carbon system also undergoes a eutectic reaction at 4.3% carbon and 1148°C. This transformation is identical, in principle, to the eutectic reactions studied in Chapter 7. Inspection of the Fe–Fe_3C phase diagram shows that the eutectic product, also referred to as *ledeburite*, is a mixture of austenite and iron–carbide (Fe_3C), and that this transformation occurs outside the composition of steels. Therefore, the eutectic product will not be involved in our discussion of steels.

Eutectoid Reaction

The eutectoid reaction was defined briefly in our introduction to the Fe–Fe_3C phase diagram (Figure 12-3). Now let us examine this very important reaction more closely.

In steel (Fe–C system) this reaction occurs under equilibrium conditions at a temperature of 727°C and composition of 0.8% carbon. If the phase rule [see equation (7-1)] is applied to the system at this point, there are zero degrees of freedom. In other words, the eutectoid reaction is invariant! We may express this reaction in steels as follows:

$$\text{austenite } (\gamma) \quad \underset{\text{cooling}}{\overset{\text{heating}}{\rightleftarrows}} \quad \text{ferrite } (\alpha) + \text{iron carbide } (Fe_3C)$$

The product of this solid-state transformation is an intimate mixture of ferrite (α) and iron–carbide (Fe_3C). This structure is commonly referred to as *pearlite*, because it sometimes resembles "mother-of-pearl" under the microscope. The micro-

Figure 12-7 Photomicrograph illustrating the laminar nature of pearlite (1125×).

structure of pearlite in a carbon steel is shown in Figure 12-7. When etched with solute-sensitive reagents, pearlite appears characteristically lamellar, containing alternating bands of ferrite (light) and iron carbide (dark).

Example 12-4

Determine the proportions of ferrite and iron carbide in pearlite from the Fe–Fe$_3$C phase diagram given in Figure 12-3.

Solution   The quantities of these phases may be calculated from the following lever-arm relationship:

$$\alpha \quad \vdash\!\!-\!\!-\!\!-\!\!-\!\!-\!\!\blacktriangle\!\!-\!\!-\!\!-\!\!-\!\!-\!\!-\!\!\dashv \quad Fe_3C$$

$$0.025 \qquad\qquad 0.8 \qquad\qquad 6.7$$

$$\% \; C$$

$$\alpha = \frac{6.7 - 0.8}{6.7 - 0.025} \times 100$$

$$= \frac{5.9}{6.675} \times 100$$

$$= 88.4\% \quad Ans.$$

The balance is, therefore, $100\% - 88.4\% = Fe_3C$.

$$Fe_3C = 11.6\% \quad Ans.$$

These values are ordinarily taken to be:

$$\text{eutectoid } \alpha = 88\%$$

$$\text{eutectoid } Fe_3C = 12\%$$

During the formation of this microstructural constituent (pearlite), carbon, which cannot be dissolved by ferrite, combines with iron to form iron carbide, as illustrated in Figure 12-8. The proportions of these two phases in the eutectoid are fixed because the reaction is invariant. Pearlite contains 88% ferrite and 12% iron carbide, as demonstrated by Example 12-4, and very often this is evidenced by the much larger width of the ferrite lamella compared to that of the carbide.

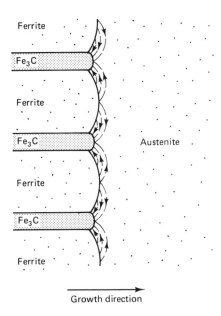

Ferrite

Fe₃C

Ferrite

Fe₃C

Austenite

Ferrite

Fe₃C

Ferrite

Growth direction

Figure 12-8 Schematic showing the growth of a pearlite colony. Carbon atoms diffuse to form Fe₃C. (Adapted from J. H. Hollomon and L. D. Jaffe, *Ferrous Metallurgical Design*, John Wiley & Sons, Inc., New York, 1947.)

To clarify the transformation and structures that we have just discussed, let us perform two examples (12-5 and 12-6) dealing with the Fe–Fe₃C phase diagram. These examples involve cooling a hypoeutectoid and heating a hypereutectoid steel, under equilibrium conditions. These processes are representative of the cooling and heating conditions experienced in large ingots and castings, which often take many hours. The alloys and their transformation microstructures are shown in Figure 12-9.

Example 12-5

Determine the transformations experienced by plain carbon steel containing 0.4% carbon (1040 steel) as it is slowly cooled from the liquid state to room temperature. This alloy is identified as A in Figure 12-9.

Solution

Point a: Solidification of bcc delta solid solution begins, with a carbon content of approximately 0.05%.

Point b: At this temperature (1495°C) the peritectic reaction takes place, transforming the delta solid solution and some liquid alloy to fcc austenite. In the range *b* to *c*, the remaining liquid solidifies as austenite.

Point c: At this temperature, the solidification process is completed and the entire system (mass) is austenite with a nominal carbon content of 0.4%. Continued cooling from *c* to *d* produces no phase changes. The microstructure of this single phase solid solution is shown in Figure 12-9. Such a photomicrograph is obtained by *hot-stage* metallographic techniques.

Point d: When alloy A reaches this temperature, bcc ferrite begins to form from the austenite. Note that this is a solid-state reaction and therefore is controlled by diffusion of carbon in the austenite. The carbon content of the ferrite formed is very low, as our diagram predicts (less than 0.025%). Thus, as this process con-

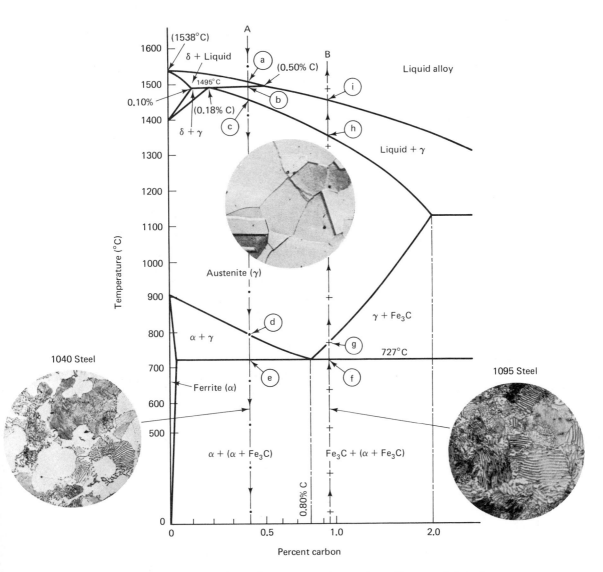

Figure 12-9 Transformation microstructures in two iron alloys, cooled and heated under near-equilibrium conditions.

tinues with gradual cooling, the concentration of carbon in the remaining austenite continually increases. (If one sketches a series of lever arms in this region for decreasing temperatures, the right end steadily moves to higher carbon contents.)

Point e: As we approach 727°C (the eutectoid temperature), the remaining austenite approaches 0.8% carbon (in spite of a nominal composition of 0.4% carbon). Consequently, at this point the eutectoid reaction occurs, transforming the austenite left, to pearlite ($\alpha + Fe_3C$), as discussed earlier.

We can determine the proportions of the microstructural *constituents* in this alloy by applying the lever arm at this point (temperature), as follows:

$$\alpha \quad \vdash\!\!-\!\!-\!\!-\!\!-\!\!\blacktriangle\!\!-\!\!-\!\!-\!\!-\!\!-\!\dashv \text{ Pearlite}$$

$$\quad 0.025 \qquad\qquad 0.4 \qquad\qquad 0.8 \quad (\alpha + Fe_3C)$$

$$\% C$$

$$\% \alpha = \frac{0.8 - 0.4}{0.8 - 0.025} \times 100$$

$$= \frac{0.4}{0.775} \times 100 = 51.6$$

$$= 52$$

(Note that this is the portion of ferrite that formed *prior* to the eutectoid reaction.) The balance of the structure, therefore, consists of 48% pearlite. These relative proportions may be observed by inspection of the photomicrograph in Figure 12-9. Often, the carbon content of a plain carbon steel may be estimated in this manner from its microstructure, especially if the material has been slowly cooled.

Example 12-6

Determine the transformations that occur during very slow heating of a 0.95% carbon steel (1095 grade) from room temperature to the liquid state. This steel is identified as alloy B in Figure 12-9.

Solution   From room temperature to point f (727°C) the structure consists almost entirely of pearlite ($\alpha + Fe_3C$) plus a small amount of *free* (pro-eutectoid) iron carbide (Fe_3C), as shown in the photomicrograph. We may calculate the proportions of these room-temperature microstructural constituents as follows:

$$\text{Pearlite} \quad \vdash\!\!-\!\!\blacktriangle\!\!-\!\!-\!\!-\!\!-\!\!-\!\!-\!\dashv \quad Fe_3C$$

$$\qquad 0.8 \quad 0.95 \qquad\qquad 6.7$$

$$\% C$$

$$\% \text{ pearlite} = \frac{6.7 - 0.95}{6.7 - 0.8} \times 100$$

$$= \frac{5.75}{5.9} \times 100 = 97.5$$

$$= 97.5$$

Therefore,

$$\% Fe_3C = 2.5$$

This is the amount of "free" iron carbide that exists in the grain boundary network.

Point f: At 727°C the system undergoes the eutectoid reaction and pearlite transforms to austenite. The amount of austenite is the same as the amount of pearlite that existed just before the transformation (97.5%). As the temperature of the alloy continues to increase (between *f* and *g*), the "free" iron carbide transforms to austenite.

Point g: The transformation of Fe_3C to austenite is complete at this temperature. As our diagram predicts, the entire system is fcc austenite, with a nominal carbon content of 0.95%.

Point h: At this temperature (solidus), austenite begins to melt and the composition of the initial liquid alloy is in excess of 2% carbon. This surprising statement may be supported if one simply pictures a lever arm in the two-phase (γ + liquid) region for this alloy. The left end (γ) is C_0 (0.95% C), while the right end (liquid) is just beyond 2% C.

Melting continues as the temperature increases between points *h* and *i*. Also, if a series of lever arms were inserted at increasingly higher temperatures, we would see that the composition of the remaining austenite decreases in carbon, while the composition of the liquid in equilibrium with it approaches C_0.

Point i: At this temperature (liquidus) the steel is completely liquid, with a nominal composition of 0.95% carbon (balance Fe).

We might point out that in industry it is common practice to take chemical analyses of steels from the liquid state while the molten alloy is in the furnace or ladle. Based on our knowledge of solidification (Chapter 7) and the solid-state transformation characteristics just discussed, we should recognize that ladle analyses may be representative of the bulk or macro composition, but may not accurately represent the localized composition of a steel (or any alloy for that matter) after it has experienced solidification and the solute redistribution that necessarily accompanies it.

FUNCTION OF ALLOYING ELEMENTS IN IRON AND STEEL

Binary Iron–Alloy Systems

Ferrous alloys generally contain, in addition to carbon, other elements that significantly affect their transformation behavior. These alloying elements may be divided into two general categories: *austenite stabilizers* (type A) and *ferrite stabilizers* (type B).

Austenite stabilizers are those elements that tend to enlarge the austenitic region of the iron–carbon phase diagram, as illustrated in Figure 12-10(a). Type A-I elements, such as manganese, nickel, and cobalt, enlarge the austenite region by depressing the ferrite (α)-austenite (γ) transformation, and raising the austenite (γ)-delta (δ) transformation. Type A-II elements are similar to A-I, with the exception of stable iron-rich compounds at compositions approaching the α and δ phases. These compounds are designated by C in Figure 12-10. Examples of this type of austenite stabilizer include carbon, nitrogen, copper, and zinc.

Conversely, ferrite stabilizers are elements that tend to reduce the temperature range over which austenite is stable, as shown in Figure 12-10(b). Type B-I elements narrow the austenite region, forming a "γ-loop" which is surrounded by a two-phase

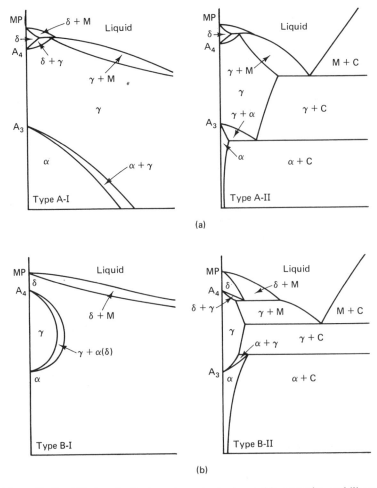

Figure 12-10 Effects of alloying elements in iron: (a) austenite stabilizers. (b) ferrite stabilizers. (From E. C. Bain and H. W. Paxton, *Alloying Elements in Steel*, American Society for Metals, Metals Park, Ohio, 1961, p. 9.)

field. Examples of this type include silicon, chromium, tungsten, molybdenum, vanadium, phosphorus, titanium, and aluminum. Similarly, type B-II elements reduce the austenite region, but include intermetallic compounds or constituents other than ferrite and austenite. Examples include boron, sulfur, tantalum, and zirconium. The effect of various alloying additions on the composition and temperature of the eutectoid reaction is summarized in Figure 12-11. For instance, additions of nickel and manganese depress the eutectoid temperature, while the other elements shown increase the temperature at which this reaction occurs. Also, as this figure illustrates, additions of these elements lower the carbon content of the eutectoid reaction. If this behavior sounds contradictory (we explained previously that the eutectoid reaction is invariant), remember that such alloy additions produce a steel which is no longer a *binary* iron–carbon alloy.

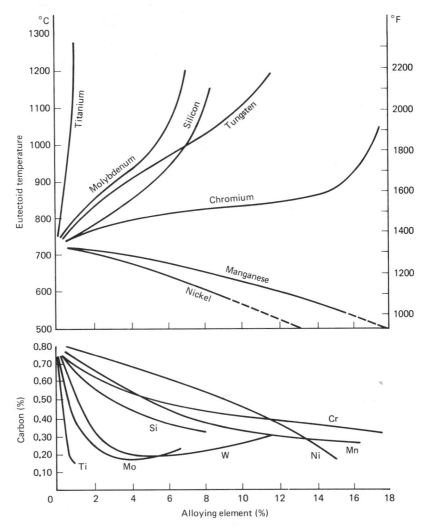

Figure 12-11 Effects of various alloying additions on the eutectoid reaction. (From E. C. Bain and H. W. Paxton, *Alloying Elements in Steel*, American Society for Metals, Metals Park, Ohio, 1961, p. 112.)

Classification of Steels

Specifications that relate to the properties and chemical compositions for carbon and alloy steels have been developed by many sources, including the Society of Automotive Engineers (SAE), the American Iron and Steel Institute (AISI), and the American Society for Testing and Materials (ASTM), as well as branches of the federal government and many private industries. These specifications serve as the basis for purchasing steels and subsequently determining their acceptability according to certain standards.

Steels are classified in a numerical system according to their chemical compositions. Such a system allows easy identification of the type or grade of steel, and also sets specific limits on the allowable chemistry. Both the SAE and the AISI have established the same system of numbers for the various types of steel, with the exception that AISI numbers are preceded by a letter indicating the melting practice, as follows:

A   basic open-hearth—alloy

B   acid Bessemer—carbon

C   basic open-hearth—carbon

D   acid open-hearth—carbon

E   electric furnace

The general classifications of carbon and alloy steels are given in Table 12-1. In this system the first digit indicates the *type* of steel (the major alloying element). The second digit indicates the approximate percentage of the major alloying element. The last two (or three in some cases) digits denote the carbon content in hundredths of a percent. For example, a C1018 steel is a plain carbon grade (10), which contains an average carbon content of 0.18% and was produced by the basic open-hearth process. This grade of steel is typically used for structural purposes. An E52100 steel, used to produce ball bearings, is a "high"-chromium steel containing approximately 1.45% Cr and 1.00% carbon. The "E" denotes that this steel was melted in an electric furnace.

We must emphasize that the steels in Table 12-1 are a basic list of *standard* grades. These series do not include the corrosion-resistant *stainless steels, tool steels*, and modifications to the standard grades which many users specify in order to achieve certain desirable properties.

Effects of Alloys in Iron and Steel

Alloying elements are added to iron and steel in order to change the crystal structure and the microstructure, thereby producing a change in the properties of these materials. Generally, steel contains small amounts of such elements as manganese, silicon, sulfur, and phosphorus, and traces of copper, tin, aluminum, and so on. These elements may be present in the initial iron charge, and may also result from recycled ferrous scrap which is contaminated with them. However, unless these elements exceed certain levels, they ordinarily are tolerated, because it is uneconomical to remove them completely, and in the case of alloy steels, manganese and silicon are usually added in greater quantities anyway.

Carbon is the chief element that affects the characteristics of steels. As discussed previously, even very small concentrations of carbon in iron can produce significant changes in microstructure, and such changes accordingly have a profound effect on the mechanical and physical behavior of the resultant alloy. A distinction is made between steels that contain carbon and small amounts of manganese plus residual sulfur and phosphorus, and steels that contain other alloying elements, such as nickel, chromium, silicon, and molybdenum. The former group is called "plain" carbon steels, while the latter are referred to as alloy steels. Alloy steels may be further clas-

TABLE 12-1 NUMBERING SYSTEM FOR WROUGHT OR ROLLED STEEL

Numerals and digits		Type of identifying elements
UNS	SAE	
		Carbon steels
G10XX0	10XX	Nonresulfurized, manganese 1.00% maximum
G11XX0	11XX	Resulfurized
G12XX0	12XX	Rephosphorized and resulfurized
		Alloy steels
G13XX0	13XX	Manganese steels
G23XX0	23XX	Nickel steels
G25XX0	25XX	Nickel steels
G31XX0	31XX	Nickel–chromium steels
G32XX0	32XX	Nickel–chromium steels
G33XX0	33XX	Nickel–chromium steels
G34XX0	34XX	Nickel–chromium steels
G40XX0	40XX	Molybdenum steels
G41XX0	41XX	Chromium–molybedenum steels
G43XX0	43XX	Nickel–chromium–molybdenum steels
G44XX0	44XX	Molybdenum steels
G46XX0	46XX	Nickel–molybdenum steels
G47XX0	47XX	Nickel–chromium–molybdenum steels
G48XX0	48XX	Nickel–molybdenum steels
G50XX0	50XX	Chromium steels
G51XX0	51XX	Chromium steels
G50XX6	50XXX	Chromium steels
G51XX6	51XXX	Chromium steels
G52XX6	52XXX	Chromium steels
G61XX0	61XX	Chromium–vanadium steels
G71XX0	71XXX	Tungsten–chromium steels
G72XX0	72XX	Tungsten–chromium steels
G81XX0	81XX	Nickel–chromium–molybdenum steels
G86XX0	86XX	Nickel–chromium–molybdenum steels
G87XX0	87XX	Nickel–chromium–molybdenum steels
G88XX0	88XX	Nickel–chromium–molybdenum steels
G92XX0	92XX	Silicon–manganese steels
G93XX0	93XX	Nickel–chromium–molybdenum steels
G94XX0	94XX	Nickel–chromium–molybdenum steels
G97XX0	97XX	Nickel–chromium–molybdenum steels
G98XX0	98XX	Nickel–chromium–molybdenum steels
		Carbon and alloy steels
GXXXX1	XXBXX	B denotes boron steels
GXXXX4	XXLXX	L denotes leaded steels
		Stainless steels
S2XXXX	302XX	Chromium–nickel steels
S3XXXX	303XX	Chromium–nickel steels
S4XXXX	514XX	Chromium steels
S5XXXX	515XX	Chromium steels
		Experimental steels
None	EX—	SAE experimental steels

Source: SAE Handbook Vol. 1, 1983, p. 107. Reprinted with permission, Society of Automotive Engineers, Inc.

sified into *low* alloy or *high* alloy, based on the total content of alloy addition as follows:

Low alloy: less than 10% alloy addition
High alloy: more than 10% alloy addition

Alloying elements perform many functions when added to iron or steel. Overall, these functions may be summarized as follows:

- The alloying addition forms a solid solution with iron, resulting in solid-solution strengthening and increased corrosion resistance. Examples include carbon, chromium, manganese, nickel, molybdenum, silicon, and cobalt.
- Alloys combine with carbon in steel to form various alloy carbides in addition to Fe_3C. Such carbides impart additional hardness and elevated temperature strength. Examples include titanium, tungsten, and vanadium.
- Certain elements combine with oxygen in the liquid steel, thereby removing dissolved oxygen from the material. Such *deoxidation* results in improved mechanical properties in the final product. Examples include aluminum, silicon, calcium, and manganese.
- Certain alloying elements remain undissolved in the steel, forming second phases which promote machineability and damping characteristics. Examples include lead, sulfur, and phosphorus.

As a commentary to this discussion of alloying elements and their function in steel, we should mention the effects of alloy additions in promoting the *hardenability* of steels. Hardenability refers to the capability of a steel to harden throughout the cross section of a bar or component during heat treatment. Since heat treatment is explained in the next section, we will defer an extensive discussion of the influence of alloying elements on hardenability and subsequent mechanical properties. However, we should note that some alloying elements, such as manganese, molybdenum, tungsten, vanadium, chromium, silicon, and nickel, promote hardenability and facilitate the nonequilibrium transformation of austenite to microstructural products other than pearlite. This type of transformation, which is accomplished by heat treatment, is the basis for hardening and strengthening steels.

HEAT TREATMENT FUNDAMENTALS

A major part of the versatility steels display stems from their response to thermal treatment or, simply, heat treatment. Recall that iron (and therefore steel) undergoes a polymorphic transformation from fcc to bcc upon cooling. Under equilibrium conditions, this transformation is described by the iron–carbon phase diagram (Figure 12-3). However, under nonequilibrium conditions, such as those commonly encountered in commercial practice, the transformation of austenite does not correspond exactly to the equilibrium phase diagram.

The result of all heat treatments to steel is an alteration of the form in which the carbon is distributed. The most pronounced changes are obtained by first heating the material to a temperature where stable austenite is formed, holding for sufficient time to obtain a fully austenitic condition at this temperature, then transforming the austenite to its decomposition products by either *isothermal* transformation or *continuous cooling* transformation.

Austenitizing Conditions

The grain size of the transformation product is directly related to the grain size of the austenite from which it forms. Small grain size in transformation structures is generally associated with good mechanical and physical properties. Thus, to obtain the smallest possible grain size in the product, it is important to start with the smallest practical austenite grain size. Although the precise austenitizing temperature and holding time depends strongly on the degree of chemical homogeneity that a particular alloy exhibits, we may use certain temperatures as guidelines. For hypoeutectoid compositions, the austenitizing temperature is generally 10 to 25°C (50 to 75°F) above the A_3 temperature (see Figure 12-3). If the temperature exceeds this range, coarsening of the austenite grains can occur, which may deleteriously affect the final mechanical properties, particularly toughness. In the hypereutectoid steels, coarsening of the austenite grains would be too extreme at temperatures above the A_{cm} line. Therefore, hypereutectoid steels are usually austenitized at a temperature between the $A_{1,3}$ and A_{cm} lines.

Although the exact length of time a steel part is austenitized depends on its composition, size and degree of chemical homogeneity, a "rule-of-thumb" is to heat for one hour per inch of diameter or maximum thickness, and to hold at the appropriate austenitizing temperature for at least $\frac{1}{3}$ of the heating time.

Isothermal Transformation

One convenient method for describing the nonequilibrium transformation of austenite during cooling is the isothermal transformation diagram, referred to more commonly as the *TTT diagram* (time–temperature–transformation). As the name implies, this method consists of transforming austenite, isothermally, at temperatures below the upper critical temperature. The upper critical temperature is the temperature where the alloy under consideration is fully austenitic (e.g., A_3 for hypoeutectoid steels and A_{cm} for hypereutectoid compositions). Correspondingly, the lower critical temperatures are A_1 and $A_{1,3}$, respectively. This type of diagram may be produced experimentally for any desired alloy in order to predict the structure of that alloy under conditions of isothermal transformation. Typical TTT diagrams for two plain carbon steels are shown in Figures 12-12 and 12-13. In the case of the hypoeutectoid and hypereutectoid steels, both the upper critical and the lower critical temperatures must be considered, because either *proeutectoid*, ferrite or Fe_3C forms prior to the eutectoid (pearlite) reaction. A eutectoid steel (0.8%), on the other hand, forms a structure consisting entirely of pearlite from austenite, because of its composition. Therefore, only the A_1 line need be considered. The shaded region indicates the portion of the diagram where transformation is occurring as a function of time for a particular

Heat Treatment Fundamentals

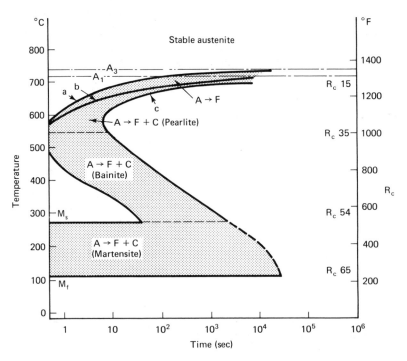

Figure 12-12 Isothermal transformation diagram for 0.6% C (1060) steel. A, austenite; F, ferrite; C, iron carbide. Transformation of austenite starts at curve a, and is complete at curve c.

(constant) temperature. For example, in Figure 12-12, at 650°C, transformation of austenite to ferrite begins, at curve (a) and continues isothermally until curve (b). At this time, austenite begins to transform to pearlite and continues this reaction until curve (c) is reached. At (c) the austenite is completely transformed and a microstructure of ferrite and pearlite results.

Overall, the TTT diagrams show the change in transformation behavior as temperature decreases below the critical temperatures, and may also be used to show differences in isothermal transformation behavior resulting from changes in carbon or alloy content. Notice that at high transformation temperatures (near A_3 or A_1) the amount of proeutectoid constituent formed approaches that found under equilibrium conditions. But at lower temperatures (near the nose of the curve), very little proeutectoid constituent forms because insufficient time is available before pearlite starts forming.

Bainite formation. At temperatures below the nose of the TTT curve, transformation of austenite produces a structure in which the ferrite and iron carbide are not lamellar. This transformation product, called *bainite*, exhibits a feathery or acicular (needle-like) microstructure, as shown in Figure 12-14. Whereas pearlite is nucleated by iron carbide and accompanied by the subsequent formation of ferrite, bainite is nucleated by ferrite, which is followed by the precipitation of iron carbide.

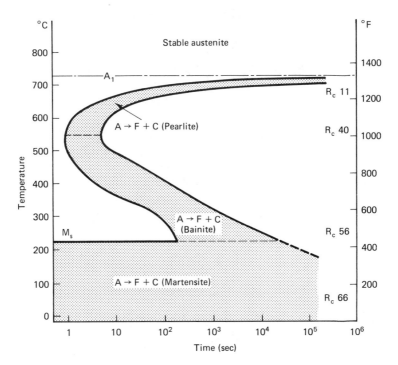

Figure 12-13 Isothermal transformation diagram for 0.8% C (1080) steel. A, austenite; F, ferrite; C, iron carbide.

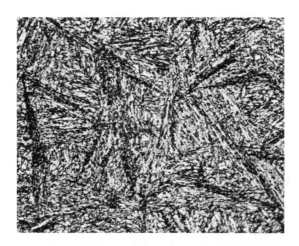

Figure 12-14 Photomicrograph showing bainitic microstructure in low alloy steel (750×). (Courtesy of T. V. Brassard, Benét Laboratories.)

This process, shown schematically in Figure 12-15, leads to a dispersion of iron carbide in a matrix of ferrite. With lower transformation temperatures, the distribution of carbide is finer and the ferrite needles are thinner, resulting in a structure that is harder than pearlite. As indicated on the TTT diagrams in Figures 12-12 and 12-13, the hardness of bainite ranges approximately from 35 to 56 on the Rockwell C (R_C) scale.

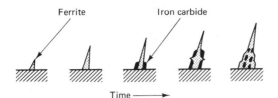

Ferrite   Iron carbide

Time ⟶

Figure 12-15 Schematic representation of bainite formation. (From D. K. Allen, *Metallurgy Theory and Practice*, American Technical Society, Alsip, Ill. 1969.)

Martensite formation. Examination of the TTT diagrams further reveals that transformation of austenite at temperatures lower than those in the bainite range occurs as a function of temperature only; transformation is independent of time. The latter process is a diffusionless transformation which starts at a temperature labeled M_s and produces a body-centered tetragonal crystal structure of iron with carbon atoms located at interstitial sites as shown in Figure 12-16. The resultant product is

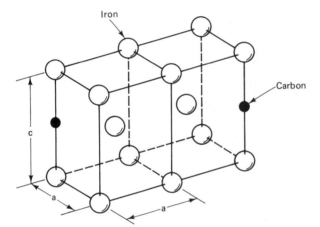

Iron

Carbon

c

a

a

Figure 12-16 Body centered tetragonal (bct) structure of martensite.

called *martensite* and usually exhibits a lath or plate-like morphology, as illustrated in Figure 12-17. Since the martensite transformation occurs at relatively low temperatures, diffusion of carbon is not involved, but rather a shearing of the austenite lattice takes place. Therefore, the resulting structure has the same chemical composition as the austenite from which it formed, and is supersaturated with carbon. As the unit cell sketch in Figure 12-16 implies, the body-centered structure is distorted, with the lattice dimensions being unequal and giving rise to internal strains and expansion of the lattice. The expansion associated with this transformation is depicted in Figure 12-18, which displays the change in length of a typical low-alloy steel as a function of temperature. Notice the prominent increase in length that occurs at the M_s temperature during cooling of this material. Such an expansion can produce significant strains in the iron–carbon lattice. These conditions—carbon supersaturation and lattice deformation from shearing—contribute to the high hardness exhibited by martensite.

Perhaps a final remark about the general features of the isothermal transformation diagram is in order. Figures 12-12 and 12-13, which illustrate the TTT behavior for carbon steel, show that increasing the carbon content increases the time before transformation starts at any temperature. In other words, the TTT curve is displaced

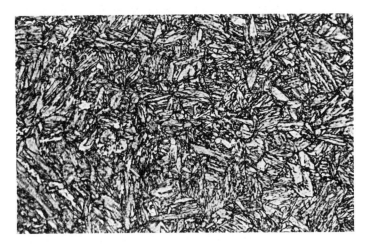

Figure 12-17 Microstructure of martensite in a tempered low alloy steel (825 ×). (Courtesy of T. V. Brassard, Benét Laboratories.)

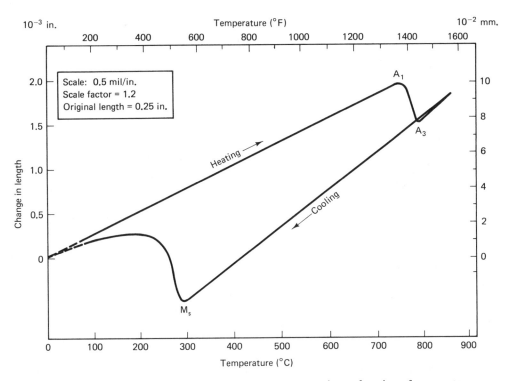

Figure 12-18 Dilation behavior of a low alloy (4335) steel as a function of temperature. (Courtesy of P. J. Cote, Benét Laboratories.)

to longer times. Although we have not discussed the effects of other alloying additions on the TTT curve, their influence is generally similar. Increasing the alloy content of a steel increases the time before transformation of austenite begins. Moreover, the martensite transformation occurs at lower temperatures. The carbon content has a significant effect on the start temperature (M_s) and finish temperature (M_f) of this reaction, as shown in Figure 12-19. Indeed, these data demonstrate that the M_f tem-

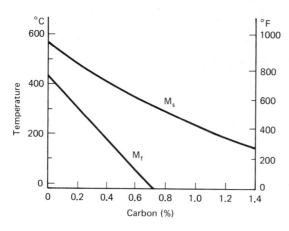

Figure 12-19 Influence of carbon content on the M_s and M_f temperatures. (From A. R. Troiano and A. B. Greninger, *Metal Progress 50*, American Society for Metals, Metals Park, Ohio, 1946, p. 303).

perature for carbon steels with compositions greater than about 0.7 %C is below room temperature! Certain steels, then, cooled to ambient temperatures may not be completely transformed, and austenite in an unstable condition would be retained.

Large percentages of *retained austenite* are usually undesirable, since it is softer than the surrounding transformation products. Furthermore, it is in an unstable state and eventually may isothermally transform to bainite during subsequent thermal operations. Such later transformation may occur when the dimensional tolerances and properties of the component will not withstand any volumetric changes and the strains that accompany them. As a result, the part may not meet the dimensional requirements imposed on it, or worse, cracks may be produced by the attendant internal stresses, and render the material useless.

In the case of retained austenite, *cold treatment* or cooling the material to below ambient temperatures may be necessary to convert the retained austenite to martensite. Such treatments, referred to as *refrigeration*, are typically used in high-carbon grades and steels containing austenite stabilizing additions (i.e., manganese, cobalt, and nickel).

Applications of Isothermal Transformation

The procedure of cooling steels rapidly enough to obtain a desired hardness, which results from the structural transformation, has one inherent disadvantage. If the material is cooled too rapidly, cracks may develop from the resulting thermal and transformational stresses. Such cracks are commonly referred to as *quench cracks*. Quench cracks can behave as stress concentrators and initiation sites for fatigue or, depending on their severity, can result in complete failure of the heat-treated part. Therefore, they should obviously be avoided.

In some instances, satisfactory hardness and attendant mechanical properties may be attained by transforming the austenite at temperatures above M_s. This process, called *austempering*, consists of cooling the steel to some intermediate temperature and holding for sufficient time to allow complete transformation of austenite to the structure or hardness level desired, as illustrated in Figure 12-13. The austempering process is pictured in Figure 12-20.

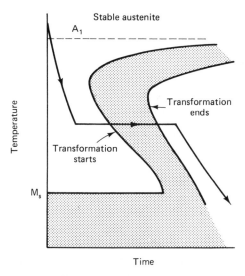

Figure 12-20 Schematic of a typical austempering process.

Another application of the isothermal transformation diagram involves cooling the material to a temperature slightly above the M_s, at a rate sufficiently fast to avoid the nose of the TTT curve. This procedure prevents transformation of the austenite to products other than martensite. However, in large sections a considerable temperature gradient may exist in the piece. If the martensite transformation occurs before this temperature gradient diminishes, sizable internal stresses can be produced and occasionally result in cracking. This potential problem can be alleviated by a procedure referred to as *martempering* or *marquenching*. As illustrated in Figure 12-21, this treatment consists of rapidly cooling the steel to just above the M_s and holding isothermally until the temperature becomes uniform throughout the cross section. Then the material is slowly cooled through the martensite transformation region, allowing the entire structure to transform to martensite in a uniform manner.

Continuous Cooling Transformation

In actual practice, the preponderance of steel heat treatments involve heating the material to an appropriate austenitizing temperature, followed by continuous cooling to room temperature or the desired finishing temperature. This type of transformation behavior is described by the use of *continuous cooling* transformation (CCT) diagrams.[1]

---

[1] It should be noted that with CCT diagrams as with TTT diagrams, each alloy grade has a different curve, which results from its different composition.

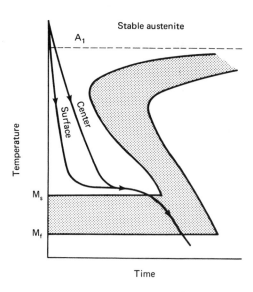

Stable austenite

A_1

Center

Surface

Temperature

M_s

M_f

Time

Figure 12-21 Schematic of a typical martempering process.

The CCT diagram for an SAE 4340 grade, low-alloy steel is shown in Figure 12-22. This grade is capable of being heat treated to very high tensile strengths and is widely used in applications where strength and toughness are required, such as aircraft landing gear and automotive axles, shafts, and rods. Continuous cooling is usually accomplished by immersion of the material (in its austenitized state) into a quenching medium. Typical quenchants include air, oil, water, and brine (5 to 10% NaCl solution). Often in order to obtain a more effective quenching process, the part and/or the quench media is agitated. This increases the rate of heat transfer between the surface of the part and the quenchant, thereby promoting a more efficient cooling process. The severity of quench (H) is an indication of how effective a specific cooling medium is from the thermodynamic standpoint. Some typical values of H are given in Table 12-2.

TABLE 12-2 QUENCHING CHARACTERISTICS OF CERTAIN COOLANTS

	H values	
Quench medium	No circulation	Good circulation
Air	0.02	0.25 (violent)
Oil	0.25	0.45
Water	1.00	1.50
Brine	2.00	5.00 (violent)
Water spray		5.00

The CCT diagram illustrated in Figure 12-22 displays certain cooling rates which define the specific regions of austenite transformation. For instance, cooling rates less than about 4°C/hr (40°F/hr) produce a structure consisting of ferrite and pearlite. Between 4°C/hr and 66°C/hr (150°F/hr) the transformation products include marten-

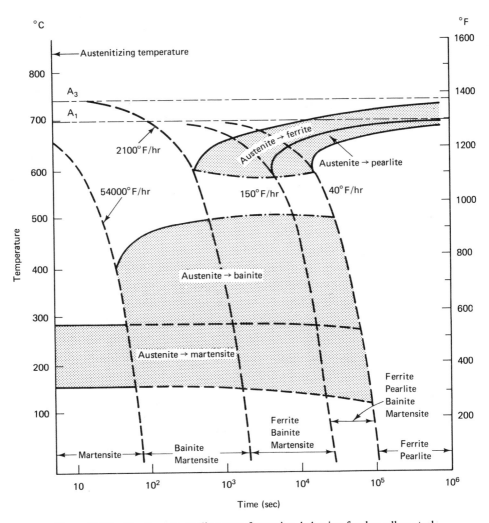

Figure 12-22 Continuous cooling transformation behavior for low-alloy steel (SAE 4340). (From H. E. McGannon, Ed., *The Making, Shaping and Treating of Steel*, U.S. Steel Corporation, 1964, p. 1049.)

site, bainite, ferrite, and pearlite. At cooling rates of 66°C/hr to 1150°C/hr (2100°F/hr) the pearlite transformation region is missed entirely and the final microstructure will consist of ferrite, bainite, and martensite. Cooling rates between 1150°C/hr and 30,000°C/hr (54,000°F/hr) will produce a structure consisting largely of bainite with a relatively small amount of martensite. In this alloy, a cooling rate of at least 30,000°C/hr is necessary to transform all the austenite to martensite. This cooling rate is called the *critical cooling rate* (CCR).

The final microstructure and therefore the resultant properties of a steel can be predicted accurately by superimposing the cooling conditions corresponding to various heat treatments on the continuous cooling transformation diagram. Furthermore, the transformation behavior of a steel and the cooling conditions can be controlled within certain limits by adjustments to chemistry and heat treatment parameters. Thus the properties of the steel can be accurately controlled. As we indicated in the introductory remarks, this aspect of steels is one of the important factors contributing to its versatility and widespread use.

Let us examine briefly some of the important heat treatment processes that are associated with continuous cooling.

Annealing. In general, annealing consists of heating a material to some specified elevated temperature for an appropriate length of time. Ordinarily, the temperature is sufficiently high enough for diffusion to take place. The purpose of annealing is to (1) relieve residual stresses induced by prior processing, such as cold working and cooling processes, and (2) soften the steel for improved machineability or formability. In some cases, annealing may involve subcritical (below A_1) heating to relieve stresses, recrystallize cold-worked material (see Figure 6-27), or to spherodize the carbides of previously transformed material. These processes are often referred to as *subcritical annealing*.

On the other hand, *full annealing* consists of heating to a temperature above the upper critical temperature (A_3) in hypoeutectoid steels, and between the $A_{1,3}$ and A_{cm} temperatures in hypereutectoid steels. This is followed by slow cooling, typically 10°C/hr (50°F/hr), so that transformation occurs only and completely in the high-temperature range. Although this transformation does not actually occur under equilibrium conditions, it can yield a reasonable approximation of the equilibrium structure predicted by the phase diagram (e.g., Figure 12-3). Plainly, the slower the cooling rate, the closer equilibrium conditions are approached. A full annealing process is shown schematically in Figure 12-23 for a hypoeutectoid carbon steel.

In commercial practice, the cooling conditions necessary to produce an annealed structure are often obtained by cooling the workpiece in the furnace. This provides accurate control of the cooling rate and allows very slow cooling conditions to be established. The microstructure of a fully annealed hypoeutectoid steel (1040) is shown in Figure 12-24. Full annealing is a relatively simple and effective thermal treatment for most steels, but it is also time-consuming and often ties up valuable furnace space during the slow cooling procedure.

We should also point out that annealing is conducted on many other engineering materials, including nonferrous alloys and ceramics, in order to develop certain desirable properties. For example, annealing is typically applied to cold-worked wire products, drawn from copper or aluminum, to soften the material and restore ductility. Many glass products, such as containers and sheet, are annealed to relieve residual stresses that may result during solidification.

Normalizing. The process of *normalizing* consists of heating the steel above the upper critical temperature (A_3 or $A_{1,3}$-A_{cm}), followed by air cooling at room temperature. Basically, normalizing has two primary purposes: (1) to alleviate chemical segregation of carbon and other alloying elements, and (2) to produce refinement of the austenitic grain size, and a more uniform, finer carbide distribution, so that carbide dissolution will be promoted during subsequent heat treatment. The cooling

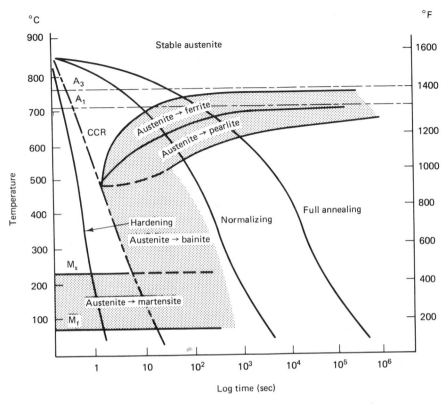

Figure 12-23 Schematic showing various heat treatment processes in a continuously cooled *hypo*eutectoid steel. CCR denotes the critical cooling rate.

Figure 12-24 Microstructure of plain carbon steel (1018) in the annealed condition. Dark regions are pearlite colonies in a matrix of ferrite (335×).

rate for normalizing is typically on the order of 100°C/hr (212°F/hr) and results in a structure with more finely laminated pearlite and less proeutectoid constituent. A typical normalizing procedure is depicted in Figure 12-23. Note in this process that transformation to products other than ferrite and pearlite is avoided.

Hardening. Hardening is the process of heating the steel to a temperature sufficient to produce a fully austenitic condition, as described earlier in this section, followed by cooling (quenching) at a rate fast enough to prevent transformation to any product other than martensite. The rate that just produces an entirely martensitic structure is the critical cooling rate (CCR). Therefore, this rate must be exceeded to ensure transformation of all the austenite to martensite. A typical hardening procedure that exceeds the CCR is illustrated in Figure 12-23.

The critical cooling rate of a steel is strongly affected by its chemical composition and also by its austenitic grain size. Increases in carbon and other principal alloying elements *decrease* the critical cooling rate, thereby making it relatively easier to obtain martensite. Essentially, these alloy additions move the "nose" of CCT curve to higher values of time, thus delaying the start of transformation. Unfortunately, austenitic grain size may be interpreted as having the opposite effect. Fine austenitic grain size, which incidentally promotes good mechanical properties, *increases* the critical cooling rate, thereby making it more difficult to obtain transformation completely to martensite.

Example 12-7

A structural component manufactured from SAE 4340 grade steel is to undergo heat treatment, consisting of annealing and hardening, respectively. What cooling rates should be used for these processes, and what microstructures would you predict for these thermal conditions?

Solution Referring to the continuous cooling transformation diagram for this particular alloy (Figure 12-22), we can easily determine by inspection that transformation to a structure of ferrite and pearlite will occur for cooling rates slower than approximately 40°F/hr. Therefore, full annealing can be accomplished at a cooling rate of about 40°F (or slower).

Hardening by definition is the transformation of austenite to martensite. Essentially, complete transformation to martensite occurs between M_s and M_f if the cooling rate exceeds 54,000°F/hr in this alloy. This rate can be considered the critical cooling rate. *Ans.*

Tempering

The hardening process we have just discussed results in a structure of untempered martensite which is very hard (typically on the order of $60\text{-}66R_c$), brittle, and contains high internal (residual) stresses. These characteristics are a result of the unique martensitic transformation behavior, as explained previously in the section on heat treatment. However, in this physical condition, the steel is not only susceptible to sudden cracking, but exhibits poor ductility and toughness. It is, at this point, generally unsuitable for use in a manufactured product.

The purpose of *tempering*, an elevated temperature thermal process, is to relieve these deleterious stresses and restore the ductility and toughness of the steel. Tempering is conducted at temperatures *below* the lower critical temperature (A_1) to avoid recrystallization of any martensite to austenite. If such recrystallization occurs, a precipitous drop in room temperature tensile strength can result. During tempering (another diffusion-controlled process) the stress relief and recovery of ductility and toughness occur through the precipitation of carbide (iron and alloy) from the supersaturated solid solution of carbon in *bct* iron. This reaction produces a fine dispersion of hard carbide particles in the ferrite matrix, which reverts to a stable bcc structure during the process. Tempered martensite therefore consists of a uniform dispersion of submicroscopic-to-microscopic carbide particles in a matrix of ferrite. Such a structure promotes good tensile strength and ductility, plus good toughness (impact and K_{Ic}) since the dispersion of small hard particles impedes dislocation motion, thereby preventing slip, while the ferrite matrix tends to plastically deform at stress concentrators and microcracks, thus raising the energy necessary to initiate and propagate cracks.

Effects of alloying elements. Generally, the presence of alloying elements in steel retards the loss of hardness and strength during tempering. Carbide-forming elements such as chromium, molybdenum, tungsten, vanadium, and titanium are particularly effective in delaying the softening process. In fact, these alloys can produce an increase in hardness when tempered at certain temperatures. This process is referred to as *secondary hardening*. The effects of molybdenum additions to a 0.35% carbon steel containing 0.5% chromium are shown in Figure 12-25. As the molybdenum content is increased, softening is retarded for any tempering temperature.

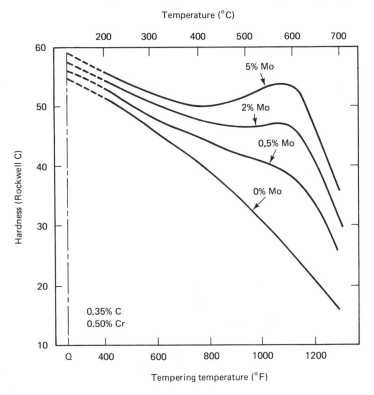

Figure 12-25 Effect of molybdenum additions on the tempering response of low-alloy steel. (From E. C. Bain and H. W. Paxton, *Alloying Elements in Steel*, American Society for Metals, Metals Park, Ohio, 1961, p. 200.)

Furthermore, in the range 450 to 650°C (842 to 1200°F), the hardness is actually recovered for molybdenum additions of 2% and 5%, as the tempering temperature is increased. This secondary hardening effect is due to the formation of a submicroscopic alloy carbide precipitate. With higher temperatures or longer times, these alloy carbides grow larger and eventually become ineffective as hardeners. In principle, this is the same process as *overaging* that we examined in the section "Precipitation hardening" in Chapter 6. Consequently, the material re-experiences softening and this process resumes at a faster rate, as shown in Figure 12-25.

As the tempering process continues and carbon diffuses to form more and more carbide, the particles grow large and eventually coalesce. The overall result is a decrease in tensile strength and hardness of the steel. The effects of tempering tempera-

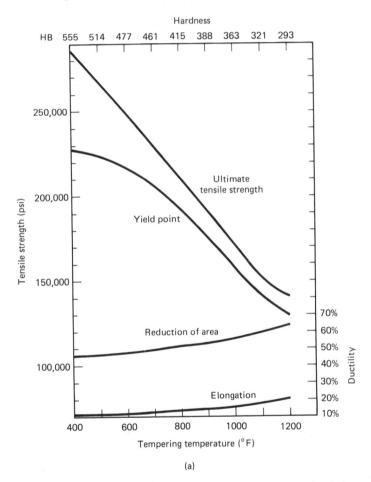

(a)

Figure 12-26 Effects of tempering temperature on selected mechanical properties. (a) Hardness and tensile properties in 4340 steel. (From *Modern Steels and Their Properties*, Handbook 3310, Bethlehem Steel Corp., Bethlehem, Pa., 1967.) (b) Impact toughness as measured by the Izod test in several alloy steels. (From E. C. Bain and H. W. Paxton, *Alloying Elements in Steel*, American Society for Metals, Metals Park, Ohio, 1961, p. 227.)

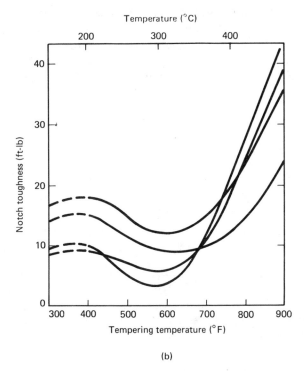

(b)   **Figure 12-26**   (Cont.)

ture on the hardness, yield strength, and impact toughness of steel are demonstrated in Figure 12-26. Since the tempering reaction is a diffusion-controlled process, it is reasonable to expect the rate of precipitation and growth of carbide particles to be greater at higher temperatures. The influence of temperature and time on the tempering process are shown in Figure 12-27. This relationship, which is typical of many

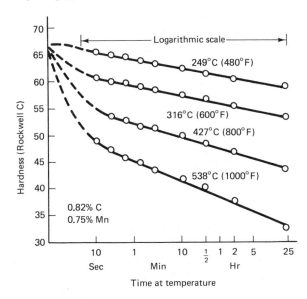

Figure 12-27   Influence of temperature and time during the tempering of a eutectoid steel. (From E. C. Bain and H. W. Paxton, *Alloying Elements in Steel*, American Society for Metals, Metals Park, Ohio, 1961, p. 185.)

steels, shows that tempering temperature is the major variable affecting the softening of the as-quenched material. Moreover, this figure shows the logarithmic relationship between hardness and the time at temperature. Such a relationship demonstrates that increased times produce a lesser effect in lowering hardness than increased temperatures, which exert a very strong influence on the tempering operation. This effect is shown by Example 12-8.

Example 12-8

A tempering operation is conducted on the eutectoid composition steel of Figure 12-27. (a) Determine the time required to reduce the hardness of this material from an as-quenched value of approximately 67R$_C$ to a hardness level of 45R$_C$, if the tempering temperature is 427°C (800°F). (b) What would be a more practical procedure for attaining this same hardness requirement?

Solution (a) Utilizing the time–temperature relationships in Figure 12-27, we can interpolate the time necessary to temper this steel to 45R$_C$ at 427°C as approximately 10 hr. *Ans.*

(b) Further inspection of this "graph" shows that a hardness level of 45 R$_C$ can also be attained at a tempering temperature of 538°C (1000°F), but in only about 1 min. However, to provide sufficient control over the tempering process, a temperature between 427 and 538°C, which will provide the desired hardness throughout the part, in approximately 1 to 2 hrs at temperature, should be established by further experimentation. *Ans.*

Comment: A tempering time of 10 hr duration is not unusual in very large parts. However, such prolonged processing times tie up valuable furnace space and consume additional energy. These factors should be considered in determining the processing parameters for a specific material.

Since continuation of the tempering reaction results in a corresponding decrease in tensile strength and hardness, the process must be arrested when sufficient ductility and toughness are recovered. The point at which tempering is stopped depends on the alloy being treated, together with the desired properties or design requirements imposed by the application.

After tempering, some steels may be air-cooled from the tempering temperature, while others exhibit a loss of toughness on slow cooling from temperatures above 538°C (1000°F) and must be rapidly cooled. This loss of toughness develops from a phenomenon generally known as *temper embrittlement*, which apparently involves a damaging precipitate at the prior austenitic grain boundaries. Alloys subject to this problem are rapidly cooled to room temperature to avoid such precipitation effects.

Hardenability

When we refer to the *hardenability* of a steel, we do not mean its ability to resist indentation or penetration (in other words, hardness); rather, we mean the relative *ease* of transforming austenite to martensite in a particular alloy. Therefore, a material that has high hardenability can be *through-hardened* with less difficulty than can one of lower hardability. A through-hardened condition results when the center of a round bar, or the midpoint of the thickest section in a more geometrically complex part, is quenched to the desired hardness level.

Fortunately, a number of techniques have been developed to predict the hardness at the center of a quenched steel part, or to specify the quenching (cooling) conditions that will through-harden a particular steel. Let us examine one of these methods for determining hardability.

Jominy end-quench test. The Jominy test consists of selectively cooling a standard size bar of steel, as illustrated in Figure 12-28. Note that only the *end* of the specimen is actually quenched, and as a circumstance of such cooling conditions, transformation of austenite begins at the water-cooled end and progresses "up" the bar. The cooling rate decreases with increasing distance from the quenched end, as

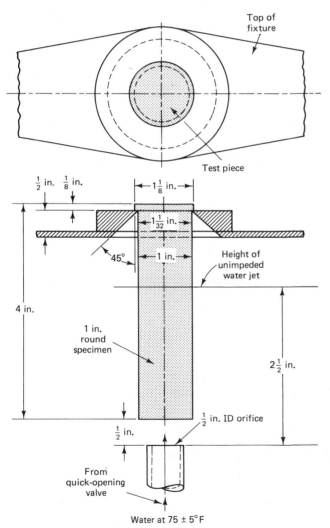

Figure 12-28 Standard Jominy end-quench test apparatus. (From ASTM Standard A255-67, *End-Quench Test for Hardenability,* American Society for Testing Materials, Philadelphia, Pa.) (Copyright, ASTM, 1916 Race Street, Philadelphia, Pa. 19103. Reprinted with permission.)

Heat Treatment Fundamentals

431

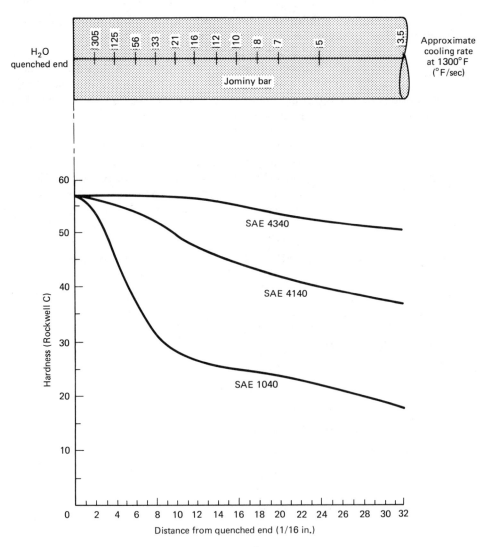

Figure 12-29 Typical Jominy hardenability curves for carbon and low-alloy steel containing 0.40% C.

shown in Figure 12-29. Therefore, the hardness of steels with relatively low hardenability decreases with distance from the water quenched end, since transformation of austenite to martensite depends on the cooling rate. This behavior is reflected by the Jominy curves for the steels in Figure 12-29. The sharp drop in hardness with distance from the quenched end is apparent for the 1040 grade, which is a plain carbon steel. Comparatively, the hardenability of 4140 and 4340 is greater because of the addition of alloying elements, a topic that will be discussed in a later section.

In Figure 12-30, the continuous cooling transformation behavior of 8630 steel is correlated with end-quench data. Examination of this figure shows that cooling rates produced near the quenched end of the Jominy bar (location *A*) exceed the

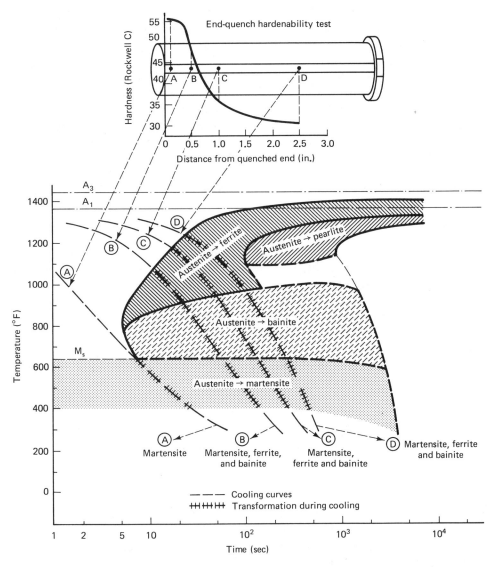

Figure 12-30 Relationship between end-quench test and the continuous cooling transformation (CCT) diagram for an 8630-type steel. (From U.S.S. Carilloy Steels, United States Steel Corp., 1948.)

critical cooling rate and result in a fully martensitic microstructure. This is reflected in the hardness value obtained in the Jominy bar at this point. Conversely, the cooling rate at location *D*, which is 2.5 in. (63.5 mm) from the quenched end, is substantially lower, resulting in transformation to ferrite, bainite, and martensite. Again, this transformation is reflected in the hardness level of the Jominy bar at this location.

Hardenability (Jominy) curves for a variety of standard grades are widely published in the metallurgical literature and heat treatment handbooks. These curves generally give the high and low limits of hardenability and therefore are usually referred to as "H"-bands. They are widely used by engineers to determine which steel should be used to make a particular component, and how it should be heat treated.

Application of end-quench results. The results of a Jominy end-quench test can be applied to various-size round bars produced from that particular steel. Simply, the hardness level at any location along the Jominy bar (the Jominy distance) is the result of a specific cooling rate. Therefore, if this cooling rate is duplicated at some location in a round bar, or for that matter any component made from this steel, the hardness corresponding to the appropriate Jominy distance should result. Of course, this relationship is strongly dependent on similar chemical compositions in the Jominy bar and the part or cross section in question.

Factors affecting hardenability. Alloying elements exert a major influence on the hardenability of steel. With the exception of cobalt, alloy additions increase hardenability by retarding transformation of austenite at temperatures above M_s. Cobalt additions in steel tend to encourage the nucleation and growth of pearlite. Certain elements also exert a stronger influence on hardenability than do others. For example, carbon, manganese, molybdenum, and chromium make a greater contribution to hardenability than silicon and nickel per unit weight of alloying element.

The austenitic grain size also has a considerable influence on hardenability. Since pearlite nucleates at the austenite grain boundaries, the greater the surface area available, the more pearlite formation is encouraged, thereby decreasing hardenability. However, it is generally not good practice to use coarse-grained austenite to obtain increased hardenability. Large austenitic grain size has frequently been associated with poor ductility, impact strength, and fracture toughness in the subsequent transformed product. Therefore, other means, such as increased alloy content, should be used if increased hardenability is necessary.

PROPERTIES AND APPLICATIONS OF FERROUS ALLOYS

Ferrous alloys are generally selected or specified for an application by one or more of the following criteria:

- Ability to meet a standard specification (e.g., ASTM, SAE, ASME)
- Ability to be fabricated into a particular configuration by means of stamping, forging, extrusion, casting, etc.
- Mechanical properties (e.g., yield strength, hardness, impact strength)
- Chemical composition, particularly when the material is to receive further processing, such as forming, heat treating, welding, or machining

Earlier in this chapter we discussed the classification of steels, which included plain carbon and alloy grades, as listed in Table 12-1. Let us now reexamine these grades together with ultra-high-strength steel and corrosion-resistant steel (stainless steel) from the standpoint of their properties and applications. Although tool steels and cast irons are also ferrous alloys, they will not be included in our discussions.

Plain Carbon Steel

Basically, a plain carbon steel consists of iron containing small amounts of carbon. In this material, the carbon content can vary from 0.008% to approximately 2.0%, as shown on the iron–iron carbide diagram in Figure 12-3. Plain carbon steels may also contain limited amounts of manganese (1.65% max.), silicon (0.60% max.) and copper (0.60% max.). The standard carbon steels are the types such as 1005 which contains approximately 0.05% carbon through 1095 which contains approximately 0.95%C. Carbon steels in the lower end of this range are generally used in applications where a soft, deformable material is needed. This includes structural shapes, rivets, nails, wire, and pipe. Steels with higher carbon contents (0.30%) are hardenable by quenching, and therefore find application where greater strength is required. Properties that are typically developed in these grades are given in Table 12-3. Applications for the mid-carbon range include gears, shafts, axles, rods, and a multitude of machine parts. Higher-carbon steels find application in hammers, chisels, drills, punches, cutters, knives, springs, wire, and dies for all purposes.

Low-Alloy Steel

Steels containing alloy additions which usually do not exceed a total of about 10% are referred to as low-alloy steels. These alloying elements, consisting primarily of manganese, silicon, nickel, chromium, and molybdenum, are added to enhance certain properties, such as strength, toughness, fatigue resistance, and wear resistance. Some of the general industries where these properties are important include transportation, agriculture, construction, and military applications.

Properties typical of heat-treatable low-alloy steels are listed in Table 12-4. In addition to the factors that we have already mentioned, selection of these alloys may also be based on hardenability. It is possible to divide standard alloy steels into *through-hardening* (higher hardenability) grades and *surface-hardening* (lower hardenability) grades.

The through-hardening alloys, including 4130, 4340, 8640, 9260, and 52100, are widely used in such applications as high-strength fasteners, connecting rods, springs, torsion bars, and ball bearings. Surface-hardening grades, such as 1320, 4620, and 8620, are used in shafts and gears as well as in many other parts that require external wear and abrasion resistance, combined with a tough interior.

Ultra-High-Strength Steel

Steels capable of developing yield strengths greater than about 1104 MPa (160,000 psi) are considered *ultra-high-strength* alloys. These materials consist of several classes of steels, including certain *stainless steels*. Discussion of the stainless steels, however, will be deferred until later in this section.

TABLE 12-3 TYPICAL PROPERTIES FOR CARBON STEELS[a]

Grade	Ultimate tensile strength		Yield strength		Reduction of area (%)	Elongation (%)	Hardness (BHN)
	MPa	psi	MPa	psi			
Water quench							
1030	607	88,000	473	68,500	69	28	179
1040	744	107,750	542	78,500	63	23	217
1050	905	131,250	636	92,250	55	20	262
1095	1138	165,000	707	102,500	41	16	311
Oil quench							
1060	942	136,500	592	85,750	48	18	269
1080	1145	166,000	714	103,500	38	15	331
1137	745	108,000	523	75,750	56	21	223
1141	760	110,200	520	75,300	57	22	217
1144	749	108,500	502	72,750	46	19	223

[a]One-inch (25.4-mm)-diameter bars tempered at 1000°F (538°C).

Source: Adapted from *Modern Steels and Their Properties*, Handbook 3310, Bethlehem Steel Corp., Bethlehem, Pa., 1967.

TABLE 12-4 TYPICAL PROPERTIES FOR LOW-ALLOY STEELS[a]

Grade	Ultimate tensile strength		Yield strength		Reduction of area (%)	Elongation (%)	Hardness (BHN)
	MPa	psi	MPa	psi			
Water quench							
4027	960	139,250	843	122,250	60	19	285
4130	997	144,500	894	129,500	62	18	293
8630	930	134,750	849	123,000	60	19	269
Oil quench							
1340	950	137,750	835	121,000	57	19	285
4140	1076	156,000	988	143,250	57	16	311
4340	1208	175,000	1145	166,000	46	14	352
5140	973	141,000	838	121,500	59	18	293
8740	1232	178,500	1133	164,250	53	16	352
6150	1197	173,500	1158	167,750	48	14	352
9255	1133	164,250	923	133,750	38	17	321

[a]One-inch (25.4-mm)-diameter bars tempered at 1000°F (538°C).

Source: Adapted from *Modern Steels and Their Properties*, Handbook 3310, Bethlehem Steel Corp., Bethlehem, Pa., 1967.

Medium-carbon low alloy steel.   These alloys consist of grades such as 4130, 4330, and 4340, which can be quenched and tempered to yield strengths on the order of 1725 MPa (250,000 psi) with corresponding ultimate tensile strengths as high as 2070 MPa (300,000 psi). It should be pointed out, however, that these alloys achieve their ultrahigh strength at the expense of a certain degree of ductility and toughness. Such steels are typically used in applications requiring high strength and high hardenability, such as in rocket motor cases; aircraft components, including bolts, pins, main landing gears, and brake housings; and a wide variety of structural and machinery parts.

Maraging steel.   This class of steel consists basically of extra-low-carbon (less than 0.03%) iron-based alloys to which a high percentage of nickel has been added, as shown in Table 12-5. These alloys are capable of attaining very high yield

TABLE 12-5 TYPICAL COMPOSITIONS OF MARAGING STEELS

Grade	Chemical composition (w/o)					
	Ni	Co	Mo	Ti	Al	Cb
20 Ni	19–20	—	—	1.3–1.6	0.15–0.30	0.30–0.50
25 Ni	25–26	—	—	1.3–1.6	0.15–0.30	0.30–0.50
18 Ni (200)	17–19	8–9	3–3.5	0.15–0.25	0.05–0.15	—
18 Ni (250)	17–19	7–8.5	4.6–5.2	0.3–0.5	0.05–0.15	—
18 Ni (300)	18–19	8.5–9.5	4.6–5.2	0.5–0.8	0.05–0.15	—
18 Ni (350)	17.5–18.5	12–12.5	3.8–4.6	1.4–1.7	0.10–0.15	—
13 Ni (400)	13	15–16	10	0.2	—	—

and ultimate tensile strengths, accompanied by comparatively good levels of ductility and toughness, as shown in Table 12-6.

The term *maraging* is derived from the fact that these alloys are strengthened by a combination of martensite transformation followed by an age hardening or a precipitation-hardening reaction. The student is referred to the elementary discussion regarding age hardening in Chapter 7. Depending on the precipitate formers, such as titanium, aluminum, and molybdenum, that may be present in a particular alloy, a submicroscopic precipitate (e.g., Ni_3Ti or Ni_3Mo) is produced in the iron–nickel martensite lattice. Strengthening results from the interaction between this extremely fine precipitate and dislocations, as we explained in Chapter 6.

Corrosion-Resistant (Stainless) Steel

Stainless steels may be divided into four categories: ferritic, martensitic, austenitic, and age-hardenable. This classification depends on their chemistry and the resultant structure or properties they display. Chromium is one of the principal alloying elements in stainless steel and a minimum of approximately 11 to 12% is necessary for adequate corrosion resistance. Although we will explore the mechanisms of corrosion in detail later (Chapter 17), it will be sufficient at this point simply to state the manner

TABLE 12-6 TYPICAL PROPERTIES OF MARAGING STEELS

Grade	Ultimate tensile strength		Yield strength		% RA	Charpy impact		Fracture toughness		Hardness (BHN)
	MPa	10³ psi	MPa	10³ psi		J	ft-lb	MPa-m$^{1/2}$	ksi-in.$^{1/2}$	
18 Ni (200)	1345–1585	195–230	1310–1550	190–225	35–67	35–68	26–50	110–176	100–160	419–717
18 Ni (250)	1690–1860	245–270	1655–1825	240–265	35–60	24–45	18–33	99–165	90–150	464–488
18 Ni (300)	1825–2105	265–305	1790–2070	260–300	30–50	16–26	12–19	88–143	80–130	500–548
18 Ni (350)	2468	358	2427	352	43	16	12	—	—	584
18 Ni (400)	2792	405	2724	395	25	—	—	—	—	628

in which stainless steels resist corrosion. Essentially, the chromium in steel reacts with oxygen in the environment to form a relatively stable oxide film (Cr_2O_3) on the steel surface. This oxide film serves to protect (insulate) the surface from further corrosive reaction (oxidation). Such a process is referred to as *passivation*, or in other words, a loss of chemical reactivity under certain environmental conditions. In our case, the stainless steels simply form a passive film on their surface which, left undisturbed, retards further corrosion.

The chemical compositions and typical properties of certain stainless steel grades are listed in Tables 12-7 and 12-8. Let us now briefly examine the important aspects of each group.

Ferritic stainless steels.   This group of stainless steels contains between 11.5 and 27% chromium as the only major alloying element in addition to a maximum of 0.25% carbon. Because these alloys are ferritic they do not respond to heat treatment. Examination of Figure 12-31 shows that these steels fall outside the "γ-loop" and consist of a solid solution of chromium in bcc iron at all temperatures up to the solidus.

Ferritic stainless steels are magnetic, and are typically employed where corrosion resistance is needed but where strength requirements are relatively moderate. Such applications include furnace parts, boiler baffles, kiln linings, stack dampers, chemical processing equipment, automobile trim, catalytic converters, and decorative purposes in general.

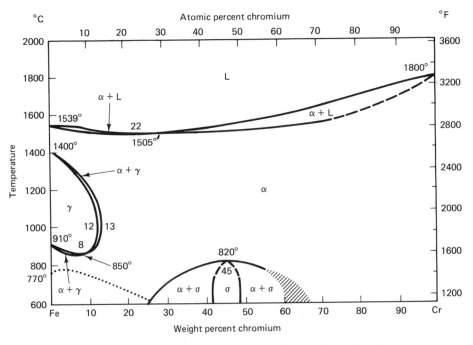

Figure 12-31   Iron–chromium equilibrium phase diagram illustrating the gamma (γ) loop. (From *Metals Handbook*, 1948 ed., American Society for Metals, Metals Park, Ohio, 1948, p. 1194.)

TABLE 12-7 TYPICAL CHEMICAL COMPOSITIONS OF STANDARD STAINLESS STEELS

AISI Type	Composition (%)			
	C	Cr	Ni	Other
Martensitic chromium steels				
410	0.15 max.	11.5–13.5	—	—
416	0.15 max.	12–14	—	Se, Mo, or Zr
420	0.35–0.45	12–14	—	—
431	0.20 max.	15–17	1.25–2.5	—
440A	0.60–0.75	16–18	—	—
Ferritic (nonhardenable) steels				
405	0.08 max.	11.5–14.5	0.5 max.	0.1–0.3 Al
430	0.12 max.	14–18	0.5 max.	—
442	0.25 max.	18–23	0.5 max.	—
446	0.20 max.	23–27	0.5 max.	0.25 N max.
Austenitic chromium–nickel steels				
201	0.15 max.	16–18	3.5–5.5	5.0–7.5 Mn, 0.25 N max.
202	0.15 max.	17–19	4–6	7.5–10 Mn, 0.25 N max.
301	0.15 max.	16–18	6–8	2 Mn max.
302	0.15 max.	17–19	8–10	2 Mn max.
302B	0.15 max.	17–19	8–10	2–3 Si
304	0.08 max.	18–20	8–12	1 Si max.
304L	0.03 max.	18–20	8–12	1 Si max.
308	0.08 max.	19–21	10–12	1 Si max.
309	0.20 max.	22–24	12–15	1 Si max.
309S	0.08 max.	22–24	12–15	1 Si max.
310	0.25 max.	24–26	19–22	1.5 Si max.
314	0.25 max.	23–26	19–22	1.5–3.0
316	0.10 max.	16–18	10–14	2–3 Mo
316L	0.03 max.	16–18	10–14	2–3 Mo
317	0.08 max.	18–20	11–14	3–4 Mo
321	0.08 max.	17–19	8–11	Ti (4 × C) min.
347	0.08 max.	17–19	9–13	Cb + Ta (10 × C) min.
Alloy 20	0.07 max.	29	20	3.25 Cu, 2.25 Mo
Age-hardenable steels[a]				
17–7 PH	0.07	17	7	1.0 Al
17–4 PH	0.05	16.5	4.25	4.0 Cu
PH 13–8 Mo	0.05 max.	14	8.5	2.5 Mo, 1 Al
AM 350	0.10	16.5	4.3	2.75 Mo
CD4M Cu	0.03	25	5	3.0 Cu, 2.0 Mo

[a]Typical compositions.

TABLE 12-8 TYPICAL MECHANICAL PROPERTIES OF STANDARD STAINLESS STEELS

Material	Condition	Tensile strength		Yield strength, 0.2% offset		Elongation (% in 2 in.)	Brinell hardness (BHN)
		psi	MPa	psi	MPa		
Type 410	Annealed	75,000	518	40,000	276	30	155
Type 410	Hardened, tempered at 600°F	180,000	1242	140,000	966	15	375
Type 410	Hardened, tempered at 1000°F	145,000	1000	115,000	794	20	300
Type 420	Annealed	95,000	656	50,000	345	25	195
Type 420	Hardened, tempered at 600°F	230,000	1587	195,000	1346	25	500
Type 440A	Annealed	105,000	724	60,000	414	20	215
Type 440A	Hardened, tempered at 600°F	260,000	1794	240,000	1656	5	510
Type 450	Annealed	75,000	518	45,000	310	30	155
Type 446	Annealed	80,000	552	50,000	345	23	170
Type 301	Annealed	110,000	759	40,000	276	60	165
Type 301	Cold worked, ½ hard	150,000	1035	110,000	759	15	320
Type 304	Annealed	85,000	586	35,000	242	55	159
CF 8	Annealed (15% ferrite)	87,000	600	47,000	324	52	150
Type 304L	Annealed	80,000	582	30,000	207	55	140
Type 310	Annealed	95,000	656	40,000	276	45	170
Type 316	Annealed	80,000	552	36,000	248	55	149
Type 347	Annealed	92,000	635	35,000	242	50	160
Alloy 20	Annealed	85,000	586	35,000	242	50	160
17–4 PH	Annealed	150,000	1035	110,000	759	10	332
	H900	200,000	1380	185,000	1276	14	420
PH 13–8 Mo	Annealed	—	—	—	—	—	363
	H950	225,000	1552	205,000	1414	12	465
17–7 PH	Annealed	130,000	897	40,000	276	35	159
	RH 950	215,000	1484	195,000	1346	9	465
	CH 900	265,000	1828	260,000	1794	2	494
PH 15–7 Mo	Annealed	150,000	1035	65,000	448	25	241
	RH 950	225,000	1552	200,000	1380	5	456
AM 350	Annealed	160,000	1104	55,000	380	40	215
	H 850	220,000	1518	190,000	1311	13	450
CD4M Cu	Annealed	105,000	724	85,000	586	20	240
	H950	140,000	966	120,000	828	15	310

Source: Adapted from M. G. Fontana and N. D. Greene, *Corrosion Engineering*, McGraw-Hill Book Company, New York, 1978.

Martensitic stainless steels.   This type of stainless steel is also primarily a chromium steel, but in contrast to the ferritic group, contains enough carbon to produce martensite by quenching. Martensitic stainless steels contain between 11.5 and 18% chromium and between 0.15 and 0.75% carbon. The influence of carbon content is instrumental in forming a martensitic structure in these alloys because it tends to expand the size of the γ-loop. If stainless steel with sufficient carbon can be heated to the gamma (austenitic) region and quenched rapidly enough, the austenite transforms to martensite. Table 12-8 shows typical properties that can be developed in the martensitic grades of stainless steel (i.e., 410, 420, 440).

Martensitic stainless steels are magnetic, exhibit good resistance to atmospheric corrosion and some chemicals, and in addition to high strength, certain grades (e.g., 410) display reasonably good toughness. Applications for these stainless steels include cutlery, surgical instruments, valves, turbine parts, pump parts, and oil well equipment.

Austenitic stainless steels.   These stainless steels are alloyed to the extent that they remain austenitic at low temperatures. The principal alloying elements added to iron are chromium and nickel, generally totaling greater than 23%. Although these steels are not hardenable by heat treatment, they may be cold worked (strain hardened) to develop a wide range of tensile strengths. For example, ultimate strengths up to 2415 MPa (350,000 psi) for cold-drawn wire and 1725 MPa (250,000 psi) for cold-rolled strip can be obtained.

Occasionally, it is helpful to our understanding of multicomponent alloys, such as stainless steel, to examine a "slice" of the ternary phase diagram. This technique is accomplished by *fixing* one of the components at a constant composition,[2] then describing the phase relationships between the two remaining components and temperature. The result is a *pseudo-binary* phase diagram of a ternary system. Figure 12-32 represents the pseudo-binary diagram for the Fe–Cr–Ni system at 70% iron. This relationship allows us to predict the phases and trends in phase changes as a function of temperature and composition under equilibrium conditions. In essence, alloys containing < about 22% Cr and > about 8% Ni (e.g., 18Cr–8Ni) are austenitic at room temperature. This composition corresponds to type 302 and 304 grades, which are also known as type "18-8" stainless steels. Note the location of type 304L[3] in Figure 12-32.

Austenitic stainless steels have greater resistance to general corrosion than do the other types. However, they are particularly susceptible to *integranular corrosion*, that is, corrosion which occurs selectively along the grain boundaries. This type of attack can take place when an austenitic stainless steel becomes *sensitized* by heating at temperatures between 427 and 871°C (800 and 1600°F) for a sufficient period of time. During this process, chromium carbide precipitates at the grain boundaries, thus depleting the chromium concentration in the matrix adjacent to the boundary.

---

[2]Basically, this involves establishing a vertical plane through the ternary diagram at the desired composition. The ternary Fe–Cr–Ni phase diagram is shown in Figure 7-27.

[3]L designates very low carbon concentration.

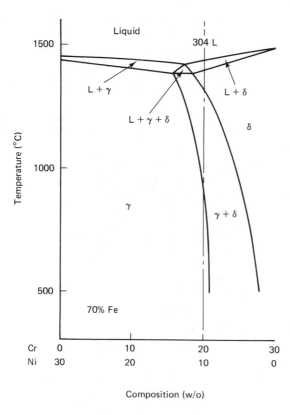

Figure 12-32 Pseudo-binary slice of the Fe–Cr–Ni ternary phase diagram at 70% Fe. γ-austenite, δ-ferrite.

If the chromium content of the steel falls below the amount necessary to provide adequate corrosion protection (approximately 12%) these regions become highly susceptible to attack. Such a sensitized condition is illustrated in Figure 12-33.

Sensitization may be avoided in stainless steels by using lower carbon contents or by the addition of certain elements which have stronger carbide-forming tendencies than chromium. Such elements as columbium, titanium, and molybdenum form a very stable carbide, minimizing the possibility of chromium carbides and thus preventing chromium depletion of the steel matrix.

Austenitic stainless steels that have experienced sensitization or carbide precipitation because of a heating process such as welding, may be restored to a uniform structure by solution heat treatment. This thermal treatment, called *stabilization*, typically consists of heating to a temperature between 1000 and 1150°C (1850 and 2100°F) for a period of time sufficient to dissolve the chromium carbides and obtain a reasonably homogeneous solid solution. The material is then cooled rapidly through the sensitizing temperature range to avoid any further carbide precipitation.

The austenitic stainless steels are used for decorative purposes; interior show cases; signs; automobile trim; aircraft fittings; food-handling equipment; surgical implantations, such as wires, screws, nails, and plates; prosthetic devices, such as hip and knee joints; and certain applications requiring corrosion resistance in conjunction with a nonmagnetic steel.

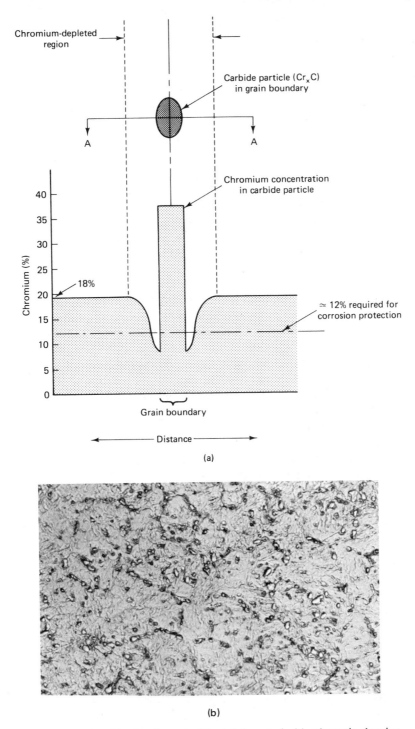

Figure 12-33 Sensitization in austenitic stainless steel: (a) schematic showing chromium depletion of matrix adjacent to a grain boundary carbide; (b) photomicrograph of carbide precipitation in the grain boundaries of stainless steel (825×).

Properties and Applications of Ferrous Alloys

Precipitation-hardening stainless steels. The last class of stainless steels we will discuss depends on precipitation hardening (aging) for the optimum development of properties. These alloys, listed in Tables 12-7 and 12-8, are utilized in applications that require very high strength together with corrosion resistance.

Basically, these steels are made either martensitic, semiaustenitic, or austenitic by varying the chromium/nickel ratio. Since chromium is a ferrite stabilizer and nickel is an austenite stabilizer, lowering the ratio of chromium to nickel tends to promote an austenitic condition, while raising this ratio promotes transformation to martensite. Small additions of aluminum, titanium, molybdenum, and copper lead to the precipitation of intermetallic compounds during thermal treatment. For example, 17–7 PH type stainless steel is solution treated at about 1050°C (1922°F) to dissolve coarse precipitates and obtain a uniform austenitic solid solution. The material is then cooled rapidly to room temperature, producing a supersaturated solid solution. This is followed by aging at 510°C (950°F), resulting in a fine, submicroscopic precipitate (intermetallic compounds) which strengthens the steel by the mechanisms described in Chapter 6.

Tensile strengths on the order of 1380 to 2070 MPa (200,000 to 300,000 psi) can be obtained by the proper selection and processing of these alloys. Generally, these materials find the widest range of applications where there is a high-performance requirement combined with potential corrosion problems. For example, 17–7 PH is routinely used for valves, bearings, gears, splines, mandrels, and valve seats. PH 13–8 Mo is used for aircraft parts, nuclear reactor components, landing gear parts, high-performance shafting, and petrochemical applications requiring stress corrosion resistance.

The corrosion resistance of the precipitation hardening stainless steels is usually less than that of the 18–8 austenitic types, except for CD4M Cu, which is better than the others listed in Tables 12-7 and 12-8, due to its exceptionally high chromium content.

STUDY PROBLEMS

12.1. Iron undergoes a polymorphic transformation from bcc to fcc at 910°C (1670°F), as shown in Figure 12-2.
  (a) What is the percent volume change associated with this transformation if the radii of the iron atoms are 1.24 Å (bcc) and 1.27 Å (fcc), respectively?
  (b) If you are heating a sample of iron, what effect does this transformation have on the size of the sample?

12.2. Based on the data shown in Figure 12-18 (e.g., A_1 and A_3 temperatures), determine the change in volume that occurs in this steel when austenite forms. How does this compare with the answer for problem 12.1?

12.3. Verify the solubility approximations of carbon in iron stated in the section on ferrite (α): Approximately, 1 carbon atom per 1000 iron atoms at 0.025% and 1 carbon atom per 2500 iron atoms at 0.008%.

12.4. An alloy consisting of 0.20% carbon, with the balance iron, is cooled very slowly to room temperature (assume equilibrium conditions). What are the proportions of microstructural constituents (ferrite and pearlite) formed? Sketch the microstructure that you would expect to observe under the microscope.

12.5. What are the percentages of the phases ferrite and iron carbide formed in an iron alloy containing 0.65% carbon if it is cooled to room temperature under near-equilibrium conditions?
Answer: $\% \, \alpha = 90.6$, $\% \, Fe_3C = 9.4$

12.6. During the microscopic examination of an iron–carbon alloy, you observe a microstructure which contains approximately 35% pearlite and 65% ferrite. Estimate the weight percentage of carbon in this material.

12.7. An alloy of iron containing 1.2% carbon is heated under near-equilibrium conditions until it becomes austenitic.
(a) What is the approximate temperature at which this transformation occurs?
(b) If this alloy is slowly cooled to room temperature, what are the percentages of the microstructural constituents formed?
(c) What are the relative proportions of phases coexisting in the room temperature product of part (b)?

12.8. In question 12.7(b), how much iron carbide is contained in the pearlite constituent?

12.9. Certain alloy additions to iron can have a pronounced effect on the eutectoid reaction.
(a) What happens to the eutectoid conditions if 17% chromium is added?
(b) What are the eutectoid parameters for an 8% nickel addition?
(c) What can you state regarding these two additions to iron?

12.10. Categorize the following elements as to their ferrite or austenite stabilizing tendency. What are the respective eutectoid temperature and carbon content for a 2% addition of each element?
(a) Si   (b) Mo
(c) Mn   (d) W

12.11. Classify the following steels according to their chemical compositions. List the SAE or AISI designation for each alloy.
(a) 0.21% C, 0.75% Mn, 0.035% P, 0.034% S, 0.25% Si, 0.50% Ni, 0.46% Cr, 0.17% Mo, balance Fe
(b) 0.41% C, 0.65% Mn, 0.033% P, 0.038% S, 0.26% Si, 1.92% Ni, 0.80% Cr, 0.25% Mo, balance Fe
(c) 0.18% C, 0.66% Mn, 0.036% P, 0.040% S, balance Fe
(d) 0.60% C, 0.85% Mn, 0.039% P, 0.028% S, 1.95% Si, balance Fe
(e) 0.21% C, 0.77% Mn, 0.030% P, 0.121% S, balance Fe
(f) 1.02% C, 0.30% Mn, 0.020% P, 0.015% S, 0.27% Si, 1.48% Cr, balance Fe
(g) 0.30% C, 0.49% Mn, 0.029% P, 0.034% S, 0.26% Si, 0.95% Cr, 0.21% Mo, balance Fe

12.12. The following ferrous alloys have been chemically analyzed and the results of these analyses are presented below. Do these materials meet the compositional requirements of their respective grades? Note any discrepancies and comment if they appear to be significant.
(a) *302—"stainless steel"*: 0.13% C, 1.8% Mn, 18.5% Cr, 9.2% Ni, balance Fe
(b) *18Ni (300 grade) marage steel*: 18.5% Ni, 9.3% Co, 4.9% Mo, 0.65% Ti, 0.15% Al, 0.15% C, balance Fe
(c) *316L—"stainless steel"*: 0.08% C, 17.2% Cr, 12.0% Ni, 1.0% Mo, balance Fe
(d) *4140—low-alloy steel*: 0.38% C, 0.80% Mn, 0.029% P, 0.030% S, 0.25% Si, 0.95% Cr, 0.25% Mo, balance Fe

12.13. Classify the following alloying elements in steel as to whether they tend to be solid-solution strengtheners or carbide formers.

(a) Ti   (b) Ni
(c) Si   (d) V
(e) Mn   (f) Cr
(g) W   (h) C

12.14. What is the percent change in volume that accompanies the transformation of austenite to martensite? The bct unit cell dimensions are $a = 2.84$ Å and $C = 2.94$ Å for a eutectoid composition (0.8% carbon), while the lattice parameter for the austenite is 3.60 Å (*Hint:* Base your solution on equal numbers of Fe atoms; i.e., fcc austenite = 4 atoms/u.c.)

Answer: $+1.65\%$

12.15. Based on the data displayed in Figure 12-18, what is the *maximum* possible change in volume accompanying the transformation of austenite to martensite at the M_s temperature if the original length of the specimen is 0.25 in.? Can you make a more realistic estimate of this volume change?

12.16. A 4340-grade steel is continuously cooled to room temperature at a cooling rate of approximately 600°F/hr. Predict the microstructure that will be produced by this cooling rate.

12.17. In a eutectoid composition steel, it is desired to produce a microstructure that contains approximately 50% martensite–50% bainite. Show how you would obtain this structure by means of a diagram.

12.18. A component made from 4340-grade steel has been oil-quenched to an approximate hardness of 600 BHN. Specify the nominal tempering temperature necessary to produce a yield strength of 160,000 psi in this material. What would be the corresponding ultimate tensile strength and ductility properties? How would you quantitatively assess the effects of this heat treatment prior to testing mechanical property specimens?

12.19. Estimate the ultimate tensile strength developed in the alloy of problem 12.17 based on the relationship between BHN and UTS given in Chapter 8. How much error is introduced by using such an estimation technique?

12.20. Jominy end-quench tests are conducted on two steels to evaluate their hardenability characteristics. Plot the Jominy curves (end-quench hardenability) for the data below.

Distance from quenched end ($\frac{1}{16}$in.)	Hardness (R_C)	
	Steel A	Steel B
2	58	60
4	57	58
6	54	55
8	52	51
12	49	47
16	47	44
24	45	38
32	42	32
40	39	29
48	37	26
64	34	21

(a) What is the cooling rate (at 1300°F) that produces the same hardness at the same location in both materials (bars)?

(b) Which steel exhibits greater hardenability?

(c) Suggest reasons for the difference in hardenability behavior between the two materials.

12.21. For each of the following applications, select a type or grade of ferrous alloy which you feel is best suited for that part or situation. Explain the reason for your choice.

(a) Automobile fender

(b) Reinforcing filaments for radial tires

(c) "Free"-machining steel

(d) Ball bearings

(e) Aircraft landing gear

(f) Surgical instruments

12.22. What is sensitization in stainless steel, and how is it prevented?

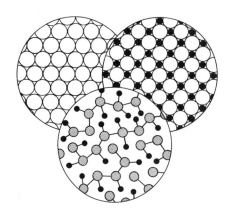

CHAPTER THIRTEEN:
Nonferrous Engineering Alloys and Their Applications

Although ferrous metals comprise the most important group of materials on a volume basis, no text on engineering materials would be complete without a discussion of nonferrous metals. Some of the important groups of metallic materials are the light metals (aluminum, beryllium magnesium, and titanium), copper and nickel-based alloys, lead, tin, refractory metals, and the precious metals. Although this list does not represent all the commercial nonferrous metals, it is sufficiently broad so that it shows the diversity of properties and applications occurring with nonferrous metals. These groups all have different attributes which make them useful as engineering materials, and proper utilization of the metals and alloys within the group requires an understanding of these properties.

ALUMINUM AND ALUMINUM ALLOYS

Aluminum, as it is used commercially, has a purity of about 99%, with silicon and iron making up the balance. It has an fcc structure and a density of 2.70 g/cm³. It is moderately corrosion resistant, except for strong alkaline solutions. Its corrosion resistance derives from the formation of a naturally occurring oxide layer. The corrosion resistance of aluminum and its alloys can be improved by *anodizing*.[1] This is an electrochemical process in which an adherent oxide layer is developed on the surface of the aluminum. The thickness and uniformity of the oxide layer is much greater

---

[1] Anodizing is simply the reverse of an electroplating reaction. In the case of anodizing, the part to be protected is made the anode during the electrochemical reaction and experiences oxidation. This results in an adherent oxide (passivating) layer on the surface of the material which inhibits corrosion.

than that occurring naturally and in addition can be dyed to produce interesting decorative effects.

Aluminum is nonsparking and nonmagnetic. Its nonmagnetic characteristics make it useful for electrical shielding purposes, while its nonsparking characteristics make it less hazardous around flammable or explosive substances. Aluminum also has excellent electrical conductivity and is used as an electrical conductor. However, because its conductivity on a volume basis is only 61% that of copper, the cross section of the conductors must be correspondingly larger (see Example Problem 10-3).

Aluminum is also essentially nontoxic and this accounts for its use in cookware, in water storage containers, and for food wrapping (aluminum foil).

One of the principal drawbacks of aluminum and its alloys is its poor wear resistance in applications where there is metal-to-metal contact. The wear resistance can be markedly improved by *hard-coating* aluminum alloys. This is an electrochemical process similar to anodizing, except that the coating obtained is much thicker and harder. It provides a relatively thick (0.05 to 0.25 mm) ceramic-like layer which is integrally bonded to the metallic substrate.

Commercially pure aluminum is used primarily for cookware, foil, wire, and as a paint pigment. Pure aluminum has a tensile strength of approximately 12,000 psi (83 MPa). This relatively low value can be increased by three principal means, which were introduced in Chapter 6:

1. Strain hardening resulting from cold work
2. Solid-solution strengthening due to alloying
3. Precipitation hardening

Aluminum and its alloys can be readily cold worked to improve their strength, although there is a corresponding decrease in ductility as the strength increases. Alloying is another mechanism that can be used to raise strength levels, and as will be seen later in this section, these mechanisms can be used either individually or in combination to achieve various property levels. The principal alloying elements used in aluminum alloys are copper, silicon, magnesium, zinc, manganese, and chromium.

Aluminum–Copper Alloys

Copper is the major alloying element in aluminum alloys. It provides the basis for precipitation hardening in both cast and wrought alloys and is also employed to decrease porosity due to shrinkage during solidification of cast alloys.

The equilibrium phase diagram for the aluminum–copper system is shown in Figure 13-1. The student should recall from Chapter 6 that decreasing solid solubility with decreasing temperature, is one of the requisites for precipitation hardening. An examination of the Al–Cu system at low concentrations of copper [Figure 13-1(b)] reveals that this is the case here.

The solubility of copper in aluminum decreases from 5.65% at 1018°F (548°C) to less than 0.25% at room temperature. A eutectic is formed at 33% copper. These

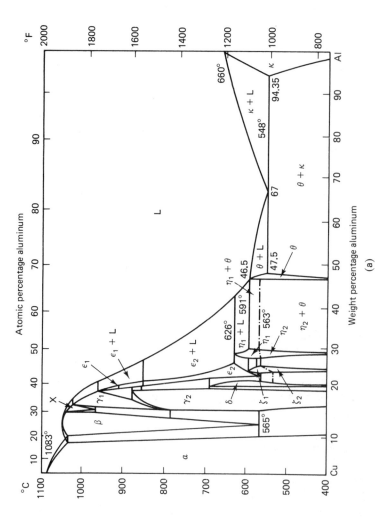

Figure 13-1 Equilibrium phase diagram for the aluminum–copper system: (a) overall diagram; (b) aluminum-rich portion. (From *Metals Handbook*, 8th ed., Vol. 8: *Metallography, Structures and Phase Diagrams*, American Society for Metals, Metals Park, Ohio, 1973, p. 259.)

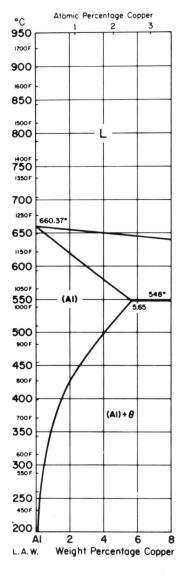

°C
Atomic Percentage Copper

L

660.37°

(Al)

548°
5.65

(Al) + θ

Al   2   4   6   8
L.A.W.   Weight Percentage Copper

(b)   **Figure 13-1** (Cont.)

alloys may be strengthened by the precipitation of the theta (θ) phase ($CuAl_2$) from the solid solution (κ) by means of an aging heat treatment. The effect of copper content and the time of aging (or overaging) on hardness is shown in Figure 13-2. The effect of time and temperature on strength is illustrated in Figure 13-3.

Copper is usually employed up to about 4% in cast alloys and 8% in wrought alloys. True Al–Cu binary alloys are not common and the commercial Al–Cu alloys usually contain various percentages of magnesium, zinc, silicon, and manganese. Table 13-1 lists the nominal compositions for several cast aluminum alloys, while the compositions for several wrought aluminum alloys are tabulated in Table 13-2.

Aluminum and Aluminum Alloys   **453**

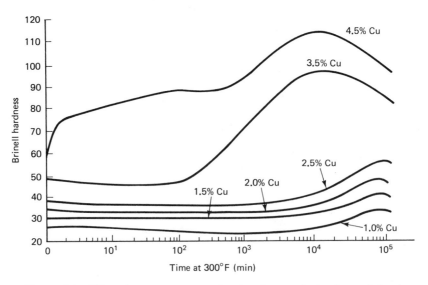

Figure 13-2 Effect of copper content and aging time on the hardness of aluminum–copper alloys. (From D. S. Clark and W. R. Varney, *Physical Metallurgy for Engineers*, D. Van Nostrand Company, Princeton, N.J., 1962, p. 422.)

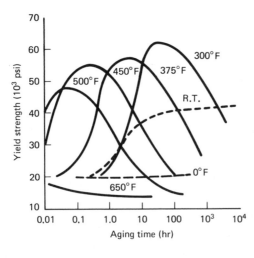

Figure 13-3 Effect of aging temperature and time on the yield strength of 2014 aluminum alloy. (From K. R. VanHorn, Ed., *Aluminum*, Vol. 1: *Properties, Physical Metallurgy and Phase Diagrams*, American Society for Metals, Metals Park, Ohio, 1967, p. 116.)

Example 13-1

2014-T4 aluminum alloy may be strengthened by aging (precipitation hardening), as illustrated in Figure 13-3. (a) Determine the difference in yield strength that would be obtained in 10 hr aging time if the process were conducted at a temperature of 300°F (149°C) versus room temperature (natural aging), and (b) estimate the aging conditions that would meet the requirements for a yield strength of 60,000 psi (414 MPa).

Solution (a) At 10 hr aging time, the yield strength is approximately 37,000 psi (255 MPa) for room temperature aging, while at 300°F, the yield strength is approximately 53,000 psi (366 MPa), a difference of 16,000 psi (111 MPa). *Ans.*

(b) For a requirement of 60,000 psi, this alloy could be aged at 300°F for approximately 30 hr.   *Ans.*

Comment: Increasing aging temperature to "speed up" the process tends to decrease the maximum attainable yield strength level. Observe what occurs for an aging temperature of 650°F (343°C). Can you think of a reason for this behavior?

TABLE 13-1 NOMINAL COMPOSITIONS FOR SEVERAL CAST ALUMINUM ALLOYS

| Alloy | Form | % Alloying Elements | | | | | |
		Cu	Fe	Si	Mg	Zn	Ni
43	SC, PM	0.1	0.8	5.0	0.05	0.2	—
108	SC	4.0	1.0	3.0	0.03	0.2	—
112	SC	7.0	1.5	1.0	0.07	2.2	0.3
113	SC, PM	7.0	1.4	2.0	0.07	2.2	0.3
122	SC, PM	10.0	1.5	1.0	0.2	0.5	0.3
142	SC, PM	4.0	0.8	0.6	1.5	0.1	2.0
195	SC	4.5	1.0	1.2	0.03	0.3	—
212	SC	8.0	1.4	1.2	0.05	0.2	—
214	SC	0.1	0.4	0.3	4.0	0.1	—
B214	SC	0.1	0.4	1.8	4.0	0.1	—
F214	SC	0.1	0.4	0.5	4.0	0.1	—
220	SC	0.2	0.3	0.2	10.0	0.1	—
319	SC	3.5	1.2	6.3	0.5	1.0	0.5
355	SC, PM	1.3	0.6	5.0	0.5	0.2	—
A355	SC	1.5	0.6	5.0	0.5	0.1	0.8
356	SC, PM	0.2	0.5	7.0	0.3	0.2	—
A612	SC	0.5	0.5	0.15	0.7	6.5	—
750	SC, PM	1.0	0.7	0.7	—	—	1.0
A750	SC, PM	1.0	0.7	2.5	—	—	0.5
B750	SC, PM	2.0	0.7	0.4	0.75	—	1.2

SC = Sand Cast
PM = Permanent Mold

Aluminum–Magnesium Alloys

Magnesium is alloyed with aluminum in amounts usually ranging from about 1 to 10%. The resulting alloys are lighter than pure aluminum, possess good mechanical properties, are easily machined, and possess good salt water and alkaline corrosion resistance.

The equilibrium phase diagram for the Al–Mg system is shown in Figure 13-4. As can be seen, this system also displays decreasing solid solubility and would appear to be a candidate for precipitation hardening. Magnesium has a maximum solid solubility of 15% at 845°F (451°C), decreasing to less than 2% at room temperature. Precipitation hardening is achieved by the precipitation of the beta (β) phase (AlMg)

Alloy	Cu	Si	Fe	Mn	Mg	Zn	Cr	Ni	Other
EC	Max. impurities 0.40%								
2EC	0.05	0.4	0.3	0.01	0.6	0.05	0.01	—	
1050	0.05	0.25	0.4	0.05	0.05	0.05	—	—	Ti 0.03
1100	0.2	Si + Fe	1.0	0.05	—	0.10	—	—	
2014	4.4	0.9	1.0	0.8	0.5	0.25	0.10	—	Ti 0.15
2018	4.0	0.9	1.0	0.2	0.7	0.25	0.10	2.0	
2024	4.5	0.5	0.5	0.5	1.5	0.25	0.10	—	
2117	2.6	0.8	1.0	0.2	0.3	0.25	0.10	—	
2218	4.0	0.9	1.0	0.2	1.5	0.25	0.10	2.0	—
2618	1.3	0.25	1.1	—	1.5	—	—	1.0	Ti 0.07
3003	0.2	0.6	0.7	1.2	—	0.10	—	—	
3004	0.25	0.3	0.7	1.2	1.0	0.25	—	—	
4032	0.9	12.2	1.0	—	1.0	0.25	0.10	0.9	
4043	0.3	5.0	0.8	0.05	0.05	0.10	—	—	Ti 0.20
4343	0.25	7.5	0.8	0.10	—	0.20	—	—	
5005	0.2	0.4	0.7	0.2	0.8	0.25	0.10	—	
5052	0.1	Si + Fe	0.45	0.1	2.5	0.10	0.25	—	
5056	0.1	0.3	0.4	0.12	5.0	0.10	0.12	—	
5086	0.1	0.4	0.5	0.5	4.0	0.25	0.15	—	Ti 0.15
5154	0.1	Si + Fe	0.45	0.1	3.5	0.20	0.25	—	Ti 0.20
5356	0.1	Si + Fe	0.50	0.12	5.0	0.10	0.12	—	Ti 0.15
5454	0.1	Si + Fe	0.40	0.75	2.7	0.25	0.12	—	Ti 0.20
5456	0.2	Si + Fe	0.40	0.75	5.0	0.25	0.12	—	Ti 0.20
5554	0.1	Si + Fe	0.40	0.75	2.7	0.25	0.12	—	Ti 0.12
5556	0.1	Si + Fe	0.40	0.75	5.0	0.25	0.12	—	Ti 0.12
5652	0.04	Si + Fe	0.40	0.01	2.5	0.10	0.25	—	
6061	0.27	0.6	0.7	0.15	1.0	0.25	0.25	—	Ti 0.15
6063	0.1	0.4	0.35	0.10	0.67	0.10	0.10	—	Ti 0.10
6151	0.35	0.9	1.0	0.20	0.62	0.25	0.25	—	Ti 0.15
6253	0.1	0.6	0.5	—	0.25	2.0	0.25	—	
6463	0.2	0.4	0.15	0.05	0.67	—	—	—	
6951	0.27	0.3	0.8	0.10	0.6	0.20	—	—	
7075	1.6	0.5	0.7	0.30	2.5	5.6	0.3	—	Ti 0.20
7076	0.65	0.4	0.6	0.5	1.6	7.5	—	—	Ti 0.20
7079	0.6	0.3	0.4	0.2	3.3	4.3	0.17	—	Ti 0.10
7178	2.0	0.5	0.7	0.3	2.7	6.8	0.3	—	Ti 0.20
7277	1.2	0.5	0.7	—	2.0	4.0	0.25	—	Ti 0.10

from the α solid solution. If silicon is added to form ternary alloys, additional strengthening can be developed through the formation of Mg_2Si.

Aluminum–Silicon Alloys

Silicon is also an important alloying element in aluminum alloys. It is used both as a solid solution strengthening agent and to improve casting qualities, such as fluidity, toughness, and soundness. The equilibrium phase diagram for Al–Si is shown in Figure 13-5. The solid solubility of silicon in aluminum decreases from a maximum of 1.65%

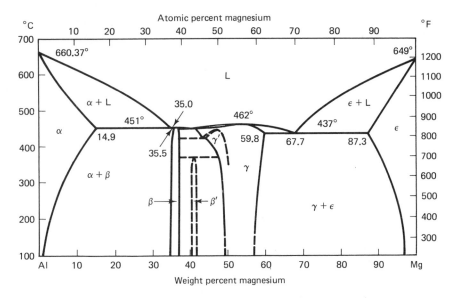

Figure 13-4 Equilibrium phase diagram for the aluminum–magnesium system. (From *Metals Handbook*, 8th ed., Vol. 8: *Metallography, Structures and Phase Diagrams*, American Society for Metals, Metals Park, Ohio, 1973, p. 261.)

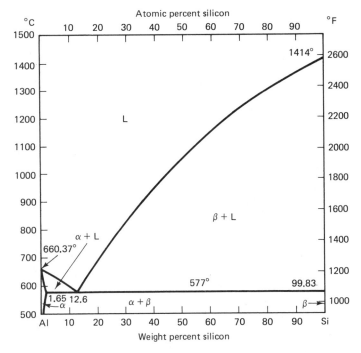

Figure 13-5 Equilibrium phase diagram for the aluminum–silicon system. (From *Metals Handbook*, 8th ed., Vol. 8: *Metallography, Structures and Phase Diagrams*, American Society for Metals, Metals Park, Ohio, 1973, p. 263.)

Aluminum and Aluminum Alloys **457**

at 1077°F (577°C) to essentially zero at room temperature. A eutectic having the composition of 12.6% Si produces the strength and hardness desired in cast Al–Si alloys. The silicon content of these alloys ranges down to 1% but rarely exceeds 14%. Other elements, such as zinc, copper, and iron, are sometimes added to the Al–Si system to form ternary alloys with age-hardening characteristics. The effect of silicon content on mechanical properties of a typical aluminum alloy is shown in Figure 13-6.

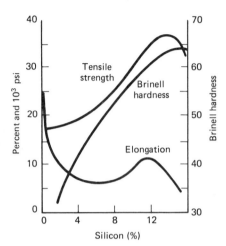

Figure 13-6 Effect of silicon content on the mechanical properties of an aluminum–silicon alloy.

The microstructure of cast 356 alloy (see Table 13-1) is presented in Figure 13-7. This alloy contains from 0.20 to 0.40% magnesium in addition to silicon, which forms Mg_2Si, the primary strengthening constituent.

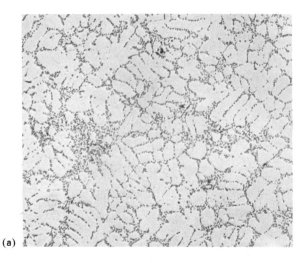

(a)

Figure 13-7 Microstruure of cast 356 aluminum alloy. (a) Solid solution of aluminum–silicon (light gray regions), outlined by an interdendritic network of Al–Si eutectic (dark gray) (80×). (b) Microstructure at 400× showing morphology of eutectic particles. Black needle-like constituent is Mg_2Si. (Courtesy of Reynolds Aluminum Co., Richmond, Va.)

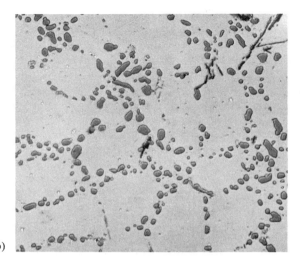

(b)

Figure 13-7 (Cont.)

Classification of Aluminum Alloys

Aluminum alloys may be grouped into those alloys which are wrought and those which are cast. A four-digit system of numerical designations is used to identify wrought aluminum and wrought aluminum alloys. The first digit indicates the alloy group as shown below. The 1xxx series is for minimum aluminum purities of 99.00% and greater; the 2xxx through 8xxx series classifies aluminum alloys by major alloying elements. The last two digits identify the aluminum alloy or indicate the aluminum purity.

	Alloy designation
Aluminum, 99.00% minimum and greater	1xxx
Major alloying element	
Copper	2xxx
Manganese	3xxx
Silicon	4xxx
Magnesium	5xxx
Magnesium and silicon	6xxx
Zinc	7xxx
Other element	8xxx
Unused series	9xxx

The composition designation for cast alloys is somewhat different and generally consists of a two or three-digit designation, such as those shown in Table 13-1. The mill condition of both cast and wrought aluminum alloys is shown by a series of letters and numerals following the alloy designation.

A letter following the alloy designation and separated from it by a hyphen indicates the basic-temper designation. The addition of a subsequent digit, where applicable, indicates the specific treatment employed to produce the basic temper. Those compositions that are hardenable only by strain hardening are given "-H" designations, whereas those hardenable by heat treatment through precipitation or a combination of cold work and precipitation are given the letter "-T" in accordance with the following classification:

-F   As fabricated
-O   Annealed, recrystallized (wrought only)
-H   Strain-hardened
     -H1   Strain-hardened only
     -H2   Strain-hardened and then partially annealed
     -H3   Strain-hardened and then stabilized
-W   Solution heat-treated–unstable temper
-T   Heat-treated to stable tempers
     -T2   Annealed (cast only)
     -T3   Solution treated and cold-worked
     -T4   Solution treated followed by natural aging at room temperature
     -T5   Artificially aged only after an elevated-temperature, rapid-cool fabrication process such as casting or extrusion
     -T6   Solution treated and artificially aged
     -T7   Solution treated and stabilized to control growth and distortion
     -T8   Solution treated, cold-worked, and artificially aged
     -T9   Solution treated, artificially aged, and cold-worked

For example, the overall designation 6063-T6 would refer to a specific aluminum alloy containing magnesium and silicon as its principal alloy elements, which was then strengthened by solution treating and artificial aging. The mechanical properties of several wrought and cast aluminum alloys are shown in Tables 13-3 and 13-4, respectively.

MAGNESIUM AND ITS ALLOYS

Magnesium is the lightest of the commercially important engineering metals, with a density of 1.74 g/cm^3 and a hcp structure. The pure metal is readily available by electrolytic reduction from seawater, a factor that may account for its increased use in the future when other metals and alloys become less available or more costly to refine.

The pure metal is used primarily as an alloying element and as an active agent in incendiary devices such as flares and fire bombs. Magnesium alloys are generally

TABLE 13-3 TYPICAL MECHANICAL PROPERTIES FOR CERTAIN WROUGHT ALUMINUM ALLOYS

Alloy, condition	Tensile strength (psi)[a]	Yield strength (psi)[a]	Elonga-tion (% in 2 in.)	Brinell hardness (500-kg load, 10-mm ball)	Shear strength (psi)[a]
EC-0	12,000	4,000		—	8,000
EC-H19	27,000	24,000		—	15,000
2EC-T6	32,000	29,000	19	—	—
2EC-T64	17,000	9,000	24	—	—
1060-0	10,000	4,000	45	19	7,000
1060-H18	19,000	18,000	10	35	11,000
1100-0	13,000	5,000	45	23	9,000
1100-H18	24,000	22,000	15	44	13,000
2011-T3	55,000	43,000	15	95	32,000
2011-T6	57,000	39,000	17	97	34,000
2011-T8	59,000	45,000	12	100	35,000
2014-0	27,000	14,000	18	45	18,000
2014-T4	62,000	42,000	20	105	38,000
2014-T6	70,000	60,000	13	135	42,000
2017-0	26,000	10,000	22	45	18,000
2017-T4	62,000	40,000	22	105	38,000
2024-0	27,000	11,000	22	47	18,000
2024-T3	70,000	50,000	20	120	41,000
2024-T36	72,000	57,000	15	130	42,000
2024-T4	68,000	47,000	19	120	41,000
2024-T81	70,000	65,000	10	128	43,000
2024-T86	75,000	71,000	8	135	45,000
6061-0	18,000	8,000	30	30	—
6061-T6	45,000	40,000	12	95	—
7075-0	33,000	15,000	16	—	—
7075-T6	83,000	73,000	11	—	—
7178-0	33,000	15,000	16	60	—
7178-T6	88,000	78,000	10	160	—

[a]Multiply by 6.9×10^{-3} to obtain MPa.

employed where its low density can be used to maximum advantage. For example, they are used in the manufacture of ladders; certain hand tools, such as levels; aircraft gearbox housings; lawn mower decks; and chain saw housings.

Magnesium and its alloys are generally considered to have poor corrosion resistance, particularly in seawater. This relatively low corrosion resistance accounts for the use of magnesium as a sacrificial anode, as you will see in Chapter 17 (see Figure 17-11). However, its corrosion resistance can be markedly improved by the utilization of electrolytic anodizing followed by the application of protective coatings, such as chromate-bearing paints and polymeric resins.

TABLE 13-4 TYPICAL MECHANICAL PROPERTIES FOR SELECTED CAST ALUMINUM ALLOYS

Alloy	Tensile strength (psi)[a]	Yield strength (psi)[a]	Elonga-tion (% in 2 in.)	Brinell hardness (500-kg load, 10-mm ball)	Shear strength (psi)[a]
43-F	19,000	8,000	8.0	40	14,000
108-F	21,000	14,000	2.5	55	17,000
112-F	24,000	15,000	1.5	70	20,000
113-F	24,000	15,000	1.5	70	20,000
122-T61	41,000	40,000	< 0.5	115	32,000
142-T21	27,000	18,000	1.0	70	21,000
142-T571	32,000	30,000	0.5	85	26,000
142-T77	30,000	23,000	2.0	75	24,000
195-T4	32,000	16,000	8.5	60	26,000
195-T6	36,000	24,000	5.0	75	30,000
195-T62	41,000	32,000	2.0	90	33,000
212-F	23,000	14,000	2.0	65	20,000
214-F	25,000	12,000	9.0	50	20,000
B214-F	20,000	13,000	2.0	50	17,000
F214-F	21,000	12,000	3.0	50	17,000
220-T4	48,000	26,000	16.0	75	34,000
319-F	27,000	18,000	2.0	70	22,000
319-T5	30,000	26,000	1.5	80	24,000
319-T6	36,000	24,000	2.0	80	29,000
355-T51	28,000	23,000	1.5	65	22,000
355-T6	35,000	25,000	3.0	80	28,000
355-T7	38,000	36,000	0.5	85	28,000
355-T71	35,000	29,000	1.5	75	26,000
A355-T51	28,000	24,000	1.5	70	22,000
356-T51	25,000	20,000	2.0	60	20,000
356-T6	33,000	24,000	3.5	70	26,000
356-T7	34,000	30,000	2.0	75	24,000
356-T71	28,000	21,000	3.5	60	20,000
A612-F	35,000	25,000	5.0	75	26,000
750-T5	20,000	11,000	8.0	45	14,000
A750-T5	20,000	11,000	5.0	45	14,000
B750-T5	27,000	22,000	2.0	65	18,000

[a]Multiply by 6.9×10^{-3} to obtain MPa.

Alloying Elements

A number of elements are utilized to improve the mechanical properties of magnesium. Among these elements are aluminum, zinc, and manganese. Occasionally, other elements, such as the rare earths, tin, and zirconium, are added to provide special properties. Iron and copper are considered impurities and are kept to a minimum to avoid problems with corrosion.

The magnesium alloys are classified by a two-letter designation representing the alloying elements present in order of their concentration, followed by numerals that represent the amounts of those elements. The letters representing the alloying elements are as follows:

A	aluminum	L	lithium
B	bismuth	M	manganese
C	copper	N	nickel
D	cadmium	P	lead
E	rare earth	Q	silver
F	iron	R	chromium
G	magnesium	S	silicon
H	thorium	T	tin
K	zirconium	Y	antimony
		Z	zinc

Based on this system, therefore, the designation AZ63 represents a magnesium alloy containing 6% aluminum and 3% zinc.

Wrought versus Cast Properties

There are both cast and wrought forms of the magnesium alloys. Table 13-5 lists a number of these alloys, together with the mechanical properties that can be attained. It should be noted, however, that magnesium, being a hcp metal, does not have as many slip systems available at room temperature as does aluminum, for example (3 versus 12). You should recall that the subject of slip and slip systems was introduced in Chapters 4 and 6 (see Table 6-3). In any event, the significance of this is that cold working of the magnesium alloys is more difficult, since cracking may occur. Consequently, hot working at temperatures of 200 to 400°C (392 to 752°F) is generally employed to shape and form the desired end products.

Many of the cast and wrought alloys can be age-hardened, in particular, the Mg–Al alloys. The chemical composition and mechanical properties for several cast and wrought magnesium alloys are shown in Table 13-5. An examination of this table reveals that certain magnesium alloys (e.g., ZK60) are capable of being heat treated (solution treated and aged) to yield strengths in excess of 40,000 psi (276 MPa).

BERYLLIUM AND ITS ALLOYS

Beryllium is a metal which has achieved commercial significance only in the last 35 years. It is a metal with a hcp structure and a density of 1.85 g/cm³. It is this low density combined with a relatively high elastic modulus [44.1 × 10⁶ psi (304, 290 MPa)] which makes beryllium and its alloys interesting and potentially valuable as engineering materials. Beryllium is relatively scarce and is generally refined only from the mineral beryl, which is obtained by hand picking beryl crystals from volcanic deposits.

TABLE 13-5 CHEMICAL COMPOSITION AND PROPERTIES OF SEVERAL CAST AND WROUGHT MAGNESIUM ALLOYS

	Nominal composition	Form and condition	Yield strength (0.2% offset) 1000 psi	Tensile strength 1000 psi	Elon. in 2 in. %	Hardness brinell
Magnesium Alloy AZ 31B	Mg—bal Al—3 Zn—1 Mn—0.2 min.	Sheet Annealed	22	37	21	56
		Hard Sheet	32	42	15	73
Magnesium Alloy AZ 80A	Mg—bal Al—8.5 Zn—0.5 Mn—0.12 min.	As Forged	33	48	11	69
		Forged and Aged	36	50	6	72
Magnesium Alloy AZ 91B	Mg—bal Al—9 Zn—0.7 Mn—0.15 min.	Die Cast	22	33	3	63
Magnesium Alloy AZ 92A	Mg—bal Al—9 Zn—2 Mn—0.10 min.	As Cast	14	25	2	65
		Solution Treated	14	40	10	63
		Solution Treated and Aged	22	40	3	81
Magnesium Alloy HK 31A	Mg—bal	Strain Hardened Partially-Annealed	29	37	8	68
	Th—3.25 Zr—0.7	Castings, Heat Treated	15	32	8	66
Magnesium Alloy HZ 32A	Mg—bal Th—3.25 Zn—2.1 Zr—0.7	Solution Treated and Aged Castings	13	27	4	55
Magnesium Alloy ZK 60A	Mg—bal Zn—5.5 Zr—0.45 min.	As Extruded	38	49	14	75
		Aged	44	53	11	82

By far the largest tonnage use of beryllium since its early development and up through the present is in the form of beryllium–copper alloys. The phase diagram for the Be–Cu system in the region of importance for the beryllium–copper engineering alloys is shown in Figure 13-8. These alloys can be age hardened to very high strength

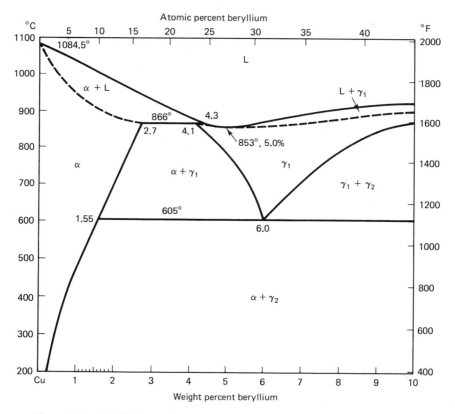

Figure 13-8 Equilibrium phase diagram for the beryllium–copper system. (From *Metals Handbook*, 8th ed., Vol. 8: *Metallography, Structures and Phase Diagrams*, American Society for Metals, Metals Park, Ohio, 1973, p. 271.)

levels. For example, the 2% beryllium–copper alloy (C17200) can be age hardened to a yield strength of 180,000 psi (1242 MPa). This is accomplished by solution treating the alloy to 800°C (1472°F) to obtain an α solid solution, which is then quenched to room temperature and aged at 315°C (599°F) for 3 hr to develop the γ_2 (CuBe) phase.

The following two characteristics of these alloys have been largely responsible for their spectacular development: (1) beryllium–copper alloys are unsurpassed in their ability to withstand fatigue and wear and at the same time conduct electrical current under high-temperature conditions, and (2) they are unique among copper-based alloys in that they can be worked in a relatively soft state and then brought to their final level of strength and hardness by simple low-temperature heat treatment (aging). These properties immediately suggested numerous applications in electrical contacts and springs, and in castings and forgings subjected to severe wear conditions.

A large application for beryllium–copper alloys is for the manufacture of nonsparking safety tools. Large quantities of these tools are used in the petrochemical industry and other industries where a spark from conventional steel tools might conceivably set off a disastrous explosion. Beryllium alloys are also used in aircraft and space applications, where their low density and high modulus make them extremely useful for such applications as brake components, bearings, bushings, and small gears. We should also mention that beryllium may be used as a moderating material in nuclear reactor applications, because of its nuclear cross-section characteristics (neutron absorption).

One of the problems with beryllium, however, is that it presents a potential health hazard as a result of its relative toxicity. Beryllium metal and beryllium compounds produce a variety of symptoms in human beings, including pneumonia and skin reactions such as dermatitis, ulcers, and benign tumors. These problems arise mainly from the inhalation of dust and fine particles or the accidental contamination of a surface wound. Therefore, the melting, processing, and machining of beryllium and its alloys must be carried out with extreme care.

TITANIUM AND ITS ALLOYS

Like beryllium, titanium is a relatively recent addition to the engineering scene, being introduced on a commercial scale about 1950. It is produced commercially by reducing titanium tetrachloride with magnesium to produce a titanium sponge. This sponge is remelted under vacuum to produce a metallic ingot. It is a lustrous white metal, with a density of 4.51 g/cm^3 and a modulus of elasticity of 16×10^6 psi (11,400 MPa).

Titanium alloys can be strengthened to the level of high-strength steel in many cases (see Example 9-2). This factor, combined with its low density and relatively high modulus, makes it a prime candidate for use in aerospace applications. Among other applications, the metal finds use in this field in wing spacers, fuselage structural members, and jet engine components, such as turbine disks and compressor blades. The latter use is based on the fact that titanium has a relatively high melting point, 3140°F (1127°C), and maintains its strength at elevated temperatures.

Titanium exhibits two polymorphic forms, alpha (α) at temperatures below 1625°F (885°C) and beta (β) above this temperature. Alpha titanium has a hcp structure, while beta is body-centered cubic. Most alloying elements decrease the α-to-β transformation temperature; however, aluminum raises the transformation temperature. The effect of aluminum on the α phase is to promote stability of this phase at higher temperatures, which makes aluminum an important element in many of the titanium alloys.

The alloying elements, iron, manganese, chromium, molybdenum, vanadium, columbium, and tantalum, stabilize the β phase, thus decreasing the α-to-β transformation temperature. Additions of columbium and tantalum produce improved strength and help in preventing the embrittlement produced by the presence of titanium and aluminum compounds. There are three general types of titanium alloys depending on the structures: α, α + β, and β. The α alloys consist of a single phase and do not undergo a polymorphic transformation. These alloys are therefore not responsive to heat treatment and hence do not develop the strength possible in other titanium alloys. The α + β alloys are heat-treatable by precipitation hardening and possess good ductility. The microstructure of Ti–6 Al–4 V, a high-strength, α + β titanium alloy, is shown in Figure 13-9. The β alloys on the other hand, have rela-

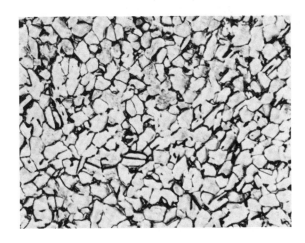

Figure 13-9 Microstructure of Ti–6 Al–4 V alloy, annealed bar (280×). Light gray equiaxed grains consist of α; dark intergranular phase is β.

tively low ductility for strengths comparable to the other alloys. The mechanical properties of several commercial titanium alloys are presented in Table 13-6.

The α alloys in general have the highest strength at elevated temperatures in the range 600 to 1100°F (315 to 595°C) and have the best resistance to oxidation in this temperature range. Their room temperature strength is not as good as that of the α + β alloys. The strongest alloys are the α + β type, which respond to heat treatment. These alloys are more formable than the α alloys, but they are not weldable. When they are heated above the α-to-β transformation temperature, they lose ductility as a result of grain growth.

Titanium and titanium alloys also have exceptional corrosion resistance and good human tissue compatability. This is because the very stable oxide of titanium that forms on the surface is extremely tenacious and readily reforms if damaged. Therefore, these materials are commonly used for surgical implants such as the hip prosthesis shown in Figure 13-10.

Table 13-7 lists the properties of several of the light metals, including titanium, on a comparative basis relative to steel. An examination of this table shows that on a weight basis, these metals and alloys exhibit a significant advantage which can be utilized in many applications.

TABLE 13-6 COMPOSITION, STRUCTURE, AND MECHANICAL PROPERTIES FOR SEVERAL TITANIUM ALLOYS

| Composition | Condition | Room temperature strengths | | Elongation (%) | Structure | Heat treatable | Typical application |
		UTS (ksi)[a]	Yield (ksi)[a]				
Unalloyed ASTM Grade 2 ASTM Grade 3 ASTM Grade 4	Annealed	59 75 95	40 62 80	28 27 25		No	Hydraulic control valves; gyrowheel structure; fittings; attachment brackets; welded ducts; complex tube shapes; skin-stringer structures
Ti–5 Al–2.5 Sn	Annealed	125	120	18	α	No	Transmission and gear housings; jet-engine compressor case assembly and stator housing
Ti–8 Al–1 Mo–1 V	1100°F (8 hr), air cooled	145	135	16	α	No	Jet-engine compressor blades, disks, and housings; inner skin and frame for jet-engine nozzle assembly; experimental sheet-stringer structures; bulkhead forgings
Ti–6 Al–4 V	Annealed 1700°F (20 min), water quench +975°F (8 hr), air cooled	135 170	120 150	11 7	$\alpha + \beta$	Yes	Jet-engine compressor blades, disks, etc.; landing-gear wheels and structures; fasteners; brackets; fittings; pressure bottles; frames; firewalls; stiffeners; hip prostheses
Ti–6 Al–6 V–2 Sn	Annealed	165	155	12	$\alpha + \beta$	Yes	Fasteners and air-intake control track; experimental structural forgings
Ti–13 V–11 Cr–3 Al	Annealed 1400°F (30 min), air cooled	130 175	125 165	16 6	β	Yes	Structural forgings; primary and secondary sheet-stringer structures; skins; frames; brackets; fittings; fasteners; and specialty uses

[a]Multiply by 6.9 to obtain MPa.

TABLE 13-7 TYPICAL PROPERTIES OF LIGHTWEIGHT METALS AND ALLOYS COMPARED WITH STEEL

Material	Density (g/cm³)	Modulus of elasticity (10⁶ psi)	Ultimate tensile strength (psi)[a]	Yield strength, 0.2% offset (psi)[a]	Elongation (% in 2 in.)	UTS/Density (10³ in.)
Magnesium	1.7	6.5	37,000	27,000	9	602
Magnesium AZ80A-T5	1.8	6.5	55,000	40,000	7	846
Beryllium-extruded flake	1.9	40	59,000–78,000	20,000–30,000	4–7	859–1136
Aluminum 1060-0	2.7	10.4	10,000	4,000	43	102
Aluminum 7178-T6	2.8	10.4	88,000	78,000	11	870
Titanium (99.9%)	4.5	16	34,000	20,000	54	209
Ti–7% Al–4% Mo (aged)	4.4	16	190,000	175,000	12	1195
Zirconium (0.2% O₂)	6.4	12	30,000	12,000	30	129
Zircaloy 2	6.5	—	69,000	45,000	22	—
Steel-SAE 1035	7.8	30	88,000	55,000	31	312
Steel-SAE 4130	7.8	30	190,000	168,000	10	674
Steel, stainless, 17–7 PH	7.9	28	230,000	217,000	7	806
Steel, stainless, 302-30% cold worked	7.9	28	175,000	140,000	15	613

[a]To obtain MPa, multiply by 6.9×10^{-3}.

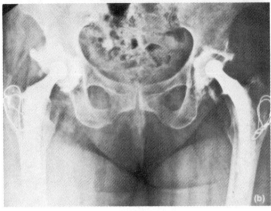

Figure 13-10 Total hip prosthesis. (a) Titanium alloy implant and polyethylene socket prior to implantation. (b) X-ray radiograph showing a double replacement.

COPPER AND ITS ALLOYS

Copper is a metal that has been in use since the dawn of ancient civilizations. It is a reddish metal with an fcc structure and a density of 8.96 g/cm³. Although most metals find their major industrial use in the form of alloys, this is not exclusively true for copper. While it is used in the alloyed condition, it also finds extensive application in its commercially pure form. Its principal use in this form is as an electrical conductor, as discussed in Chapter 10. In fact, it ranks second only to silver in electrical conductivity. Most copper for electrical applications contains a small amount of oxygen (0.02 to 0.05%) which is combined in the form of Cu_2O (cuprous oxide). This is termed electrolytic tough pitch copper (ETP) and has a slightly greater electrical conductivity (101.6%) than does pure copper. The microstructure of ETP copper is shown in Figure 13-11. In general, however, the presence of small amounts of other elements tends to decrease the electrical conductivity, as shown in Figure 13-12. Regardless of this fact, however, the reduced conductivity is permitted in some applications where higher strength is required. ETP copper should not be utilized in a reducing atmosphere above 400°C (750°F) because the gases react with the oxide particles at the grain boundaries, causing embrittlement.

There is a special type of copper in which the last traces of Cu_2O are reduced during refining by a charcoal treatment, then processed in a protective atmosphere to exclude oxygen. This oxygen-free high conductivity copper (OFHC) has exceptional

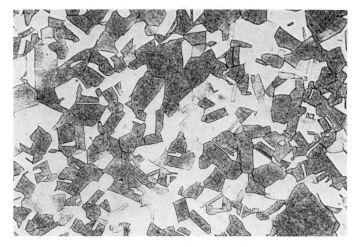

Figure 13-11   Microstructure of electrolytic tough pitch (ETP) copper (200×).

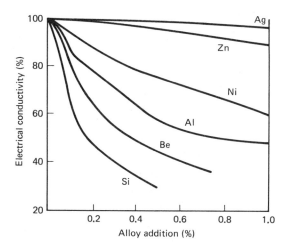

Figure 13-12   Change in electrical conductivity as a result of alloying additions to copper.

plasticity, good electrical properties and weldability, and can be used safely in reducing atmospheres.

The thermal conductivity of copper ranks second only to silver (see Appendix 1). However, copper is generally favored for most applications, because of economic reasons. Applications that take advantage of this characteristic (in conjunction with good corrosion resistance) include automotive radiators, solar collector panels, certain cookware, heat sinks in solid-state electronic circuits, and heat-exchanger elements.

It is important to recognize, however, that the comparatively high values of electrical and thermal conductivity are not in themselves sufficient criteria for copper's enormous economic value. Rather, it is these properties in conjunction with good corrosion resistance and ease of formability that makes copper a very practical choice for many engineering applications.

Copper exhibits excellent plasticity and can be extruded into tubes, rolled into shapes, or drawn into wire. In the annealed condition, copper is very soft and ductile. Its tensile strength is approximately 35,000 psi (242 MPa), with an elongation of about 40% in 2 in.

The properties of copper can be altered appreciably by cold working. The tensile strength may be raised to about 50,000 psi with an elongation of about 4% in 2 in., by rolling or drawing. Copper has good corrosion and weather resistance and is used for roofing and other architectural applications, and as tubing for hydraulic fluids, refrigeration, water and sanitary drainage.

Copper is also used extensively in the alloyed condition to improve its strength and hardness, and to develop special properties. The elements that are commonly alloyed with copper are zinc, tin, nickel, silicon, aluminum, cadmium, and beryllium. In the following sections we examine several of these systems and study the unique properties exhibited by such copper alloys.

Copper–Zinc Alloys

The alloys of copper and zinc are commonly classified as *brasses*; however, the term *commercial bronze* may also be used for some compositions of copper and zinc. Since the terms used in the brass industry may be misleading to the student, the alloys of copper will be discussed from the standpoint of both composition and name, rather than from name alone. The equilibrium diagram of the copper–zinc system is shown in Figure 13-13. The region of the α solid solution, which has an fcc structure, extends from 0% to about 39% zinc. With increased amounts of zinc, a second solid solution, β, is formed. This phase has a bcc structure. The brasses may be classified as (1) α brasses or (2) $\alpha + \beta$ brasses. Alpha brasses containing from 5 to 20% zinc are called red brasses because of their reddish-copper color, while those containing 20 to 36% zinc are called yellow α brasses. The microstructure of yellow (cartridge) brass was displayed in Figure 6-21. Alpha + beta brasses contain 36 to 45% zinc.

The copper–zinc alloys are the most important of the copper alloys, due to their desirable properties and relatively low cost. The mechanical properties of alloys in this system are very closely related to the phases obtained. The tensile strength and ductility of the copper–zinc alloys increase with increasing amounts of zinc up to about 30%. The mechanical properties of these α-brasses depend on the zinc content and the degree of cold working they receive. The effect of cold drawing on the properties of a typical brass alloy is illustrated in Figure 13-14. With the appearance of the β solid solution, the strength continues to increase, but the ductility begins to decrease. The β and γ solid solutions are not as ductile as the α solid solution; therefore, very few alloys containing more than about 40% zinc are of commercial importance, unless they are for special applications. If the structure of the alloy contains an appreciable amount of the γ phase, the alloy will be quite brittle and of little engineering value.

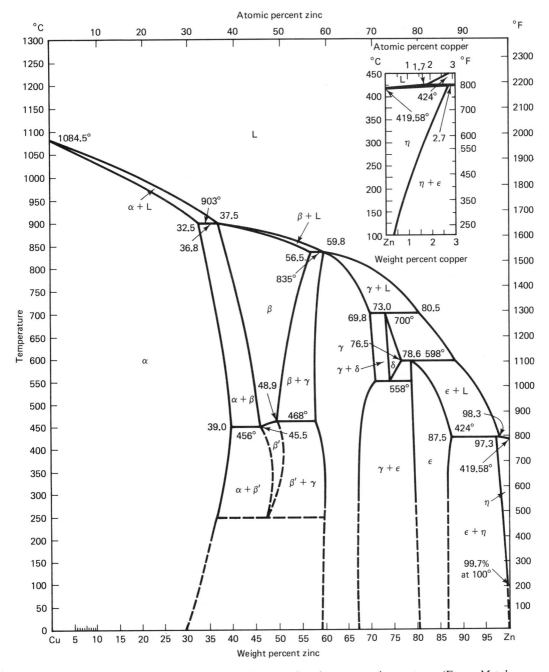

Figure 13-13 Equilibrium phase diagram for the copper–zinc system. (From *Metals Handbook*, 8th ed., Vol. 8, *Metallography, Structures and Phase Diagrams*, American Society for Metals, Metals Park, Ohio, 1973, p. 301.)

Copper and Its Alloys

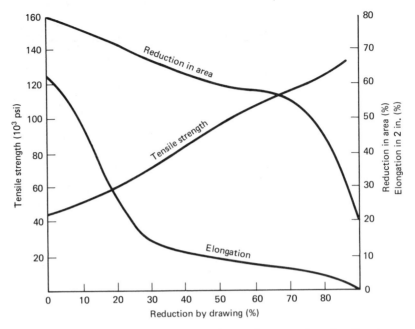

Figure 13-14 Effects of cold work on the tensile properties of yellow brass (65 Cu–35 Zn).

Example 13-2

A yellow brass (65% Cu–35% Zn) rod has been reduced in diameter from 10 mm in its annealed condition to a final diameter of 4 mm by cold drawing. (a) Calculate the tensile strength and the ductility (percent elongation) of the finished rod. (b) What would the mechanical properties of the finished rod be if it were annealed?

Solution   (a) Cross sectional area of original rod:

$$\frac{\pi d^2}{4} = \frac{\pi (10 \text{ mm})^2}{4} = 78.53 \text{ mm}^2$$

Cross section of finished rod:

$$\frac{\pi d^2}{4} = \frac{\pi (4 \text{ mm})^2}{4} = 12.56 \text{ mm}^2$$

$$\% \text{ reduction by drawing} = \frac{\text{original area} - \text{final area}}{\text{original area}} \times 100$$

$$= \frac{78.53 - 12.56}{78.53} = \frac{65.97}{78.53} \times 100$$

$$= 84\%$$

Using Figure 13-13, we have

$$\text{tensile strength} = 130,000 \text{ psi}$$

$$\text{elongation} = 2.5\% \qquad Ans.$$

> (b) After annealing, the 4-mm bar would have the following properties:
>
> $$\text{tensile strength} = 40{,}000 \text{ psi (276 MPa)}$$
>
> $$\% \text{ elongation} = 62\% \quad Ans.$$
>
> These would be the same properties as the original rod before cold reduction.

Those alloys that contain only α solid solution can be easily hot or cold worked, whereas those alloys that contain both α and β cannot withstand an appreciable amount of cold working without rupture and therefore must be formed while hot. Alloys that contain mostly β solid solution can be hot rolled, forged, or extruded easily. When the γ solid solution is present, both cold and hot working are difficult. Consequently, wrought forms of these alloys are rarely found. Figure 13-15 illustrates the effect of composition on the strength and ductility of copper–zinc alloys. Table 13-8 lists the mechanical properties for several important copper–zinc alloys and indicates the change in properties with increasing cold work.

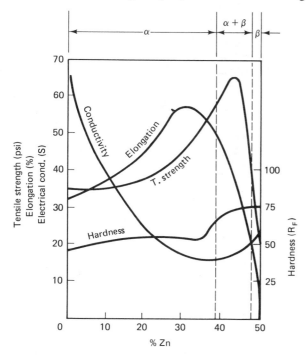

Figure 13-15 Influence of increasing zinc content on the properties of typical copper–zinc alloys. Note the location of the respective single and dual phase regions.

Copper–Tin Alloys

The term "bronze" was originally used to describe copper–tin alloys, but is now used for any copper alloy that contains up to 25% of the principal alloying element (except copper–zinc alloys). The name "bronze" conveys the idea of a higher-class alloy than brass, although it has been incorrectly applied to some alloys which are really special brasses. Alloys containing principally copper and tin are referred to as the *true* bronzes, although other elements are also frequently present to improve the charac-

Common name and compo-sition (%)	Condition	UTS (psi)[a]	Y.S. (psi)[a]	Elong. 2 in. (%)	Some uses
Gilding metal	0.035 mm G.S.	35,000	11,000	45	Coins, fuse caps,
	$\frac{1}{4}$ hard	42,000	32,000	25	emblems,
	Hard	56,000	50,000	5	jewelry
	Spring temper	64,000	58,000	4	
Commercial bronze	0.035 mm G.S.	38,000	12,000	45	Grillwork,
	$\frac{1}{4}$ hard	45,000	35,000	25	screen cloth,
	Hard	61,000	54,000	5	marine hard-
	Spring temper	72,000	62,000	3	ware, primer caps, costume jewelry
Cartridge brass	0.035 mm G.S.	49,000	17,000	57	Radiator cores,
(70 Cu, 30 Zn)	$\frac{1}{4}$ hard	54,000	40,000	43	lamp fixtures,
	Hard	76,000	63,000	8	springs
	Extra spring temper	99,000	65,000	3	
Yellow brass	Annealed	49,000	17,000	57	Architectural
(65 Cu, 35 Zn)	$\frac{1}{4}$ hard	54,000	40,000	43	grillwork
	Hard	74,000	60,000	8	
	Spring temper	91,000	62,000	3	
Architectural bronze (57 Cu, 40 Zn, 3Pb)	As extruded	60,000	20,000	30	Trim and hard-ware
Naval brass	Rod soft ann.	57,000	25,000	47	Condenser plates,
(60 Cu, 39$\frac{1}{4}$ Zn,	Rod $\frac{1}{4}$ H (8%)	69,000	46,000	27	welding rod,
3/4 Sn)	Rod $\frac{1}{2}$ H (20%)	75,000	53,000	20	marine hard-ware, propeller shafts, valve stems
Manganese bronze	Rod soft ann.	65,000	30,000	33	Pump rods,
(58.5 Cu, 39 Zn,	Rod $\frac{1}{4}$ H (10%)	77,000	45,000	23	valve stems,
1.4 Fe, 1 Sn, 0.1 Mn)	Rod $\frac{1}{2}$ H (20%)	84,000	60,000	19	welding rod
Aluminum brass (76 Cu, 22 Zn, 2 Al)	Tube 0.025 G.S.	60,000	27,000	55	Condenser tubes

[a]Multiply by 6.9×10^{-3} to obtain Mpa

teristics of the plain two component system. The equilibrium phase diagram for the Cu–Sn system is shown in Figure 13-16.

Tin increases the hardness, strength, and wear resistance of copper to a greater extent than does zinc. True bronzes also have good corrosion resistance, including saltwater corrosion resistance. Additions of tin from 0% up to about 16% form a homogeneous face-centered cubic solid solution with copper. At room temperature, the solubility of tin in copper theoretically is nil. However, in commercial practice, the reduction of solubility rarely occurs. In all but alloys annealed for extremely long times after severe cold working, the solubility below 968°F (520°C) can be considered

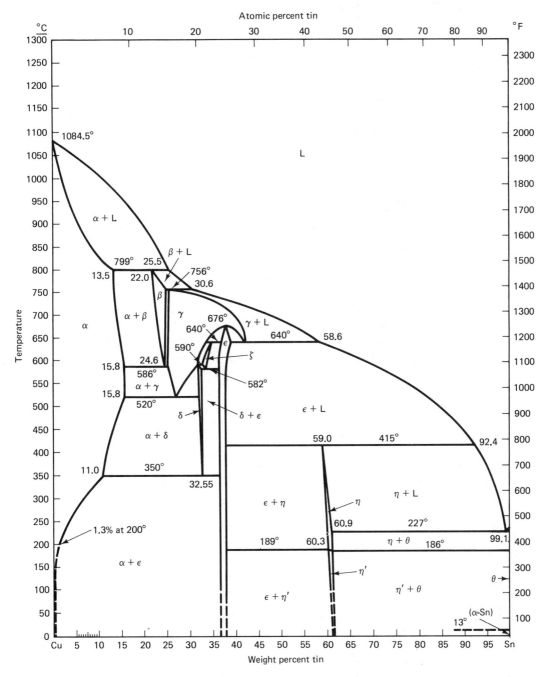

Figure 13-16 Equilibrium phase diagram for the copper–tin system. (From *Metals Handbook*, 8th ed., Vol. 8: *Metallography, Structures and Phase Diagrams*, American Society for Metals, Metals Park, Ohio, 1973, p. 299.)

Copper and Its Alloys

as constant at 15.8%. Similarly, the decomposition of ϵ (Cu_3Sn) below 662°F (350°C) occurs extremely slowly. Thus, for most practical purposes, the ($\alpha + \epsilon$) region can be considered as extending from 15.8% to 32.6%.

The copper–tin alloys may be divided into four groups, depending on the percentage of tin present.

1. *Up to 8% tin:* used for sheets, wire, coins; readily cold worked to increase strength; typical parts include high-strength springs, clips, snap switches, electric sockets and plug contacts, fuse clips, and flexible tubing
2. *Between 8 and 12% tin:* used for gears, machine parts, bearings, and marine fittings
3. *Between 12 and 20% tin:* used considerably for bearings and bushings
4. *Between 20 and 25% tin:* used primarily for bells and cymbals; very hard and brittle; used in the cast condition

However, as stated previously, not many of the true bronzes are pure binary copper–tin alloys. These bronzes may contain phosphorus, lead, zinc, and nickel as alloying elements. Tables 13-9 and 13-10 list the nominal composition and mechanical properties for several of the bronze alloys.

Phosphor bronze. Phosphor bronzes are tin bronzes containing from 1 to 10% copper and from 0.03 to 0.3% phosphorus added as a deoxidizing agent during melting. Hardness and strength are increased considerably, due to the presence of phosphorus, since it forms a hard precipitate (Cu_3P) distributed in the matrix of α

TABLE 13-9 NOMINAL COMPOSITION AND FORM OF SELECTED BRONZE ALLOYS

Common name	Form	Cu	Sn	P	Pb	Zn
Phosphor bronze (5 Sn)	Wrought	Bal.	3.5–5.8	0.03–0.35	—	—
Phosphor bronze (8 Sn)	Wrought	Bal.	7.0–9.0	0.03–0.25	—	—
Phosphor bronze (10 Sn)	Wrought	Bal.	9.0–11.0	0.03–0.25	—	—
Leaded tin bronze	Cast	86–90	5.5–6.5	—	1.0–2.0	3.0–5.0
Leaded tin bearing bronze	Cast	85–89	7.5–9.0	—	1.0	3.0–5.0
Gun metal	Cast	87–89	7.5–10.5	—	0–2.0	1.5–4.5
High-leaded tin bronze (bearing)	Cast	81–85	6.25–7.5	—	6.0–8.0	2.0–4.0
High-leaded tin bronze (bushing)	Cast	78–82	9.0–11.0	—	8.0–11.0	—
High-leaded tin bronze (anti-acid)	Cast	75–79	6.25–7.5	—	13.0–16.0	—
High-leaded tin bronze (semi-plas.)	Cast	68.5–73.5	4.5–6.0	—	22.0–25.0	—
85–5–5–5	Cast	84–86	4.0–6.0	—	4.0–6.0	4.0–6.0

TABLE 13-10 TYPICAL MECHANICAL PROPERTIES OF SELECTED BRONZE ALLOYS

Common name	Form	Section (in.)	Tensile strength (psi)	Yield strength (psi)	Elongation (%)	Hardness
Phosphor bronze	Flat annealed	0.040	47,000	19,000	64	26 R$_B$
(5% Sn)	Flat hard	0.040	81,000	75,000	10	87 R$_B$
Phosphor bronze	Flat annealed	0.040	55,000	—	70	75 R$_F$
(8 Sn)	Flat hard	0.040	93,000	72,000	10	93 R$_B$
Phosphor bronze	Flat annealed	0.040	66,000	—	68	55 R$_B$
(10 Sn)	Flat hard	0.040	100,000	—	13	97 R$_B$
Leaded tin bronze (Navy M)	Sand cast	0.505	38,000	16,000[a]	35	66 Brinell[b]
Leaded tin bronze (bearing bronze)	Sand cast	0.505	36,000	18,000[a]	30	68 Brinell[b]
Gun metal	Sand cast	—	40,000	20,000[a]	30	75 Brinell[b]
High-leaded tin bronze (bearing bronze)	Sand cast	0.505	34,000	17,000[a]	20	60 Brinell[b]
High-leaded tin bronze (bushing and bearing bronze)	Sand cast	—	32,000	17,000	12	65 Brinell[b]
High-leaded tin bronze (anti-acid bronze)	Sand cast	—	30,000	16,000	15	55 Brinell[b]
High-leaded tin bronze (semiplastic bronze)	Sand cast	—	21,000	—	10	48 Brinell[b]
85–5–5–5 (ounce metal)	Sand cast	—	34,000	17,000	25	60 Brinell[b]

[a] 0.5% elongation under load.
[b] 500-kg load.

phase. The phosphorus also increases the fluidity of the molten metal, thereby increasing the ease of casting and aiding in the production of sounder castings.

Leaded bronze. This is a term used to describe a mechanical mixture in which lead is dispersed in a copper alloy. Lead does not alloy with copper, but may be mixed with molten copper, and under suitable conditions the mixture may be cast into a mold. The solidified alloy has lead well distributed throughout the casting in small particles. This type of immiscibility was discussed in Chapter 7 (see Figure 7-25). Lead may be added to both bronze and brass alloys to improve machineability and bearing properties.

The lead particles, with their softness and low shear strength, reduce friction in parts subjected to sliding wear, such as in a bearing, or to machining tool wear, as at the tool–chip interface. Lead can be a source of weakness and is usually kept below 2%, although some bearing bronzes may contain higher percentages.

Copper–Aluminum Alloys

Alloys of copper containing between 5 and 11% aluminum are known as aluminum bronzes [see Figure 13-1(a)]. They may also contain other elements, such as iron, silicon, manganese, and nickel, for increased strength. The choice of aluminum bronze over tin bronze is usually based on properties other than corrosion resistance, such as superior wear resistance and mechanical properties at room and elevated temperatures. The addition of aluminum to copper up to about 11% causes a progressive increase in tensile strength; about 90,000 psi (621 MPa) can be reached. Elongation values decrease with increasing aluminum content. At about 12% aluminum, elongation values drop to 1 or 2%.

Above 9.5% aluminum, aluminum bronzes may be heat treated to improve mechanical properties. The hardness and strength of aluminum bronzes may be markedly improved by heat treatment. Reference to the copper-rich end of the copper-aluminum equilibrium diagram [Figure 13-1(a)] shows how the heat treatment is accomplished. An alloy containing approximately 10% aluminum may be solution treated by heating it to 1650°F (900°C) followed by water quenching. During the solution treatment, the α phase is transformed to the β phase. The β phase is retained, at room temperature in part, by the rapid cooling. A subsequent aging treatment at a temperature of 700 to 1100°F (371 to 593°C) causes the β phase to become unstable; it then undergoes a transformation to form a fine structure of α and γ phases. Such a transformation causes an increase in tensile strength of from 80,000 psi (552 MPa) in the annealed condition to 100,000 psi (690 MPa) in the solution-treated and aged condition. Hardness is increased from 90 R_B to 25 R_C, while elongation is decreased from 22% to about 2 to 6%.

The family of aluminum bronze alloys offers high strength and hardness, excellent corrosion resistance, good wearing qualities, good fatigue resistance, and is well suited for service at elevated temperatures. The properties of alloys within certain composition ranges can be improved further by heat treatment.

Because of their unusual range of properties, aluminum bronzes are used for special engineering applications such as dies, valve seats, slides, worm gears, cams, screw-down nuts, and nonsparking safety tools. Tensile strength after heat treatment may be as high as 120,000 psi (828 MPa) and hardness may reach 37 R_C or higher with certain compositions.

Copper–Silicon Alloys

The alloys made by adding silicon to copper generally have excellent properties: corrosion resistance almost equivalent to copper, a wide range of mechanical properties and good castability. A portion of the copper–silicon equilibrium diagram is shown in Figure 13-17. Silicon is soluble in copper up to about 5.3% at a temperature of 1550°F (843°C). This solubility decreases to less than 4% at room temperature. Although these alloys are not hardenable by heat treatment, they may be cold worked to achieve strengths as high as 45,000 psi (310 MPa).

Most of the commercial silicon–bronze casting alloys contain from 3 to 5% silicon, with tin, manganese, iron, and zinc added to enhance properties. Mechanical properties range from 45,000 to 65,000 psi (310 to 449 MPa) tensile strength, 15,000 to 40,000 psi (104 to 276 MPa) yield strength, and 15 to 75% elongation, depending on the proportions of alloying ingredients.

Silicon–bronze alloys exhibit excellent corrosion resistance to organic acids, sulfite solutions, and seawater. Silicon bronzes find application in electrical fittings,

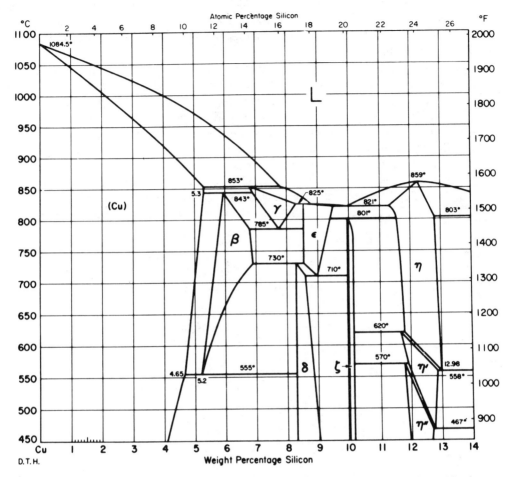

Figure 13-17 Equilibrium phase diagram for the copper–silicon system. (From *Metals Handbook*, 8th ed., Vol. 8: *Metallography, Structures and Phase Diagrams*, American Society for Metals, Metals Park, Ohio, 1973, p. 298.)

marine hardware, boilers, pumps, and shafting. These alloys are also cast as ingots and may then be worked either hot or cold. They are often used as lower-cost substitutes for the Cu–Sn alloys and combine high strength with good corrosion resistance and ease of welding.

Copper–Nickel Alloys

An examination of the copper–nickel system (see Figure 7-12) reveals that copper and nickel are totally soluble in both the liquid state and the solid state. The addition of nickel to copper produces several important effects. Corrosion resistance generally increases with increasing nickel content. Copper alloys containing between 2 and 30% nickel are called *cupronickels*. These alloys have moderately high strength in the cold-worked condition, combined with excellent corrosion resistance. For

Copper and Its Alloys

example, 30% cupronickel has a yield strength of 70,000 psi (483 MPa) and an elongation of 15% when cold worked. These alloys are used in systems requiring good corrosion resistance, such as condenser tubes, desalination piping, and other marine components.

Another change that occurs with nickel additions is the appearance. The color of the alloys lightens until at 20% nickel and above, the alloys are silver in appearance. There are a group of alloys called *nickel silver*, which contain copper, nickel, and zinc. The compositions vary widely, but the alloys generally contain approximately 60% copper, 10 to 30% nickel, and from 5 to 25% zinc. As the nickel content is increased, the zinc content is usually decreased. These alloys are pleasing in appearance and possess moderately good corrosion resistance. Such alloys are used as the base material for silver plate, costume jewelry, plumbing fixtures, zippers, and eyeglass frames.

The electrical characteristics of copper also change markedly with the addition of nickel. The electrical resistivity changes dramatically, as shown in Figure 13-18.

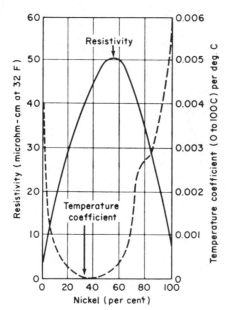

Figure 13-18 Effect of nickel content on the resistivity of copper. Note the value of temperature coefficient that results at approximately 40% Ni.

An alloy called *Constantan* utilizes this behavior. It contains about 45% nickel and 55% copper. The alloy has high resistivity, combined with a low temperature coefficient of resistivity, which makes it useful for wire-wound resistors and for thermocouples.

NICKEL AND ITS ALLOYS

Nickel is one of the most important of the major industrial metals. It is a light gray metal with an fcc structure and a density of 8.90 g/cm³. Commercially pure nickel has a tensile strength of 45,000 psi (310 MPa) in the annealed condition, but a very low yield strength, which is on the order of 8500 psi (58.7 MPa). Consequently, it is used in its commercially pure form, primarily where resistance to corrosion is desired, rather than strength characteristics. Table 13-11 lists the composition and applications for several of the commercially pure forms of nickel, together with several nickel-based alloys. Commercially pure nickel is available in several grades, depending on its impurities or minor alloying elements. A nickel is a commercially pure nickel (99.4%), with the balance being principally cobalt. Its use is mostly in the chemical industry and in electroplating. D nickel contains 4.5% manganese to improve resistance to attack by sulfur compounds at temperatures up to 1000°F (538°C). E nickel, which contains 2% manganese, is very similar to D nickel, while L nickel, with a low carbon content, is used in forming operations requiring substantial plastic deformation.

Another nickel alloy, called Z nickel, contains about 4.5% aluminum and is subject to precipitation hardening. Maximum yield strength of cold-worked and age-hardened strip approaches 230,000 psi (1587 MPa), with a hardness of 46 R_C. This alloy is used principally for pump rods, springs, and shafts that demand high strength and high resistance to corrosion.

Nickel is also used as an alloying element in stainless steels (discussed in Chapter 12), copper-based alloys, and *high-temperature* alloys. The latter alloys are vitally important in elevated temperature applications and are discussed later in this section.

Nickel–Copper Alloys (Monels)

The Monels are nickel–copper alloys in which nickel is the predominant element. There are five of these alloys, as shown in Table 13-11, containing approximately 65% nickel and 30% copper with varying amounts of the other elements. The Monels are useful alloys in that they possess excellent corrosion resistance, combined with moderately high strength, at lower cost than pure nickel. Monel alloys can be cast, forged, rolled, and welded with relative ease. Monels can achieve a tensile strength of 80,000 psi (552 MPa) with 45% elongation in the annealed condition, and 100,000 psi (690 MPa) with 25% elongation when cold worked.

The R-Monel has the same general characteristics as Monel, but it is a free-machining alloy, containing 0.025 to 0.050% sulfur, intended for processing in automatic screw-making machines. K-Monel is a precipitation-hardenable Monel which contains about 3% aluminum, and its strength is similar to that of a heat-treated steel. The heat treatment is as follows: solution treated at 1600°F (870°C) for $\frac{1}{2}$ hr, followed by a water quench. This is followed by heating to a precipitation temperature of 1100°F (595°C) for 8 to 16 hr, and then furnace cooling.

S-Monel is used primarily in castings and contains about 4% silicon. This alloy is also responsive to precipitation hardening. A hardness of about 37 R_C makes it suitable for use when resistance to galling and erosion is important, as in valve seats, and where sliding contact is involved under corrosive conditions. This alloy is solutionized by heating to a temperature of 1600°F (870°C) for 1 hr, followed by cooling in air to 1200°F (650°C) and then quenching in oil or water. Hardening is accomplished after the softening (solutionizing) treatment by heating to 1100°F (595°C) for 4 to 6 hr, followed by furnace cooling.

TABLE 13-11 COMPOSITION AND APPLICATIONS FOR COMMERCIAL NICKELS AND NICKEL ALLOYS

Alloy	Composition (w/o)											Typical properties and uses
	Ni	C	Mn	Fe	S	Si	Cu	Cr	Al	Ti	Other	
Commercially pure nickel												
A nickel	99.4	0.1	0.2	0.15	0.005	0.05	0.1					Chemical industry, electroplating
D nickel	95.2	0.1	4.5	0.15	0.005	0.05	0.05					Resistance to sulfur attack
E nickel	97.7	0.1	2.0	0.10	—	0.05	0.05					Similar to D nickel
L nickel	99.4	0.02	0.2	0.15	0.005	0.05	0.1					Severe plastic forming operations
Z nickel	94.0	0.16	0.25	0.25	0.005	0.40	0.05			0.33		Age-hardenable, springs, pump rods, shafts
Nickel and copper alloys												
Monel	66.0	0.12	0.90	1.35	0.005	0.15	31.50					Corrosion resistance, toughness, high strength
R-Monel	66.0	0.18	0.90	1.35	0.050	0.15	31.50					Free machining
K-Monel	65.3	0.15	0.60	1.00	0.005	0.15	29.50		2.80	0.50		Age-hardenable, non-magnetic
H-Monel	65.0	0.1	0.9	1.5	0.015	3.0	29.50					Age-hardenable, machinable
S-Monel	63.0	0.1	0.9	2.0	0.015	4.0	30.0					Age-hardenable, anti-galling, impellers, pump liners
Nickel-iron alloys												
Invar	36.0			63.0								Very low expansion coefficient, length standards, tuning forks
Platinite	46.0			54.0								Glass seal
Permalloy	78.5			21.5								High magnetic permeability, submarine telegraph cables

Nickel–Iron Alloys

Nickel and iron are totally soluble in the liquid state, but form various solid solutions upon solidification. Alloys that contain up to 6% nickel are ferritic. With increasing nickel content, the alloys become increasingly subject to air hardening. Alloys containing between 6 and 28% nickel form martensite upon rapid cooling. After slow cooling or reheating, these alloys decompose into α and γ phases.

Alloys containing greater than 34% nickel are entirely austenitic, nonmagnetic, and even at extremely low temperatures are structurally stable. Alloys of nickel and iron have rather unique properties of thermal expansion, magnetism, thermal conductivity, and modulus of elasticity. These properties can be varied considerably by changes in the percentage of nickel present in the alloy. For example, an iron–nickel alloy containing 36% nickel (with minor amounts of Mn, Si, and C) has a coefficient of thermal expansion (α) so low that its length is almost invariable for ordinary changes in temperature, hence it is called Invar. The very low value of α for Invar $[\alpha = (0.877 + 0.00127 T^\circ) \times 10^{-6}$ in./in./$^\circ$C] makes it ideally suited for mechanisms which must maintain critical dimensions, such as the balance wheel in watches and clocks, dimensional standards, and tuning forks.

Other nickel–iron alloys are used in the production of glass-to-metal seals. These alloys have a coefficient of expansion which approximates glass and provides a transition between a metal component and the glass in which it is embedded.

Nickel–Base Superalloys

The nickel-base superalloys (Tables 13-12 and 13-13) are initially strengthened by solid-solution hardening from the addition of cobalt, iron, chromium, molybdenum, tungsten, and aluminum. Further strengthening results from a precipitation hardening reaction which produces a fine intermetallic precipitate of Ni_3Al and Ni_3Ti. This unique precipitate is referred to as *gamma prime* (γ') and is shown in Figure 13-19. In contrast to the strengthening mechanisms that we discussed in Chapter 6 which apply to other particles, remarkably, the strength (hardness) of γ' increases with increasing temperature.

Another method to improve elevated-temperature properties is to produce a fine "mechanical" dispersion of *insoluble* particles within a metal matrix. Mechanical dispersion in this sense implies that the particles are exogenous to the alloy system and do not result from a precipitation reaction within the alloy. Rather, they are mechanically mixed into the matrix material. This is usually accomplished in powder form. For example, TD-nickel (see Table 13-13) consists of 2% thoria (ThO_2) uniformly dispersed in a matrix of 98% nickel. This material is, in fact, superior to most other nickel-based alloys at temperatures above approximately 2000°F (1094°C).

Nickel-base superalloys are used extensively in applications involving aircraft and high temperature engines, and also land-based power generating systems. For example, the sections of gas and jet turbines which are subjected to both high stresses and high temperatures utilize these alloys for disks, shafts, bolts, blades (buckets), and so on. Reciprocating engines employ nickel-base alloys in superchargers (turbochargers) and exhaust valves. Even the metalworking industry utilizes these alloys

TABLE 13-12 NOMINAL COMPOSITION OF SELECTED CAST SUPERALLOYS

Alloy Designation	Composition (w/o)															Tempera-ture (°C)[a]
	C	Mn	Si	Cr	Ni	Co	Mo	W	Nb	Fe	Ti	Al	B	Zr	Other	
MC-102	0.04	—	—	20.0	bal.	—	6.0	2.5	6.5	—	—	—	—	—	—	850
GMR 235-D	0.15	—	—	15.5	bal.	—	5.0	—	—	4.5	2.5	3.5	0.050	—	—	927
MAR-M 421[b]	0.15	—	—	16.0	bal.	9.0	2.0	3.8	2.0	—	1.8	4.3	0.015	0.05	—	975
Alloy 713LC	0.05	—	—	12.0	bal.	—	4.5	—	2.0	—	0.6	5.9	0.010	0.10	—	980
Alloy 713C	0.12	—	—	12.5	bal.	—	4.2	—	2.0	—	0.8	6.1	0.012	0.10	—	985
IN-738	0.17	—	—	16.0	bal.	8.5	1.7	2.6	0.9	—	3.4	3.4	0.010	0.10	1.8 Ta	901
B-1900	0.10	—	—	8.0	bal.	10.0	6.0	—	—	—	1.0	6.0	0.015	0.10	4.0 Ta	1000
IN-100	0.18	—	—	10.0	bal.	15.0	3.0	—	—	—	4.7	5.5	0.014	0.06	1.0 V	1000
IN-731	0.15	—	—	10.0	bal.	10.0	2.5	—	—	—	4.5	5.6	0.014	0.06	1.0 V	1000
MAR-M 200[b]	0.15	—	—	9.0	bal.	10.0	—	12.5	1.0	—	2.0	5.0	0.015	0.05	—	1020
X-40	0.50	0.75	0.75	25.5	10.5	bal.	—	7.5	—	—	—	—	—	—	—	—
FSX-414	0.25	—	—	29.0	10.0	bal.	—	7.5	—	1.0	—	—	0.010	—	—	—
MAR-M 302[b]	0.85	—	—	21.5	—	bal.	—	10.0	—	—	—	—	0.005	0.20	9.0 Ta	—

[a]Temperature capability °C for 100 hr of life at 137 MPa (20,000 psi).
[b]Trademark of Martin Marietta Corporation.

TABLE 13-13 NOMINAL COMPOSITION OF SELECTED WROUGHT SUPERALLOYS

Alloy designation[a]	Composition (w/o)															Temperature (°C)[b]
	C	Mn	Si	Cr	Ni	Co	Mo	W	Nb	Fe	Ti	Al	B	Zr	Other	
A-286	0.05	1.35	0.50	15.0	26.0	—	1.3	—	—	bal.	2.0	0.2	0.015	—	—	780
Inconel alloy 718	0.04	0.20	0.30	18.6	bal.	—	3.1	—	5.0	18.5	0.9	0.4	0.005	—	—	802
IN-120	0.04	—	—	21.0	bal.	14.0	4.0	—	2.0	—	2.5	0.25	0.005	0.05	—	802
Incoloy alloy 901	0.05	0.10	0.10	12.5	42.5	—	5.7	—	—	bal.	2.8	0.2	0.015	—	—	826
Nimonic 80A	0.06	0.10	0.70	19.5	bal.	1.1	—	—	—	—	2.5	1.3	—	—	—	826
Inconel alloy X-750	0.04	0.70	0.30	15.0	bal.	—	—	—	0.9	6.8	2.5	0.8	—	—	—	835
D-979	0.05	0.25	0.20	15.0	bal.	—	4.0	4.0	—	27.0	3.0	1.0	0.010	—	—	843
Nimonic 90	0.07	0.50	0.70	19.5	bal.	18.0	—	—	—	—	2.4	1.4	—	—	—	843
René 41	0.09	—	—	19.0	bal.	11.0	10.0	—	—	—	3.1	1.5	0.005	—	—	881
M-252	0.15	0.50	0.50	20.0	bal.	10.0	10.0	—	—	—	2.6	1.0	0.005	—	—	884
Waspaloy	0.08	—	—	19.5	bal.	13.5	4.3	—	—	—	3.0	1.3	0.006	0.06	—	894
Inconel alloy 700	0.12	0.10	0.30	15.0	bal.	28.5	3.7	—	—	0.7	2.2	3.0	—	—	—	905
Nimonic alloy 105	0.20	1.0	1.0	14.6	bal.	20.0	5.0	—	—	2.0	1.2	4.7	—	—	—	925
Udimet 500	0.08	—	—	18.0	bal.	18.5	4.0	—	—	—	2.9	2.9	0.006	0.05	—	927
Nimonic alloy 115	0.15	—	—	15.0	bal.	15.0	3.5	—	—	—	4.0	5.0	—	—	—	950
Udimet 700	0.08	—	—	15.0	bal.	18.5	5.2	—	—	—	3.5	4.3	0.030	—	—	960
TD Nickel	—	—	—	—	98.0	—	—	—	—	—	—	—	—	—	2.0 ThO$_2$	824
Haynes Alloy 188	0.10	1.25	0.40	22.0	22.0	bal.	—	14.0	—	3.0	—	—	—	—	0.08 La	—
L-605	0.10	1.50	0.50	20.0	10.0	bal.	—	15.0	—	—	—	—	—	—	—	—

[a]Trademarks: Inconel, Incoloy, and Nimonic—INCO family of Companies; René 41—Teledyne Allvac; Waspaloy—United Technologies Corporation; Udimet—Special Metals Corporation; Haynes—Cabot Corporation.
[b]Temperature capability °C for 100 hr of life at 137 MPa (20,000 psi).

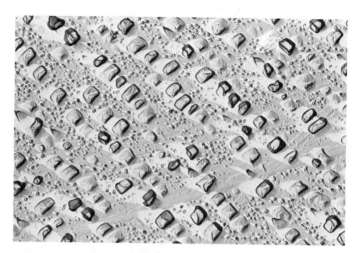

Figure 13-19 Transmission electron micrograph (replica) of a nickel-base superalloy showing gamma prime precipitate (cubical) 11,500×. (Courtesy of L. J. McNamara, Benét Laboratories).

in hot-work tool and die applications, together with heat-treatment equipment and fixtures.

LEAD AND ITS ALLOYS

Lead is another common metal which finds wide application as an engineering material. It is bluish gray in appearance and has an fcc structure with a density of 11.36 g/cm³, the highest of the common metals. There are a number of applications for lead which are based on its density. For example, it is used to produce ballast, sinkers for sport and commercial fishermen, and bullets.

Upon exposure to the atmosphere, lead quickly acquires a protective film of oxide or sulfate. Consequently, lead has good resistance to atmospheric corrosion, to fresh water, and to many acids and gases, including chlorine. In its pure state it is used as a coating for steel sheets. This coated steel is called *terne* metal and is used in the fabrication of roofs, gas tanks, and other products requiring atmospheric corrosion resistance.

Lead, from the time of the Romans up to the early twentieth century, had been used as a material to produce pipes. In fact, the term "plumber" derives from the Latin word, *plumbum*, meaning lead. Although the use of lead pipe has been supplanted by copper and plastic piping, countless installations still exist in older residences and structures.

Other applications for lead include its use as a cable sheathing material, as a radiation shield, and as a sound deadener. Lead is also used as an additive in free-machining brasses, bronzes, and stainless steels, where it forms a dispersed phase which provides lubricity and chip-breaking characteristics. This was mentioned earlier in this chapter in the section "Leaded Bronze" and in Chapter 7 in the section "Monotectic Alloys" (see Figure 7-25).

Lead–Calcium Alloys

The major use for lead is for the manufacture of grids (plates) in lead–acid batteries (secondary batteries). Although a lead–antimony alloy has historically been the most widely used battery grid alloy, the recent development of lead–calcium alloys has led to their widespread use in *maintenance-free* storage batteries.

The calcium alloying addition strengthens the lead grids so that they are sufficiently strong to resist sagging and trauma due to road vibration while eliminating much of the undesirable electrochemical production of hydrogen and loss of water associated with the presence of antimony. With concentrations of calcium between 0.01 and 0.10%, strengthening occurs over time by the precipitation of very fine Pb_3Ca particles (i.e., age hardening).

The addition of tin to the lead–calcium alloys increases the strength, as shown in Figure 13-20, and also improves the electrochemical characteristics. It has been

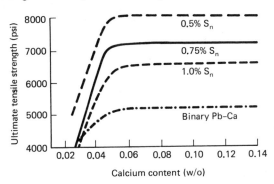

Figure 13-20 Effects of calcium and tin additions on the ultimate tensile strength of lead. A very small Ca addition (0.05%) produces an optimum increase in the strength of lead. (Courtesy of J. A. Passmore, St. Joe Lead, Company, Herculaneum, Missouri.)

proposed that the tin promotes a dual precipitation mechanism in which Sn_3Ca precipitates in addition to Pb_3Ca. The Pb_3Ca precipitates early in the reaction but causes only moderate lattice distortion, while the Sn_3Ca precipitates more slowly but creates significantly more lattice distortion.

Lead–Tin Alloys

Lead–tin alloys are commonly used as low-temperature brazing materials (soft solders). A review of the equilibrium phase diagram for the lead–tin system presented in Chapter 7 (see Figure 7-18) shows that lead and tin form a eutectic at 61.9% Sn. However, solder compositions range from 95% Pb–5% Sn to 50% Pb–50% Sn. Except for the eutectic composition, the Pb–Sn solders freeze over a range of temperatures, during which the material forms a slushy mixture which is more amenable to wiping operations than those which require the material to flow readily. The Pb–Sn solders are used in such operations as the joining of electrical wires, plumbing fittings, and filling in seams in automotive bodies.

Ternary Alloys

Type metal. Lead, combined with tin and antimony, forms a class of metals known as *type* metals. The low melting point deriving from the lead base facilitates casting. Antimony provides hardness and wear resistance, while tin adds fluidity and improves the structure, making it finer and more capable of reproducing detail, the latter property being an essential attribute for a type metal. While a eutectic exists at 84% Pb, 4% Sn, and 12% Sb, the exact composition varies with the typesetting process.

Babbitt. Lead-based babbitts are ternary alloys composed of lead, antimony, and tin which take advantage of the low coefficient of friction of lead. A typical composition for a lead-based babbitt would be 75% Pb–10% Sb–10% Sn. These alloys are used as bearing materials for light to moderate loads. With small additions of arsenic (1 to 3%), they are used in heavier bearing applications, such as journal bearings in steamships, and automotive and diesel engines.

Low-melting alloys. This term is generally applied to metals that melt below 450°F (233°C). This is the upper range of temperature for these alloys, and alloys that melt as low as 117°F (46.7°C) have been developed. Many of the alloys have lead, tin, and bismuth as their primary elements, while others include cadmium and bismuth as well. Some of these alloys form eutectics, whereas others are non-eutectic. The eutectic alloys melt and yield at a specific temperature, while the non-eutectic alloys melt over a temperature range, as might be expected. These low-melting alloys are used primarily as *fusible links* in such devices as electrical fuses, sprinkler systems, smoke release vents, and fire detection apparatus.

THE PRECIOUS METALS

The precious metals are generally considered to be gold, silver, and platinum, as well as their alloys. On a tonnage basis, they do not account for significant production, but they have unique properties which make them useful as engineering materials. Let us look briefly at this group of nonferrous metals and examine some of the properties that make the precious metals and alloys unique engineering materials.

Gold

Gold is a soft, lustrous yellow metal with a density of 19.3 g/cm³ and an fcc structure. It has been known to man since antiquity and has been used in jewelry and for decoration since before recorded history. Pure gold is termed "24 karats." Therefore, 18-karat gold contains 75% pure gold. In fact, most "gold" jewelry is made from 14- to 18-karat gold alloyed with copper because the pure metal is much too soft and will not exhibit adequate hardness and wear characteristics. Gold has several unique properties that make it important industrially. It is exceptionally malleable and ductile. It can be drawn into wire finer than a human hair and a single Troy ounce (31 g) can be hammered into a sheet measuring (300 ft²)!

It is remarkably corrosion resistant and is not attacked by ordinary corrosive media, even many acids. It is for this reason that it is often found free and uncombined in nature. From an engineering standpoint, this freedom from oxidation and tarnish is the basis for the making of gold electrical contacts in applications requiring high reliability and integrity and also its use in dental restorations.

Gold also has good thermal conductivity and reflectivity. It has been used in thin coats (electrodeposited) to provide thermal protection on satellites and other aerospace applications.

Silver

Silver is a brilliant white metal with an fcc structure and a density of 10.49 g/cm^3. Like gold, silver is used as a store of value and, historically, has been used as coinage, in jewelry, and in table service. However, pure silver is also too soft for most uses; therefore, it is alloyed with other metals to increase its hardness and strength. Indeed, many jewelry applications consist of alloys from the silver–copper system illustrated in Figure 7-20. For example, *sterling silver* contains 92.5% Ag and 7.5% Cu. Silver also has properties that make it important as an engineering material. For example, it displays high electrical conductivity, as explained in Chapter 10 (see Table 10-2). Although its cost makes ordinary conductor applications prohibitive, silver is used in certain cases, such as transvenous leads in heart pacemakers.

The photosensitive properties of silver, in the form of silver halides, also make it the basis of the photographic industry. Silver chloride or silver bromide is precipitated in gelatin which is spread on cellulose acetate sheet. The action of light radiation (photons) on this material produces decomposition of the silver halide as illustrated by the following reaction:

$$AgBr \longrightarrow Ag + Br$$

During subsequent development, the silver salt which has been exposed to light is further reduced and metallic silver (Ag°) is precipitated as a fine black powder. In this manner, the latent image is converted into a visible image (negative) on the film. As you can see from an examination of Table 13-14, photographic materials are the largest single use for silver.

TABLE 13-14 CONSUMPTION OF SILVER IN THE UNITED STATES IN A TYPICAL YEAR CLASSIFIED BY END USE

End use	Millions of ounces
Electroplated ware	9.5
Sterling ware	21.0
Jewelry	9.5
Photographic materials	53.0
Dental and medical supplies	1.6
Mirrors	4.0
Brazing alloys and solders	13.8
Electronic and electronic products: batteries	3.5
Catalysts	11.0
Coins, medallions, and commemorative objects	8.5
Miscellaneous	0.5
Total industrial consumption	167.5

The Precious Metals

Silver is also an important ingredient in the silver–tin–mercury amalgam commonly used in dental work to restore cavities (dental caries) in teeth.

Recently, a dispersed phase alloy which contains spheres of silver-copper eutectic added to the conventional amalgam has been developed. The composition of this alloy is 70% Ag, 16% Sn, 13% Cu, and 1% Zn. During amalgamation, the copper attracts and combines with the free tin, thereby eliminating the tin-rich phase responsible for corrosion and producing dental restorations that tend to be more resistant to crevice corrosion.

Platinum Group Metals

The six metals in the platinum group are platinum, palladium, iridium, rhodium, ruthenium, and osmium. These metals are grouped together because they are found in the same ores. What makes these metals important is their superior resistance to corrosion and oxidation. Their high intrinsic value is due to the fact that, like gold, they are relatively scarce. Since platinum is the least rare and most widely used metal in this group we will limit our discussion to this metal.

Platinum. Platinum is a silvery white metal with an fcc structure and a density of 21.43 g/cm³. It is produced primarily as a by-product of nickel refining. Platinum is malleable and ductile when pure and is used in jewelry, for thermocouple wire, for laboratory equipment such as electrochemical anodes and crucibles, for decorative purposes on dishware and crystal, and for electrical contacts. Additionally, one of its major uses is as a catalytic agent in the petrochemical industry. Its effectiveness as a catalyst in the cracking and refining of petroleum is well known. More recently, it has found use as the primary catalyst in automotive catalytic converters, where it reduces noxious emissions.

REFRACTORY METALS

Refractory metals become increasingly important as high-temperature materials when temperatures approach about 1000°C (1832°F). The high-temperature strength of these materials is due largely to their very high melting points, as shown in Table 13-15. Molybdenum, for example, alloyed with titanium, and with titanium plus zirconium (TZM), exhibits elevated-temperature strengths superior to other commercial superalloys. But the utilization of this material is severely restricted by its poor oxidation resistance. Molybdenum displays a very high rate of oxidation, and molybdenum oxide volatilizes rapidly at temperatures well below 1000°C (1832°F) as well.

The short-time tensile strength of certain refractory materials is displayed in Figure 13-21. Note the tensile strengths exhibited by these materials at temperatures in excess of 2000°F (1093°C). As we indicated above, the refractory alloys display considerably better high-temperature-strength capability than that of the nickel-based alloys. However, their commercial applications as high-temperature materials will be limited until the aspects of chemical reactivity and processing difficulties, such as their relative brittleness, are resolved.

TABLE 13-15 SELECTED PHYSICAL AND MECHANICAL PROPERTY DATA FOR REFRACTORY METALS

Metal	Melting temperature °C	Melting temperature °F	Density gm/cm³	Thermal conductivity $\left(\dfrac{\text{cal}}{\text{cm}^2\text{sec}}\right)\big/\left(\dfrac{°\text{C}}{\text{cm}}\right)$	Tensile strength[a] at 25°C 10³ psi	Tensile strength[a] at 25°C MPa	Modulus of Elasticity at 25°C 10⁶ psi	Modulus of Elasticity at 25°C MPa
Tungsten	3410	6170	19.3	0.40	100–500	689–3445	59	406,500
Tantalum	3030	5486	16.6	0.13	35–70	241–482	27	186,000
Molybdenum	2620	4748	10.2	0.35	120–200	827–1378	46	316,900
Rhenium	3180	5756	21.0	0.17	280	1929	67	461,600
Columbium	2468	4474	8.57	0.12	30–60	207–413	15	103,400

[a]Ranges are given because the tensile strength of these materials vary significantly with their form and processing history.

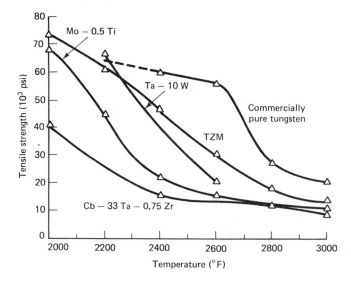

Figure 13-21 Elevated temperature tensile strength of selected refractory sheet materials, tested in protective atmosphere. (From R. W. Gilbert, Jr. and J. V. Houston, Jr., *Metal Progress*, November 1962.)

STUDY PROBLEMS

13.1. Aluminum is used to manufacture containers for many beverages. What attributes of aluminum make it suitable for such an application? Explain your reasons in detail.

13.2. Several problems have been encountered with pure aluminum conductors used in residential housing. One of these problems has been the increased corrosion where aluminum has been connected to copper wire in the system. Based on the properties of aluminum, what other problems might you expect with aluminum wire, resulting from ordinary residential construction methods?

13.3. If 1 oz of gold is hammered into a sheet measuring 5.27 m per side, how thick will the sheet be? What property of gold does this illustrate?

13.4. A cast aluminum highway guard post having a yield strength of 52,000 psi (358.8 MPa) can be substituted for a steel guard post with a yield strength of 100,000 psi (690 MPa). (a) Assuming a uniform stress distribution, how much larger must the cross section of the new post be? (b) What is the relative weight of the new system? (c) What is the percentage difference in weight?
Answer: (a) 1.92 times larger

13.5. With reference to problem 13.4, list the advantages of the new system. Do you anticipate any serious problems with using the new system? (*Hint:* Examine mechanical properties other than strength and ductility.)

13.6. Indicate which of the following metals and alloys are hardenable by cold work, heat treatment, or both: lead, Be–Cu, Ti, K-Monel, 70 Cu–30 Zn, Ti–6 Al–4 V, 80 Cu–20 Ni.

13.7. Examine the phase diagram of the Cu–Ni and Al–Mg systems. Based on the diagrams which of the following alloys would you expect to exhibit age hardening?
(a) 5 Cu–95 Ni
(b) 8 Al–92 Mg
(c) 90 Ni–10 Cu
(d) 95 Al–5 Mg

13.8. Using Table 13-7, compare the following alloys to steel (SAE1035) on a yield strength-to-weight (density) ratio and on a modulus-to-weight basis: magnesium AZ80-T5, aluminum 7178-T6, Ti–7 Al–4 Mo, and beryllium.

   (a) Which metal or alloy has the highest modulus-to-density ratio?

   (b) Which metal or alloy has the highest yield strength-to-density ratio?

13.9. Examine Figure 13-1(b).

   (a) Indicate the range of copper content most suitable for precipitation hardening.

   (b) Explain, using diagrams, what happens to the microstructure when an alloy containing 3.5% Cu is quenched from 500°C to room temperature.

   (c) Similarly, show what happens when the alloy is cooled from 500°C under equilibrium conditions.

13.10. A batch of aluminum components are age hardened at 300°F for 16 hr. Hardness values taken on two specimens taken from the batch were 70 BHN and 37 BHN. What is the most likely explanation?

13.11. In many applications, such as a floor or deck, a structural material is limited not by its strength but by the stiffness of the structure. The deflection for a cantilever beam is proportional to the density divided by the square root of the modulus. Calculate these values for the metals listed in Table 13-7.

13.12. What metal is the most likely alloying element for copper if the sole attribute was to minimize the loss in conductivity?

13.13. An annealed brass bar (70 Cu–30 Zn) 1 in. square has a tensile strength of 42,000 psi.

   (a) Determine what percentage reduction in cross-sectional area it must be given by cold working to raise its tensile strength to 100,000 psi.

   (b) What would the final width of the bar be, assuming that the reduction was uniform?

   Answer: (b) 0.685 in

13.14. In problem 13.13, if the original 1 in. sq. bar had a tensile strength of 60,500 psi, what additional amount of cold working would be necessary to raise the tensile strength to 97,000 psi? What would be the ductility properties of the final bar?

13.15. Are Cu–Zn alloys comprised primarily of β phase used commercially to any extent? Explain your answer.

13.16. Match up the word or phrase in one column with the most appropriate term in the opposite column.

   (a) Highest thermal conductivity
   (b) Electrical fuses
   (c) Surgical implant material
   (d) Resistance strain gage
   (e) 7xxx series contain
   (f) High modulus
   (g) Low density
   (h) Low thermal expansion
   (i) Catalytic agent
   (j) Sterling silver

   (1) Be
   (2) Cu–Ag
   (3) Platinum
   (4) Constantan
   (5) Ag
   (6) Titanium
   (7) ternary Pb
   (8) Mg
   (9) Zn
   (10) 36 Ni–Fe

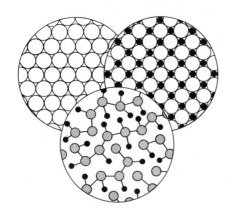

CHAPTER FOURTEEN:
Polymeric Engineering Materials and Their Applications

POLYMERS

Polymers are large, high-molecular-weight macromolecules constructed from a repeating series of smaller structural units. Polymers may be natural in origin and derived from plant, animal, or mineral substances, such as cellulose, wool, or asbestos, or may be synthetically created from petrochemicals or silicones. It is also convenient to classify polymers into various subgroups depending on their molecular weight and compositional basis. *High polymers* are those having molecular weights ranging from 20,000 to several million, while *low polymers* are those with molecular weights of several thousand up to 20,000.

Organic and Inorganic Polymers

As stated previously, polymers can be classified as to the size of the molecule; however, they can also be classified as to whether the polymer structure is organic, that is, based on a carbon chain, or inorganic, i.e., based on some other element such as silicon.

The ability of the carbon atom to form macromolecules by covalently bonding with other carbon atoms as well as oxygen, sulfur, and nitrogen forms the basis for the wide variety of organic polymers which exist. While organic polymers do exhibit many unusual properties, they also have several deficiencies: in general they do not have good high-temperature resistance. At prolonged heating above 150°C, they tend to melt and decompose, because of the reaction of the carbon atoms with atmospheric oxygen. Most organic polymers dissolve or swell when immersed in hot organic fluids, a factor that minimizes their utilization as seals and gaskets in jet engines and other propulsion systems. In addition, relatively few of the organic polymers exhibit sufficient flexibility over a wide enough temperature range to be useful at both elevated and cryogenic temperatures. In military and aerospace applications, this is a serious limitation. However, the carbon atom is not unique in the capability to form polymers; other elements, such as silicon, can also form long-chain molecules. These inorganic polymers frequently exhibit properties which are quite different from those displayed by organic polymers. For example, these inorganic polymers display remarkable high-temperature resistance, as will be seen later in this chapter when silicone polymers are discussed.

Monomers, Dimers, and Trimers

A monomer is a chemical compound that can be converted (polymerized) into a polymer. The individual repeating units are often referred to as *mers*. The polymerization of a polymer frequently occurs in a sequential manner; that is, two monomers react to form a *dimer*, which may then react with a third monomer to form a *trimer*. The process may continue indefinitely until it is stopped by the reaction of another chemical species called a *terminator*, which satisfies the bond requirements at the ends of the molecule. Thus long macromolecules of high molecular weight can result.

The reaction cited below exemplifies the relationship between monomers, dimers, and trimers for glycolic acid.

STRUCTURAL CONSIDERATIONS

Basic Structure

This term describes the basic configuration of the polymer, that is, whether it is composed of straight chains or forms a branched or network structure.

Linear polymers. A linear polymer is one in which the skeletal structure consists of a long chain of atoms to which the side groups are attached. Examples of linear polymers are polyvinyl chloride (PVC), polyethylene, and polymethyl methacrylate (PMMA). The structure of polyvinyl chloride is shown below.

$$
\begin{array}{cccccccccc}
 & H & Cl & H & Cl & H & Cl & H & Cl & & H \\
 & | & | & | & | & | & | & | & | & & | \\
-\!\!&C\!\!-\!\!&C\!\!-\!\!&C\!\!-\!\!&C\!\!-\!\!&C\!\!-\!\!&C\!\!-\!\!&C\!\!-\!\!&C\!\!-& \cdots & C- \\
 & | & | & | & | & | & | & | & | & & | \\
 & H & H & H & H & H & H & H & H & & H
\end{array}
$$

It should be noted that the molecules of a linear polymer are not "stick straight," but rather resemble the form of a rope section which is loosely piled with considerable coiling and bending, as shown in Figure 14-1(a).

Branched polymers. This type of structure can occur with linear polymers as well as with other types. In essence, it consists of side branches of similar structure attached to the main chain, as shown in Figure 14-1(b).

Cross-linked polymers. A cross-linked polymer is one in which chemical bonds exist between chains, as depicted in Figure 14-1(c). This cross-linking can proceed in three dimensions to form a network. The increased cross-linking increases the strength and toughness of the system and reduces the ability of the molecules to slide. Stress, heat, and pressure facilitate the formation of cross-links in many polymers and the resultant material becomes permanently rigid or *thermosetting*. An example of such a material is the phenolic resin, in which phenol is reacted with formaldehyde to form a network phenolic structure as follows:

three-dimensional network

Ladder polymer. A ladder polymer is one in which two linear polymers are linked in a regular sequence as depicted in Figure 14-1(d). Ladder polymers exhibit a more rigid structure than do ordinary linear polymers and frequently show good

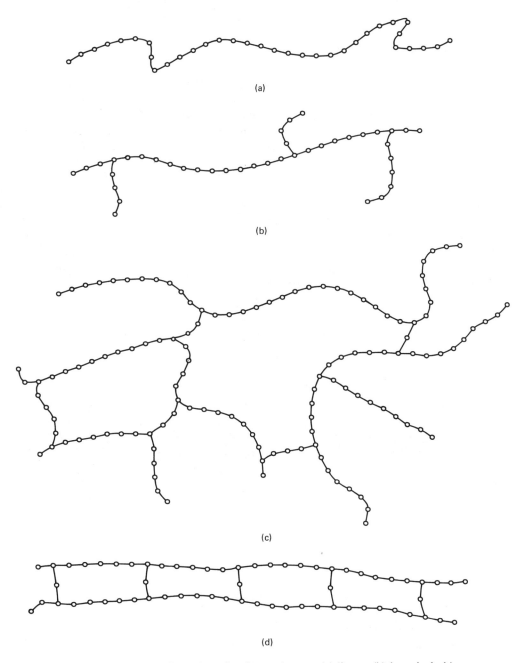

(a)

(b)

(c)

(d)

Figure 14-1 Basic configuration of various polymers: (a) linear; (b) branched; (c) cross-linked; (d) ladder.

Structural Considerations

499

thermal resistance, because two bonds must be fractured at each site in order to fragment the molecule.

Bond Angle and Molecular Length

If one reexamines the carbon tetrahedron as described in Chapter 2 (Figure 2-17), it is found that the bond angle is approximately 109°; however, the exact angle is dependent on the atoms which are forming the bonds with the carbon atom. In a linear chain, this permits the carbon atoms to form a zigzag orientation in space, with the angle between the C—C bonds being approximately 109°, as shown in Figure 14-2(a).

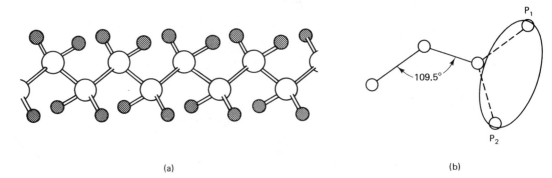

(a)   (b)

Figure 14-2   (a) Construction of a linear polymer chain (white spheres: carbon; black spheres: hydrogen or substitutional atoms); (b) rotation of atoms about C—C bond.

The bond length for a C—C bond is 1.54 Å. However, this is an idealized presentation, since in actuality the carbon atoms can rotate about the single C—C bond and occupy any position such as P_1 or P_2 in the circular locus described in Figure 14-2(b). This rotation of the bond produces kinking and shortening of the molecule much like that occurring if a necklace or jewelry chain is kinked. The most probable length ($\bar{L}$) can be calculated statistically according to the relationship

$$\bar{L} = ln^{1/2} \qquad (14\text{-}1)$$

where n = number of bonds
l = bond length

Example 14-1

(a) Calculate the maximum molecular length of a polyethylene molecule having 420 carbon atoms in the chain if the atoms are arranged in a "zigzag" pattern as shown in the attached figure. (b) What would be the most probable length?

Solution   (a) The number of C—C bonds in the chain is one less than the number of carbon atoms, as shown in the adjacent figure. The maximum length would be the sum of the straight-line distances between the carbon atoms in the axial direction of the chain.

$$x = 1.54 \text{ Å (sin 54.5)}$$
$$= 1.253 \text{ Å}$$
$$L = (420 - 1)(1.253 \text{ Å})$$
$$= 525 \text{ Å} \qquad Ans.$$

(b) The most probable length would be

$$\bar{L} = l n^{1/2}$$
$$= (1.54 \text{ Å})(419)^{1/2} = 31.5 \text{ Å} \qquad Ans.$$

Spatial Configuration

The spatial orientation or steric configuration of a polymer exerts a considerable influence on the degree of molecular packing that occurs. If the structure has a regular steric structure, the chains will be able to achieve a high degree of packing, fitting together and aligning themselves in a parallel mode. Intermolecular attraction will be maximized and the material will be capable of crystallization. Conversely, if the structure is not spatially regular, the degree of intermolecular attraction is lower and the material will tend not to crystallize. There are three primary classifications of a polymer structure, based on its spatial configuration with respect to substitutional groups. These configurations are presented in Figure 14-3, where R represents a radical group, such as a methyl group (CH_3) or an ethyl group (CH_2—CH_3).

Isotactic polymers. Isotactic polymers are characterized by always having the substitute groups on the same side of the main chain. Isotactic polymers tend to crystallize easily, because the regular array of the substitute groups along one side of the molecule permits the chain to form a helical shape and allows adjacent chains to nest together in an ordered manner.

Syndiotactic polymers. With syndiotactic polymers, the substitute groups occupy alternating positions on the chain, as shown in Figure 14-3(b). The regular configuration of syndiotactic polymers also favors close packing and the formation of a structure with a high degree of crystallinity.

Atactic polymers. Atactic polymers, presented schematically in Figure 14-3(c), display a random distribution of the substitute units with respect to the chain. Atactic polymers, because of the random positioning of the side groups, do not tend to crystallize readily.

Trans–Gauche Orientation

A second type of structural isomerism[1] has to do with the position of substitutional groups relative to each other as a result of twisting of the C—C bond. An examination

---

[1] An isomer is a substance having the same molecular formula as another, but having a different structural arrangement.

Structural Considerations

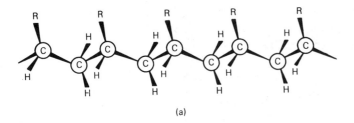

(a)

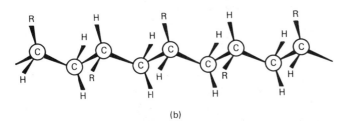

(b)

(c)

Figure 14-3 Schematic representation of (a) isotactic, (b) syndiotactic, and (c) atactic polymers. C, carbon; R, radical group; H, hydrogen.

of the ethane molecule (C_2H_6) using a "hard-sphere" molecular model presented in Figure 14-4(a) shows that the molecule can be twisted about the C—C bond so that the hydrogen atoms, when viewed on end, can assume either eclipsed positions (hydrogen atoms are in alignment) or staggered positions (hydrogen atoms are intermediately placed). Energy measurements have shown that the eclipsed conformation exhibits the highest integral energy, as shown in Figure 14-4(b). If a Cl atom is substituted for one of the hydrogen atoms at each carbon atom to form ethylene chloride ($C_2H_4Cl_2$), the results are generally similar, but the angular positions are not all equivalent and different energies result. There is one eclipsed position (0°), one *trans* position (180°), and two *gauche* positions (60°, 300°), as illustrated in Figure 14-5.

Copolymers

Up to this point we have limited our discussion to homopolymers, which are comprised of only one type of repeating unit:

-A-A-A-A-A-A-A-A-

Copolymers are the result of reacting two or more polymers to achieve a structure in

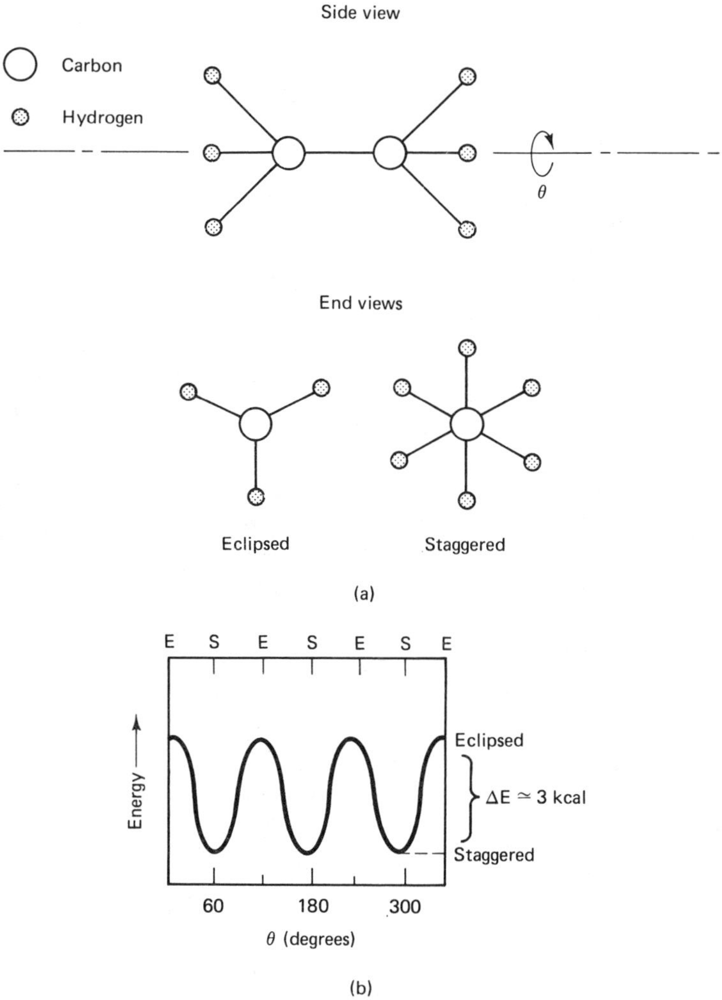

Side view

Carbon

Hydrogen

θ

End views

Eclipsed   Staggered

(a)

E   S   E   S   E   S   E

Eclipsed

$\Delta E \simeq 3$ kcal

Staggered

60   180   300

θ (degrees)

(b)

Figure 14-4 (a) Trans and gauche conformations of ethane; (b) energy versus angular position (θ) for ethane.

which the repeating units are varied in their position. The arrangement of the repeating units within the copolymer structure can vary markedly; however, there are four basic arrangements.

Alternating copolymer. The repeating units A, one type of polymer, and B, a second type of polymer, occupy alternating positions in the structure:

-A-B-A-B-A-B-A-B-A-B-

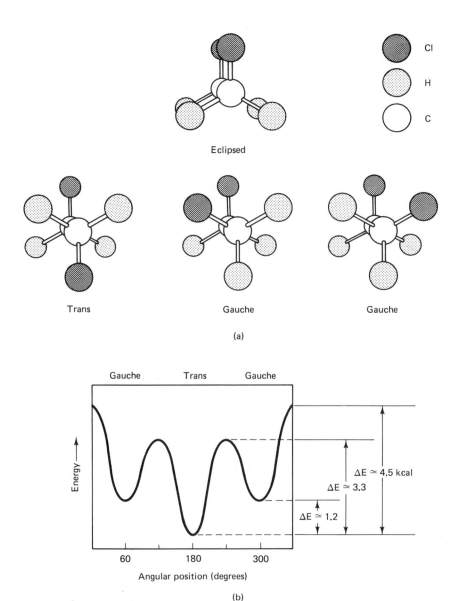

Figure 14-5 (a) Trans and gauche conformations for ethylene chloride; (b) energy versus angular position.

Random copolymers. The structural units A and B are located in random positions in the chain:

-B-A-B-A-B-A-A-A-A-B-B-A-B-B-A-

Graft copolymers. One repeat unit makes up the main chain, while the second comprises short side chains which are attached to the main trunk:

```
                                                B
                                                |
                                                B
                                                |
                                                B
                                                |
  -A-A-A-A-A-A-A-A-A-A-A-A-A-A-A-A-A-A-A-
        |                           |
        B                           B
        |                           |
        B                           B
        |                           |
        B                           B
        |                           |
        B                           B
        |                           |
        B                           B
```

Block copolymers. The repeating units appear in blocks of random length in an alternating fashion:

$$\text{-A-A-A-A-B-B-A-A-A-A-B-B-B-A-A-B-B-B-A-B-B-}$$

These copolymers containing mixtures of two or more repeating units in various combinations frequently have differing physical and mechanical properties than either homopolymer. They have less orderly structures and therefore tend to be less crystalline and are tougher and more flexible than the corresponding homopolymers. The copolymers of the ABS (*a*crylonitrile–*b*utadiene–*s*tyrene) family are primary examples of copolymerization in which the properties of the polymer can be altered by varying the proportions of the monomers. The process of copolymerization thereby permits the manufacture of a group of products possessing a wide range of physical, chemical, and mechanical properties from a limited number of raw materials. For instance, the ABS system is used to produce rigid pipe for plumbing and drainage purposes, while synthetic rubber can be produced by the copolymerization of styrene and butadiene as follows:

styrene   butadiene   styrene-butadiene rubber

Very often, combinations of polymers can result in materials with properties superior to those displayed by the individual polymers. A good case in point is the blend of polymers and synthetic rubber used in automobile bumpers. This material has the formability of "plastic" and the elasticity of rubber. By absorbing light impacts and then returning to its original shape, this material helps reduce minor collision damage.

POLYMERIZATION REACTIONS

There are two fundamental methods of reacting monomers to achieve long chain molecules of high molecular weight. These are *addition* polymerization and *condensation* polymerization.

Addition Polymerization

This type of reaction is characterized by the addition of successive unsaturated monomer units which create a product that contains saturated multiples of the original monomer unit. Saturation, you will recall, was discussed with respect to covalent bonding in Chapter 2. The reaction starts by the creation of an active initiation site. This active site is produced by the addition of an *initiator* or catalyst to the monomer. The initiator may be an active radical (R) with a free electron or an ionized group (e.g., Cl^-, OH^-, CH_3 or benzene rings (C_6H_6). The free electron acts on the double carbon bond by breaking it. The reaction then proceeds by the successive addition of the monomers to the active site in a chain reaction. The driving force for this process is the decrease in potential energy that accompanies the formation of the two single carbon bonds (see Table 2-6). These reactions are notable for their extreme rapidity, and reaction rates of 10^4 units/sec have been reported. Termination of the reaction occurs when the active ends of two rapidly growing chains collide or by their collision with a *terminator* radical. This reaction can be depicted as follows:

Initiation:

$$R_1 + \begin{array}{c} H \;\; H \\ | \;\; | \\ C=C \\ | \;\; | \\ H \;\; H \end{array} \longrightarrow \begin{array}{c} H \;\; H \\ | \;\; | \\ R_1-C-C- \\ | \;\; | \\ H \;\; H \end{array}$$

Propagation:

$$R_1-\begin{array}{c} H \;\; H \\ | \;\; | \\ C-C- \\ | \;\; | \\ H \;\; H \end{array} + \left[\begin{array}{c} H \;\; H \\ | \;\; | \\ C=C \\ | \;\; | \\ H \;\; H \end{array}\right]_n \longrightarrow R_1-\begin{array}{c} H \;\; H \\ | \;\; | \\ C-C \\ | \;\; | \\ H \;\; H \end{array} \cdots\cdots \begin{array}{c} H \;\; H \\ | \;\; | \\ C-C- \\ | \;\; | \\ H \;\; H \end{array}$$

Termination:

$$R_1-\begin{array}{c} H \;\; H \\ | \;\; | \\ C-C- \\ | \;\; | \\ H \;\; H \end{array} \cdots\cdots \begin{array}{c} H \;\; H \\ | \;\; | \\ C-C- \\ | \;\; | \\ H \;\; H \end{array} + R_2 \longrightarrow R_1-\begin{array}{c} H \;\; H \\ | \;\; | \\ C-C- \\ | \;\; | \\ H \;\; H \end{array} \cdots\cdots \begin{array}{c} H \;\; H \\ | \;\; | \\ C-C-R_2 \\ | \;\; | \\ H \;\; H \end{array}$$

Examples of polymers that are created by addition polymerization are polyethylene, polystyrene, and polyvinylchloride as illustrated below:

$$\begin{array}{c} H \;\; H \\ | \;\; | \\ C=C \\ | \;\; | \\ H \;\; H \end{array} \longrightarrow \left[\begin{array}{c} H \;\; H \\ | \;\; | \\ C-C \\ | \;\; | \\ H \;\; H \end{array}\right]_n$$

ethylene   polyethylene

$$\underset{\text{styrene}}{\overset{\displaystyle H \quad H}{\underset{\displaystyle \overset{|}{\underset{|}{C}}}{}}} \longrightarrow \underset{\text{polystyrene}}{}$$

styrene → polystyrene

vinyl chloride → polyvinylchloride

Condensation Reaction

Condensation reactions occur by the reaction of two molecules with the formation and loss of a small molecular by-product such as water or ammonia. The reactions proceed in a stepwise fashion, with each step requiring initiation and reaction with the existing molecule. This condition is considerably different from that occurring with addition reactions and accounts for the much slower reaction rates that occur with condensation reactions.

The reacting molecules are usually of different types, such as the reaction of dicarboxylic acid and a dihydroxyalcohol to form a polyester.

$$\underset{\text{dicarboxylic acid}}{HO-\overset{O}{\overset{||}{C}}-R-\overset{O}{\overset{||}{C}}-OH} + \underset{\text{dihydroxy alcohol}}{H\ O-R-OH}$$

↓ catalyst

$$H\left[O-\overset{O}{\overset{||}{C}}-R-\overset{O}{\overset{||}{C}}-O-R\right]_{n}-OH + nH_2O$$

polyester

Functionality

The concept of functionality deals with the number of active sites at which chemical reactions can occur. If a monomer has only two possible sites, such as polyethylene or polyvinyl chloride, it is considered to be bifunctional; if it has three possible sites, it is trifunctional; and so on. Bifunctional monomers tend to form linear polymers, because the presence of two connecting sites promotes the formation of long chains, whereas trifunctional monomers tend to form branched and network (cross-linked) polymeric structures.

CRYSTALLINITY IN POLYMERS

It was discovered in the early part of this century that polymers exhibited X-ray diffraction patterns (Chapter 3). Unlike metals, which displayed a well-defined pattern, the polymers produced only a few broad diffraction peaks superimposed on a diffuse background. Studies of these crystallites have shown that they are created by the polymer chain being folded back and forth on itself, as shown schematically in Figure 14-6. A view of some actual polyethylene crystals is shown in Figure 14-7. The crystals

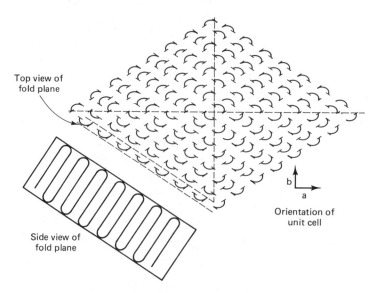

Top view of
fold plane

Side view of
fold plane

Orientation of
unit cell

Figure 14-6 Representation of folding and packing in a polyethylene single crystal.

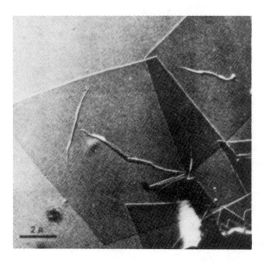

Figure 14-7 Single crystals of linear polyethylene. (From D. H. Reneker and P. H. Geil, *J. Appl. Phys. 31*, 1960, p. 1916.)

in this case exhibit a rhombohedral shape. These crystals are packed in a three-dimensional array and are approximately 100 to 400 Å thick.

Although crystallization does occur in polymers, because of the size and complexity of polymer molecules it does not occur easily. The tendency to crystallize is related to the structure, regularity, and polarity of the molecules. Molecules possessing a high degree of regularity without side groups and branches show a strong tendency to crystallize; those that are irregular or possess a high degree of side branching do not. Figure 14-8 shows the crystal structure which has been established for nylon 66. Note

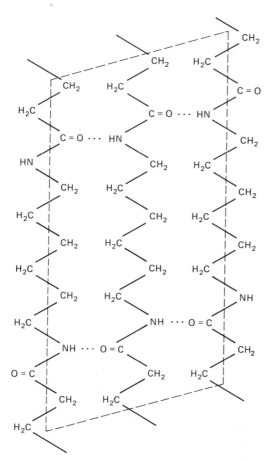

Figure 14-8 Relationships of atoms in adjacent chains of a polyamide (nylon 66) polymer crystal.

the affinity and alignment of the polar groups in the structure. The C=O group produces an electric dipole, a negative charge associated with the oxygen atom. This strongly attracts the hydrogen atoms in adjacent molecules which have exposed protons on the end of a covalent bond. When polar groups are spaced regularly and symmetrically, polymer crystallization is improved. On the other hand, polyethylene, which has long paraffinic molecules, exhibits relatively little attraction and although it can crystallize, the polymer chains can move readily by one another when a stress is applied.

The presence of crystallinity therefore affects the mechanical behavior of polymers. The cross-links formed stiffen and toughen the polymer. Semicrystalline polymers are generally tougher than amorphous polymers. They resist impact better and are less affected by solvent penetration and temperature changes. In addition, crystallinity affects the electrical behavior of polymers. The electrical properties of polymeric materials were discussed in Chapter 10.

Polymer crystallization can be increased by thermal processing, such as annealing at elevated temperatures or slow cooling from the melt temperatures. Similarly, the degree of crystallization can be increased by mechanical working. When the polymer product is initially cast, only a few randomly oriented crystallites are formed. If the fiber is placed under tension during subsequent coiling and stretching operations, the stretching aligns the adjacent chains into a parallel orientation, increasing packing and promoting crystallization. This can be illustrated by examining what happens when rubber is stretched. As the strain is increased, the degree of crystallinity also

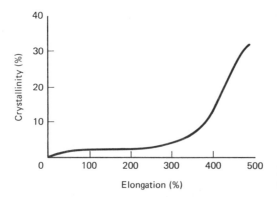

Figure 14-9 Increase in rubber crystallinity as a result of stretching.

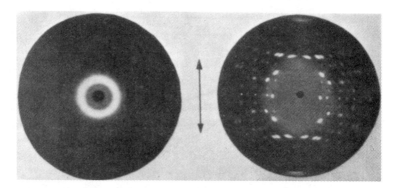

Figure 14-10 X-ray diffraction patterns: (a) unstretched polyisobutylene; (b) polyisobutylene stretched to the maximum. Arrow indicates fiber axis. (From C. S. Fuller et al., *J. Am. Chem. Soc. 62*, 1940, p. 1905.)

increases, as shown in Figure 14-9. This can be verified by X-ray crystallographic techniques. For example, Figure 14-10 illustrates the change in structure, observable by X-ray diffraction, which occurs as the result of straining in polyisobutylene. The X-ray pattern changes from one showing no regularity to one exhibiting a regular array of diffraction spots, indicating the periodicity associated with a crystalline lattice.

MOLECULAR WEIGHT DISTRIBUTION

During polymerization, the polymers being formed are subjected to a sequence of random events, such as contact with an initiator, other molecules, or a terminator. Consequently, all polymer molecules do not grow to the same length, and the final product contains a series of polymers with a range of molecular weights. The distribution of molecular weights can be determined experimentally by a variety of means and defined in various ways. However, we shall confine our discussion to two methods of definition: number-average molecular weight ($\bar{M}_n$) and weight-average molecular weight ($\bar{M}_w$). The number-average molecular weight ($\bar{M}_n$) is a numerical mean in which the total weight of the polymer sample is divided by the number of molecules. This number average equals the average degree of polymerization ($\overline{DP}_n$).

$$\bar{M}_n = \overline{DP}_n = \frac{w}{\sum n_i} = \frac{\sum n_i M_i}{\sum n_i} \qquad (14\text{-}2)$$

where w = total weight of sample
    n_i = number of molecules with molecular weight, M_i
    M_i = mean molecular weight of the i fraction

The weight-average molecular weight ($\bar{M}_w$) is the weight fraction of each polymer group multiplied by the corresponding mean molecular weight of the fraction (M_i), divided by the total weight of the sample. $\bar{M}_w$ can be calculated from the following equation:

$$\bar{M}_w = \frac{\sum w_i M_i}{\sum w_i} = \frac{\sum n_i M_i^2}{\sum n_i M_i} \qquad (14\text{-}3)$$

where w_i = weight fraction of molecules with molecular weight, M_i
    n_i = number of molecules with molecular weight, M_i

These molecular weight distributions are important because it is necessary from the standpoint of manufacturing process control to be able to define the end product resulting from a polymerization process and because the mechanical and physical properties of the product are related very directly to the mix of molecule sizes in the product. The lower-molecular-weight polymers are softer and more flexible, while the higher-molecular-weight polymers are harder and tougher. The method of calculating the molecular weight distribution by the two methods is shown in Example 14-2.

Example 14-2

A high polymer was analyzed using chromatographic methods and found to have the characteristics corresponding to the data shown in columns 1 and 2 of the accompanying table. From these data, calculate (a) the weight-average molecular weight ($\bar{M}_w$) and (b) the number-average molecular weight ($\bar{M}_n$).

Solution (a) The mean value of the molecular weight in each classification group is taken by calculating the midpoint of the classification range. These data are shown in column 3. Then using equation (14-3), we determine the values of $M_i w_i$ by multiplication to give the values shown in column 4. Summing columns 2 and 4, we obtain

$$\bar{M}_w = \frac{\sum w_i M_i}{\sum w_i} = 412.05 \text{ kg/mol} \qquad Ans.$$

(b) The number-average molecular weight is then established by determining the number of molecules in each group (n_i) and multiplying this by the mean molecular weight of each group (M_i). The number of molecules in each group is the number of moles (w_i/M_i) multiplied by Avogadro's number:

$$n_i = \frac{w_i}{M_i} A_0$$

where A_0 is Avogadro's number. Using a sample of 1000 g, the number of molecules in each group can be calculated as shown in column 5. Similarly,

$$n_i M_i = \frac{w_i}{M_i} A_0 (M_i) \qquad n_i M_i = w_i A_0$$

as shown in column 6. The number-average molecular weight can then be calculated as being:

$$\bar{M}_n = \frac{(1000)A_0}{(30.885 \times 10^{-4})A_0} = 323.78 \text{ kg/mol} \qquad Ans.$$

These results demonstrate the mean molecular weights obtained by both methods.

MECHANICAL BEHAVIOR OF POLYMERS

Viscoelastic Behavior

In discussing the mechanical behavior of polymers, the concepts of stress and strain described in Chapter 6 are directly applicable. Polymers can be subjected to mechanical testing to determine tensile strength, percent elongation and impact strength using procedures outlined in Chapter 8. There are, however, some differences that should be discussed. Many materials, such as metals, for example, display mechanical behavior resembling that of an ideal elastic solid; that is, stress is proportional to strain [see equation (6-13)] and independent of time, so that

$$\sigma = Ee$$

At the other extreme from the perfectly elastic solid is the perfectly viscous fluid,

(1) Molecular weight groups (10^{-3} kg/mol)	(2) Weight fraction, w_i	(3) Mean value (M_i) of molecular weight in each group (10^{-3} kg/mol)	(4) $M_i w_i$	(5) Number of molecules in each group[a] ($\times 10^{-4}$)	(6) $n_i M_i$
10,000–99,999	0.01	55×10^3	0.55×10^3	$\dfrac{10\text{ g}}{5.5 \times 10^4} = 1.82 A_0$	$10 A_0$
100,000–199,999	0.08	150×10^3	12×10^3	$\dfrac{80\text{ g}}{15 \times 10^4} = 5.33 A_0$	$80 A_0$
200,000–299,999	0.22	250×10^3	55×10^3	$\dfrac{220\text{ g}}{25 \times 10^4} = 8.80 A_0$	$220 A_0$
300,000–399,999	0.23	350×10^3	80.5×10^3	$\dfrac{230\text{ g}}{35 \times 10^4} = 6.57 A_0$	$230 A_0$
400,000–499,999	0.17	450×10^3	76.5×10^3	$\dfrac{170\text{ g}}{45 \times 10^4} = 3.77 A_0$	$170 A_0$
500,000–599,999	0.13	550×10^3	71.5×10^3	$\dfrac{130\text{ g}}{55 \times 10^4} = 2.36 A_0$	$130 A_0$
600,000–699,999	0.08	650×10^3	52×10^3	$\dfrac{80\text{ g}}{65 \times 10^4} = 1.23 A_0$	$80 A_0$
700,000–799,999	0.05	750×10^3	37.5×10^3	$\dfrac{50\text{ g}}{75 \times 10^4} = 0.661 A_0$	$50 A_0$
800,000–899,999	0.02	850×10^3	17×10^3	$\dfrac{20\text{ g}}{85 \times 10^4} = 0.235 A_0$	$20 A_0$
900,000–1,000,000	0.01	950×10^3	9.5×10^3	$\dfrac{10\text{ g}}{95 \times 10^4} = 0.105 A_0$	$10 A_0$
	1.00 (100%)		412.05×10^3	$30.885 A_0$	$1000 A_0$

[a] Based on 1000 g of sample.

which is characterized by shear stress (τ) being proportional to the rate of strain ($d\gamma/dt$) as follows:

$$\tau = \eta \frac{d\gamma}{dt} \tag{14-4}$$

The proportionality constant (η) in this expression is the viscosity, and this relationship is Newton's law of viscous flow.

Polymers are the largest class of materials whose mechanical properties exhibit characteristics of *both* elastic solids and viscous liquids. These materials are termed *viscoelastic*. Viscoelastic behavior can be demonstrated by a series of experiments. If a polymer rod of unit length is loaded axially and held briefly at a constant stress, a corresponding strain results, as shown in Figure 14-11. If the specimen were unloaded

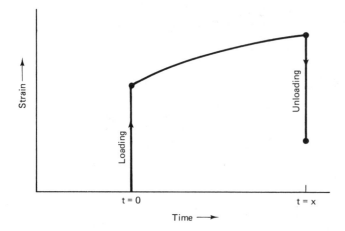

Figure 14-11 Increase in strain with time in a specimen held at constant stress.

at this point, the specimen would behave elastically and there would be no permanent change in length. However, as the stress is maintained on the rod from $t = 0$ to some time $t = x$, a further increase in strain results. When the rod is unloaded, the final increment of strain remains as permanent (plastic) deformation. This phenomenon is sometimes referred to as *creep*. Creep is the continued deformation of a material under constant stress.

A similar experiment can be utilized to show the effect of time on the stress required to produce a given amount of strain. If the rod described earlier were stressed so as to produce a given amount of strain corresponding to Δl, the stress required to hold and maintain this strain would decrease with time as shown in Figure 14-12. This is referred to as *stress relaxation*. For many polymers the relationship between stress and time can be expressed by

$$\sigma_t = \sigma_0 e^{-t/\lambda} \tag{14-5}$$

where σ_0 = initial stress
$\quad\quad \sigma_t$ = stress at time t
$\quad\quad t$ = time
$\quad\quad \lambda$ = relaxation time

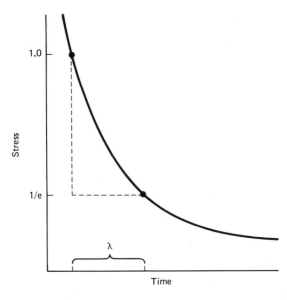

Figure 14-12 Stress relaxation
illustrating the decrease in stress
occurring with time.

where relaxation time is the time necessary to reduce the stress to $1/e$ of its initial value. Stress relaxation, then, is the continued decrease in stress required to maintain a given deformation.

Example 14-3

A stress of 10.5 MPa produces an elongation of 40 cm in a 100-cm-long elastic tie-down cord. After 13 weeks of service, the stress in the cord is only 5.1 MPa. (a) What is the relaxation time for the material? (b) If the cord requires at least 2 MPa to be effective, how long will it be before the cord is not suitable for service?

Solution. (a) Utilizing equation (14-5), we can determine the relaxation time (λ) as follows:

$$\sigma_t = \sigma_0 e^{-t/\lambda}$$

$$\frac{\sigma_t}{\sigma_0} = e^{-t/\lambda}$$

$$\ln \frac{\sigma_t}{\sigma_0} = \frac{-t}{\lambda}$$

$$\ln \frac{5.1}{10.5} = -\frac{91}{\lambda} \text{ days}$$

$$-0.722 = -\frac{91}{\lambda} \text{ days}$$

$$\lambda = \frac{91}{0.722} \text{ days}$$

$$\therefore 126 \text{ days} \quad Ans.$$

$$\ln \frac{\sigma_t}{\sigma_0} = -\frac{t}{\lambda}$$

$$\ln \frac{\sigma_t}{\sigma_0} \lambda = -t$$

$$-t = (-1.66)(126 \text{ days})$$

$$t = 209 \text{ days} \qquad Ans.$$

As one might expect, the ability of a polymer to relax is dependent on molecular movement, which, like atomic diffusion, increases as temperature increases; therefore, stress relaxation times are shortened at higher temperatures in accordance with the classical Arrhenius equation:

$$\frac{1}{\lambda} = e^{-Q/RT}$$

where T = temperature ($^\circ$K)
R = constant = 1.98 (cal/mol-$^\circ$K)
Q = activation energy (cal/mol)

Glass Transition in Polymers

It is important to recognize that although the polymer molecule is large and has a high molecular weight, it must still respond to temperature changes. At high temperatures, segments of the long polymer molecules, even though entwined and entangled, exhibit vibrational, rotational, and translational motion which results in a continuous rearrangement of the molecule and the creation of unoccupied "space" between molecules. This range of motion also provides a mechanism of absorbing energy and thereby imparts toughness to the polymer. As the temperature is reduced, the molecular motion is diminished, the volume of unoccupied space is reduced, and the stress on intermolecular bonds is reduced, allowing them to play a greater role in fixing the position of the molecular segments. The random liquid structure is retained; however, the viscosity increases. For an amorphous polymer at temperatures below the *glass transition temperature* (T_g), rotation and translational motion ceases and only vibration around fixed positions remains. This means that the atoms cannot rearrange themselves to achieve better packing and there is a change in slope of the volume versus temperature curve. This point defines the glass transition temperature (T_g) and is illustrated in Figure 14-13. Although it is not crystalline, the polymer becomes more rigid and brittle. This can be demonstrated by taking a rubber ball, immersing it in liquid nitrogen (-195.8°C), and showing that it shatters on impact.

In semicrystalline polymers, a similar phenomenon occurs in the amorphous zone above the melting point (T_m); however, below this temperature, the crystalline regions would remain intact. The higher the degree of crystallinity, the smaller the effect of the glass transition on the mechanical properties of the material. This behavior is depicted in Figure 14-14. Note the smaller differences in slope between *EF* and *FG* compared to *BC* and *CD*.

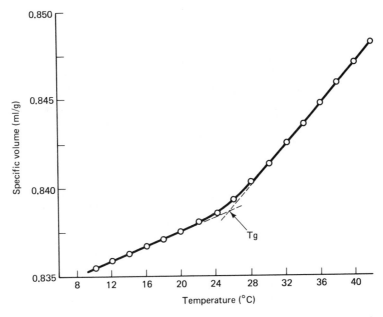

Figure 14-13 Specific volume as a function of temperature illustrating the glass transition temperature (T_g) for polyvinyl acetate.

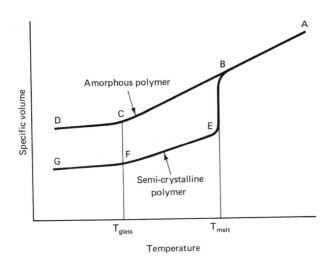

Figure 14-14 Specific volume–temperature curves for an amorphous polymer (AD) and a semicrystalline polymer (AG). A–B, liquid region; B–C, rubbery behavior; C–D, glassy behavior; E–F, crystallites in a rubbery matrix; F–G, crystallites in a glassy matrix.

Service Temperatures Relative to Transition Temperatures

The temperatures at which polymers are to be used relative to their transition temperatures are very important since mechanical behavior is directly affected by temperature. Polymers can be grouped into various classes according to their behavior relative to their transition temperatures.

1. *Elastomers:* These materials must be used well above the glass transition temperature (T_g) in order to retain the molecular mobility necessary to achieve elasticity.

2. *Amorphous polymers:* These polymers, such as polymethyl methacrylate, are utilized for their structural rigidity and consequently must be used at temperatures below T_g.

3. *Crystalline polymers:* These polymers are used at temperatures well below T_m, since changes in crystal structure can occur as T_m is approached. Typically, the glass temperature is not important, since it represents only a minor change in behavior.

These three classes serve only to illustrate the effect of transition temperature, and there are other intermediate classes with varying behavior. Table 14-1 cites transition temperatures for several common polymers.

TABLE 14-1 TRANSITION TEMPERATURE FOR SEVERAL POLYMERS

Polymer	T_g (°C)	T_m (°C)	T_g/T_m (°K)
Polyisoprene	−70	28	0.67
Polyvinylidene fluoride	−39	210	0.48
Polyvinylchloride	82	180	0.78
Polystyrene	100	230	0.75
Polyethylene	−68	135	0.50
Polypropylene	−18	176	0.57
Nylon	47	225	0.64

Elastomers

Elastomers are polymers that exhibit large elastic strains when stressed, often in the vicinity of several hundred percent, as shown in Figure 14-15. The rubber materials

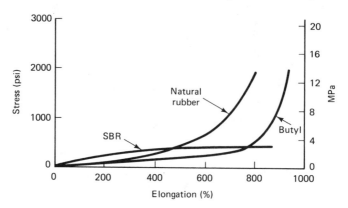

Figure 14-15 Typical stress versus elongation behavior of vulcanized rubbers.

that we encounter in engineering applications are members of this group. They are useful for gaskets, seals, pneumatic bladders, flexible tubing, and a variety of other uses. Any polymer can exhibit this behavior provided that the structure is long and irregular, so that thermal agitation will produce a coiled conformation. In addition, as stated previously, the material must be above its glass transition temperature in order that the thermal energy is sufficient to keep the polymer molecules in a mobile condition. An examination of Figure 14-15 illustrates that the initial application of stress to a specimen produces large strains compared to subsequent stresses. The reason for this is simply that the action of the stress is uncoiling and aligning the elastomer molecules. As the molecules become straighter, and greater alignment is achieved, proportionally greater stress is required to produce the same strain. Table 14-2 illustrates the chemical and mechanical properties obtained with a number of common elastomers.

CHEMICAL STABILITY

Polymers can exhibit a remarkable resistance to many forms of chemical attack; however, they can also be degraded in service. Each class of polymer species has areas in which it is resistant and others in which it is sensitive to attack. The following areas are a partial listing of the type of degradation that can occur.

Solvent Attack

Thermoplastic polymers can be dissolved by various organic solvents. In general, the solubility of the polymer is inversely proportional to its molecular weight. In a polymer that is comprised of a distribution of fractions with various molecular weights, the low-molecular-weight fraction is subject to attack and extraction by the solvent. Cross-linking in a polymer keeps the individual molecules from dissolving and tends to make the polymer insoluble.

Swelling

Swelling can be considered to be a form of solvent attack. It is manifested by dimensional growth of the affected polymer. This is of particular importance for materials used in gaskets and seals.

Oxidation

Oxygen, particularly in the form of ozone (O_3), can permeate the surface of polymeric materials reacting with the structure, producing cross-links and a resultant decrease in flexibility. This is particularly evident in elastomers, where surface cracking can be seen after prolonged exposure to air.

Ultraviolet Radiation

Exposure to ultraviolet light such as sunlight can furnish enough energy to fracture the C—C bonds which comprise the backbone of the polymer molecule. This produces

TABLE 14-2 PROPERTIES AND APPLICATIONS OF SELECTED ELASTOMERS

Type →	Natural isoprene	Polybutadiene	Styrene-butadiene	Chloroprene	Polysulfide	Ethylene/acrylic	Polyurethanes
ASTM Designation	NR	BR	SBR	CR	PTR		AU, EU
PHYSICAL PROPERTIES							
Specific Gravity	0.92–0.93	0.91	0.94	1.23–1.25	1.35	1.08–1.12	1.02–1.25
Ther Cond, Btu/hr/sq ft/F/ft	0.082	—	0.143	0.11	—	—	0.09–0.10
Coef of Ther Exp (cubical), 10^{-5} per °F	37	37.5	37	34	—	—	5–25
Colorability	—	—	Good	Fair	Fair	—	Good–Excellent
MECHANICAL PROPERTIES							
Hardness, Durometer	30A–100A	45A–80A	30A–90D	30A–95A	20A–80A	64A	10A–80D
Ten Str, 1000 psi	3.5–4.5	2.5	2.5–3.0	0.5–3.5	0.5–1.5	1.95	0.8–8.0
Modulus (100%), psi	—	300–1500	300–1500	100–3000	—	800	25–5000
Elongation, %	500–700	450	450–500	100–800	210–450	450	250–800
Compression Set, Method B, %	10–30	10–30	5–30	20–60	29–38	—	10–45
Resilience, %							
Yerzley (ASTM 945)	80	50–90	20–90	50–80	—	—	5–75
Rebound (Bashore)	—	—	10–60	50–80	—	20	20–65
Hysteresis Resistance	Excellent	Good	Fair-Good	Very good	—	—	Fair-Good
Flex Cracking Resistance	—	Excellent	Good	Very good	—	Excellent	Good-Excellent
Tear Resistance	Excellent	Good	Fair	Good	Poor-Fair	Excellent	Outstanding
Abrasion Resistance	Excellent	Excellent	Excellent	Excellent	Poor-Fair	Excellent	Exc-Outstand
Impact Resistance	Excellent	Good	Excellent	Excellent	Poor-Fair	—	Exc-Outstand
ELECTRICAL PROPERTIES							
Vol Res, ohm-cm	—	—	$5.0–8.4 \times 10^{13}$	2.0×10^{13}	5×10^{13}	1.9×10^{12}	$0.3 \times 10^{10}–4.7 \times 10^{13}$
Dielectric Str, V/mil	400–600	400–600	600–800	400–600	—	7.30	330–700
Dielectric Constant							
60 Hz	—	—	—	8.0	7.3	—	4.7–9.53
1 MHz	2.9	3.3	—	6.7	6.8	—	5.9–8.51
THERMAL PROPERTIES							
Service Temperature, F							
Min for Cont Use	−70	−150	−75	−60	−50	−30	−65
Max for Cont Use	250	200	250	225	>250	400	250

TABLE 14-2 (CONT.)

Type →	Natural isoprene	Polybutadiene	Styrene-butadiene	Chloroprene	Polysulfide	Ethylene/ acrylic	Polyurethanes
ASTM Designation	NR	BR	SBR	CR	PTR		AU, EU
ENVIRONMENTAL RESISTANCE							
Ozone	Poor	Poor	Poor	Very good	Excellent	Outstanding	Excellent
Oxidation	Good	Good	Good	Very good	Excellent	Excellent	Excellent
Weathering	Fair	Fair	Fair	Very good	Excellent	Excellent	Good
Water	Excellent	Excellent	Excellent	Good	Good	Excellent	Good-Excellent
Radiation	Fair-Good	Poor	Good	Good	Good	—	Good-Excellent
Alkalies	Fair-Good	Fair-Good	Fair-Good	Excellent	Excellent	Excellent	Poor-Fair
Aliphatic Hydrocarbons	Poor	Poor	Poor	Good	Excellent	Good	Excellent
Aromatic Hydrocarbons	Poor	Poor	Poor	Fair	Fair-Good	Good	Fair-Good
Halogenated Hydrocarbons	Poor	Poor	Poor	Poor	Very good	Good	Poor-Fair
Alcohol	Good	Good	Good	Good	Excellent	Fair	Poor-Good
Animal, Vegetable Oils	Poor-Good	Poor-Good	Poor	Good	Excellent	Excellent	Excellent
Acids							
Dilute	Fair-Good	Fair-Good	Fair-Good	Excellent	Good	Excellent	Fair
Concentrated	Fair-Good	Fair-Good	Fair-Good	Fair-Good	Poor	Poor	Poor
Synthetic Lubricants (diester)	Poor-Fair	Poor-Fair	Poor	Poor	Good	—	Poor-Good
Hydraulic Fluids							
Silicates	Poor-Good	Poor-Good	Poor-Good	Poor-Good	Poor-Good	Good	Fair
Phosphates	Poor-Good	Poor-Good	Poor-Good	Poor	Poor-Fair	Good	Poor
Permeability to Gases	Low	Low	Low	Low-Medium	Very low		Medium
Limiting Oxygen Index	—	—	—	38-45	—	48	15-20
USES	Pneumatic tires, tubes; power transmission belts; gaskets; shock absorption, seals against air, moisture, sound, dirt; sponge stock; heels, soles	Pneumatic tires; heels, soles; gaskets, seals, belting, sponge stocks; used in blends with other rubbers for better resilience, abrasion resistance, low temp properties	Same as natural rubber and polybutadiene	Wire and cable; belts, hose, extruded goods, coatings; molded and sheet goods; adhesives, automotive gaskets, seals; petroleum, chemical tank linings	Seals, gaskets, diaphragms, valve seat disks, flexible seat mountings, hose in contact with solvents, balloons, boats, life vests, rafts	Automotive ignition wire jackets, spark plug boots, coolant and power steering hose, motor mounts, timing belts, transmission seals	Fork lift truck and airplane tail wheels; back-up wheels for turbine blade grinders; spinning cotts for glass fiber, hydraulic vacuum seals; rolls, gaskets, seals, mechanical goods

Source: Materials Engineering, Dec. 1983, 4-1.

smaller molecules and changes the molecular weight distribution. Mechanical properties and viscosity are thereby affected. Ultraviolet irradiation also facilitates oxidative attack and the two factors, when combined, provide a severe test of a polymer's weathering ability.

Moisture and Humidity

Moisture can be entrapped during the polymerization process or be absorbed during storage and service. The presence of excessive moisture can affect dimensional stability and can alter mechanical properties, as shown in Figure 14-16.

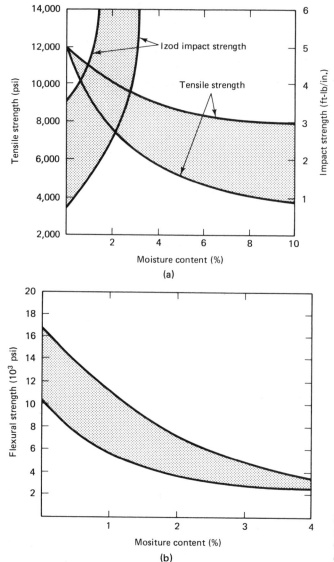

Figure 14-16 Effect of moisture on mechanical properties of nylon 66: (a) tensile and impact strength; (b) flexural strength.

FORMS OF POLYMERS

An examination of the world around us reveals that polymers can be and are manufactured in a large variety of forms in addition to solid shapes, a fact that greatly enhances their utility. In this section we show a number of the forms available and the characteristics of each group. Table 14-3 cites the properties obtainable and typical applications for several of the more common thermoplastics, and Table 14-4 shows similar data for a number of thermosetting resins.

Fibers

Polymer fibers are the basis of the entire textile industry. Natural polymers such as silk, cotton, and wool have recently been augmented by the synthetic fibers formed from the polyesters, nylon, polyethylene, and polypropylene, among others. The fibers are generally produced by extruding the molten polymer through a spinneret, which is a die containing fine holes, so that threads are produced as the material solidifies. After extrusion, the fibers are drawn under tension at temperatures between T_g and T_m to produce a favorable orientation to the crystallites. The advantage of synthetic fibers is that the properties are predictable and can be controlled within relatively narrow limits. One disadvantage of synthetic fibers relative to natural fibers is that the latter frequently have a more complex shape, which can impart improvements in some attributes, such as thermal insulation (e.g., wool) and absorbency (e.g., cotton). The morphology of some natural and synthetic polymeric fibers is shown in Figure 14-17. However, as fiber technology progresses, some of the disadvantages of synthetic fibers are gradually being eliminated. For instance, *hollow* synthetic fibers are commercially available which exhibit very good thermal insulating properties combined with low weight, a desirable feature for certain cold weather garments.

Films

Films are produced by the extrusion of a polymer through a wide slit in a die. The resulting sheet is then further reduced in thickness by rolling. The end product is a thin film which can be used for packaging, sealing, backing for photographic film and audio tapes, vapor barriers, general construction, and a wide variety of other uses. Such films have completely changed the packaging industry. Heat-shrinkable plastic films are an example of such an application. Heat-shrinkable films are manufactured by placing the film in tension and stretching it longitudinally and transversely while it is being produced. The product to be packaged is wrapped in the film, which is then heated with hot air. The application of heat causes the retraction of the molecular structure and the films shrink by as much as 25%, encasing the product.

Adhesives

Polymeric materials form an important class of adhesives, which are materials that are used to bond other materials together. An adhesive must have several characteristics in order to function properly. It must wet the surfaces to be joined, and it

TABLE 14-3 MECHANICAL, PHYSICAL, AND CHEMICAL PROPERTIES OF THERMOPLASTIC POLYMERS

Name and structure	Tensile strength (psi)[a]	Elongation (%)	Rockwell hardness	Izod impact (ft-lb/in.)	Tensile modulus (10^3 psi)	Specific gravity	Coefficient of expansion [$°F^{-1} \times 10^{-6}$ ($°C^{-1} \times 10^{-6}$)]	Maximum service temperature (°C)	Chemical resistance	Typical applications
Polyethylene High density Low density	4,400 1,400–2,500	15–100 500–725	R40 R10	1–12 4–14	120 25	0.96 0.92	70 (120) 100 (180)	Softens 80 125	Excellent resistance to acids and bases at room temperature, except oxidizing acids	Films, wire insulation, ski bottoms, bottles, surgical implants
Polypropylene	5,000	30–200	R90	0.4–2.2	200	0.91	50 (90)	57° at 264 psi	Resistant to most acids and alkalis even at higher temperatures	Rope, housings, fibers
Polystyrene	5,000–10,000	1–2	M72	0.3	460	1.05	38 (68.5)	105° at 264 psi	Not resistant to aromatic and chlorinated hydrocarbons	Containers, thermal insulated bead board
Polyvinylchloride (rigid)	5,000–10,000	2–30	R110	1	400	1.40	30 (54)	77° at 264 psi	Resistant to wear, acids, and alkalis; affected by ketones, esters, and aromatic hydrocarbons	Floors, coated fabrics, pipe, phono records
Polytetrafluoroethylene (PTFE)	2,500–6,500	250–350	D52	4	65	2.2	55 (99)		High resistance to most corrosive chemicals and organic solvents	Chemical ware, seals, bearings, gaskets, pipes, linings
ABS, acrylonitrile–butadiene–styrene copolymer	4,000–7,000	5–20	R108	2–4	350	1.06	50 (90)	85° at 264 psi	Resistant to dilute acids and alcohols; damaged by concentrated sulfuric; soluble in esters, ketones	Pipe, tubing, appliance housings

TABLE 14-3 (CONT)

Polyamides (nylon 66) $-N-C-$ H O	Excellent resistance to most organic solvents; do not use with strong acids	11,800	5–25	R113	1–2	475	1.14	55 (90)	105° at 264 psi	Textiles, rope, gears, bearings, machine parts
Acrylics (type II) $-C-C-C-O-CH_3$ CH₃ O	Resists weak alkalis, acids; attacked by esters ketones, chlorinated hydrocarbons	8,000	4	R100	0.4	450	1.19	40 (72)	60°	Aircraft enclosures, signs, lighting fixtures, windows, surgical implant adhesive
Acetals (standard)	Excellent resistance to most organic solvents; do not use in strong acids	10,000	25 / 12	R94	1.4	5.2	1.42	45 (81)	124° at 264 psi	Appliance parts, hardware, gears, bushings, aerosol bottles
Cellulose acetate	Dissolved by acetone and ethyl acetate	7,000–8,000	10	R115	3	500–4,000	1.29	75 (135)	63° at 264 psi	Blister packaging, films, tool handles
Polycarbonates $-O-C-O-R-$ O	Resists wear, acids; attacked by fuels and organic solvents	8.5–10,000	10–116	M85	16	350	1.25	25 (45)	127° at 264 psi	Light globes, machine parts, propellers, lenses, sporting goods
Polyesters $-C-O-$ O	Resists oils, alcohols, and esters; attacked by strong acids and bases	8,000	300	R117	>1	340	1.35	33 (60)	55° at 264 psi	Gears, bearings, valves, pump parts

[a]Multiply psi by 6.9×10^{-3} to obtain MPa.

TABLE 14-4 MECHANICAL, PHYSICAL, AND CHEMICAL PROPERTIES OF THERMOSETTING POLYMERS

Name and structure	Chemical resistance	Tensile strength (psi)	Elongation (%)	Rockwell hardness	Impact izod (ft-lb/in.)	Tensile modulus (10^3 psi)	Specific gravity	Coefficient of expansion [$°F^{-1} \times 10^{-6}$] ($°C^{-1} \times 10^{-6}$)]	Max service temperature (°C)	Typical applications
Phenolics (phenol-formaldehyde) [structure]	Severely attacked by strong acids and alkalis	7,500	nil	115R	0.5	1000	1.46	45 (81)	150–180	Electrical devices, heat resistant handles (pots, irons, etc.)
Urea-melamine [structure]	Resistant to weak acids, alkalis, solvents, greases, oils	7,000	nil	110R	0.3	1000	1.50	20 (36)	100	Dishware, wear-resistant laminates
Polyesters [structure]	Attacked by strong acids, alkalis, ketones, solvents	4,000–10,000	2	100R	0.4	150	1.12	45 (75.5)	120–180	Electrical components, fiberglass composites, coatings
Epoxies [structure]	Highly resistant to water alkalis, less resistant to acids and oxidizers	10,000	nil	106M	0.5	450	1.15	33 (72)	230–260	Adhesives, fiberglass composites, coatings

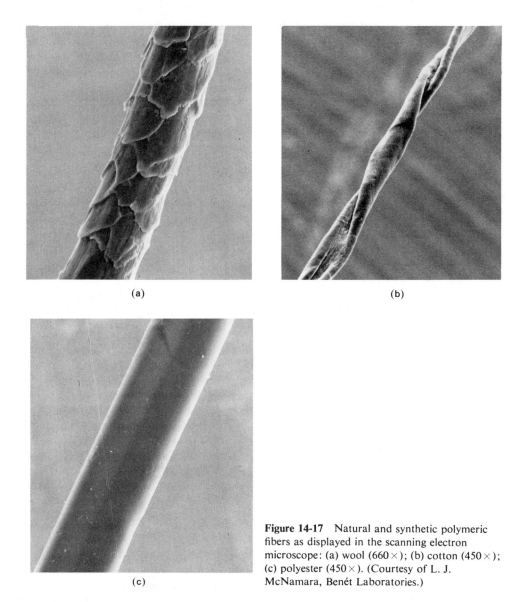

(a)

(b)

(c)

Figure 14-17 Natural and synthetic polymeric fibers as displayed in the scanning electron microscope: (a) wool ($660\times$); (b) cotton ($450\times$); (c) polyester ($450\times$). (Courtesy of L. J. McNamara, Benét Laboratories.)

should have high shear strength in its final state. Polymeric adhesives can generally be grouped into one of the following categories:

Solvent-based systems. These materials consist of a thermoplastic or elastomeric resin which has been cut (diluted) with a solvent. After application, the solvent evaporates, allowing the resin to cure. An example of such a material is polystyrene model cement.

Thermoplastics. These are adhesive materials which are heated to a temperature at which they flow quite readily, then injected or sprayed onto the joint, then allowed to cool to room temperature. This process is favored by modern manufacturing for joining operations because it does not require extensive cure time.

A variation of this type adhesive employs polymethylmethacrylate (PMMA), described as an acrylic in Table 14-3, for joining prosthetic devices to bone in surgical implants. This adhesive consists of solid PMMA and a PMMA-polystyrene copolymer. Liquid monomer containing a catalyst is provided for use by the surgeon. The liquid, when combined with the powder, causes polymerization of the monomer, bonding spherical particles of the copolymer together in a PMMA matrix in about 10 to 12 min.

Thermosetting adhesives. These are materials, such as epoxy resins, which are in the partially polymerized state and are mixed with an activator and then applied. The polymerization is completed *in situ*, forming the bond. High-performance, high-strength adhesives fall into this category; however, one of the problems is that reactions often produce a condensation product, the removal of which must be considered in the application.

Coatings

A significant percentage of polymer use is in the form of coatings. These coatings are applied to other materials to supplement and enhance their appearance, to provide color, to increase the corrosion resistance and weatherability, or to modify the surface behavior (e.g., to electrically insulate components). The application of a polymer coating can beneficially alter the suitability of a material for a particular application. For example, magnesium alloys have many desirable characteristics, such as light weight and ease of fabrication; however, these alloys generally have poor corrosion resistance in many applications. The use of polymeric coatings such as epoxies and polyurethanes greatly enhances the corrosion resistance of these alloys and permits their use in applications where the corrosion rate would normally be too severe for unprotected magnesium.

Similarly, coatings can be used to improve wear resistance. Tetrafluoroethylene-impregnated anodized coatings are used to improve the wear resistance of titanium, and molybdenum disulfide-filled epoxy coatings can be used to impart wear resistance to a number of materials, including steels, bronzes, and aluminum alloys. Of course, the student may be more familiar with coatings of polytetrafluoroethylene (Teflon), which are frequently applied to cooking utensils to prevent foods from sticking and to facilitate cleanup. The use of these coatings expands the number of applications where various materials can be used by minimizing some of the problem characteristics that the base material might have.

Two general categories of plastic coating exist: those in which the polymer is dissolved in a solvent or suspended in an emulsion, and those which are formed by applying a molten layer of the polymer.

The polymer coatings of the solvent type contain a volatile solvent which evaporates, permitting the resin molecules to come together and polymerize. In addition, the coating frequently contains pigments which provide color and may also modify the hardness and oxidation resistance.

Foams

Plastic foams might properly be discussed as a composite material, since they consist basically of two phases, gas bubbles dispersed in a polymer matrix. The gas can be injected during solidification or be produced by the reaction of a specific foaming agent. The entrapment of large volumes of gas in the final product produces changes in structure which markedly affect certain properties. Strength, ductility, heat and sound transmission, and density are among the properties that are affected. These new properties, which result from foaming, permit the polymeric materials to be used in applications for which they would be otherwise unsuited. Some of these applications are padding in bedding and upholstery, flotation in boats and life jackets, thermal insulation, and as packaging materials. Table 14-5 cites the mechanical properties for a number of foamed polymers.

The properties that result for a particular foam are dependent on certain geometrical factors. The gas in a polymer foam is distributed in voids which are referred to as cells. The polymer phase encloses these cells and comprises the cell walls. Total characterization of the foam would require complete knowledge of the size, shape, and distribution of the cells as well as the composition and thickness of the cell walls. In addition, one should be aware of the degree to which the structure is open-celled or closed-celled. A foam whose cells are noninterconnecting is called *closed cell*, whereas one where the cells are interconnected is referred to as *open-celled*. Examples of these structures are shown in Figure 14-18. Usually, the fraction of open and closed cells in a structure is expressed as a percentage of the total volume. This factor is

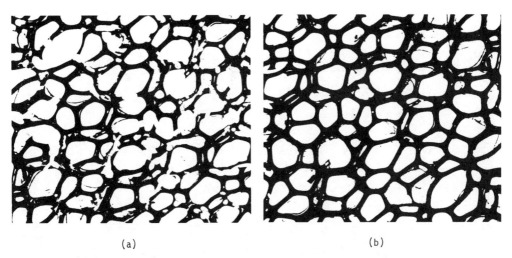

(a)   (b)

Figure 14-18 Cross section showing the appearance of (a) an open-celled foam and (b) a closed-cell foam.

TABLE 14-5 PROPERTIES OF COMMERCIALLY AVAILABLE FOAMS

	Polystyrene Extruded			Polystyrene Molded			Polyurethane Polyether board	Polyurethane FIP[a]	Polyurethane Polyester FIP[a]	Epoxy	Phenol-formaldehyde			Polyethylene			Urea-formaldehyde	Silicone		Cellulose acetate
Density, lb/ft³:	1.9	2.9	4.4	1.0	2.0	4.0	2.3	2.5	2.1	2.3	2.0	4.0	8.0	2.0	29[b]	30[c]	1.8	3.5	14	6–7
Mechanical properties at 75°F																				
Compressive strength (psi)	35	65	130	20	35	70	50[d]	32	37	25	25	55	140				8	6.2		125
Tensile strength (psi)	70	105	178	20	45	85	60	30	47	40	15	30	70	25	670	1800	17	200		170
Flexural strength (psi)	70	80	160	20	60	120	30[e]	55	60		45	90	205							147
Shear strength (psi)	40	58	88									25	45							140
Compressive modulus (psi × 10³)	1.0	3.0	5.05	0.25	0.75	1.75	1.04			0.57							0.7			
Flexural modulus (psi × 10³)	2.5	2.0	2.95	2.0	2.4	6.6	1.0													
Shear modulus (psi × 10³)	0.9	1.8	2.95				0.5[e]													
Thermal properties																				
Thermal conductivity (initial) (Btu-in. °F^{-1}ft^{-2}hr^{-1})	0.26			0.16	0.16	0.16	0.12	0.110	0.110	0.11	0.20	0.20	0.27					0.281	0.3	
Thermal conductivity (equil.) (Btu-in. °F^{-1}ft^{-2}hr^{-1})	0.26			0.260	0.240	0.243	0.165	0.150	0.157	0.15				0.035			0.23			0.31
Coefficient of thermal expansion (in. in.$^{-1}$ °F^{-1} × 10^{-4})	3.5			3.3 to 3.5			2.7													2.5
Flammability[f]				Burns—can be made FR			FR				1-3 FR	1-3 FR	1-3 FR	Burns			FR	FR		Burns
Electrical properties																				
Heat distortion temperature (°F)	170	170	170	175	175	175	250			300	250			160			120	650	700	350
Dielectric constant at 10⁹ cps	<1.05	1.07	1.07	<1.017	1.03	1.06	1.04							1.05	1.50	1.55		1.09	1.25	1.12
Dissipation factor at 10⁹ Hz (× 10⁻⁴)	<4.0	<4.0	<4.0	<1.0	7.0		13							2.0	3.3	40.0		10.2		20
Chemical properties																				
Water absorption (10-ft head) (lb/ft²)	0.08	0.08	0.08	nil	nil	nil	<0.04	0.06	0.04	0.03				0.4				0.284		
Water absorption (vol. %)				<1.0	<1.0	<1.0	<2.0				100			4.0				2.3		4.5
Moisture-vapor transmission (perm-inch)	1.5	1.5	1.5	2.0	2.0	2.0	<2.5	1.7	1.0	1	{0.4[g] / 21[h]}							41.2		
Specific heat (Btu/lb)	0.29											0.38	0.38				0.40			

[a] FIP, foamed-in-place.
[b] Prepared from low-density polyethylene.
[c] Prepared from high-density polyethylene.
[d] Load parallel to thickness dimension.
[e] Load perpendicular to thickness dimension.
[f] FR, flame retardant.
[g] With skin.
[h] Without skin.

Source: E. Baer, Ed., *Engineering Design for Plastics*, Van Nostrand Reinhold Company, Inc., New York, 1964.

important in several applications, such as thermal insulation and flotation, because it affects the permeability of the foam with respect to the surrounding gases and fluids. Obviously, a foam with a high degree of open cells will take in water and would not be suitable for a flotation device, nor would it be efficient for thermal insulation under these conditions.

ADDITIVES

Many objects which are commonly considered to be entirely polymeric contain other substances in addition to the basic polymer. These additives are of various types and generally fall into one of the following categories.

Fillers

Fillers are fine particles or fibers which are added to polymers to modify the mechanical properties of the material. Carbon black is an important example of such a material. It is added to rubber used for tire stock and greatly improves the strength and wear resistance. Other fillers, such as zinc oxide and titanium oxide, are used as pigments to impart color to the basic polymer. In some cases, such as wood flour (fine sawdust) or clay, the function of the filler is to provide bulk and reduce cost by reducing the amount of resin in the product.

Fillers can also impart special characteristics to a polymer such as improved heat resistance or increased friction. The addition of asbestos and metallic powders to phenolic resins in the manufacture of brake linings is an example of such usage.

Plasticizers

These are materials which are added to thermoplastics to increase their flexibility and reduce brittleness. They are usually materials which would be classified as high-molecular-weight liquids, such as dioctyl phthalate. These materials function by increasing the distance between adjacent chains and reducing the effects of intermolecular forces and the tendency toward crystallization, thereby facilitating the relative mobility of the polymer molecule. As stated, these materials can markedly alter both the mechanical and physical properties of the basic polymer. Figure 14-19 illustrates the effect of plasticizer concentration on the properties of cellulose acetate.

Special-Purpose Additives

There are a variety of materials that can be added to a polymer to achieve special purposes, such as oxidation resistance, resistance to thermal degradation, and ultraviolet resistance. Phenols, for example, are used to retard oxidation of polymers. Carbon black, which was described previously as a filler material in tires, is also used to block the ultraviolet penetration and degradation of polyethylene. Dyes can be added to change the color of the product. Flame retardants can be added to various polymers

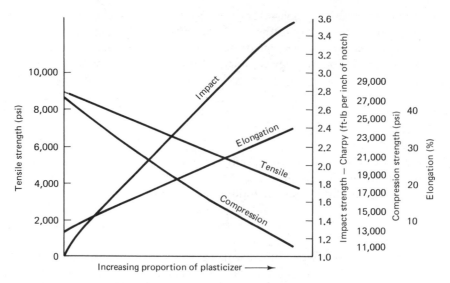

Figure 14-19 Effect of plasticizer content on mechanical properties of cellulose acetate.

to minimize their flammability and to make them self-extinguishing. Graphite can even be added to certain polymers, thus making them electrically conductive.

SILICONE POLYMERS

Theoretically, any inorganic material that contains long-chain molecules should behave as a polymer, and indeed there are a number of inorganic polymers which do exist. As you will see in Chapter 15, glass may be considered a polymeric material, since the structure is comprised of ring and chains based on repeating silicate units. Similarly, asbestos is characterized by a double-silicate, ladder structure, as indicated in Figure 14-20. However, among the most important inorganic polymer groups are

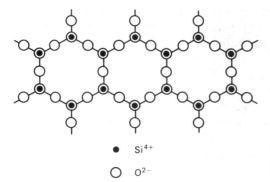

● Si^{4+}

○ O^{2-}

Figure 14-20 Schematic of the double silicate chain (ladder) structure of asbestos.

the polyorganosiloxanes or *silicones*. These materials are produced by the polymerization of silanes.

The synthesis of silicone polymers is a multistep chemical process in which the first step is the reaction of elemental Si with methyl chloride at high temperatures to produce dimethyl chlorosilane,

$$Si + 2CH_3Cl \xrightarrow[Cu]{heat} \begin{array}{c} CH_3 \quad Cl \\ \diagdown \diagup \\ Si \\ \diagup \diagdown \\ CH_3 \quad Cl \end{array}$$

which can hydrolyze to form octamethyl cyclotetrasiloxane plus HCl. This silane can be polymerized by heating above 100°C in a slightly acid solution to form the polymer according to the following reaction:

$$\begin{array}{cc} CH_3 & CH_3 \\ | & | \\ CH_3-Si-O-Si-CH_3 \\ | & | \\ O & O \\ | & | \\ CH_3-Si-O-Si-CH_3 \\ | & | \\ CH_3 & CH_3 \end{array} \xrightarrow[heat]{acid} \begin{array}{ccc} CH_3 & CH_3 & \left[\begin{array}{c} CH_3 \\ | \\ \end{array}\right. & CH_3 \\ | & | & | & | \\ -O-Si-O-Si- & \left|O-Si-\right| & O-Si-O- \\ | & | & | & | \\ CH_3 & CH_3 & \left.CH_3\right]_n & CH_3 \end{array}$$

Silicones may also be polymerized (cured) at room temperature by the use of a catalyst which promotes cross-linking. The catalytic reaction proceeds as follows:

$$CH_3SiX_3 + \begin{array}{c} CH_3 \\ | \\ HO-Si-O- \cdots \\ | \\ CH_3 \end{array} \longrightarrow \begin{array}{cc} X & CH_3 \\ | & | \\ R-Si-O-Si-O- \cdots + HX \\ | & | \\ X & CH_3 \end{array}$$

$$\text{catalyst} \qquad \text{silane}$$

In the reaction above, X represents a moisture hydrolyzable species such as an acetoxy or alkoxy group.

The silicon-oxygen repeating backbone of the dimethyl silicone polymer provides the inherent chemical stability, environmental stability, and extremely good high-temperature performance characteristics of these inorganic structures.

The silicone bond linkage is similar to bond linkages found in other high-temperature-resistant materials, such as quartz, glass, and sand, hence the outstanding high-temperature properties of silicone. The organic methyl side groups provide the flexibility characteristics of organic materials, but because the methyl group is the smallest organic unit, it also contributes to the thermal stability of the structure as a whole.

Silicone polymers have relatively long bond lengths within the polymer and low intermolecular forces between adjacent polymer chains. These characteristics result in relatively low-viscosity materials for a given molecular weight in comparison to organic polymers and relatively high vapor transmission properties in comparison to organic polymers. This particular factor, high vapor transmission, is vitally important in certain applications, such as gas-permeable membranes. For example, contact lenses made with or containing silicone allow sufficient quantities of oxygen to reach the cornea, which needs on the order of 2 to 4% oxygen to prevent tissue inflammation.

The properties obtainable for one type of silicone, a room-temperature vulcanizing (RTV) rubber, are listed in Table 14-6. Note the range of temperatures in which this material is useful. Indeed, RTV silicone rubber is widely used by many industries (e.g., automotive) to produce gaskets and seals in situ. In such applications, a bead of uncured rubber is simply applied to the surfaces to be sealed. The gasket is formed by the subsequent assembly of the parts, and curing takes places under ambient conditions by the condensation mechanism.

TABLE 14-6 PROPERTIES OF A TYPICAL SILICONE PRODUCT: A ROOM-TEMPERATURE VULCANIZING (RTV) RUBBER

Description: Easily pourable product that cures to a tough transparent rubber. Product is used for electronic potting, where see-through feature is desired for easy component identification and repair. Also useful where clear rubber is required, such as solar cell potting and optical instrument applications.

Typical physical properties		*Typical thermal properties*	
Viscosity (cP)	4000	Thermal conductivity	0.00045 (0.11)
Specific gravity	1.02	cal/sec/cm/cm^2/°C	
Hardness, Shore A durometer	45	(Btu/hr/ft/ft^2/°F)	
Tensile strength kg/cm^2, (psi)	63 (900)	Coefficient of thermal expansion	27×10^{-5} (1.5×10^{-4})
Elongation (%)	150	cm/cm/°C	
Tear resistance kg/cm, (lb/in.)	4 (25)	(in./in./°F)	
Useful temperature °C, (°F)	−60 to 204 (−75 to 400)	*Typical electrical properties*	
Linear shrinkage (%)	< 0.2	Dielectric strength kV/mm, (V/mil)	19.7 (500)
		Dielectric constant at 100 Hz	3.0
		Dissipation factor at 100 Hz	0.001
		Volume resistivity (Ω-cm)	1×10^{15}

The silicone rubbers are also used in plastic and reconstructive surgery. For example, a silicone rubber ear implant is shown in Figure 14-21. This device is used as an underlying framework in reconstructive plastic surgery. Although the silicone rubbers have relatively poor mechanical properties and tear easily without fillers such as silica to strengthen them, their consistency is such that it resembles cartilaginous structures and they are well tolerated by tissue while not being degraded by the body environment.

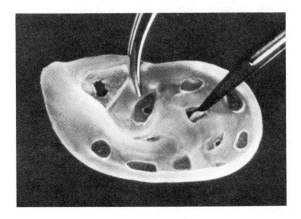

Figure 14-21 Silicone rubber ear implant. Perforations permit tissue in-growth and adhesion.

STUDY PROBLEMS

14.1. Polypropylene exists in three isomeric forms: isotactic, syndiotactic, and atatic.
(a) Sketch schematically the resultant structures.
(b) Discuss the effect of the structures on crystallinity and properties.

14.2. ABS is a copolymer produced by grafting of acrylonitrile and styrene onto a butadiene main chain. Construct a schematic of the ABS structure showing the elemental atoms.

14.3. Polyvinylchloride occurs via an addition reaction of a monomer (C_2H_3R) into a polymer. How much energy is given off when 100 g of C_2H_3Cl reacts?
Answer: 32.7×10^3 cal

14.4. Urea-formaldehyde is produced by a condensation reaction between urea (NH_2-$CONH_2$) and formaldehyde (CH_2CO). Sketch the expected reaction.

14.5. A copolymer is comprised by weight of 13% polyvinyl acetate and 87% polyvinyl-chloride. What is the ratio of polyvinylchloride mers to polyvinyl acetate mers in the structure?
Answer: 9 : 1

14.6. A low-density polyethylene has a density of 0.916 g/cm³, while a totally crystalline polyethylene has a density of 1.030 g/cm³. If polyethylene in the amorphous state has a density of 0.90 g/cm³, what is the percent crystallinity of the low-density polyethylene?

14.7. High-density polyethylene is used as a contact material for hip-socket implants and for the running surface of skis.
(a) Write the chemical formula for polyethylene showing the monomer group.
(b) Would you expect the structure to exhibit greater or lesser crystallinity than that of conventional low-density polyethylene?

14.8. Commercial polyethylene is produced in two forms, a low-density variety and a high-density variety.
(a) Which form will exhibit greater crystallization?
(b) Which is more common?
(c) What are some of the applications of each type?

14.9. In examining the molecular distribution, the weight-average molecular size ($\bar{M}_w$) is always larger than the number-average molecular size ($\bar{M}_n$). Explain.

14.10. Polyvinylchloride has the unit structure or mer (C_2H_3Cl). The mean mass per molecule is 47,000 amu.
  (a) What is the mass of each mer?
  (b) What is the degree of polymerization?
  Answer: (a) 62.5 amu (b) 752 mer/molecule

14.11. A polypropylene polymer molecule has a degree of polymerization of 550.
  (a) What is the longest molecular length possible while maintaining the 109° carbon bond angle?
  (b) The most probable?

14.12. A specimen of polymer is composed of two polyvinylchloride polymers which are blended together. The samples have the following characteristics:

Sample	Weight (g)	DP_n	DP_w
X	20	300	350
Y	40	600	700

Calculate the number-average molecular weight ($\bar{M}_n$) and the weight-average molecular weight ($\bar{M}_w$) of the blend.

14.13. Can oxygen react with the double bonds in butadiene? If so, what would you expect the resultant structure to look like? What do you believe the effect would be on the mechanical properties of the polymer?

14.14. Isoprene (C_5H_8) is vulcanized with sulfur. Assuming that all the sulfur reacts, what weight percent of sulfur is required to completely react all the unsaturated bonds? Answer: 32%

14.15. What fraction of the isoprene is linked if the resultant polymer contains 9.5 w/o sulfur?

14.16. A rubber that contains 62 w/o butadiene, 30% isoprene, is vulcanized with 8% sulfur. What percentage of the available cross-links have been utilized?

14.17. Ski racers depend heavily on the use of waxes which are applied hot to minimize ski/snow friction and to facilitate turning. The adage in skiing is: "The wax you don't see is what does the job." Explain.

14.18. A rubber shock cord is fastened in position with a stress of 17 MPa. Two months later the cord exerts a stress of only 9 MPa. What is the relaxation time? Answer: 94 days

14.19. An elastomeric strapping band used for shipping has cross-sectional dimensions of 0.5 in. wide $\times$ 0.032 in. thick and is loaded with a force of 40 lb. The relaxation time for the elastomer is 160 days. What is the stress on the band after 1 month? After 1 year?

14.20. Was there a good choice of material in problem 14.19 if the requirements are that it retain 90% of its retentive ability after 1 year? What should the relaxation time of the proper material be?

14.21. Calculate the activation energy at 20°C for the polymer used in problem 14.20. Answer: 4743 cal/mole

14.22. What wavelength of light is necessary to break the C—C bond? Refer to the section "Photo conductors" in Chapter 10.

14.23. A plastic foam having a density of 0.0302 g/cm³ is produced from a polymer having a specific gravity of 1.035. What is the percent expansion during foaming?

14.24. Styrofoam produced from a polymer having a specific gravity of 1.10 has a density of 0.063 g/cm³ and 7.3% interconnected voids (open cells).

(a) What is the final density if the foam is saturated with water?

(b) What is the suitability of this foam as a thermal insulation material?

14.25. A polymeric foam has an initial density of 0.072 g/cm³. The foam undergoes a 1000% weight increase when immersed in water.

(a) Calculate the percentage of interconnected voids (open cells).

(b) What would be the suitability of the foam for a floatation device?

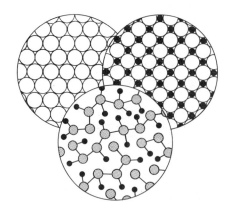

CHAPTER FIFTEEN:
Ceramic Engineering Materials and Their Applications

We have previously introduced two of the basic classes of engineering materials: metals and polymers. In this chapter we deal with the third basic class, ceramics. Unfortunately, many people, including students, have a preconceived belief that all ceramics are hard, brittle materials which are used exclusively for making pottery, sinks, water closets, and dishware. Admittedly, certain ceramics are traditionally utilized in such products. Indeed, with the exception of wood, ceramics are perhaps the oldest known engineering material, with crude pottery being developed about 10,000 B.C. and building bricks existing as early as 4000 B.C. Furthermore, it is also believed that glass was first produced in ancient Egypt about 1600 B.C.

These types of ceramics are referred to as the *traditional* ceramics and consist principally of silicate glasses, cements, and clays. The traditional category of ceramics still constitutes the bulk of the ceramics industry today, with examples such as glass products, cements, whitewares (pottery, porcelains), porcelain enameled ware, refractories, brick, tile, and abrasives.

However, a combination of technological developments, such as discoveries of new ceramic compounds with unique properties, new and better methods of processing and manufacturing ceramics, and a demand for more energy-efficient engines and high-temperature systems, have spurred the development and application of more modern ceramic materials. This category of ceramics is often referred to as the *new* ceramics, and consists primarily of pure and mixed oxide ceramics, plus carbide, nitride, boride and silicide compounds.

Although the production of these new generation ceramics is not nearly as large as the traditional, their applications are very widespread and unique. For instance, new ceramics are a vital part of the electronics and magnetics industries, as indicated in Chapters 10 and 11. They are also in strong demand for applications such as high-temperature engines, where increased efficiency is a prime consideration; machining hard metallic alloys at high cutting rates; fuels for nuclear reactors; coatings that require high-temperature resistance combined with chemical stability; solid electrolytes for fuel cells; plus a number of optical uses, including coloring pigments, fluorescent screens, lasers, and iridescent films.

Let us now examine the fundamental structure–property relationships of ceramic materials and see how they influence the behavior and thus the application or utilization of these engineering materials.

CRYSTALLINE CERAMICS

In Chapter 2 we reviewed the various types of atomic bonding as related to the basic classes of engineering materials, such as metals, ceramics, and polymers. Ceramic materials are generally crystalline (with the principal exception of glasses) and primarily exhibit ionic bonding, although a significant amount of covalent bonding may also take place. Therein lies a basic difference between ceramics and metals; that is, ceramics do not have delocalized valence electrons, whereas metallic materials do. Not only does this condition significantly affect the strength and chemical stability of ceramics, but it also heavily influences their electrical and magnetic characteristics, as explained in Chapters 10 and 11.

Basic Structural Relationships

Ionic bonding in ceramics involves electron transfer between the atoms. Consequently, there must be definite ratios of the different kinds of ions present in the solid, depending on their respective oxidation states. For example, there are equal numbers of positively charged ions (cations) and negatively charged ions (anions) in a binary ceramic compound such as magnesium oxide (MgO), since the charge on each ion is the same number (2) and of opposite sign. Ceramic compounds are generally stoichiometric; that is, they exhibit fixed integer ratios of cations to anions such as beryllia (BeO), thoria (ThO_2), alumina (Al_2O_3), and spinel ($MgAl_2O_4$). However, there are certain exceptions where the compounds do not have an exact integer ratio of ions. These nonstoichiometric compounds may involve substantial solid solutions of similar ions, or crystalline defects, such as an iron ion vacancy in FeO, resulting in $Fe_{1-x}O$ (e.g. $Fe_{0.96}O$). Crystalline defects in ionic materials were discussed in Chapter 4 (see Figures 4-2 and 4-3).

Another important characteristic of ceramic structures is the relative size of the ions involved. For example, oxygen and metal ions form the basis for a large number of ceramics. In this case, the metal ion is the cation, while the oxygen ion is the anion. Moreover, the oxygen ion is usually larger, since it gains electrons during ionization. Accordingly, the metal ion is relatively smaller since it transfers the electrons to the oxygen atom. As a consequence, cations tend to be located in the interstices[1] produced by the crystalline arrangement of anions.

[1] The student is referred back to our discussion of interstitial impurities in Chapter 4 (see Figure 4-5).

The relative size of ions in a ceramic compound also strongly influences their crystalline arrangement. According to Pauling's first rule,[2] the coordination number (CN) of anions around a cation is determined by the geometry necessary for the cation to remain in contact with each anion. In essence, this is the equilibrium separation that we discussed in Chapter 2 (shown in Figure 2-3). The particular geometry of a cation–anion combination is fixed by the radius ratio of the cation to the anion (i.e., r_c/r_A). Coordination number, as you will recall from Chapter 3, refers to the number of nearest neighbors an atom or ion has in the crystal structure. For example, consider the compounds beryllia (BeO) and magnesia (MgO). The radius ratio of BeO is $r_c/r_A = 0.35$ Å$/1.40$ Å $= 0.250$. Beryllia consists of a beryllium ion (Be^{2+}) located in the tetrahedral interstice produced by four oxygen ions (O^{2-}). This tetrahedral or fourfold (4-f) coordination is depicted in Figure 15-1(a). Thus each Be^{2+} is surrounded

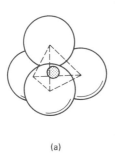

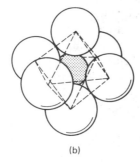

(a)   (b)

Figure 15-1   Typical coordination of cations (shaded) in ceramics: (a) fourfold coordination (tetrahedral); (b) sixfold coordination (octahedral).

by four oxygen ions and each oxygen ion is correspondingly surrounded by four beryllium ions. The coordination number of both anion and cation in this structure is therefore CN = 4. However, in MgO, the radius ratio $r_c/r_A = 0.66$ Å$/1.40$ Å $= 0.471$, and each Mg^{2+} ion is surrounded by six O^{2-} ions. Therefore, the coordination number of both cation and anion in this compound is CN = 6. Such an octahedral (sixfold) coordination is illustrated in Figure 15-1(b).

The relationship between relative ion sizes (radius ratio), coordination number (CN), and ionic arrangement is presented in Table 15-1. Examination of these data shows that as the radius ratio of cation to anion (r_c/r_A) increases, the CN increases, until $r_c/r_A = 1$ (for same size cation and anion) and CN = 8.

Oxide Structures

A large number of ceramic materials can be understood on the basis of simple cubic, close-packed cubic, and hexagonal close-packed, oxygen arrangements as shown in Table 15-2. For a more complete description of the interstitial sites in the cubic and hexagonal structures, the student is referred back to Figure 4-5 and Table 4-1. In the following section we examine some of these structures, which are representative of many commercially important ceramics.

[2]L. Pauling, *Nature of the Chemical Bond*, Cornell University Press, Ithaca, N.Y., 1945.

TABLE 15-1 RELATIONSHIP BETWEEN RADIUS RATIOS AND COORDINATION NUMBER FOR VARIOUS IONIC STRUCTURES

Radius ratio, r_c/r_A	Coordination number	Anion arrangement	Example of structure
0–0.155	2	Linear	Molecular chains
0.155–0.225	3	Corners of triangle	Boron nitride
0.225–0.414	4	Corners of tetrahedron	ZnS (see Figure 15-1a)
0.414–0.732	6	Corners of octahedron	NaCl (see Figure 15-1b)
0.732–1.00	8	Corners of cube	CsCl (see Figure 15-2)

Simple cubic. In this structure illustrated in Figure 15-2(a), the anions are located at the corners of a cube, with the cation situated in the interstitial site produced by these anions. Although this structure may appear to be bcc, it actually consists of

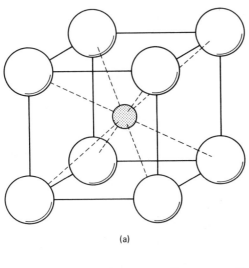

(a)

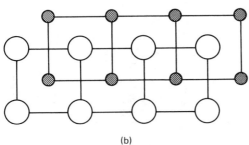

(b)

Figure 15-2 Simple cubic structure. (a) Cation (dark) centrally located within the interstice produced by eight anions—eightfold coordination. (b) Plane view of interpenetrating simple cubic cells.

TABLE 15-2 SUMMARY OF SIMPLE CERAMIC STRUCTURES

Anion packing	Coordination number of M^a and O	Sites occupied by cations	Generic structure name	Examples
Cubic close-packed	$6:6$ MO	All oct.	Rock salt	NaCl, KCl, LiF, KBr, MgO, CaO, SrO, BaO, CdO, VO, MnO, FeO, CoO, NiO
Cubic close-packed	$4:4$ MO	$\frac{1}{2}$ tet.	Zinc blende	ZnS, BeO, SiC
Cubic close-packed	$4:8$ M_2O	All tet.	Antifluorite	Li_2O, Na_2O, K_2O, Rb_2O, sulfides
Distorted cubic close-packed	$6:3$ MO_2	$\frac{1}{2}$ oct.	Rutile	TiO_2, GeO_2, SnO_2, PbO_2, VO_2, NbO_2, TeO_2, MnO_2, RuO_2, OsO_2, IrO_2
Cubic close-packed[b]	$12:6:6$ ABO_3	$\frac{1}{4}$ oct. (B)	Perovskite	$CoTiO_3$, $SrTiO_3$, $SrSnO_3$, $SrZrO_3$, $SrHfO_3$, $BaTiO_3$
Cubic close-packed	$4:6:4$ AB_2O_4	$\frac{1}{8}$ tet. (A), $\frac{1}{2}$ oct. (B)	Spinel	$FeAl_2O_4$, $ZnAl_2O_4$, $MgAl_2O_4$
Cubic close-packed	$4:6:4$ $B(AB)O_4$	$\frac{1}{8}$ tet. (B), $\frac{1}{2}$ oct. (A, B)	Spinel (inverse)	$FeMgFeO_4$, $MgTiMgO_4$
Hexagonal close-packed	$4:4$ MO	$\frac{1}{2}$ tet.	Wurtzite	ZnS, ZnO, SiC
Hexagonal close-packed	$6:6$ MO	All oct.	Nickel arsenide	NiAs, FeS, FeSe, CoSe
Hexagonal close-packed	$6:4$ M_2O_3	$\frac{2}{3}$ oct.	Corundum	Al_2O_3, Fe_2O_3, Cr_2O_3, Ti_2O_3, V_2O_3, Ga_2O_3, Rh_2O_3
Hexagonal close-packed	$6:6:4$ ABO_3	$\frac{2}{3}$ oct. (A, B)	Ilmenite	$FeTiO_3$, $NiTiO_3$, $CoTiO_3$
Hexagonal close-packed	$6:4:4$ A_2BO_4	$\frac{1}{2}$ oct. (A), $\frac{1}{8}$ tet. (B)	Olivine	Mg_2SiO_4, Fe_2SiO_4
Simple cubic	$8:8$ MO	All cubic	CsCl	CsCl, CsBr, CsI
Simple cubic	$8:4$ MO_2	$\frac{1}{2}$ cubic	Fluorite	ThO_2, CeO_2, PrO_2, UO_2, ZrO_2, HfO_2, NpO_2, PuO_2, AmO_2
Connected tetrahedra	$4:2$ MO_2	—	Silica types	SiO_2, GeO_2

[a] M is the metal ion.

[b] A and B are metal ions with different valence.

Source: W. D. Kingery, H. K. Bowen, and D. R. Uhlmann, *Introduction to Ceramics*, John Wiley & Sons, Inc., New York, 1976, p. 62.

interpenetrating simple cubic cells (Figure 15-2b). Equivalent lattice positions are related only by integer translations of a, the lattice parameter. The ions in this structure have $CN = 8$, as given in Table 15-1; therefore, *each* ion is surrounded by eight nearest neighbors.

The lattice parameter (a) can be related to the ionic radii of this structure by employing the same analysis that we used in Chapter 3 for bcc structures (see Figure 3-9). In the present case, however, we must substitute a relatively smaller cation in the center of the unit cell and the relationship is expressed as follows:

$$a = \frac{2}{\sqrt{3}}(R + r) \tag{15-1}$$

where R = radius of anion
r = radius of cation

Example 15-1

Cesium chloride (CsCl) exhibits a simple cubic structure with Cs^+ ions located within the eightfold interstitial sites produced by the Cl^- ions. (a) Determine the lattice constant for this structure. (b) What is the density of this compound?

Solution (a) The lattice parameter (a) is related to the ionic radii by equation (15-1). From inside back cover, the ionic radii are found to be, $r_{Cs^+} = 1.67$ Å, $R_{Cl^-} = 1.81$ Å. Substituting these values in equation (15-1), we obtain

$$a = \frac{2}{\sqrt{3}}(1.81 \text{ Å} + 1.67 \text{ Å})$$

$$= \frac{2}{\sqrt{3}}(3.48 \text{ Å})$$

$$= 4.0 \text{ Å or } (4.0 \times 10^{-8} \text{ cm}) \quad \textit{Ans.}$$

(b) The density can be calculated based on the weight of ions per unit cell [similar to equation (3-4)] as follows:

$$D = \frac{(\text{no. cations/u.c.})(\text{Wt. of cation}) + (\text{no. anions/u.c.})(\text{Wt. of anion})}{\text{Vol. of unit cell}}$$

Since we have one cation (Cs^+) and the equivalent of one anion (Cl^-) per unit cell, we may state

$$D_{CsCl} = \frac{(1)(\text{Atomic Wt. Cs}/A_0) + (1)(\text{Atomic Wt. Cl}/A_0)}{a^3}$$

$$= \frac{(1)(132.91 \text{ amu/ion}) + (1)(35.76 \text{ amu/ion})}{(6.02 \times 10^{23} \text{ amu/g})(4.0 \times 10^{-8} \text{ cm})^3}$$

$$= \frac{168.67 \text{ g}}{(6.02 \times 10^{23})(6.4 \times 10^{-23} \text{ cm}^3)}$$

$$= 4.38 \text{ g/cm}^3 \quad \textit{Ans.}$$

This type of crystal structure, simple cubic with eightfold coordination, is not frequently observed in ceramic materials, the reason being that a radius ratio (r_c/r_A) of 0.73 (Table 15-1) or greater is difficult to obtain, since cations tend to be much smaller than anions. Compounds forming this structural arrangement include CsCl, CsBr, and CsI.

Cubic close-packed. This type of crystalline arrangement basically consists of oxygen ions in an fcc structure with metal ions situated in some or all of the interstitial sites produced by the oxygen ions. It is important to recognize that two types of interstices are developed in an fcc structure. Recall from the section "Interstitial Impurity Atom" in Chapter 4 that both fourfold interstices (tetrahedral voids) and sixfold interstices (octahedral voids) exist in this structure [see Figure 4-5(a) and (b)].

In the *rock salt* (NaCl) type structure, which is typical of many ceramics, the anions are arranged in fcc packing and *all* the octahedral interstices contain cations. This structure is illustrated in Figure 15-3(a). When the ions are in their equilibrium positions (touching), as shown in Figure 15-3(b), the lattice parameter (*a*) is related to the ionic radii by the following expression:

$$a = 2R + 2r \qquad (15\text{-}2)$$

Since each ion has sixfold coordination, the minimum radius ratio for the formation of this type structure is 0.414 (see Table 15-1). Many ceramic compounds

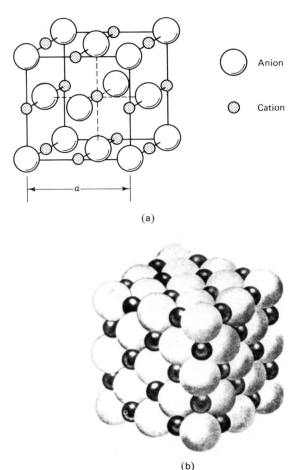

◯ Anion

◉ Cation

(a)

(b)

Figure 15-3 NaCl-type structure: (a) anions in fcc positions with cations located in all octahedral interstices; (b) NaCl structure with ions touching.

exhibit this structure, including MgO, CaO, FeO, NiO, MnO, and BaO. Furthermore, all the alkali halides, with the exception of the three we mentioned under the simple cubic arrangement, and all the alkaline earth sulfides, form this structure.

Another close-packed cubic structure which commonly occurs in ceramics is known as the *zinc blende* (ZnS) structure. The anions in this arrangement are located in fcc positions just like the NaCl structure. The cations, however, are situated in the tetrahedral interstices producing fourfold coordination. A further difference between the ZnS type structure and the NaCl type structure involves the number of filled interstices. Recall from Chapter 4 that there are eight tetrahedral interstices per unit cell in the fcc structure. But only half the sites (4) contain cations and this configuration produces maximum cation separation. The ZnS type structure is shown in Figure 15-4(a).

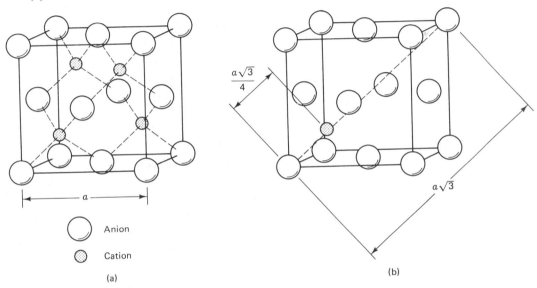

(a)

(b)

Figure 15-4 Unit cell of zinc blende (ZnS) structure: (a) anions in fcc arrangement, four cations located in tetrahedral interstices; (b) main diagonal (dashed line) through cube showing intersection with corner anion and adjacent cation.

The relationship between the lattice parameter (a) and the ionic radii is derived in a manner similar to that we used for bcc in Chapter 3 (see Figure 3-9). A body diagonal through the cube equals $a\sqrt{3}$, and one-fourth of this distance contains R from a corner ion (anion) plus r from the cation in the adjacent tetrahedral interstice, as shown in Figure 15-4(b). Therefore, we can write

$$\frac{a\sqrt{3}}{4} = R + r$$

$$a = \frac{4(R + r)}{\sqrt{3}} \qquad (15\text{-}3)$$

A number of ceramic compounds form the zinc blende structure, such as SiC, BeO, CdS, and AlP. This structure is also favored by certain semiconductive compounds, such as GaP, GaAs, InP, that were discussed in Chapter 10 (see Table 10-5).

Example 15-2

Determine the density of the compound BeO (one of the best electrical insulators in existence) if it crystallizes in the zinc blende structure.

Solution Examination of Figure 15-4(a) shows that this structure contains four anions and four cations per unit cell. Also, since it is cubic, the volume of the unit cell is a^3.

First, let us calculate the volume of the BeO unit cell. From equation (15-3), the lattice parameter is

$$a = \frac{4(R+r)}{\sqrt{3}}$$

From inside the back cover we find the values of ionic radii for Be^{2+} and O^{2-}. Therefore,

$$a = \frac{(4)(1.32 + 0.35)}{\sqrt{3}}$$

$$= \frac{(4)(1.67)}{\sqrt{3}} = 3.86 \text{ Å or } (3.86 \times 10^{-8} \text{ cm})$$

Now we can calculate the density of this compound as follows:

$$D_{BeO} = \frac{(\text{no. ions/unit cell})(\text{weight of ions})}{(\text{volume of unit cell})(\text{Avogadro's no.})}$$

$$= \frac{4(9.01 + 16.00) \text{ amu}}{(6.02 \times 10^{23} \text{ amu/g})(3.86 \text{ Å})^3}$$

$$= \frac{100.04 \text{ g}}{(6.02 \times 10^{23})(5.75 \times 10^{-23}) \text{ cm}^3}$$

$$= \frac{100.04}{34.62} \text{ g/cm}^3$$

$$= 2.89 \text{ g/cm}^3 \quad \textit{Ans.}$$

The *fluorite* structure (CaF_2) is an exception to the general situation where anions are arranged in a close-packed lattice with cations located in the resultant interstices. Instead, this structure may be viewed as a close-packed cubic lattice of cations with anions occupying the eight available tetrahedral interstices as illustrated in Figure 15-5. Several important refractory oxides, such as UO_2, ThO_2, CeO_2, and ZrO_2, form the fluorite structure. For example, UO_2 produces a dimensionally stable nuclear reactor fuel because of the unoccupied interstices in the center of the unit

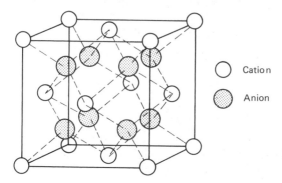

○ Cation

◉ Anion

Figure 15-5 Fluorite-type structure showing cations in fcc packing and simple cubic arrangement of the anions.

cell and the midpoints of the unit cell edges. Such vacant interstitial sites evidently provide ample collection points for the products of fission.

Hexagonal close-packed. We should also realize that not all the close-packed structures are necessarily cubic. In fact, the zinc sulfide (ZnS) compound we have just previously examined can also crystallize in hcp arrangement. When ZnS forms a hcp arrangement, it is called the *wurtzite* structure. In Chapter 3 we determined that the packing factor and CN for the hcp structure was identical to fcc. Ceramic compounds that crystallize in this fashion contain anions in the hexagonal close-packed positions with cations located in the tetrahedral or octahedral interstices. In addition to the wurtzite structure, many ceramic compounds form a crystal structure with the anions in hcp positions, including nickel arsenide (NiAs) and corundum (Al_2O_3), as previously given in Table 15-2.

Corundum is an important industrial ceramic used widely in abrasives, refractories, and cutting applications. However, you are probably more familiar with corundum in another fashion. When 1-2% Cr^{3+} is substituted for Al^{3+} in corundum, the gemstone known as ruby is produced. If the substitution consists of iron and titanium ions, sapphire results.

Additional oxide structures. In addition to the simple cubic and close-packed arrangements we have discussed thus far, many ceramics are oxide compounds containing more than one type of cation. For example, a number of oxides have the formula AB_2O_4, where A^{2+} and B^{3+} are the divalent and trivalent cation species, respectively. The structure of $MgAl_2O_4$ (*spinel*) is in this category and can be viewed as a combination of the cubic rock salt and zinc blende structures with the oxygen ions in fcc packing. The spinel structure was illustrated in Chapter 11, where we discussed the ceramic materials with magnetic applications (see Figure 11-15). As shown in Table 15-2, two types of spinel structures can occur. In *normal* spinels, the A^{2+} (divalent) ions are located in tetrahedral interstices and the B^{3+} (trivalent) ions take up octahedral sites. Examples of this structure include the compounds $ZnFe_2O_4$, $CoAl_2O_4$, and $NiAl_2O_4$, which find applications in magnetic recording tape (γ-Fe_2O_3 + polymer binder), television deflection coils, and transformer cores. In *inverse* spinels, all the A^{2+} ions and half the B^{3+} ions are in octahedral sites, while the remaining B^{3+} ions occupy tetrahedral sites. This latter spinel structure, written as $B(AB)O_4$ and commonly referred to as "ferrite," is found in the ceramics, which exhibit important magnetic properties, as alluded to in Chapter 11. Examples of ferrites include $FeNiFeO_4$, Fe_3O_4, and $FeTiFeO_4$.

The *perovskite* ($CaTiO_3$) structure is another important ceramic structure, which we introduced for barium titanate ($BaTiO_3$) in Chapter 10 during our discussion of ferroelectric behavior. This arrangement occurs when large cations are present in the compound and can form a close-packed structure in conjunction with the oxygen ions. This is the situation we illustrated in Figure 10-29, where the Ba^{2+} and O^{2-} ions are in fcc packing and the smaller Ti^{4+} ions are located in octahedral interstices. Ceramic materials that form this structure exhibit unique electromagnetic properties (ferroelectric) and also display piezoelectric properties which make them useful for electrical transducers, phonograph cartridge pickups, and ultrasonic energy applications.

The complex oxide structures we have just discussed encompass many important ceramic compounds with quite unique properties and applications. They are, however, by no stretch of the imagination, representative of all the multiple ceramic compounds. These examples simply serve to acquaint the student with a few of the more important complex structures and their applications.

Silicate Structures

Many important ceramic engineering materials are based on silica (SiO_2). In fact, most of the traditional ceramics referred to in our introduction (i.e., pottery, porcelain, brick, and tile) contain silicate units as their principal building block.

The silicon atom contains four valence electrons and has a radius ratio (r_c/r_A) with oxygen of 0.29. Therefore, silicon forms a tetrahedral arrangement with four oxygen atoms, as shown in Figure 15-6(a). Examination of the bonds in this ionic

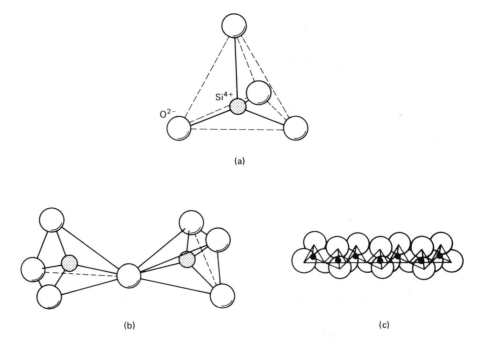

(a)

(b)   (c)

Figure 15-6 Schematic example of silicate structures: (a) silicate tetrahedron (SiO_4); (b) "oxygen bridge" connecting silicate tetrahedra; (c) connected silicate tetrahedra forming a chain-like network.

arrangement reveals that the four valence electrons from silicon are shared with the surrounding four oxygens. Correspondingly, each oxygen ion in this arrangement shares electrons with two silicon ions. The SiO_4 tetrahedra join in compounds such that the oxygen ion receives an electron from another metal ion or they link with other silicate tetrahedra as illustrated in Figure 15-6(b). In this manner, *chains* of tetrahedral silicate groups are formed [Figure 15-6(c)] and the shared oxygen ions constitute a *bridge* between the units. Groups of silicate structures exhibit several general arrangements, as given in Table 15-3.

TABLE 15-3 STRUCTURES OF THE GENERAL SILICATE GROUPS

Group	Arrangement of Si–O tetrahedrons	Typical mineral	Structural configuration
Orthosilicates	Independent tetrahedrons sharing no oxygens	Forsterite Mg_2SiO_4	
Pyrosilicates	Independent pairs of tetrahedrons sharing one oxygen	Akermanite $Ca_2MgSi_2O_7$	
Metasilicate chains	Continuous single chains of tetrahedrons sharing two oxygens	Diopside $CaMg(SiO_3)_2$	
Metasilicate chains	Continuous double chains of tetrahedrons sharing alternately two and three tetrahedrons	Tremolite $H_2Ca_2Mg_5(SiO_3)_8$	
Metasilicate rings	Closed independent rings of tetrahedrons each sharing two oxygens	Benitoite $BaTiSi_3O_9$ Beryl $Al_2Be_3Si_6O_{18}$	
Disilicates	Continuous sheets of tetrahedrons each sharing three oxygens	Muscovite $Al_4K_2(Si_6Al_2)O_{20}(OH)_4$	
Silica	Three-dimensional network of tetrahedrons each sharing all four oxygens	Quartz SiO_2 Orthoclase $KAlSi_3O_8$	(Three-dimensional structure not shown)

Source: F. H. Norton, *Elements of Ceramics*, Addison-Wesley Publishing Company, Inc., Reading, Mass., 1952, p. 7.

The capability of the silicates to form chain-like groupings with one another and with other ions or groups accounts for the cementitous characteristics of this ceramic. For example, the composition of natural and portland cements, which bind together the constituents of concrete, consist principally of silicates (see Table 16-7). The various clays and glazes also contain silicates, which significantly affect the bonding behavior of these materials.

Clay minerals. The chief elements found in clays are oxygen, silicon, and aluminum. These elements combine to form the ceramic compounds known as *aluminosilicates*. The basic structure of the common clays consist of a layer of SiO_4 tetrahedra $(Si_2O_5)_n$, linked at the corners with an $AlO(OH)_2$ layer of alumina octahedra. For instance, consider the formation of *kaolinite*, which is the most common clay mineral. Feldspar, a mineral found in virtually all igneous[3] rock, has the general formula $K_2O \cdot Al_2O_3 \cdot 6SiO_2$. Natural weathering and environmental chemical reactions combine to produce kaolinite $(Al_2O_3 \cdot 2SiO_2 \cdot 2H_2O)$ from this mineral. Pure kaolinite is white and is the basic ingredient in fine china, but ordinarily, kaolinite contains some

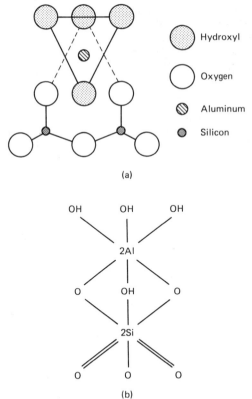

Hydroxyl

Oxygen

Aluminum

Silicon

(a)

OH   OH   OH

2Al

O   OH   O

2Si

O   O   O

(b)

Figure 15-7 Schematic showing the layered structure of kaolinite sheet. (a) Top portion of structure consists of $AlO_2(OH)_4$ octahedra, while the bottom portion consists of SiO_4 tetrahedrons. (b) Representation of the coordination between cations and anions in the layers.

[3] Igneous rock is formed by solidification or precipitation from a molten silicate solution, as opposed to sedimentary rock, which results from consolidation of sediments, and metamorphic rock which is produced by the alteration of pre-existing rock through the application of pressure and heat.

iron oxide contamination from the soil, which gives it the familiar reddish color of common clay.

The structure of kaolinite essentially consists of silicate tetrahedra, arranged in sheets where each tetrahedron shares three of its oxygen ions with three other tetradrons, combined with aluminate sheets (AlO—OH). These-layered groups are depicted in Figure 15-7, forming kaolinite sheet—$Al_2(Si_2O_5)(OH)_4$.

Kaolinite crystals are extremely small, thin platelets which tend to have an irregular hexagonal shape, as illustrated in Figure 15-8. These fine platelets readily slide past

Figure 15-8 Electron micrograph of kaolinite crystal (note hexagonal platelet morphology). (From F. H. Norton, *Elements of Ceramics*, Addison-Wesley Publishing Company, Inc., Reading, Mass., 1952, p. 8.)

one another mixed with water, accounting for the plastic-like behavior of ordinary clays, when moist.

The compositions of several commercially important traditional ceramics (whiteware) are listed in Table 15-4. Kaolin and ball clay provide fine particle sizes and good plasticity, which assists forming and firing the ceramic shapes. Feldspar helps form a viscous liquid at the firing temperature and also promotes vitrification. The flint is simply an inexpensive filler material that is unreactive at low temperatures, but forms a viscous liquid at high temperatures.

TABLE 15-4 COMPOSITION OF SELECTED WHITEWARES

Material	Composition[a] (%)				
	Kaolin	Ball Clay	Feldspar	Flint	Other
Vitreous sanitary ware	28	20	32	20	
Electrical insulators	21	25	34	20	
Vitreous wall tile	27	29	33	11	
Hotel china	34.8	7	22	35	$1.2CaCo_3$
Dental porcelain	5	—	81	14	
Refractory porcelain	50	10	10	5	$25Fe_2O_3$

[a]Typical constituents:
  Kaolin: Al_2O_3, SiO_2, H_2O
  Ball clay: Al_2O_3, SiO_2, Fe_2O_3, H_2O
  Feldspar: SiO_2, Al_2O_3, K_2O, Na_2O
  Flint: SiO_2, Al_2O_3, TiO_2, Fe_2O_3, H_2O

Polymorphism

In Chapter 3 we introduced the concept of polymorphism when we discussed the fundamentals of crystalline structures. As you will recall, polymorphism refers to the ability of a material to exist in more than one crystalline structure even though no change occurs in its chemical composition. In Chapters 12 and 13 we noted polymorphism in certain metals. Not surprisingly, this type of transformation also occurs in many crystalline ceramic compounds.

Polymorphic transformation usually occurs as a result of temperature change, but can also be produced by very high hydrostatic pressure or a combination of high pressures and severe cold work (mechanical deformation). The polymorphic form that is stable at a particular temperature depends on the free energy (G) associated with that structure. The free-energy relationships that we examined in Chapter 7 (see Figure 7-6) for alloying also apply to the solid ceramic compounds now under consideration, as illustrated in Figure 15-9. In this hypothetical system, the crystalline structure

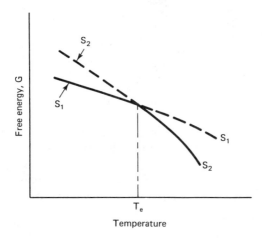

Figure 15-9 Schematic relationship between Gibbs free energy (G) and temperature for a hypothetical ceramic undergoing polymorphic transformation at T_e.

represented by S_1 has lower free energy below the equilibrium temperature T_e and therefore is the more stable form. However, above T_e, the structure S_2 has lower free energy and thus is the more stable polymorph.

Important ceramics displaying polymorphic behavior include silica (SiO_2), titania (TiO_2), zinc sulfide (ZnS), calcium titanate ($CaTiO_3$), alumina (Al_2O_3), and zirconia (ZrO_2). Consider, for example, the silicate structure we examined in the preceding section. The three fundamental crystalline structures of silica—quartz, tridymite, and cristobalite—can exist in two or three structural modifications depending on temperature, as illustrated in Figure 15-10.

Since polymorphic transformations involve solid-state rearrangement of the ions in the crystal, volume changes result in the bulk material. When these volume changes are large, the expansion (or contraction) of the lattice can produce significant strains, thereby resulting in serious consequences in ceramic materials. For example, zirconia (ZrO_2) has certain desirable high-temperature properties such as high melting point (2770°C) and low thermal conductivity, and therefore appears to be a useful

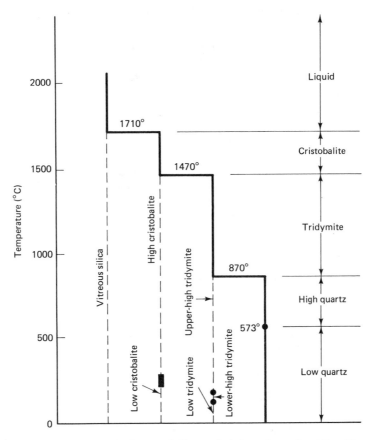

Figure 15-10 Polymorphic forms of silica and their regions of stablity. Heavy lines indicate stable phase. (Adapted from F. H. Norton, *Elements of Ceramics*, Addison-Wesley Publishing Company, Inc., Reading, Mass., 1952, p. 132.)

refractory material. However, ZrO_2 undergoes a polymorphic change at 1000°C which results in a significant volume decrease. Under these circumstances, pure ZrO_2 would crack during the heating and cooling cycles to which a refractory material is generally subjected in service.

Properties and Applications

In addition to the traditional applications that we referred to in the introduction, many current applications for crystalline ceramics, especially *new* ceramics, depend primarily on their high temperature, electrical, and magnetic characteristics. Our discussion will be somewhat restricted to the properties in these areas that tend to be structure insensitive. Mechanical properties tend to be strongly structure sensitive and therefore depend intimately on the particular ceramic compound and its processing history. Consequently, we will only briefly comment on the general nature of mechanical behavior in ceramics.

Thermal properties. The term "ceramic" is virtually synonymous with high-melting-temperature materials. Indeed, this characteristic of ceramics—high melting point—is the basis for their use in many elevated-temperature applications, such as refractory bricks and mortars (cements), furnace liners, thermal insulators, and heat-resistant coatings. The melting temperatures of selected ceramic materials, together with their respective theoretical density, are given in Table 15-5. Overall,

TABLE 15-5 MELTING TEMPERATURE AND DENSITY OF SELECTED CERAMICS

Material	Average melting temperature		Theoretical density (g/cm³)
	°C	°F	
Borides			
TiB_2	2938	5320	4.50
ZrB_2	3038	5500	6.10
TaB_2	3100	5610	12.60
Carbides			
α-SiC	2700	4892	3.50
TiC	3193	5780	4.92
TaC	3877	7010	14.37
WC	2600	4710	15.77
Nitrides[a]			
BN (s)	2732	4950	2.25
AlN (d)	2232	4050	3.05
Si_3N_4 (s)	1870	3400	3.40
TiN	2950	5340	5.43
Oxides			
BeO	2550	4620	3.01
MgO	2827	5120	3.57
α-Al_2O_3	2000	3630	3.98
ZrO_2	2766	5010	6.10
ThO_2	3300	5970	9.69
UO_2	2816	5100	10.96

[a]s, sublimes; d, decomposes.
Source: Compiled primarily from J. E. Hove and W. C. Riley, Eds., *Modern Ceramics: Some Principles and Concepts*, John Wiley & Sons, Inc., New York, 1965.

these data demonstrate the relatively high-melting behavior of ceramics, compared to the other two primary classes of engineering materials, metals and polymers. This particular characteristic of ceramics will obviously continue to be exploited in future engineering applications as the quest for higher operating temperatures in heat engines and power production progresses.

The applications of ceramic materials which involve elevated temperatures, cyclic temperature conditions, or thermal gradients must also take into consideration the *thermal expansion* and *thermal conduction* properties. First, let us consider thermal expansion. This is the basic property of a material to change its specific volume with temperature. The general increase in volume with increased temperature is due primarily to increased amplitude of atomic or ionic vibrations about a mean lattice position. In cubic crystals, the expansion coefficients along different crystallographic axes are equal. Therefore, the linear coefficient of expansion (α) is the same in the x, y, and z directions, and is expressed as follows:

$$\alpha = \frac{\Delta l}{l_0 \Delta T} \qquad (15\text{-}4)$$

where α equals the fractional expansion per degree change in temperature (e.g., mm/mm °C). The average coefficient of volume expansion (β) for isotropic materials is usually taken (for limited temperature ranges) to be 3α.

The typical values for coefficients of linear thermal expansion (α) are given for selected ceramics in Table 15-6. Note that these data treat the coefficient as an average

TABLE 15-6 MEAN THERMAL EXPANSION COEFFICIENTS FOR SELECTED CERAMICS (0–1000°C)

Material	Linear expansion coefficient, α ($10^{-6}/°C$)	Material	Linear expansion coefficient, α ($10^{-6}/°C$)
Al_2O_3	8.8	ZrO_2 (stabilized)	10.0
BeO	9.0	Fused silica glass	0.5
MgO	13.5	Soda–lime–silica glass	9.0
Mullite	5.3	TiC	7.4
Spinel	7.6	Porcelain	6.0
ThO_2	9.2	Fireclay refractory	5.5
UO_2	10.0	Y_2O_3	9.3
Zircon	4.2	TiC cement	9.0
SiC	4.7	B_4C	4.5

Source: W. D. Kingery, H. K. Bowen, and D. R. Uhlmann, *Introduction to Ceramics*, John Wiley & Sons, Inc., New York, 1976, p. 595.

(constant) value ($\bar{\alpha}$) between 0 and 1000°C. Two important points must be emphasized here; (1) not all ceramics are isotropic, and consequently, (2) the expansion characteristics are not crystallographically uniform. In fact, for crystals that are strongly anisometric,[4] the expansion coefficient in one direction may be negative! The coefficient of thermal expansion is a function of temperature, and treatment of this parameter as a constant can, in some cases, result in serious design errors. The influence of temperature on the linear thermal expansion coefficient (α) is illustrated in Figure 15-11 for several oxide compounds.

[4]Anisometric crystals have different dimensions in different directions (i.e., noncubic).

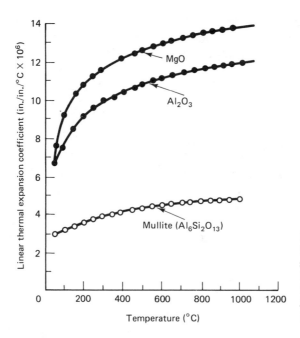

Figure 15-11 Effect of temperature on thermal expansion coefficient for several ceramics. (From W. D. Kingery, H. K. Bowen, and D. R. Uhlmann, *Introduction to Ceramics,* John Wiley & Sons, Inc., New York, 1976, p. 593.)

Example 15-3

Stabilized zirconia (ZrO_2 + CaO) is heated from room temperature (27°C) to 300°C. (a) Determine the linear strain (e) produced by thermal expansion. (b) If this ceramic material is constrained from expanding, calculate the stress that would result from this temperature change.

Solution (a) Based on equation (15-4), we can determine the engineering strain (e) produced in the material as follows:

$$\alpha = \frac{\Delta l}{l_0 \, \Delta T}$$

$$\alpha \, \Delta T = \frac{\Delta l}{l_0} = e$$

From Table 15-6, α for stabilized zirconia[5] equals 10^{-5}/°C; therefore,

$$e = (10^{-5}°C)(300°C - 27°C)$$

$$= (10^{-5}/°C)(273°C)$$

$$= 2.73 \times 10^{-3} \quad Ans.$$

(b) From Hooke's law [equation (6-13)] we may calculate the resultant stress as follows:

$$\sigma = Ee$$

[5]Pure ZrO_2 experiences a polymorphic transformation at 1000°C which produces a large volume change. This volume change renders zirconia virtually useless in refractory applications. However, the addition of lime (CaO) produces a solid solution which undergoes no transformation, and the resultant material is called *stabilized* zirconia, a valuable refractory (see Figure 7-21).

where E for stabilized ZrO_2 is found in Table 15-8 as 22×10^6 psi (151,800 MPa). The stress therefore is

$$\sigma = (151,800 \text{ MPa})(2.73 \times 10^{-3})$$
$$= 414.4 \text{ MPa or } (60,058 \text{ psi}) \text{ "compressive"} \quad Ans.$$

Since ceramics are generally electrical insulators, electronic conduction is very poor. Therefore, heat energy transport must take place by alternative means, such as a lattice wave mechanism. Essentially this type of energy transport consists of transferring the heat energy from atom to atom through the crystal lattice. The thermal conductivity of ceramics is an important property from the design and application standpoint. For instance, if high-temperature thermal insulation is required, a ceramic with relatively low conductivity may be appropriate. Conversely, if thermal shock resistance (material's ability to resist sudden temperature changes) is a consideration, a ceramic material with relatively high conductivity might be in order, so that steep temperature gradients will be alleviated.

Some typical thermal conductivity values for selected ceramics are presented in Table 15-7. The thermal conductivity of ceramics may also depend heavily on temperature. This temperature dependence is illustrated in Figure 15-12. Examination of these data reveals that materials with a relatively high conductivity at low temperature tend to exhibit significant decreases in conductivity with increased temperature, and vice

TABLE 15-7 THERMAL CONDUCTIVITY DATA FOR SELECTED CERAMICS

Material	Thermal conductivity$\left(\dfrac{\text{cal-cm}}{\text{sec cm}^2 \text{ °C}}\right)$[a]	
	100°C	1000°C
Al_2O_2	0.072	0.015
BeO	0.525	0.049
MgO	0.090	0.017
$MgAl_2O_4$	0.036	0.014
ThO_2	0.025	0.007
Mullite	0.014	0.009
UO_2	0.024	0.008
Graphite	0.43	0.15
ZrO_2 (stabilized)	0.0047	0.0055
Fused silica glass	0.0048	0.006
Soda–lime–silica glass	0.004	—
TiC	0.060	0.014
Porcelain	0.004	0.0045
Fireclay refractory	0.0027	0.0037
TiC cermet	0.08	0.02

[a]To obtain (Btu-ft/hr ft^2 °F) multiply by 242.

Source: W. D. Kingery, H. K. Bowen, and D. R. Uhlmann, *Introduction to Ceramics*, John Wiley & Sons, Inc., New York, 1976, p. 642.

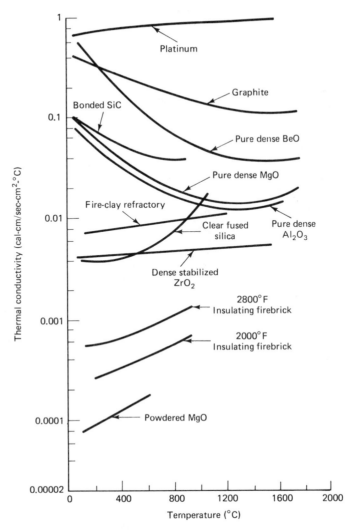

Figure 15-12 Effects of temperature on thermal conductivity (log scale) for selected ceramics and platinum. (From W. D. Kingery, H. K. Bowen, and D. R. Uhlmann, *Introduction to Ceramics*, John Wiley & Sons, Inc., New York, 1976, p. 643.)

versa. Also note the position of platinum with regard to the ceramics. Platinum metal at 0°C exhibits roughly one-third the thermal conductivity of aluminum and one-sixth the conductivity of copper, which ranks it as one of the poorer metals in terms of thermal conductivity. Clearly, the thermal conductivity of ceramics in general is far lower than most metals and metallic alloys.

Electrical and magnetic properties. As a class, ceramic materials exhibit high electrical resistivity at low temperatures and therefore have traditionally been utilized for electrical insulators. Electrical and magnetic properties of engineering materials were discussed in detail in Chapters 10 and 11, respectively. Reference was made under the pertinent topics to the particular ceramics that find applications in these fields, and the student is referred to these chapters for a more complete discussion. Our present analysis of the electrical and magnetic properties will simply summarize the important contributions being made by ceramics.

Ferroelectric ceramics such as barium titanate ($BaTiO_3$) and other similar *perovskite* structures have extremely high dielectric constants. These materials are used to manufacture smaller capacitors which have larger capacitance than components made from conventional capacitor materials. This feature has led to improved, more efficient circuitry. Another unique property of ferroelectric ceramics is their ability to change shape (dimensions) in the presence of an applied electrical field. This conversion of electrical energy to mechanical energy is characteristic of *piezoelectric* materials which are used as transducers in microphones, phonograph record players (pickup cartridges), and ultrasonic equipment used in nondestructive examination of opaque materials and structures, sonar, medical sonograms, and so on.

Electro-optic ceramics such as lithium niobate ($LiNbO_3$), and lead zirconate titanate modified with lanthanum (PLZT) are materials that experience a change in optical dielectric properties when an electric field is applied. Basically, the index of refraction (see Figure 15-18) changes under the influence of the applied field. This effect may then be used to control the transmission of light through the material. In essence, the ceramic material can be made transparent or opaque, depending on such factors as crystallographic orientation and direction of the applied electric field. Applications for these unique materials include optical oscillators, voltage-controlled switches in optical equipment (e.g., lasers) and modulators in optical communication systems.

Magnetic ceramics include a number of spinel structures commonly referred to as *ferrites*. The ferrites (e.g., $MgFe_2O_4$, Fe_3O_4, $CoFe_2O_4$, $CuFe_2O_4$) find widespread application in magnetic recording tape and disks, electron beam deflection coils, transformer cores, and so on. Magnetic ceramics are also used in the memory units of large computers. For example, thin films of magnetic garnets are deposited on nonmagnetic substrates and produce tiny magnetic domains with spin up, separated by regions with spin down, which appear as *bubbles* under polarized light. Such materials are currently being termed bubble memory materials. Other spinel ferrites (e.g., nickel–zinc ferrites) are used in microwave and high-frequency low-loss magnetic applications.

Mechanical properties. As indicated earlier in this chapter, ceramic materials as a rule tend to be hard and brittle. This is a direct consequence of their structure: the strong interatomic bonds between ions and the efficient, orderly arrangement of the ions in a compound. Also, there are fewer slip systems in these compounds than in the metallic materials studied previously. Thus in addition to relatively low dislocation densities, dislocation motion is impeded and the failure of these engineering materials occurs with little or no plastic deformation. Generally, the Charpy impact strength (energy absorbed) of ceramics is very low (typically on the order of 1 in.-lb or less) and their application where impact loading is a consideration should be avoided. In addition, as we saw in Chapter 9, ceramics typically exhibit low resistance to fracture. These materials are extremely sensitive to stress concentrators, such as surface flaws (e.g., notches, scratches, cracks, etc.) and microstructural defects (e.g., inclusions, porosity, microcracks). A comparison of the fracture toughness (K_{Ic}) of certain ceramics with metals can be made in Table 9-2.

On the other hand, many ceramic materials display relatively high values of compressive strength and therefore can be satisfactorily employed in compressive loading situations. For instance, the electrical insulators incorporated in the support "guy" wires of utility poles are uniquely designed to place the ceramic material in compression rather than tension. But perhaps the best example of compressive applications is the use of cements and concrete in foundations, piers, and footings for huge structures and heavy equipment. For instance, the CN tower in Toronto, Can., the world's tallest structure, at 553 m (1815 ft), is composed primarily of reinforced concrete. Typical values of room-temperature mechanical properties are given in Table 15-8 for selected bulk ceramic materials of commercial importance. These data include the modulus of rupture,[6] which was developed by bending tests in this case.

TABLE 15-8 TYPICAL ROOM-TEMPERATURE MECHANICAL PROPERTIES FOR SELECTED CERAMICS[a]

Material	Modulus of elasticity (10^6 psi)	Modulus of rupture (psi)	Tensile strength (psi)	Compressive strength (psi)
Al_2O_3	53	30,000–50,000	—	250,000–400,000
BeO	45	20,000–40,000	14,000	120,000
MgO	30.5	25,000–45,000	19,000	—
ZrO_2[b]	22	20,000–35,000	21,000	85,000–190,000
AlN	50	38,500	—	300,000
BN	12	7,000–14,000	8,000	16,000–46,000
B_4C	65	47,000–50,000	—	400,000
SiC (dense)	68	10,000–60,000	20,000	200,000
SiC (bonded)	50	2,000	5,000	25,000
TiC	45	125,000	35,000–40,000	200,000–400,000

[a]To obtain MPa multiply psi by 6.9×10^{-3}.

[b]Stabilized with 10% CaO.

Source: Compiled from J. F. Lynch, C. G. Ruderer, and W. H. Duckworth, *Engineering Properties of Ceramics*, Technical Report AFML-TR-66-52, June 1966.

Certain ceramic materials, including Al_2O_3, BeO, B_4C, SiC, and graphite, do display high values of tensile strength when they are tested in the form of filaments (fibers) and whiskers.[7] These materials are used as the reinforcement phase in the production of certain composite materials, as you will see in Chapter 16.

[6]Modulus of rupture is defined as the ultimate strength or the breaking load per unit area of a specimen tested in torsion or in bending. In tension, it is the ultimate tensile strength.

[7]Whiskers are filamentary crystals that are relatively free of crystalline defects, such as dislocations and grain boundaries.

Since most mechanical properties are very structure sensitive and vary considerably with the processing history of the material, the structural applications of ceramics should generally be considered on an individual basis, rather than depending on published literature values.

GLASS

Although a lay person might not ordinarily consider glass an engineering material, this material is one of the most important products of the ceramic industry. Unquestionably, glass science is rapidly becoming a distinct branch of materials science, because of the unique properties exhibited by glasses and glass-like ceramics, and their unprecedented applications in the fields of optics and electronics. Our treatment of glass in this section will be limited to the more common commercial glasses and their composition–structural relationships.

The Glass State

Glasses may be generically defined as inorganic solid materials that ordinarily do *not* crystallize when they are cooled from the liquid state. These materials consequently do not display long-range periodicity of atomic structure and are frequently referred to as "supercooled liquids" because of their amorphous nature. For this reason, glasses are often compared with polymeric materials in regard to their structure and behavior. In contrast to the crystalline metals and ceramics that we studied previously, glasses are formed from certain liquids which become very viscous near their melting temperature. In these materials, the formation and growth of crystal nuclei in a supercooled liquid are prohibited even when the cooling rate approaches equilibrium conditions. During solidification, these materials, unable to crystallize, simply super-cool to a rigid amorphous (glass) state. The typical solidification characteristics of glass and a crytalline substance are compared in Figure 15-13. This relationship

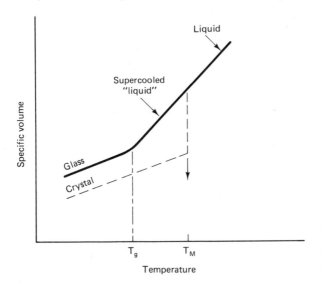

Figure 15-13 Temperature–volume relationship depicting the solidification characteristics of glass versus a crystalline material.

portrays the change in volume with temperature of a liquid that forms a crystalline structure, compared to one that forms a glass state.

In the case of crystal formation, a precipitous decrease in volume occurs at a specific temperature, the melting (freezing) point. This change is associated with the ordering and relaxation of atoms as they assume crystallographic positions in the solid. Notice that this event does *not* occur for glass formation. Rather, the volume continues to decrease, at approximately the same rate below the freezing point, until a temperature is reached where the structure undergoes no further changes in the arrangement of atoms or molecules. This temperature is called the *glass transition temperature* (T_g). Recall that a similar condition occurs in amorphous polymers, and this was also referred to as the "glass" transition temperature (see Figure 14-13).

The rate of cooling in a glass-forming material is instrumental in determining the final structure of the glass. Because the viscosity of glass-formers increases substantially in the vicinity of the freezing point, their atomic and molecular configurations lag the temperature change. Therefore, at any given time, the structure in a supercooled glass actually corresponds to the equilibrium structure for a much higher temperature. When T_g is reached, the random structural configuration corresponding to a much higher temperature is "locked in" the glass. Moreover, as the cooling rate increases, the higher the temperature to which these locked-in configurations will correspond. In other words, faster cooling rates produce higher glass transition temperatures (T_g) and result in more random or amorphous structures. This behavior is illustrated in Figure 15-14.

Recall that in Chapter 7 during our discussion of solidification we mentioned *rapid solidification processes* (RSP). It is noteworthy that these processes (RSP) have even been utilized to produce glassy (amorphous) layers on the surface of certain

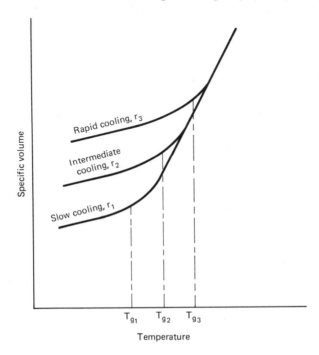

Figure 15-14 Influence of cooling rate (r) on the glass transition temperature (T_g) where $r_3 > r_2 > r_1$.

crystalline metals and ceramics. This is accomplished with *lasers* that can momentarily heat a small region of the surface above the melting point, then allow the resulting liquid to be "quenched" by the surrounding bulk material at cooling rates as high as $10^6 °C/sec$. Such glassy layers on crystalline materials can have profound effects on their mechanical and physical properties, especially resistance to corrosion and environmental attack.

Glass Structure

When glass-forming materials are cooled, a continuous random network of the constituents results. In most glasses, the principal constituent is silica with the basic unit of structure being the SiO_4 tetrahedron, which we illustrated in Figure 15-6.

X-ray diffraction analyses of silica glass have verified that no long-range periodic structure occurs, but the arrangement of the silica tetrahedron persists, resulting in fourfold coordination for the Si^{4+} ions. Each O^{2-} ion is bonded to two Si^{4+} ions and the adjacent tetrahedra are joined by an oxygen *bridge* (shared O^{2-} ion). However, the orientation of adjacent tetrahedra changes throughout the solid. This randomized structure is schematically compared with an ordered crystalline lattice in Figure 15-15.

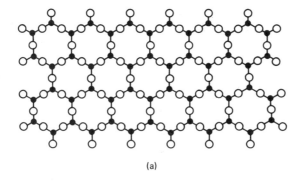

(a)

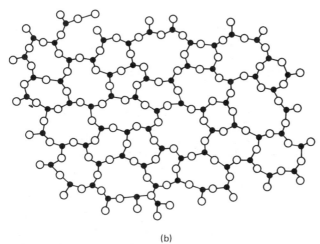

(b)

Figure 15-15 Illustrative comparison of possible structures resulting from same composition: (a) ordered crystalline structure; (b) random glassy network.

Composition. A limited number of oxide compounds are capable of forming glasses. The common glass formers include oxides of silicon, boron, and phosphorus. The chemical compositions of the commercially important glasses are listed in Table 15-9. Examination of these data shows that although silica (SiO_2) is the chief ingredient in these glasses, substantial amounts of other oxides may also be included in certain glasses. For example, soda–lime window glass contains 12 to 15% Na_2O (soda), 10 to 12% CaO (lime), and 1 to 4% MgO, in addition to the silica. These additives are called *fluxing* oxides or modifiers, and tend to disrupt the silica network at points where the tetrahedra are linked. Thus some of the O^{2-} ions are now strongly bonded to only a single metal ion and have weaker ties with one or more of the flux ions. The structure of glass containing flux is illustrated in Figure 15-16. Such additives are used extensively in the production of glass to control certain properties: e.g., lower the melting point, inhibit crystallization, and increase fluidity.

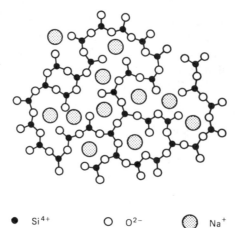

● Si^{4+} ○ O^{2-} ◉ Na^+ **Figure 15-16** Sodium modified, silicate glass structure.

Effects of temperature. The viscosity of glass is important not only during the formation of the network structure, but also for the subsequent processing of glass products. The viscosity of common soda–lime glass is shown for example in Figure 15-17. This graph shows the regions that are significant to the glass manufacturer or processor.

Annealing (thermal treatment with slow cooling of glass) is usually performed to relieve deleterious residual stresses which result from rapid or nonuniform cooling. The glass composition and the size of the object determine the annealing conditions, with very large glass structures, such as lenses for astronomical telescopes, taking as much as weeks or months to anneal properly.

In contrast to annealing, some applications benefit by controlled *quenching*. This procedure involves rapidly cooling the surface (e.g., with an air blast), which results in residual compressive stresses in the surface layers of the glass. Glass produced by this method is referred to as *tempered* glass and is commonly utilized in automotive applications. In service, any applied tensile stresses must first "overcome" the residual compressive stresses in the surface before the tempered glass experiences tension at the very regions which primarily influence failure—the surface layers. This technique

TABLE 15-9 APPROXIMATE CHEMICAL COMPOSITIONS OF SELECTED COMMERCIAL GLASSES

Material	SiO_2	Na_2O	K_2O	CaO	MgO	BaO	PbO	B_2O_3	Al_2O_3
Weight percent									
1 Silica glass (fused silica)	99.5 +								
2 96% silica glass	96.3	< 0.2	< 0.2					2.9	0.4
3 Soda-lime—window sheet	71–73	12–15		8–10	1.5–3.5				0.5–1.5
4 Soda-lime—plate glass	71–73	12–14		10–12	1–4				0.5–1.5
5 Soda-lime—containers	70–74	13–16		10–13		0–0.5			1.5–2.5
6 Soda-lime—electric lamp bulbs	73.6	16	0.6	5.2	3.6				1
7 Lead-alkali silicate electrical	63	7.6	6	0.3	0.2		21	0.2	0.6
8 Lead-alkali silicate high-lead	35		7.2				58		
9 Aluminoborosilicate (apparatus)	74.7	6.4	0.5	0.9		2.2		9.6	5.6
10 Borosilicate—low-expansion	80.5	3.8	0.4				Li_2O	12.9	2.2
11 Borosilicate—low-electrical loss	70.0		0.5				1.2	28.0	1.1
12 Borosilicate—tungsten sealing	67.3	4.6	1.0		0.2			24.6	1.7
13 Aluminosilicate	57	1.0		5.5	12			4	20.5

Source: E. B. Shand, *Glass Engineering Handbook*, McGraw-Hill Book Company, New York, 1958, p. 4.

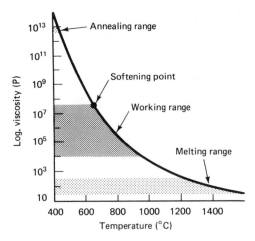

Figure 15-17 Viscosity of soda–lime–silica glass as a function of temperature. (From F. H. Norton, *Elements of Ceramics*, Addison-Wesley Publishing Company, Inc., Reading, Mass., 1952, p. 151.)

is similar to the residual compressive stresses induced in metal components to prevent fatigue failures, as we discussed in Chapter 9 (see Figure 9-31). Overall, the strength of the product is greatly increased and in many cases, such as window glass, eyeglass lenses, and so on, this results in a safer, more reliable product.

Properties and Applications

Since glasses are ceramic materials which do not ordinarily form crystalline structures, they exhibit certain useful optical and electrical properties. The fundamental reason for such characteristics can be explained in terms of their atomic bonding, which produces no delocalized valence electrons, and their random network-type structure. The optical properties of glass that make it a useful engineering material include its transparency or ability to transmit light, and its refractive properties. Glass also displays certain strength and formability characteristics which make it a desirable material, particularly as a reinforcing phase in polymers (structural glass-reinforced plastics). Finally, the chemical resistance of glass makes it an excellent container material for literally all chemicals (the only substance that rapidly attacks glass is HF).

Optical properties. When light strikes the surface of glass, some of this incident energy is reflected while the balance passes into the glass interior. The light entering the glass experiences *refraction*; that is, the direction of the light rays are changed. This change in direction depends on (1) the angle the incident light makes with the surface, and (2) the density of the glass. Refraction of light by a glass plate is depicted in Figure 15-18. The refractive index (*n*) is determined as follows:

$$n = \frac{\sin \theta}{\sin \phi} \qquad (15\text{-}5)$$

where θ = angle of incidence
  ϕ = angle of refraction

The refractive index (*n*) is in essence a measure of a glass's ability to "bend" light, with

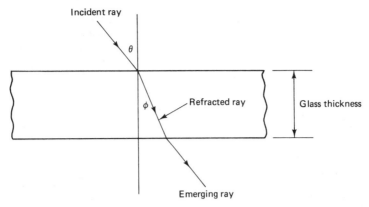

Figure 15-18 Schematic illustration of refraction and transmission of light by glass.

larger values of n indicating greater bending or refraction. The refractive indices of silicate glasses range from 1.458 for silica glass to 2.00 for very dense optical glasses. Values of n for some common glasses are listed in Table 15-10.

The proportion of light which is reflected from the surface of a glass also depends on n, the refractive index. The fraction of light reflected (R) by a single surface is expressed as follows:

$$R = \left(\frac{n-1}{n+1}\right)^2 \tag{15-6}$$

TABLE 15-10 VALUES OF REFRACTIVE INDICES (n) AND THERMAL EXPANSION COEFFICIENTS (α) FOR GLASSES LISTED IN TABLE 15-9

	Material	Refractive index (sodium D)	Coefficient of thermal expansion, α $(°C^{-1})$ (0 to 300°C)
1.	Silica glass (fused silica)	1.458	5.5×10^{-7}
2.	96% silica glass, 7900	1.458	8×10^{-7}
2a.	96% silica glass, 7911	1.458	8×10^{-7}
3.	Soda–lime—window sheet		85×10^{-7}
4.	Soda–lime—plate glass	1.510–1.520	87×10^{-7}
5.	Soda–lime—containers		85×10^{-7}
6.	Soda–lime—electrical lamp bulbs	1.512	92×10^{-7}
7.	Lead–alkali silicate—electrical	1.539	91×10^{-7}
8.	Lead–alkali silicate—high-lead	1.639	91×10^{-7}
9.	Aluminaborosilicate—apparatus	1.49	49×10^{-7}
10.	Borosilicate—low-expansion	1.474	32×10^{-7}
11.	Borosilicate—low-electrical loss	1.469	32×10^{-7}
12.	Borosilicate—tungsten sealing	1.479	46×10^{-7}
13.	Aluminosilicate	1.534	42×10^{-7}

Source: B. B. Shand, *Glass Engineering Handbook*, McGraw-Hill Book Company, New York, 1958, p. 17.

Glass

Thus, as the refractive index increases, the amount of reflection also increases and less light enters the glass.

Glass also absorbs energy from the light it transmits. Therefore, as light progresses through a glass, its intensity diminishes with distance. The overall transmittance (T) of light in glass is given by

$$T = Ke^{-\beta t} \qquad (15\text{-}7)$$

where $K =$ the reduction in intensity due to reflection $(1 - R)^2$
$\qquad \beta =$ the absorption coefficient (cm^{-1})
$\qquad t =$ glass thickness

In essence, transmittance is the ratio of intensity of emerging light to intensity of incident light. Values of T therefore may vary from 0 to 1, with $T = 1$, for example, indicating 100% transmission. Transmittance is particularly important in optical applications, including ophthalmic lenses, signal glasses, colored glasses, filters, photographic lenses, and optical fibers.

Transmittance of light in glasses can also be influenced by impurities. Although reduced transmission is undesirable in some cases, other applications depend on this condition. For example, *photochromic* lenses intentionally contain silver ions (Ag^+) in the form of silver halides. When this glass is exposed to ultraviolet light, the incident photons produce electrons which reduce Ag^+ to silver atoms (Ag^0) resulting in a reduction of light transmission. Thus the lenses darken, depending on the intensity of incident light. When the incident light diminishes, the electronic process reverses and the ions recombine, therefore the glass becomes more transparent.

Electrical properties. As we indicated in Chapter 10, glasses generally display relatively high values of electrical resistivity and therefore are generally considered as insulators. The electrical conductivity in these materials is very low and depends on migration of ions (usually sodium ions), as opposed to electron conduction in metallic conductors. Conductivity in glasses, then, tends to increase with increasing sodium ion concentration and temperature, in a classical diffusion manner [see equation (5-14)]. Certain electrical characteristics of glass, such as photoconductivity, resistivity, and dielectric properties, are discussed in Chapter 10 under the appropriate topics, and the student is referred to these sections for further details.

Mechanical properties. Glasses exhibit elastic behavior in much the same fashion as crystalline engineering materials. Therefore, if glass is elastically deformed, it will return to its original dimensions upon release of the applied loads. However, glass does *not* display any significant plastic deformation and fracture occurs before any permanent deformation takes place. The values of elastic modulus for selected commercial glasses are listed in Table 15-11, and a value of approximately 10×10^6 psi (69,000 MPa) is generally accepted for most common glasses.

Glass invariably fails from a tensile component of the applied stresses. This is true even when glass is loaded in compression, due to Poisson's ratio effects [see equation (6-7)]. There is a pronounced *size effect* on the strength characteristics of this material. Basically, this means that very small-diameter fibers can display much greater

	Material	Elastic modulus	
		10^6 psi	MPa
1.	Silica glass (fused silica)	10	69,000
2.	96% silica glass, 7900	9.7	66,930
2a.	96% silica glass, 7911	9.7	66,930
3.	Soda–lime—window sheet	10	69,000
4.	Soda–lime—plate glass	10	69,000
5.	Soda–lime—containers	10	69,000
6.	Soda–lime—electrical lamp bulbs	9.8	67,620
7.	Lead–alkali silicate—electrical	9.0	62,100
8.	Lead–alkali silicate—high-lead	7.6	52,440
9.	Aluminaborosilicate—apparatus	—	—
10.	Borosilicate—low expansion	9.8	67,620
11.	Borosilicate—low-electrical loss	6.8	46,920
12.	Borosilicate—tungsten sealing	8.7	60,030
13.	Aluminosilicate	12.7	87,630

Source: E. B. Shand, *Glass Engineering Handbook*, McGraw-Hill Book
Company, New York, 1958, p. 17.

strength values than that of larger sections (bulk specimens), even though the composition is identical. This trait is due to the brittle nature of glass and its relatively small critical crack size. The improvement occurs mainly from the elimination of flaws (especially surface discontinuities) in the finely drawn fibers. Such stress concentrators seriously affect the strength of glass, as discussed in Chapter 9, and promote failure. The influence of fiber size (diameter) on the tensile strength of drawn fibers is illustrated in Figure 15-19, which demonstrates the unusually high values of tensile strength displayed by very small-diameter glass fibers. The specific strength values of glass fibers are given in Chapter 16, where they are used extensively in glass-fiber-reinforced composites.

Summary. Since the preponderance of commercially utilized glasses fall into six basic categories (listed in Table 15-9), let us summarize this section on properties and applications by commenting briefly on each category.

Fused silica glass consists virtually of 100% silica and the addition of any other compounds places the glass in another category. This type of glass exhibits the maximum resistance to thermal shock as well as the highest permissible operating temperatures [i.e., 900°C (1652°F) for extended periods and 1200°C (2200°F) for short periods]. Although fused silica glass is superior to the other commercial glasses, its use is limited by the costs associated with producing "pure" silica and the available shapes in which it can be manufactured. Fused silica glass is therefore used only when the requirements of a particular design or application cannot be compromised. For example, high-purity fused silica glass is utilized in fiberous form to fabricate the heat-resistant tiles for the space shuttle's thermal protection system, as illustrated in Figure 15-20. These tiles are used in areas of the spacecraft such as forward fuselage,

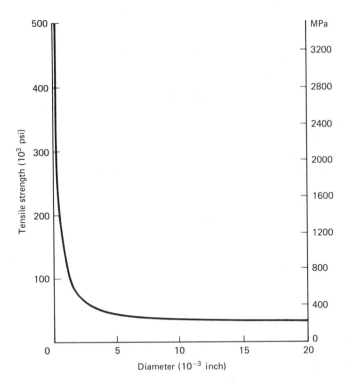

Figure 15-19 Tensile strength of typical glass fibers as a function of fiber diameter (Griffith curve).

lower wing, vertical tail, and so on, where temperatures may range between 648 and 1260°C (1200 and 2300°F) during reentry to the earth's atmosphere.

Pure silica glass is also used to produce *fiber optic* waveguides with very low loss attenuation characteristics. These fibers are utilized in communication systems for information transport, image transport, and laser transmission. For example, a *fiberscope*, illustrated in Figure 15-21, utilizes optical fibers which are produced by the rod and tube method. This process consists of slipping a rod of optical-quality silica glass inside a commercial-grade glass tube of lower refractive index. The assembly is then heated and drawn into a very thin, thread-like fiber. During use, the lower refractive index *cladding* reflects light back into the fiber core, thus preventing losses and consequent diminishment of the light beam. Fiberscopes are particularly useful in certain industrial applications which require inspections in severe or hazardous environments, difficult-to-reach areas, and nonsurgical internal examinations of human beings and animals for medical purposes.

Ninety-six percent silica glass is the designation given to silica glass containing a small percentage of B_2O_3 and other compounds. It can be used at continuous operating temperatures of 900°C (1652°F) and intermittently at 1200°C (2192°F). This glass also exhibits good resistance to thermal shock; therefore, it is applied to such items as furnace sight glasses (ports), drying trays, and space vehicle windows.

The most common type of glass is, of course, *soda–lime* (silica) glass, which is used in a wide variety of applications, such as windows, bottles, jars, lamp bulbs, and ophthalmic lenses. This type of glass is economical to produce and is readily fabricated into shapes, but does not exhibit the desirable properties of the silica glasses mentioned

(a)

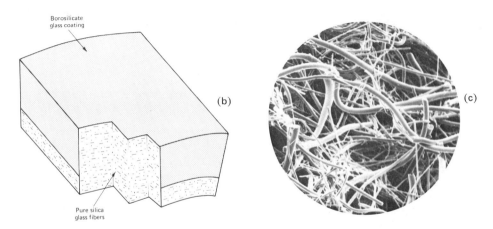

(b)

(c)

Borosilicate
glass coating

Pure silica
glass fibers

Figure 15-20   Fused silica glass applied to heat resistant tiles on space shuttle. (a) Shuttle craft showing location of tiles (dark areas). (b) Schematic of tile depicting borosilicate glaze covering pure silica fibers. (c) Fibrous nature of the tile as illustrated in the SEM (320×).

previously. For instance, soda–lime glass has poor resistance to high temperature and thermal shock, and its chemical resistance is only fair.

Lead–alkali glass is produced by substituting PbO for the lime in soda–lime glass. When the lead content is high, these glasses exhibit good dielectric properties and are used for capacitors and for the absorption of X-rays. They are also used for thermometer tubing, because of their brillance and workability.

In *borosilicate* glasses, the lime of the soda–lime glass is replaced with boric oxide (B_2O_3). This type of glass is more expensive than the soda–lime varieties and exhibits appreciable resistance to thermal shock because of its relatively low coefficient of thermal expansion (see Table 15-10). It will also withstand higher operating temperatures than either lead or lime glasses and has superior resistance to chemical

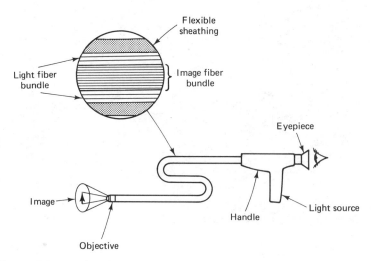

Figure 15-21 Typical fiberscope with self-contained light source. Cross section of fiber optics shaft (inset) shows image bundle surrounded by the light bundle, contained in a flexible, protective sheathing.

attack. Applications for borosilicate glasses include laboratory glassware, industrial piping, high-temperature thermometers, gauge sight glasses, and household ovenware (Pyrex).

The final category consists of *aluminosilicate* glass, which is similar to the borosilicates in behavior but with the ability to tolerate higher operating temperatures. These glasses are considerably more expensive and harder to fabricate than the borosilicates. However, they exhibit high softening temperatures and relatively low thermal expansion coefficients, which makes them especially suited for high-temperature applications such as for cookware intended to be used *directly* over flames or heating units.

STUDY PROBLEMS

15.1. Determine the radius ratio for the following compounds and predict the cation coordination number for each material based on Table 15-1. Compare your prediction with Table 15-2 and comment on any discrepancies.
  (a) SiO_2 **(b)** CsBr
  (c) FeS   **(d)** BeO
  (e) MgO   **(f)** UO_2

15.2. What is the lattice parameter (a) for cesium iodide (CsI)? What is the density of this compound?
  Answer: (a) 4.47 Å   (b) 4.839 cm³

15.3. The density of MgO is 3.6 g/cm³. Based on this information, calculate the lattice parameter (a) for this compound.

15.4. What is the density of NiO?
  Answer: 7.0 g/cm³

15.5. If a solid solution is formed from FeO and KCl, containing 10% oxygen, 25% Fe, 20% K, and 45% Cl by weight, how many cation vacancies will be present in 500 unit cells of this material?

15.6. If the concentration of vacancies in MgO at 1500°C is 100 times greater than the concentration at 1200°C, what is the energy of vacancy-pair formation in this compound? [*Hint:* See equation (4-3).]
Answer: 160 kcal/mole

15.7. What is the lattice constant (a) of the compound CdS? What is the density of this material?

15.8. ZnO forms a Wurtzite structure (see Table 15-2) and exhibits a density of 3.90 g/cm^3. Based on this information, what is the volume of a unit cell for this compound? Express your answer in Å. (*Hint:* Hcp structures have octahedral and tetrahedral interstices in the same ratio of sites to atoms as fcc structures.)

15.9. The CaF$_2$ structure is illustrated in Figure 15-5.
  (a) Sketch the arrangement of ions that intersect the (110) plane in this structure.
  (b) Based on your arrangement in part (a), what is the lattice constant (a) for CaF$_2$?
Answer: (b) 5.36 Å

15.10. The lattice constant for CeO$_2$ is 5.41 Å. Based on the structure illustrated in Figure 15-5, what is the density of this oxide compound?

15.11. Uranium dioxide (UO$_2$) forms a fluorite structure, as shown in Figure 15-5, and has a lattice constant of 5.47 Å.
  (a) Estimate the radius of the hole ($8 - f$) in the center of this unit cell.
  (b) How many such $8 - f$ holes are associated with a unit cell of UO$_2$?

15.12. Li$_2$O forms an antifluorite structure (see Table 15-2). Sketch the arrangement of ions in a unit cell of this compound. What is the coordination number of the anions in this structure?

15.13. Strontium titanate (SrTiO$_3$) forms a perovskite structure (see Figure 10-29). How many ions of each species are associated with a unit cell of this compound?

15.14. Cubic BaTiO$_3$ is stable above 120°C. Below 120°C a bct structure results, causing an electric dipole. If an ionic displacement of 0.03 Å along the c axis causes a strain of 0.75%, what is the lattice constant for the cubic arrangement?
Answer: 4.00 Å

15.15. Several important refractory products come from the Al$_2$O$_3$–SiO$_2$ system (see Figure 7-24).
  (a) For a mixture of 35 w/o Al$_2$O$_3$ and 65 w/o SiO$_2$ in equilibrium at 1650°C, how much of each phase is present?
  (b) In a "high-aluminum" brick composed of 85 w/o Al$_2$O$_3$ and 15 w/o SiO$_2$, cooled under equilibrium conditions, what are the proportions of the phases in coexistence at 1800°C?

15.16. Traditional whiteware bodies contain certain standard constituents, as indicated in Table 15-4. Give reasons for the wide differences in composition between dental porcelain and refractory porcelain.

15.17. A fire-clay refractory lining is heated from room temperature (22°C) to 566°C.
  (a) If this material is free to expand, what is the strain produced by this temperature change?
  (b) Suppose that by some oversight in design or fabrication, this material was constrained from expanding. What stress could be produced by this temperature change? ($E = 14 \times 10^6$ psi.)

(c) Comment on the type of stress (i.e., tensile or compressive) that results in the condition in part (b) and the consequences to the refractory lining.

Answer: (b) 41,888 psi or 288 MPa

15.18. MgO bricks 8 in. in length are used to line a high-temperature vessel.

(a) If 10 bricks are joined in a particular course (row), how much space will be needed to contain the bricks if they are heated from 50°C to 1650°C?

(b) If the thermal expansion is to be absorbed by properly spaced joints so that no stress is produced in the bricks, what is the minimum distance between bricks?

15.19. A cylindrical refractory liner fabricated from Al_2O_3 has dimensions 254 mm inside diameter, 280 mm outside diameter, by 305 mm long. In service, this material experiences a temperature change from 100°C to 1000°C.

(a) What is the volume change produced by this temperature increase?

(b) What percentage change occurs in the volume of the ceramic system?

Answer: (b) 2.4%

15.20. A melting furnace is lined with a cylindrical crucible made from stabilized ZrO_2 (see Table 15-6). If this vessel is used to melt nickel-based "superalloys" and experiences a temperature change from 72°F to 2500°F, what percent dimensional change would occur in the radial direction if the crucible was unrestrained? If the crucible is restrained from expanding by the surrounding furnace, what stress could develop in the liner? Comment on your answer.

15.21. The resistance of a refractory ceramic to spall or crack during heating and cooling can be related to some basic physical properties.

(a) Suggest properties that would affect spalling and comment on how such properties influence this form of failure.

(b) Formulate a relative spalling resistance index for refractory ceramics based on the properties in part (a).

15.22. A cylindrical alumina (Al_2O_3) liner, 20 in. OD, $\frac{7}{8}$ in. thick, 23 in. long, is heated to a temperature of 2000°F inside. The outside surface of this liner is maintained at 1000°F during service. How much heat energy is transferred to the surroundings in 1 hr under these conditions? (*Hint:* Heat transfer is expressed by

$$\frac{dQ}{dt} = kA\frac{dT}{dx}$$

where Q = heat energy

$\quad t$ = time

$\quad k$ = thermal conductivity

$\quad A$ = area

dT/dx = temperature gradient)

15.23. What is the heat flux in a stabilized zirconia crucible if the wall thickness equals 1.5 cm and the steady-state temperatures are 1000°C inside and 500°C outside? (*Hint:* See problem 15.22.)

Answer: 1.833 cal/sec-cm²

15.24. (a) What is the glass transition (T_g) that occurs in noncrystalline materials (including polymers)?

(b) At what temperature do these amorphous materials become solid?

15.25. (a) Explain how T_g is affected by the rate at which glasses are cooled.

(b) The random structure of many glasses can have beneficial effects on their mechanical and physical properties. If you wished to encourage randomness in a glass structure, how would you cool the system? Why?

15.26. Can metals and ceramics that ordinarily form crystalline structures be produced with amorphous or glass-like structures? Comment on why, or why not, this is possible.

15.27. Sketch a two-dimensional representation of a silicate chain (linked silicate tetrahedra) showing the bonds established between the participating ions. Compare the bond lengths and energies of this structure with those of a carbon chain. Does the latter information confirm your experience (or intuition) regarding materials composed of these chains? (*Hint:* See Table 2-7.)

15.28. Why are oxide compounds such as Na_2O, CaO, Al_2O_3, and MgO added to silica glass? Sketch a two-dimensional silicate chain containing additions of Na_2O, showing the bonds established between the adjacent ions.

15.29. A beam of light strikes a sheet of soda–lime window glass at an angle of 30° to the normal. At what angle will this incident light exit the other side of the glass sheet?

15.30. How much greater (relative percent) is the loss due to reflection of incident light for fused silica glass versus high-lead-type lead–alkali silicate glass?

15.31. A soda–lime glass with $n = 1.515$ is used for ophthalmic lenses. What is the transmittance for a 1-mm-thick lens if the absorption coefficient for this particular glass is 0.01 cm^{-1}?
Answer: 92%

15.32. A certain photographic lens glass has a refractive index of 1.458 and an absorption coefficient equal to 0.002 cm^{-1}. What is the percentage of incident light transmitted through a 2.5-mm-thick lens of this glass?

15.33. What is the percent linear expansion produced by heating an incandescent electrical light bulb from 27°C to 250°C? (See Table 15-10 for lamp bulb glass.)

15.34. A soda–lime plate glass is used in an application that subjects the glass to a temperature change from 0°C to 300°C. If this glass is restrained from expanding, what magnitude stress can develop?
Answer: 25,500 psi

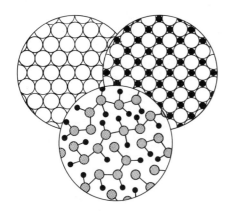

CHAPTER SIXTEEN:
Composite Materials
and Their Applications

The subject of composite materials has received considerable attention in recent years because of the unique properties and characteristics displayed by this group of engineering materials. This is especially true in structural applications requiring *high strength* and *light weight*. These materials have been highly successful in such areas as aerospace, transportation, recreational equipment, military, agriculture, and various other industrial applications.

Although many composites are often viewed as state-of-the-art materials, we would be sorely remiss if we did not point out that the fundamental concepts of composites were recognized by the Egyptians as early as about 1500 B.C. Ancient craftsmen utilized the principles of *fiber reinforcement* and *lamination* in such timely applications as brick making, long bows, armor plate, Damascus steel, and Samurai swords. Even at this early stage of our discussion it becomes clear that composites can encompass a very broad category of materials. In fact, one may ask what actually constitutes a composite material?

In order to define a composite material from the engineering standpoint, we will exclude discussions on the atomistic or molecular level and limit our explanation to a scale that extends from the microscopic to the macroscopic level. For example, on a microscopic scale, wood is a cohesive combination of *cellulose* fibers and *lignin*, as we discuss later in this chapter. Bone, as shown in Figure 16-1, is composed of *compact* tissue which is dense (ivory-like) and *cancellous* tissue which consists of slender fibers having a lattice-work structure. Many metals, such as wrought iron (see Figure 12-4), are combinations of various microstructural constituents. Concrete is a mixture of various aggregates (sand and stone) and cement which binds the aggregate particles together. Viewed on a macroscopic level, the materials above may all seem homogeneous or uniform in appearance.

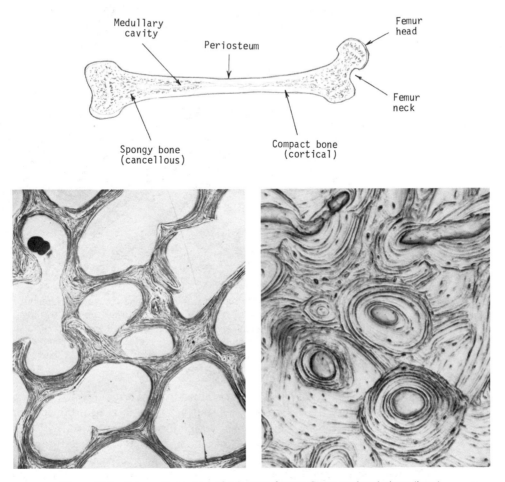

Figure 16-1 Schematic section of a human femur. Cross sectional views (inset) of cancellous bone tissue (40×) and compact bone tissue (100×). (Photomicrographs courtesy of J. W. Pugh, Hospital for Joint Diseases, New York.)

Other types of composite materials may display more obvious inhomogeneities at the "macro" level. For instance, alpine skis usually consist of multiple layers of plastic, metal, and wood, as shown in Figure 16-2. Automobile tires can contain steel or glass filaments in combination with rubber and fabric plies. Fiber-reinforced plastics, such as the *fiberglass* used in boat hulls, consist of glass filaments in a polymeric matrix. Even concrete may contain steel reinforcing bar or wire. The latter materials are perhaps more recognizably inhomogeneous. Both types of examples will be included in our definition of composite materials.

For the purpose of this text, *composite materials* will be considered as intimate

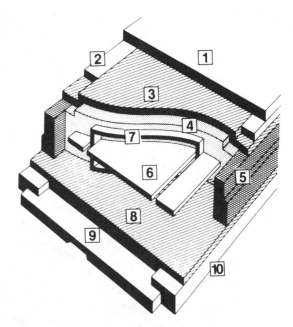

1. ABS top surface
2. ABS top edge
3. Bi-directional woven stratified glass epoxy
4. Polyurethane foam core
5. Phenolic sidewall
6. Phenolic binding insert
7. Braided fiberglass sock ± 45° bias
8. Bi-directional woven stratified glass epoxy
9. Polyethylene base
10. Hidden steel edge

Figure 16-2 Cutaway view of a high-performance alpine ski. Note the laminated construction of this "composite" product. 1, ABS top surface; 2, ABS top edge; 3, bidirectional woven stratified glass epoxy; 4, polyurethane foam core; 5, phenolic sidewall; 6, phenolic binding insert; 7, braided fiberglass sock ±45° bias; 8, bidirectional woven stratified glass epoxy; 9, polyethylene base; 10, hidden steel edge. (Courtesy of Rossignol Ski Co.)

combinations of materials differing in composition or form, in which the various constituents retain their separate identities, but act in concert to produce properties in the composites that are not attainable by the individual constituents.

Overall, the properties exhibited by a composite material are strongly influenced by the properties of the individual constituents, their respective amounts (percent), shape, orientation, and distribution, plus any synergistic interaction between these constituents when they are combined in composite form. Even though the resultant behavior of the composite material is necessarily related to the properties of the individual constituents, the composite may be far superior to the individual constituents in a particular characteristic or an application.

PRINCIPLES OF FIBER REINFORCEMENT

The mechanical properties of a composite can depend on many factors, as we alluded to in the introduction. Basically, these factors include the specific properties of the matrix and the reinforcing phase; their respective volume fractions; the shape, size, distribution, and orientation of the reinforcing phase; and the bond between the reinforcement and its matrix. Let us examine in more detail the relationships between these factors and their influence on the mechanical behavior of composite materials.

Stress–Strain Relationships

In order to analyze the stress–strain behavior of a composite, we will consider a simple situation where a matrix containing continuous cylindrical fibers is deformed in a direction parallel to the long axis of the fibers, as illustrated in Figure 16-3. If we can

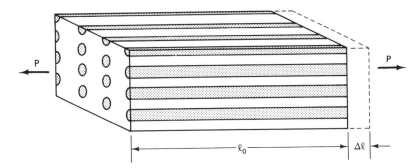

Figure 16-3 Axial deformation due to the load (P) in a composite reinforced with continuous cylindrical fibers.

assume that no slippage occurs at the interface between the fibers and the matrix, the elongation (Δl) is the same in both phases. Therefore, the engineering strain [see equation (6-1)] is identical in both phases and the strain (e_c) in the composite may be expressed as

$$e_c = e_f = e_m = \frac{\Delta l}{l_0} \tag{16-1}$$

where e_f = strain in fibers
$\quad e_m$ = strain in matrix

Furthermore, when these strains are elastic, the engineering stress in the respective constituents can be determined by applying Hooke's law [see equation (6-13)] as follows:

$$\sigma_f = E_f e_f \tag{16-2a}$$

$$\sigma_m = E_m e_m \tag{16-2b}$$

where σ_f = stress in fiber
$\quad \sigma_m$ = stress in matrix
$\quad E_f$ = elastic modulus of fiber
$\quad E_m$ = elastic modulus of matrix

In most cases, the modulus of the fiber is greater than that of the matrix, by design. Thus, for a given strain, the stress in the fibers is greater than the matrix stress. Although this is a very simplified analysis, it serves to illustrate the extremely important concept of *fiber reinforcement*. Generally, the fibers (or for that matter, the reinforcing phase in most composites) are more highly stressed than the matrix when a load is applied to the composite. This mechanism permits transfer of the stresses from the matrix to the reinforcing phase. As a result, the composite can sustain greater stresses than the unreinforced matrix material. This concept of utilizing the advantages of one

Principles of Fiber Reinforcement

material such as high modulus or high strength, in conjunction with the advantages of another, such as low density or corrosion resistance, is the very essence of composite materials.

We may summarize the stages of deformation in a composite uniaxially reinforced with continuous fibers, in the following manner:

Stage 1: Both fibers and matrix deform elastically.

Stage 2: Fibers deform elastically, matrix deforms plastically.

Stage 3: Both fibers and matrix deform plastically.

Stage 4: Fibers fracture, followed by failure of the composite.

Volume Fractions

In the simple mechanical circumstances that we have introduced, certain properties of the composite material are related to the properties of the individual constituents by the *rule of mixtures*. This relationship, which is based on the sum of the products of the individual properties and their respective volume fractions, is expressed for composites as follows:

$$E_c = E_f V_f + E_m V_m \tag{16-3a}$$

$$\sigma_c = \sigma_f V_f + \sigma_m V_m \tag{16-3b}$$

where E_c = elastic modulus of composite

σ_c = stress in the composite

V_f = fiber volume fraction

V_m = matrix volume fraction

In composites with simple geometric shapes and uniform fiber cross section, the fraction of the total composite volume occupied by the fibers (or reinforcement phase) is often known or determinable. Thus the volume fraction occupied by matrix material (V_m) is simply the total volume fraction minus the fiber volume fraction as follows:

$$V_m = 1 - V_f \tag{16-4}$$

Therefore, the expressions given in equations (16-3a) and (16-3b) may also be written exclusively in terms of V_f.

Since the load carrying capability of the composite depends principally on the modulus and strength of the reinforcing fibers, the greater the volume fraction of this portion, the more efficient is the composite from the strength standpoint. For example, consider the tensile strength behavior of copper reinforced with continuous tungsten wire 0.005 in. (0.1 mm) in diameter as shown in Figure 16-4. This figure clearly shows that increasing the amount of fibers increases the tensile strength of the composite. Essentially, the composite behaves similarly to the matrix at low fiber volume fractions, while at high fiber volume fractions, the composite behaves more like the fiber phase. Therefore, at very low values of V_f, the properties of the matrix may determine composite behavior. Such a condition is depicted in Figure 16-4, where the composite tensile strength is matrix controlled below a minimum value of V_f, but is fiber controlled above this value ($V_{f_{min}}$).

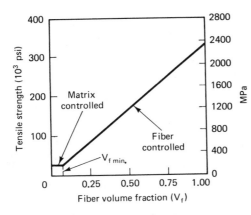

Figure 16-4 Effect of fiber concentration on the tensile strength of copper reinforced with 0.005 inch (0.1 mm) diameter tungsten wire. Source: W. R. Hibbard, Jr., "Fiber Composite Materials," American Society for Metals, Metals Park, Ohio, 1965, p. 3.

The tensile strength of the composite material illustrated in Figure 16-4 can be determined from equation (16-3b) for $V_f > V_{f_{min}}$. However, for values of $V_f < V_{f_{min}}$, the matrix properties govern this behavior and the strength may be expressed as follows:

$$\sigma_c = \sigma_{m_{UTS}}(1 - V_f) \tag{16-5}$$

where: $\sigma_{m_{UTS}}$ is the ultimate tensile strength of the matrix.

If the strength of the composite material is to exceed the ultimate tensile strength of the matrix, it is necessary that $V_f > V_{f_{min}}$. This minimum value of V_f increases as the degree of strain hardening in the matrix increases and as the matrix UTS nears that of the fibers. Consequently, when a matrix with high UTS is reinforced with fibers of slightly higher strength, large fiber volume fractions will be required to produce any significant strengthening.

Furthermore, from a practical standpoint, the tensile properties of fiber-reinforced composites generally begin to decrease when V_f exceeds about 0.80. This degradation is due to poor bonding between fibers and matrix, fiber-to-fiber contact, and voids in the composite which result from the high percentage of reinforcing phase.

Example 16-1

A composite material is produced from continuous tungsten wires ($E = 350,000$ MPa) embedded in a matrix of copper ($E = 110,400$ MPa). The composite contains 60% fibers by volume and is subjected to an axial strain (e) of 0.1%. If both the fibers and the matrix deform elastically (stage 1 deformation), determine the stress developed in the composite.

Solution The stress in this composite can be determined from equations (16-3b) and (16-4), as follows:

$$\sigma_c = \sigma_f V_f + \sigma_m(1 - V_f)$$

Since both materials deform elastically, $e_f = e_m = e_c$, we can restate the rule of mixtures using equations (16-2a) and (16-2b) as follows:

$$\sigma_c = (e_c E_f)V_f + (e_c E_m)(1 - V_f)$$

Substituting the known data, we have

$$\sigma_c = (0.001)(350{,}000 \text{ MPa})(0.6) + (0.001)(110{,}400 \text{ MPa})(0.4)$$
$$= 210 \text{ MPa} + 44 \text{ MPa}$$
$$= 254 \text{ MPa or } (36{,}812 \text{ psi}) \quad \textit{Ans.}$$

Fiber Considerations

Fiber properties. Since the fiber-reinforced composite material relies on the fiber for its strength and stiffness, it is essential that this constituent possess high strength and high modulus compared to the matrix. Then with the appropriate volume fraction accompanied by control of fiber orientation and fiber dimensions, the mechanical behavior, such as strength, toughness, and stiffness, can be optimized.

Presently, there are large numbers of fiber reinforcement materials available for composite design and fabrication. These reinforcing materials consist of both continuous and discontinuous fibers, which typically display properties as given in Table 16-1. Examination of these data quickly reveals that very high tensile strengths and/or high elastic modulus can be obtained in these fibers compared to the same material conventionally produced in bulk form. For instance, the tensile strength of the glass fibers ranges from 500,000 to 850,000 psi (3450 to 5865 MPa), while that of steel is 600,000 psi (4140 MPa). Other classes exhibit very high values of elastic modulus; the ceramic compounds, such as B_4C, SiC, and TiB_2, have moduli on the order of 70 million psi (483,000 MPa). In addition to strength and modulus other properties often play a significant role in a particular design. For example, high-temperature resistance is sometimes an important criterion for fibrous composites. In this case, materials such as carbon or graphite fibers, tungsten, and graphite whiskers may be the appropriate reinforcing phase, since they exhibit relatively high resistance to softening and melting. For example, reinforced carbon–carbon composites are used to protect the wing leading edge and the nose cap of the space shuttle orbiter, where temperatures during reentry can exceed 1260°C (2300°F).

Since the student is likely to be familiar with certain material properties at this stage, one may question the strength values presented here. For instance, laboratory tensile tests of steel ordinarily produce ultimate tensile strengths on the order of 49,000 to 300,000 psi (338 to 2070 MPa), as shown in Table 16-1 and as we indicated in Chapter 12 (see Tables 12-3 and 12-4). How, then, can strengths of 600,000 psi (4140 MPa) be attained in steel fibers? The explanation for this apparent enigma deals essentially with eliminating flaws that act as stress concentrators.[1] In very thin fibers and wires, the probability of a flaw is considerably less than in a bulk material. Also, the drawing process from which wires are produced assists in alleviating the severity of surface imperfections and may result in a favorable orientation of the crystal structure, or molecules in the case of polymeric fibers.

Whiskers, on the other hand, are very pure, unidirectional single crystals which have been grown under almost ideal conditions. These fibers typically are only a few

[1] We alluded to this process in Chapter 9 in the sections "Background and Griffith's Energy Criteria."

TABLE 16-1 TYPICAL PROPERTIES OF SELECTED FIBERS AND CONVENTIONAL ENGINEERING MATERIALS[a]

Material	Density, ρ (g/cm^3)	Tensile strength, σ_u MPa	Tensile strength, σ_u psi	Specific strength, σ_u/ρ (10^7 cm)	Tensile modulus, E MPa	Tensile modulus, E 10^6 psi	Specific modulus, E/ρ (10^7 cm)
Fibers							
E-glass[b]	2.54	3500	507,500	1.40	72,400	10.5	29.1
S-glass[c]	2.48	4600	667,000	1.89	85,500	12.4	35.2
Graphite (high modulus)	1.90	2100	304,500	1.13	390,000	56.6	209.4
Graphite (high strength)	1.90	2500	362,500	1.34	240,000	34.8	128.8
Boron	2.63	2800	406,000	1.08	385,000	55.8	149.3
Silica	2.19	5800	841,000	2.70	72,400	10.5	33.7
Tungsten	19.30	4200	609,000	0.22	414,000	60.0	21.9
Beryllium	1.83	1300	188,500	0.72	240,000	34.8	133.8
Aramid polymer	1.50	2800	406,000	1.90	130,000	18.8	88.4
Steel	7.74	4140	600,000	0.55	200,000	29.0	26.4
Conventional (bulk form)							
Steel	7.8	340–2100	49,300–304,500	0.04–0.27	210,000	30.4	27.5
Aluminum alloy	2.7	140–620	20,300–89,900	0.05–0.23	70,000	10.2	26.4
Glass	2.5	700–2100	101,500–304,500	0.28–0.86	70,000	10.2	28.6
Tungsten	19.30	1100–4100	159,500–594,500	0.06–0.22	350,000	50.8	18.5
Beryllium	1.83	700	101,500	0.39	300,000	43.5	167.2

[a]For a more detailed compilation of fiber properties, see W. D. Sutton, in *Modern Composite Materials*, L. J. Broutman and R. H. Krock, Eds., Addison-Wesley Publishing Company, Inc., Reading, Mass., 1967, p. 414.

[b]A high electrical resistivity glass based on the eutectic in the CaO-Al$_2$O$_3$-SiO$_2$ system (Fig. 16-24).

[c]A commerical glass fiber with higher strength and modulus than "E" glass, from the MgO-Al$_2$O$_3$-SiO$_2$ system.

Source: Adapted from B. D. Agarwal and L. J. Broutman, *Analysis and Performance of Fiber Composite Materials*, John Wiley & Sons, Inc., New York, 1980, p. 8.

microns in cross section and contain virtually none of the crystallographic defects found in ordinary polycrystalline materials.

Recall our discussion in the section "Theoretical Cohesive Strength" in Chapter 4 (see Figure 4-10). The fundamental reason commercial polycrystalline materials do not attain their theoretical strength is that they contain substantial numbers of dislocations. It is these crystalline imperfections that permit permanent deformation and failure at relatively low applied stresses. Since whiskers are comparatively free of such defects, they exhibit very high strengths, although they still do not achieve their maximum theoretical cohesive strength.

Variability in strength of fibers is also a concern in composite materials, since some fibers are inherently stronger than others. For example, consider the situation where bundles of fibers are axially strained, such as depicted in Figure 16-3. The weaker fibers fail first, thereby increasing the load on the remaining fibers. Even though the surviving fibers are stronger, their number is decreased and as the load increases, they continue to fail in a progressive manner. If this process continues, eventually the effective strength of the composite is insufficient to support the applied load and the composite material fails.

Effective fiber length. Thus far we have considered composite materials with continuous fibers embedded in a matrix. However, some composites contain discontinuous fibers. Such reinforcement may be by design, or perhaps the result of fiber fractures during loading. In any event, the composite properties no longer obey the rule of mixtures under these circumstances, and the reinforcement may be ineffective unless the fibers are considerably longer than a certain *critical* length (l_c).

If we examine the stresses occurring at the ends of a hypothetical fiber embedded in a matrix as illustrated in Figure 16-5, the tensile stress in the fiber drops from the nominal fiber stress ($\sigma_{f_{nom}}$) to zero at the fiber end. Correspondingly, the interfacial shear stress increases from the nominal shear stress ($\tau_{m_{nom}}$) to a maximum value near the fiber end. The portion of fiber over which this process occurs is denoted $l_c/2$, and represents the length of fiber which is *ineffective* with regards to tensile loading, because the stresses carried by end portion of the fiber are lower than the nominal stress ($\sigma_{f_{nom}}$). Since any fiber must be comprised of two ends, a reinforcing fiber must be longer than l_c in order to be effective. In other words, discontinuous fibers in a composite must be greater than l_c if the load-carrying capability of that composite is to be maximized.

Although many factors, such as fiber shape, end geometry, respective elastic and shear moduli of fiber and matrix, and interfacial bond strength, affect these stresses, the effective fiber length, when the matrix behaves as an ideal plastic material (i.e., no strain hardening), can be expressed as

$$\frac{l_c}{d_f} = \frac{\sigma_{f\max}}{2\tau_{m_y}} \qquad (16\text{-}6)$$

where l_c = critical fiber length
d_f = fiber diameter
$\sigma_{f\max}$ = maximum fiber stress
τ_{m_y} = shear strength of the matrix

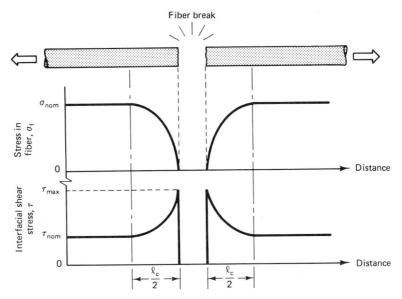

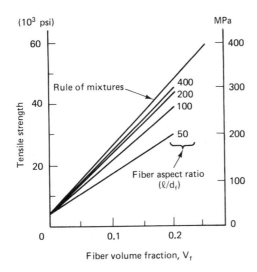

Figure 16-5 Schematic of composite stress at the ends of a fiber, when the matrix exhibits elastic deformation (l_c is the critical length).

The value l_c/d_f is referred to as the *critical aspect ratio* and is independent of volume fraction for matrices that behave plastically. In cases where the matrix deforms elastically, the critical aspect ratio decreases for increasing values of V_f. Thus, for a specific fiber diameter, the critical length (l_c) decreases as the fiber volume fraction increases. The effects of l/d and V_f on tensile strength are shown in Figure 16-6 for steel fibers embedded in an aluminum matrix.[2] It is clear from these data that the

Figure 16-6 Effect of l/d and V_f on composite tensile strength of aluminum reinforced with steel fibers. (From D. Cratchley, *Powder Met. 11,* 1963, p. 59.)

[2]Since one does not simply drill holes in the aluminum and insert steel fibers, the student is asked to contemplate possible methods for actually embedding steel fibers in a matrix of aluminum.

tensile strength of the composite increases as V_f increases, and furthermore, as the fiber aspect ratio (l/d_f) increases, this behavior approaches that predicted by the rule of mixtures [equation (16-3b)].

Example 16-2

A composite material is produced from alumina whiskers embedded in a silver matrix. The average diameter of the whiskers is 0.001 in. (0.025 mm), and their fracture strength $\sigma_{f_{max}}$ is 1,000,000 psi (6900 MPa). If the shear strength (τ_y) of the silver is 8000 psi (55 MPa) and the matrix behaves like an ideal plastic material, determine the *critical* length (l_c) of the alumina whiskers.

Solution Employing equation (16-6), we can calculate the critical length for this condition as follows:

$$l_c = \frac{(\sigma_{f_{max}})(d_f)}{(2)(\tau_y)}$$

$$= \frac{(1 \times 10^6 \text{ psi})(0.001 \text{ in.})}{(2)(8000 \text{ psi})}$$

$$= 0.062 \text{ in. (1.6 mm)} \qquad Ans.$$

Comment: This value is independent of volume fraction (V_f). But if the matrix behaved elastically, the critical length would depend on V_f and decreases for increasing V_f (at a constant d_f).

Fiber orientation.   The alignment of fibers in fiber-reinforced composites is also a very important factor. Misalignment of fibers occurs during fabrication simply because short fibers and slender filaments are difficult to align. The orientation of the fibers with respect to the direction of applied loading affects the tensile strength of a composite as shown in Figure 16-7. For small angles of misorientation (up to about

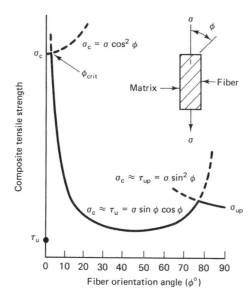

Figure 16-7   Effects of fiber orientation of composite strength. γ_u, ultimate shear strength of matrix; σ_{up}, ultimate tensile strength of matrix in plane strain; τ_{up}, ultimate shear strength of matrix in plane strain. [From A. Kelly and G. J. Davies, *Metallurgical Reviews 10* (37), 1965, p. 1.]

8°), the strength of the composite is unaffected. But, when the fibers are misaligned at angles greater than this critical value (ϕ_{crit}), the strength of the composite decreases precipitously.

Interfacial bond. In addition to the aspect ratio (l/d) and the orientation of the fibers, the strength of the bond between the fiber and matrix is a principal factor in determining the mechanical behavior of a composite. A strong bond between these constituents is necessary for them to act together during loading or service. To develop this bond, it is essential that the matrix *wet* (spread freely) the surface of the fiber during fabrication. Such wetting is necessary in obtaining good molecular or atomic interaction between the constituents.

If the matrix material does not readily wet the fiber, special finishes called *coupling agents* can be used to promote bonding between the matrix and its reinforcement. Such finishes are usually applied to the fiber reinforcement prior to fabrication of the composite, during the process referred to as *sizing*. For example, organometallic compounds of cl. omium, and organosilane compounds are typically used for the treatment of glass fibers in polymeric matrices. Although the specific interaction between the coupling agents and composite phases is not precisely known, it has been proposed that the resinophilic molecules (R) interact with the hydrocarbon matrix while the silane groups interact with sites on the glass surface. Such a process is depicted in Figure 16-8.

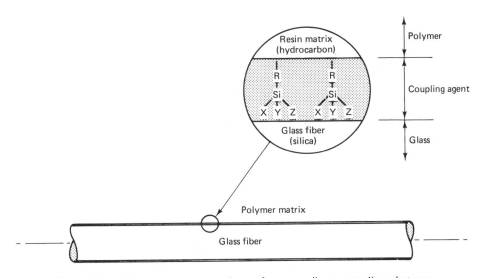

Figure 16-8 Schematic representation of organosilane coupling between polymeric matrix and glass fiber. R: resinophilic group (vinyl, amino, or ethoxy); X, Y, Z: hydrophilic substituents (halo, alkoxy, or aryloxy).

Surface roughness of the fiber phase is another important consideration in promoting bonding between the constituents of the composite. If the surface is very rough, containing many minute asperities, the matrix material may not penetrate the depressions and a poor interfacial bond will result.

Principles of Fiber Reinforcement

In the case of metallic composite constituents, only a few metals are completely insoluble in each other. Therefore, reaction of the fiber with the matrix may occur at their interface, especially if elevated service temperatures are encountered. Such reactions may increase the strength of the bond (e.g., solid solution alloying between the metals). On the other hand, metallurgical or chemical interaction between the components may also degrade the interfacial bond strength by forming weak or brittle phases, and by producing surface asperities in the fibers which act as stress concentrators. A common method for avoiding these problems is to electroplate the fiber with an unreactive metal. Such a coating prevents adverse chemical reactions between the fiber and matrix.

Properties and Applications

Fiber-reinforced plastics. The bulk of fiber-reinforced composite materials consists of plastics reinforced with continuous fibers. The reinforcement phase in these materials may be in the form of bundles of very fine glass filaments or other fibers, such as graphite, SiC, boron, metallic, and aramid polymers. Glass fibers are prepared by extruding molten glass through a small orifice, then drawing the resulting fiber to a very fine diameter, on the order of 3 to 20 μm (1.2 to 7.9 $\times$ 10^{-4} in.). Individual filaments (fibers) of glass are usually bundled into one larger strand containing a large number of filaments. A typical glass strand containing multiple fibers is shown in Figure 16-9. The glass fibers may be finished (sized) and preimpregnated with an appropriate matrix material during production.

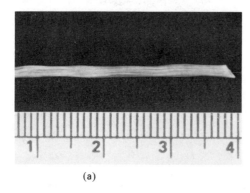

(a)

Figure 16-9 Example of glass fibers: (a) one glass strand; (b) close-up showing the large number of glass filaments contained in the strand (50$\times$).

(b)

Applications for glass-reinforced plastic (GRP) are quite varied. Hence this material is fabricated in several commercial forms. Continuous strand fibers and *roving*, which resemble ribbon or tape, are used to wind (filament winding process) such components as pipe, tanks, and pressure vessels. *Chopped* fiber (6 to 20 mm long) is utilized in manufacturing molded GRP shapes with randomly oriented fibers. The chopped fiber material can be sprayed in place or blended with other resins to produce a low-cost molding compound which facilitates its use in making panels and other shapes for among others, the automobile and aerospace industry. Woven fabrics and nonwoven mats are also important types of GRP materials and are produced in various forms, such as sheets, cloth, and roving. These composites find particular application to structural shapes, such as building shells and panels, molded GRP boats, and canopies.

Typical properties of commercial fiber reinforced plastic composite materials are given in Tables 16-2 and 16-3. Inspection of these data clearly shows the strength

TABLE 16-2 TYPICAL PROPERTIES OF GLASS-REINFORCED PLASTICS (DISCONTINUOUS FIBERS)

Material	Glass content (vol %)	Tensile strength		Elastic modulus		Specific gravity
		10^3 psi	MPa	10^6 psi	MPa	
Nylon 66	20	22	152	1.2	8,280	1.31
	40	29	200	1.6	11,040	1.41
	70	30	207	3.1	21,390	—
Polycarbonate	20	15.5	107	0.9	6,210	1.31
	40	19	131	1.5	10,350	1.44
Polystyrene	30	14	97	1.2	8,280	1.28
Polyester	35–45	15–25	104–173	0.8–1.8	5520–12,420	1.5–1.6
Epoxy	60	25	173	3	20,700	1.8

TABLE 16-3 TYPICAL PROPERTIES OF CONTINUOUS-FIBER-REINFORCED EPOXY RESINS

Material	Fiber content (vol %)	Specific gravity	Fibers parallel to load (%)	Tensile strength (psi)[a]	Tensile modulus (10^6 psi)	Compression strength (psi)[a]
E-glass/epoxy	73	2.17	100	238,000	8.1	—
E-glass/epoxy	56	1.97	100	149,000	6.2	87,000
E-glass/epoxy	56	1.97	54	75,000	3.5	71,000
E-glass/epoxy	56	1.97	0	5,000	1.5	20,000
S-glass/epoxy	72	2.12	100	275,000	9.6	200,000
S-glass/epoxy	72	2.12	0	8,000	—	20,000
Al_2O_3/epoxy	57	2.75	100	75,000	30.0	340,000
	57	2.75	0	8,000	3.0	175,000
Graphite/epoxy	63	1.61	100	250,000	23.0	198,000
	63	1.61	0	6,000	1.6	33,000
Aramid/epoxy	62	1.38	100	190,000	12.0	41,000
	62	1.38	0	5,700	0.8	20,000

[a]To obtain MPa values, multiply psi by 6.9×10^{-3}.

Principles of Fiber Reinforcement

advantage of continuous fiber reinforcement compared to discontinuous fibers, especially when the composite is loaded *parallel* to the fibers. Moreover, the aspect of mechanical property anisotropy is also vividly demonstrated by the data in Table 16-3. For example, as the percent of glass reinforcement oriented parallel to the applied load decreases for a particular composite system, the tensile strength of the composite drops sharply. In the case of epoxy reinforced with 56% E-glass ($V_f = 0.56$), this decrease amounts to 144,000 psi as the percent of glass parallel to the load changes from 100% to 0%. This orientation aspect was discussed earlier in the section on principles of fiber reinforcement. In fact, when the fibers are oriented perpendicular to the applied tensile load, the composite's tensile strength is approximately the same as that of the matrix material. Our point is that proper utilization of the continuous fiber-reinforced composite is strongly dependent on the orientation of the fibers with respect to applied loads. Any structural application of these materials must necessarily take their *anisotropic* properties into consideration.

Fiber-reinforced metals. The attractiveness of this group of composites is due to their high strength-to-weight ratios and relative resistance to softening under elevated temperature service conditions. A variety of reinforcing fibers are available for strengthening metals, including glass fibers, metal fibers, whiskers, and ceramic fibers. For example, a composite system consisting of alumina (Al_2O_3) fibers in an aluminum matrix is shown in Figure 16-10.

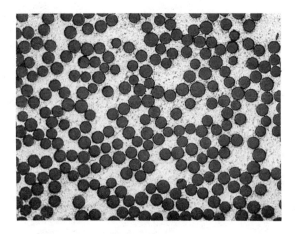

Figure 16-10 Photomicrograph showing the cross section of alumina (Al_2O_3) fibers embedded in an aluminum matrix (250×).

Typical properties for a number of fiber-reinforced metal systems are given in Table 16-4. These properties depend on the factors we discussed earlier in the section on principles of fiber reinforcement. The data presented in Tables 16-4 and 16-1 include *specific strength* (σ_u/ρ_c) and *specific modulus* (E_c/ρ_c) for a very good reason. This parameter allows the designer quickly to assess the strength-to-weight ratio or modulus-to-weight ratio of a particular material. In some instances, these aspects may be critical to the optimum performance and efficiency of a component produced from a composite material.

TABLE 16-4 TYPICAL STRENGTH PROPERTIES OF FIBER-REINFORCED METALS AT ROOM TEMPERATURE

Fiber	Matrix	V_f	Composite tensile strength, σ_u (psi)[a]	Specific strength, σ_u/ρ_c (10^3 in.)
Graphite		0.40	180,000	
Al_2O_3		0.50	90,000	769
SiO_2		0.48	126,000	1,500
Steel		0.25	173,000	1,210
B		0.10	43,000	453
Be		0.04	33,000	344
E-glass	Al	0.50	45,000	
Al_2O_3*		0.35	161,000	1,420
Al_3Ni*		0.10	48,000	470
$CuAl_2$*		0.16	39,000	307
B_4C*		0.10	29,000	302
Stainless steel*		0.11	26,000	
B		0.08	384,000	1,470
W	Ni	0.40	161,000	344
Al_2O_3*		0.19	171,000	600
B		0.35	70,000	—
Al_2O_3	Mg	0.50	75,000	742
Ta		0.32	29,000	—
Al_2O_3*	Fe	0.36	237,000	1,020
W	Ni–Cr	0.22	73,000	185
Al_2O_3*	Ni–Cr	0.09	255,000	870
Steel	Ag	0.44	65,000	191
Al_2O_3*	Ag	0.24	232,000	720
Si_3N_4*	Ag	0.15	40,000	119
Mo*	Ti	0.20	96,000	457
Graphite	Cu	0.65	115,000	—
W	Cu	0.77	255,000	420
Al_2O_3*	Al–10 Si	0.15	40,700	395
Ta_2C	Ta	0.29	155,000	267
W	Co	0.30	107,000	244
Mo	Co	0.17	52,000	156
W	316 stainless	0.18	58,600	175
E-glass	Pb	0.08	15,000	—
Al_2O_3	Pb	0.45	27,000	112
Graphite	Pb	0.35	126,000	—

[a]To obtain MPa values, multiply psi by 6.9×10^{-3}.

*Short, discontinuous fibers

Although the applications of metal-matrix composites are not presently as widespread as the reinforced plastics, there certainly are instances where metal-matrix composites are more suitable. For example, superconducting cables such as wires of niobium–tin (Nb_3Sn) intermetallic compound in a copper matrix are capable of generating very intense magnetic fields at liquid-helium temperatures. Filamentary conductors for transvenous insertion to heart pacemakers utilize wires fabricated from a nickel alloy (MP35N) embedded in a silver matrix for the appropriate combination of corrosion resistance, electrical conductivity, and mechanical properties. Table 16-5 lists the potential applications of a number of fiber-reinforced metal composite systems.

TABLE 16-5 APPLICATIONS FOR METAL-MATRIX COMPOSITE MATERIALS

Fiber	Matrix	Potential applications
Graphite	Aluminum	Satellite, missile, and helicopter structures
	Magnesium	Space and satellite structures
	Lead	Storage battery plates
	Copper	Electrical contacts and bearings
Boron	Aluminum	Compressor blades and structural supports
	Magnesium	Antenna structures
	Titanium	Jet-engine fan blades
Borsic	Aluminum	Jet-engine fan blades
	Titanium	High-temperature structures and fan blades
Alumina	Aluminum	Superconductor restraints in fusion power reactors
	Lead	Storage battery plates
	Magnesium	Helicopter transmission structures
Silicon carbide	Aluminum	High-temperature structures
	Titanium	High-temperature structures
	Superalloy (Co-based)	High-temperature engine components
Molybdenum	Superalloy	High-temperature engine components
Tungsten	Superalloy	High-temperature engine components

Source: C. T. Post, *Iron Age*, July 2, 1979, p. 52.

PRINCIPLES OF PARTICLE REINFORCEMENT

The subject of particle reinforcement was actually introduced in Chapter 6, during our discussion of strengthening mechanisms. There we examined the mechanism of precipitation hardening wherein a fine distribution of coherent precipitate particles was produced by aging a supersaturated solid solution. In the case of precipitation hardening and also dispersion hardening (mechanical mixture of dispersoid in a matrix), the particles impede dislocation motion in the matrix, thereby increasing the overall strength of the composite system.

Particulate reinforcement is generally divided into two categories, *dispersion-strengthened* and *particle-strengthened*, based on the following factors:

1. Dispersion-strengthened:

    Particle diameter: < 0.1 μm
    Matrix mean free path: 0.01 to 0.3 μm
    Volume fraction: < 0.15

2. Particle-strengthened:

    Particle diameter: $> 1 \mu$m
    Matrix mean free path: > 1 μm
    Volume fraction: > 0.25

Dispersion Strengthening

In contrast to the fiber-reinforced composites just discussed, the matrix in dispersion-reinforced composites is the principal load-carrying constituent. Perhaps this sounds contradictory, but the reinforcing particles actually impede slip on a very localized scale, thus increasing the resistance to plastic deformation. The degree of matrix strengthening therefore depends on how effective the dispersed particles are as a barrier to slip.

The effectiveness of dispersion reinforcement may be analyzed by considering the particle size (d), the volume fraction of particles (V_p), and the mean free matrix path between particles (mfp). In some cases, the interparticle spacing (D_p) is used in place of the matrix mean free path. The relationship between these variables is expressed as follows:

$$\text{mfp} = \frac{2d}{3V_p}(1 - V_p) \qquad (16\text{-}7)$$

As discussed in Chapter 6, if the applied shear stress (τ) is sufficient, dislocations in the matrix will bend or bow around the particles (see Figure 6-18). The stress necessary to produce such *bowing* is given as

$$\tau = \frac{G_m \bar{b}}{D_p} \qquad (16\text{-}8)$$

where G_m = shear modulus of matrix
$\bar{b}$ = Burger's vector of dislocation

Thus the strengthening effect of particle dispersions is inversely related to interparticle spacing (D_p). Moreover, the loops of slipped crystal surrounding the dispersed

Principles of Particle Reinforcement

particles effectively reduce the interparticle spacing, making it more difficult to move other dislocations through the dispersion. The higher applied stresses necessary to continue dislocation motion (slip) demonstrates the essential function of dispersion reinforcement (i.e., dispersed particles strengthen the matrix).

Particle Strengthening

Previously we distinguished between particle strengthening and dispersion strengthening on the basis of particulate size, volume fraction, and the mean free path of the matrix. But particle reinforcement also differs from dispersion reinforcement in that the matrix and the particles *share* the applied load. Particle strengthening basically is an extension of dispersion reinforcement in that particle size is larger, the volume fraction is greater, and so on.

Strengthening in this type (particle strengthened) of composite is similar to the dispersion-strengthened materials because the particles initially impede deformation of the matrix. Certain combinations of ceramic particles in a metal matrix (*cermets*), and cemented carbides such as tungsten carbide (WC) in a cobalt matrix, as shown in Figure 16-11, exhibit this behavior because the reinforcing particles are very hard and do not deform under load. Rather, these particles fail when the applied

Figure 16-11 Microstructure of a "cemented" carbide. Angular particles are tungsten carbide (WC) in a cobalt (Co) matrix (1000×).

shear stress exceeds their fracture strength. Cracks nucleate at the failed particles and the composite yields.

The yield strength of composites strengthened with particles that do not deform during loading has been shown to be inversely proportional to the square root of the interparticle spacing (D_p), as follows:

$$\sigma_{c_y} \propto \frac{1}{\sqrt{D_p}}$$

This behavior is shown in Figure 16-12 for plain carbon steel, which contains iron carbide (Fe_3C) particles dispersed throughout a ferrite matrix.

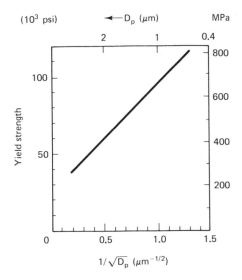

Figure 16-12 Room-temperature yield strength of plain carbon steel as a function of carbide particle spacing (D_p). (From G. S. Roberts, R. C. Carruthers, and B. L. Averbach, *Trans. ASM 44*, 1952, p. 1150.)

Particulate-Strengthened Polymers

Many polymeric materials, such as epoxies and polyesters, exhibit brittle behavior which limits their use in applications that are subject to fatigue or where fracture toughness (particularly at low temperatures) may be a critical factor. This is unfortunate, since these engineering materials possess certain desirable properties, such as low density, good corrosion resistance, and high dielectric constants.

The crack growth resistance of these polymers can be increased however, by the addition of certain particulate materials, sometimes called "fillers." For example, at room temperature, cross-linked epoxy resin can be made significantly more crack resistant by a suspension of fine *elastomeric* particles. Low-molecular-weight carboxy-terminated butadiene–acrylonitrile copolymers (CTBN), in concentrations up to 10 pph,[3] have been especially effective in increasing the toughness of an epoxy matrix. In a manner similar to "aging," the elastomer is soluble in the epoxy above a certain temperature, but below this temperature it forms precipitate particles. The nature of the precipitate, its particle size, and its concentration are important factors in determining the degree of toughening. For instance, Figure 16-13 shows the increase in toughness, as measured by the change in fracture surface work (energy required to propagate a crack), due to the increase in concentration of CTBN in an epoxy matrix.

Inorganic *fillers* are also utilized for a variety of engineering purposes in polymeric materials. These purposes include increased surface hardness, fire retardancy, coloration, and modification of thermal and electrical properties. One of the largest applications of this technology is the addition of particulate fillers to rubbery polymers such as those used in automotive tires. The typical analysis of a "rubber" tire is presented in Table 16-6. Depending on the particular tire application, the rubber may be natural or synthetic (styrene–butadiene). Carbon in the form of *carbon*

---

[3] pph, parts per hundred.

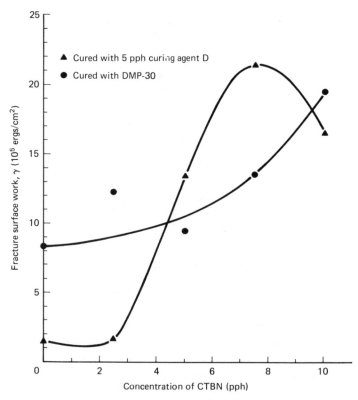

Figure 16-13 Variation of fracture surface work in an epoxy matrix containing CTBN particles. (From J. N. Sultan, R. C. Laible, and F. J. McGarry, *Applied Polymer Symposium*, No. 16, John Wiley & Sons, Inc., New York, 1971, p. 127.)

TABLE 16-6 TYPICAL ANALYSIS OF A RUBBER TIRE

Material	Concentration (w/o)
Rubber	45
Sulfur	1
Carbon/silica	23
Oil	11
Chemical additives	9
Cord (polymer)	7
Bead wire (steel)	4

black,[4] and *silica* are the reinforcing fillers which enhance the strength of the rubber matrix. The effectiveness of various carbon blacks in increasing the tensile strength of styrene–butadiene rubber (SBR) is illustrated in Figure 16-14.

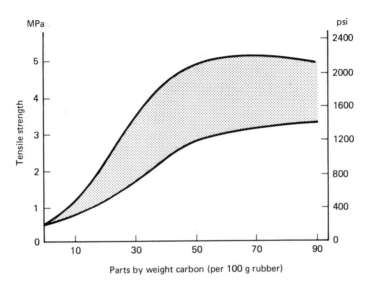

Figure 16-14 Increase in tensile strength of SB rubber by the addition of various carbon blacks. (Adapted from D. Parkinson, *Reinforcement of Rubbers*, Lakeman and Company, London, 1957.)

MATRIX CONSIDERATIONS

Matrix materials utilized in most commercial composites can be divided into two general categories: (1) polymeric, which include a number of thermosetting and thermoplastic resins, and (2) metallic, consisting of pure metals and alloys.

The function of the matrix in a composite material is usually multifold. Matrices are designed to protect the reinforcing phase from structural damage, corrosive attack, and reactions that would degrade the reinforcement properties. The matrix phase also serves to transmit applied stresses to the reinforcing constituents and stabilize them against buckling in situations where compressive stresses are axially applied to long cylindrical fibers.

In other cases, the matrix may share the load-carrying responsibility with the reinforcing phase. This aspect is important in particulate strengthened composites, as we explained in the preceding section. Furthermore, the matrix may be selected for its physical properties, such as density, thermal and electrical conductivity (or electrical resistivity), thermal expansivity, melting or softening temperature, and translucency or opacity.

[4]Carbon black is a form of impure carbon (soot) produced by chilling a smoky natural gas flame.

Although the subjects of metallic (nonferrous) alloys and polymers were covered in detail in Chapters 13 and 14, let us briefly re-examine some of the more widely used polymers and metals with respect to their application as matrix materials.

Thermosetting Polymers

This group of polymers is characterized by three-dimensional networks. Such non-linearity occurs when there are more than two reactive sites per monomer and *cross-linking* takes place between linear chains (see Figure 14-1). Indeed, this type of polymer is initially cured or hardened by the application of heat, hence the term *thermoset*.

These matrix materials are generally used in liquid formulations, so they can be cast into any desired shape. In addition to castability, the initial liquid form facilitates infiltration of the reinforcing phase and enhances wetting of the fibers.

An important benefit of the curing aspects of the thermosetting resins is that these materials can be stored in the *semicured* state. Therefore, these resins and the reinforcing fibers can be preformed in the appropriate volume fractions, as sheets, tapes, filaments, mats, and so on. In this stage, the composite material is referred to as preimpregnated or "prepregged." These products are subsequently formed into desired shapes by molding or pressing and heating (e.g. an autoclave), to obtain the final properties and configuration.

Thermosetting polymers were the first matrix materials to be used with glass fiber reinforcement and presently are principally reinforced with continuous glass fibers in the form of fabric, roving, or mats. The thermosetting resins commonly used as matrix materials include epoxies, polyesters, phenolics, and melamines. These polymers were discussed in Chapter 14 and their typical properties were presented in Table 14-4.

Thermoplastic Polymers

This type of polymer consists of linear molecular chains which are *not* cross-linked (see Figure 14-1). Instead, the attractive forces between the chains are the weaker secondary bonding forces; i.e., van der Waals forces. Thus they are not as rigid as the thermosets, and exhibit softening upon heating.

Essentially, the thermoplastic resins are heated until soft and flowable, then they are simply molded into the desired shape and allowed to cool. They are generally reinforced with short or chopped fibers and the composite material is prepared as cylindrical *pellets*. Subsequently, the composite pellets are processed by a fabricator in the conventional manner for thermoplastics, such as injection or compression molding.

The thermoplastics commonly used as matrix materials for fiber-reinforced plastics include polycarbonates, polyamides (nylon 66), polystyrene, polyethylene, and polypropylene. These polymers were also discussed in Chapter 14 and their typical properties listed in Table 14-3.

In general, the polymeric materials display low tensile strengths compared to metals and ceramics. However, as a group they also exhibit relatively low specific gravity, which makes them attractive candidates for lightweight matrix applications.

Metal Matrix Materials

Metallic matrices serve the same basic function in composites as the polymeric materials we have just discussed. For instance, they protect the reinforcing phase from structural damage and environmental degradation. They also transmit stress to the reinforcement phase and stabilize the composite against buckling in compressive loading situations.

The selection of matrix material depends on several intrinsic factors, such as strength, elastic modulus, plasticity, density, and melting temperature. In addition, the interaction between the potential metallic matrix and its reinforcement must be considered if the composite is to function properly. Factors including the wetting and bonding characteristics between fiber and matrix are extremely important, as well as chemical reactivity between these constituents after the composite material is fabricated and put in service.

Clearly, in any system the matrix material must thoroughly infiltrate the reinforcement to promote good wetting and bonding. This may be accomplished by *melting* the matrix and drawing it through the reinforcement; using the metal matrix in *powder* form and mixing it with the reinforcement, followed by pressing and sintering[5]; *electroplating* the matrix material on the reinforcement; or using the metal matrix in the form of sheets or foil and diffusion bonding alternate layers of the constituents.

Some of the matrix materials commonly employed in metal-matrix composites include aluminum, magnesium, nickel, copper, silver, and titanium. The typical properties and characteristics of these metals were discussed in Chapter 13; therefore, we will not comment on their behavior here. However, it is important to recognize that the integrity and performance of a metal-matrix composite depends strongly on the interfacial relationship between the matrix and the reinforcement phase, just as it did for polymeric resin matrix composites. Unfortunately, chemical and metallurgical reactions can take place between the constituents in the former type of composite after it is in service. For instance, if the composite is used for an elevated-temperature application, diffusion of one component into the other may occur. Such a reaction can substantially weaken the components, or form another phase in the vicinity of the interface, which is deleterious with respect to the design or function of the composite material. Such a reaction is illustrated for nickel reinforced with boron fibers in Figure 16-15. Reactions between components in this type of composite may be prevented by coating the reinforcing phase with a material that will not react with the matrix.

---

[5]Sintering refers to an elevated-temperature process typically applied to powdered metals or ceramics to consolidate these powders (by diffusion) into a solid.

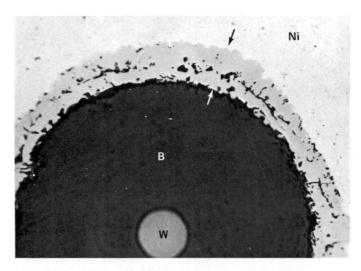

Figure 16-15 Reaction zone (between arrows) at the fiber–matrix interface in a nickel–boron composite (780×). W, tungsten filament (core); B, boron fiber; Ni, nickel matrix.

LAMINAR COMPOSITE MATERIALS

During the introductory section of this chapter we alluded to the lamination concepts utilized by ancient craftsmen to produce certain products such as swords. In a laminated product, the constituents are arranged in a series of alternating layers. Laminates may be considered yet another form of composite because they are either micro or macro combinations of materials which result in improved or different properties compared with their respective constituents. Laminated composite materials are utilized for such diverse purposes as strength, corrosion and wear resistance, controlled distortion, decorative purposes, and safety and protection.

However, the combination of different engineering materials in laminar form also introduces problems that are not experienced in single-phase materials and the composites we studied previously. The individual plies or layers in a laminated material may consist of the same basic ingredient, such as fiber-reinforced plastic sheets, but with different orientations of the fibers in alternating plies. Or the various layers may be entirely different materials altogether, such as metal and polymer. An example of a commercial structural laminate composed of several materials is shown in Figure 16-16. Regardless of the specific combination, some general principles apply to the mechanical behavior of laminates. Let us briefly examine these precepts.

Stresses in Laminates

When loads are applied to a laminated material, the resultant stresses in the lamina are proportional to their respective elastic moduli and shear moduli. However, inter-

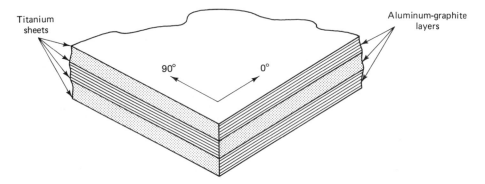

Figure 16-16 Laminate composed of alternating layers of aluminum (containing graphite fibers) separated by titanium sheets. Note that alternating Al–graphite layers are oriented at 90° with respect to each other. (Courtesy of Fiber Materials, Inc., Biddeford, Maine)

nal stresses may also be developed in directions for which no load is directly applied. These stresses result from differential contraction or expansion of the lamina, due to the dissimilar Poisson's ratios of the various constituents. In addition to the "transverse" stresses in the lamina, shear stresses also occur at the interface between the respective laminations.

Consider a relatively simple example where sheets of the same *isotropic* metal are adhesively bonded together. If a tensile load is applied normal to the plane of the sheets, the metal and the adhesive initially deform according to their elastic moduli, which may differ significantly. Correspondingly, the two materials contract differently in the transverse direction because of dissimilar Poisson's ratios. This differential contraction produces shear stresses between the adhesive and the metal. Furthermore, the shear stresses are not uniform over the interfacial area, but increase to a maximum at the edges. Such shear stresses can rupture the interfacial bond between the two materials and cause failure of the composite.

Laminates composed of anisotropic materials, such as continuous fiber-reinforced composites, and alloys with heavy microstructural banding or "fibering" may also experience problems with shear stresses when the applied loads are parallel to the laminate plane. Figure 16-16 illustrated a laminate with orthotropic layers. In this example, the alternating layers are oriented differently (at 90°) with respect to their "fiber" axes. If this material is loaded parallel to either the 0° or the 90° direction shown, the resultant tensile stress will produce *interlaminar* shear stresses.

In addition to the magnitude of the applied tensile load, the severity of this shear stress also depends strongly on the dissimilarities between the respective lamina. The major factors include differences in orientation of the lamina with respect to each other and with respect to the direction of loading, differences in elastic modulus and shear modulus, and Poisson's ratio of the respective lamina. Interlaminar shearing is a typical service failure mode for this type of composite material.

Laminar Composite Materials

Bending in Laminates

When a laminated material is subjected to bending or *flexure*, the corresponding stresses are proportional to the elastic properties of the respective lamina and their position within the laminate. Differences in properties between lamina can result in crack initiation *within* the laminate beam rather than at the extreme outer surface as one would expect in an isotropic material. Furthermore, bending a laminate in one direction may produce different bending in other directions, consequently *warping* the structure.

Applications of Laminates

Laminated materials are clearly a unique form of composites, extending from such mundane applications as furniture veneers and countertops to the highly complex structural *sandwiches* and honeycombs used in aircraft and aerospace vehicles. Let us briefly examine just a few applications of laminates.

Structural sandwich materials are a form of laminate used frequently in building construction and aerospace applications; the obvious advantage being the combination of an external material exhibiting properties such as high strength, hardness, corrosion resistance, and high temperature resistance, with a core that exhibits other desirable properties: low density, low thermal conductivity, and low acoustical transmission.

Many sandwich applications take the form of *honeycombs* in which the core consists of hexagonal grids, as illustrated in Figure 16-17. These grids are bonded to the facing material by such joining methods as resistance welding, brazing, and adhesive bonding, depending on the particular combination of materials. Honeycomb structures are used as stiffeners on such varied products as skis, aircraft fuselage panels, and doors.

Bimetallic strips also fall into the category of laminates. These components, which are an integral part of temperature-regulating thermostats, consist of two metals with different coefficients of thermal expansivity, such as brass ($\alpha = 20 \times 10^{-6}/°C$) and Invar[6] ($\alpha = 1 \times 10^{-6}/°C$). When the temperature of this laminate changes, the difference in expansivities causes deflection of the strip, thereby making or breaking an electrical contact.

Safety glass is a laminar composite consisting of a polymer (polyvinyl butyral) laminated between two sheets of soda–lime glass. The polymeric layer prevents glass fragments from detaching when the composite "glass" is fractured by impacts. "Bulletproof" vests utilize layers of various materials, such as glass, Kevlar, nylon, and high-strength fabrics, laminated together to form a tough, penetration-resistant composite textile.

---

[6]Invar is an alloy containing 65% Fe and 36% Ni, which exhibits a very low coefficient of expansion at room temperature. In fact, below temperatures of about 120°C, the coefficient of expansion is virtually invariable, hence the term *Invar*.

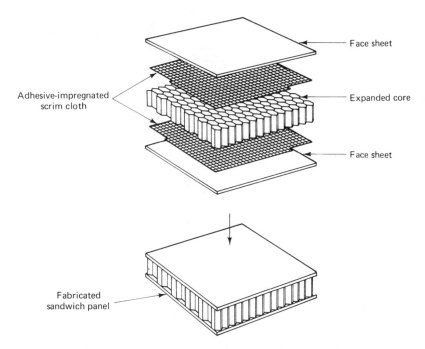

Figure 16-17 Honeycomb sandwich components. (From D. V. Rosato, *Handbook of Fiberglass and Advanced Plastics Composites*, Van Nostrand Reinhold Company, Inc., New York, 1969.)

Face sheet

Adhesive-impregnated scrim cloth

Expanded core

Face sheet

Fabricated sandwich panel

DIRECTIONALLY SOLIDIFIED EUTECTICS

As we studied in Chapter 7, a eutectic structure forms by the transformation of liquid alloy to two solid solutions. Also recall that this reaction is invariant; it takes place at a fixed composition and a constant temperature. Ordinarily, the eutectic structure produced during solidification tends to be random with respect to distribution of the microstructural constituents (see Figures 7-5 and 7-17). But if the solidification process of a eutectic alloy is precisely controlled, for example by removing heat from the liquid at a constant rate in one direction, and a planar solid–liquid interface can be maintained, the resulting structure will be aligned in the direction of heat flow. This procedure, referred to as *unidirectional solidification*, is illustrated schematically in Figure 16-18.

Mechanical Behavior

A unidirectionally solidified eutectic thus exhibits a microstructure resembling a fiber-reinforced metal-matrix composite, as demonstrated by Figure 16-19. Depending on the particular alloy system involved, one of the constituents may behave as the

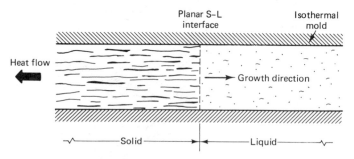

Figure 16-18 Unidirectional solidification of a eutectic alloy. The solid contains microstructural constitutents aligned parallel to the growth direction.

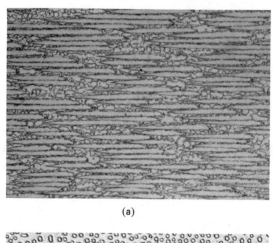

(a)

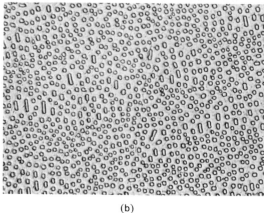

(b)

Figure 16-19 Microstructure of a directionally solidified eutectic (Ni–Mo–Al system): (a) section parallel to growth direction showing longitudinal orientation of eutectic (425×); (b) section perpendicular to growth direction (425×). (Courtesy of N. S. Stoloff, Rensselaer Polytechnic Institute.)

reinforcement while the other acts as the matrix. The principles of strengthening are essentially the same as those we discussed earlier in this chapter in the section "Fiber reinforcement." Furthermore, it is possible to determine the composition of the solid phases in the eutectic and their volume fractions if the densities of these phases are known. With this information and other pertinent data regarding the constituents, such as modulus and tensile strength values, the rule of mixtures can be used to calculate composite strength.

For instance, the Al–Ni system forms a eutectic consisting of 10 vol percent ($V_f = 0.10$) Al$_3$Ni embedded in an Al matrix. In the conventionally solidified condition, this alloy has a tensile strength of about 13,000 psi (90 MPa) and total elongation in excess of 15%. However, if this alloy is directionally solidified (growth rates $\simeq$ 2 to 10 cm/hr), the strength increases to approximately 43,000 psi (297 MPa) with 2% elongation. This behavior is shown in Figure 16-20.

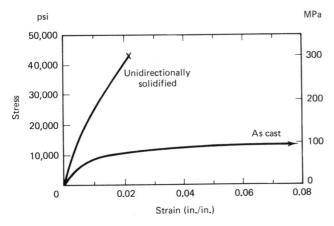

Figure 16-20 Comparison of the tensile behavior in Al–Al$_3$Ni eutectic alloy. (From F. D. Lemkey, R. W. Hertzberg, and J. A. Ford, *Trans. AIME 233*, 1965, p. 334.)

Orientation of the reinforcement phase with respect to the direction of applied stresses in this type of composite affects mechanical behavior in a manner similar to other fiber-reinforced composites. For example, the flexure strength of directionally solidified Al–CuAl$_2$ decreases noticeably, as the "fiber" axis is rotated from a position parallel to the applied stress to a position perpendicular to the applied stress, as shown in Figure 16-21.

One exceptionally important aspect of this type of composite material is the nature of the bond between reinforcing phase and surrounding matrix. Studies of the deformation experienced in these materials have been conducted by both light and electron microscopy. Such examinations have revealed that load transfer from the matrix to the "fiber" is very efficient, and that typically the fiber or whisker fails rather than the interface. This observation strongly suggests that an excellent bond is obtained between the microstructural phases, which is what might be expected, since these phases grow concurrently during solidification. Furthermore, subsequent chemical reactivity between the phases is also minimized in this type of composite system.

Applications

In general, applications for directionally solidified eutectic composites are limited compared to other fiber-reinforced materials. One reason for their low utilization is related to the *fixed* eutectic composition; therefore, no variation of constituent volume

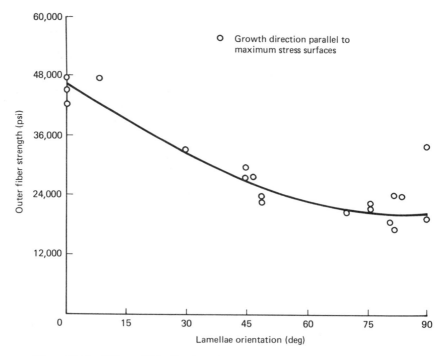

Figure 16-21 Effect of "fiber" orientation on the bending (flexure) strength of a directionally solidified Al–CuAl$_2$ eutectic. (From R. W. Hertzberg, F. D. Lemkey, and J. A. Ford, *Trans. AIME 233*, 1965, p. 342.)

fractions is possible. Also, they are relatively costly to produce, and result in the lowest melting temperature material of the alloy system under consideration.

However, certain eutectic alloy systems display unique electronic, magnetic, and optical properties when they are directionally solidified. For example, the aligned eutectic of the InSb–NiSb system is used for *magnetoresistive*[7] devices such as contactless variable resistors. In the directionally solidified condition, 60% of the theoretical magnetoresistance is obtained, while only 6% is realized for the unaligned condition.

In addition, directional solidification of eutectic alloys offers improved *creep* strength and resistance to thermal fatigue, due to the absence of transverse grain boundaries. Creep, as we have previously stated in several chapters, is permanent deformation that occurs over a period of time in components subjected to a constant stress; typically below the room-temperature elastic limit. Such deformation is also temperature dependent and increases with increasing temperature. Because of such improved thermal properties, this type of composite has been applied to aircraft engine turbine buckets or blades, as illustrated in Figure 16-22. These components are blade-shaped airfoils in turbines, which "catch" the hot gases of combustion and

[7]Magnetoresistivity refers to a material's ability to change its electrical resistance in the presence of a magnetic field.

Figure 16-22 Turbine blades manufactured by three different processes: (a) conventional casting—equiaxed grains; (b) directional solidification—columnar grains; (c) single crystal—one grain. (Courtesy of United Technologies—Pratt and Whitney Aircraft.)

essentially power the engine. Accordingly, these parts are subjected to very severe operating conditions (i.e., high temperature, stress, and corrosive atmospheres).

CONCRETE

Concrete is one of the most widely used structural materials, and as we pointed out in the introduction, this engineering material falls into the category of a composite. Just what makes concrete such a useful construction material? Well, there are many reasons that concrete is utilized to such a great extent, among them: availability of the raw materials that constitute concrete, favorable economics related to its production and placement, compressive strength, resistance to corrosion and biological attack, nonflammability, and its ability to be cast into various shapes, both "in-plant" and at remote job sites.

Basically, concrete is made from a mixture of *cement*, *aggregate*, and *water*. The aggregate is usually composed of fine material such as sand and coarse material such as washed gravel or crushed stone. The macrostructure of concrete, illustrating the composite nature of its constituents, is shown in Figure 16-23. Furthermore, most concrete applications incorporate steel reinforcing wire or bar to assist the structure in carrying tensile stresses, because as you will soon discover, concrete exhibits very low tensile strength.

Let us examine the materials that are used to produce concrete, and then introduce the fundamental properties and behavior of this composite material.

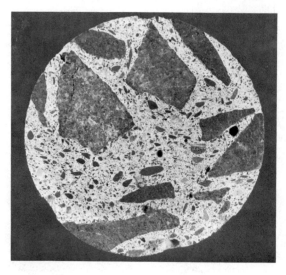

Figure 16-23 Structure of concrete showing various sized aggregates in a matrix of cement ($1\frac{1}{2}\times$).

Cement

Generically, *cementing* materials cover a very broad range of ceramic compounds, which can be divided into two categories: (1) simple cements and (2) complex cements. The simple cementing materials, such as *plaster of paris* and gypsum plasters, owe their hardening properties to the reabsorption of water. On the other hand, complex cementing materials, such as *portland* cement, harden due to the formation of new chemical compounds (silicates) during their processing or in service. The phase diagram for the $CaO-SiO_2-Al_2O_3$ system is illustrated in Figure 16-24. This diagram shows the compositional regions and compounds that are essential to cementitious materials.

Portland cement, which is usually specified for most general concrete construction, is produced by firing (heating) an appropriate mixture of calcium carbonate-

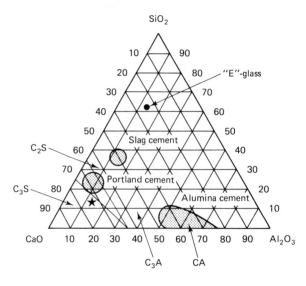

Figure 16-24 Equilibrium phase diagram for the $CaO-SiO_2-Al_2O_3$ system. Shaded regions denote compositional fields for various cements. C_2S, dicalcium silicate; C_3S, tricalcium silicate; C_3A, tricalcium aluminate; CA, calcium aluminate. (From K. N. Derucher and C. P. Heins, *Materials for Civil and Highway Engineers*, Prentice-Hall, Inc., Englewood Cliffs, N.J., 1981, p. 113.)

bearing materials, such as limestone, chalk, sea shells, slate, shale, and clay. This process is conducted in a rotary kiln at a temperature of 1550°C (2822°F), and produces portland cement *clinker*, which are fused lumps (about the size of peas) containing calcium silicate, aluminates, and relatively small amounts of iron oxide, magnesia, and alkalis. The essential constituents of portland cement are *lime*, *silica*, and *alumina* in the following approximate proportions:

Lime (CaO)	60–65%
Silica (SiO_2)	20–25%
Alumina (Al_2O_3) Iron oxide (Fe_2O_3)	7–12%

The partially sintered clinker is subsequently ground to a very fine powder and combined with a *retarder*. The retarder is added in the form of 2 to 3% gypsum ($CaSO_4 + 2H_2O$) and sufficiently delays the cement's setting reaction to enable it to meet commercial requirements. Cement may be packaged in 94-lb (43-kg) bags or bulk stored.

Five types of portland cement are commonly used in construction, depending on specific designs or applications. These types, designated in ASTM Standard C150, and their typical compositions are given in Table 16-7.

TABLE 16-7 COMPOSITION OF PORTLAND CEMENT

Type	Description	$3CaO \cdot SiO_2$[a]	$2CaO \cdot SiO_2$[b]	$3CaO \cdot Al_2O_3$[c]	$4CaO \cdot Al_2O_3 \cdot Fe_2O_3$	MgO	CaO	$CaSO_4$
				Compound (%)				
I	General purpose	45	27	11	8	2.9	0.5	3.1
II	Moderate heat	44	31	5	13	2.5	0.4	2.8
III	High early strength	53	19	11	9	2.0	0.7	4.0
IV	Low heat	28	49	4	12	1.8	0.2	3.2
V	Sulfate resisting	38	43	4	9	1.9	0.5	2.7

[a] Tricalcium silicate.
[b] Dicalcium silicate.
[c] Tricalcium aluminate.

Hardening reaction. Portland cement experiences a *setting* reaction usually within a few hours after mixing with water. This initial stiffening involves the tricalcium aluminate ($3CaO \cdot Al_2O_3$), which is the fastest setting component. *Hardening*, on the other hand, is the relatively slow development of strength which involves *hydration* of the clinker compounds. The main cementitious product that forms during this reaction is a noncrystalline calcium silicate *gel*. As water reacts with the clinker particles, a progressively thicker layer of gel is produced around the clinker. Eventually, if this process continues, the cementitious gel forms a continuous network, including any unreacted clinker. The reaction of portland cement clinker with water is schematically shown in Figure 16-25.

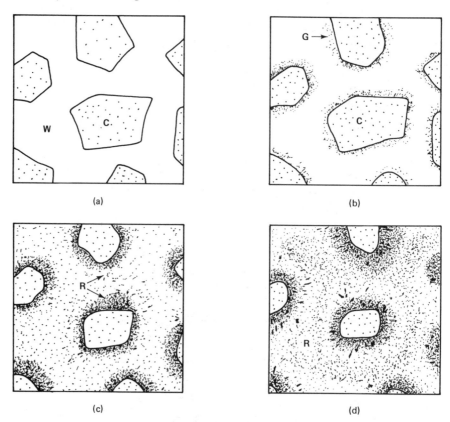

Figure 16-25 Schematic of the progressive hardening reaction in cement: (a) clinker particles (C) dispersed in water (W); (b) hydration reaction begins after a few minutes, forming siliceous gel (G) around clinker particles; (c) after a few hours reaction products (R) begin to impinge, thus "setting up" or stiffening the mix; (d) after sufficient time has elapsed, the reaction products (R) are continuous and hardening is under way. (Adapted from H. F. W. Taylor. *The Chemistry of Cements*, Vol. 1, Academic Press, Inc., London, 1964.)

Properties of Portland cement.   The various compounds in portland cement harden at different rates, as indicated by Figure 16-26. For example, $3CaO \cdot SiO_2$ reaches a compressive strength of 6000 psi (41.4 MPa) in approximately 7 days, while its counterpart, $2CaO \cdot SiO_2$, attains this same strength level in about 120 days. It is evident from these data that the hydration reaction proceeds very slowly and that it may take more than a year to reach completion (maximum strength).

If unreacted water evaporates from the cement mixture, voids can remain in the "hardened" material. Water in excess of the amount necessary to react with the cement tends to increase subsequent porosity in the hardened product. Porosity (especially interconnected pores) weakens the cement and increases its susceptibility to damage by frost and sulfate penetration.

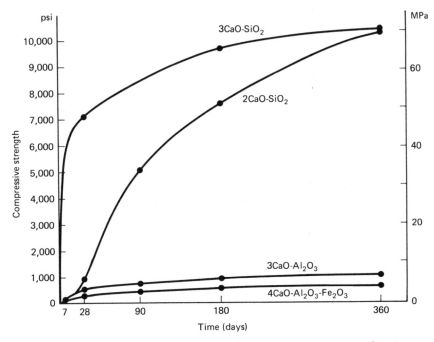

Figure 16-26 Comparison of compressive strength in cement constituents.

Aggregate

In terms of construction materials, *aggregate* refers to the inert materials, such as sand, gravel, and stone, which are mixed with cement. When fine aggregate such as sand is mixed with cement and water, the resulting product is *mortar*. A mixture of sand, cement, water, and coarser aggregate such as crushed stone or gravel is termed *concrete*.

The primary function of aggregates in concrete is to provide a *filler* material. Ordinarily, aggregate materials make up about 60 to 80% of the total volume, thereby reducing the amount of cement and water necessary in the mixture. From this standpoint, the aggregate particles do not strengthen the concrete, as did particulate matter (reinforcement phase) in composites we previously studied. However, the aggregate, particularly the coarser particles, certainly makes a significant contribution to the abrasion resistance of concrete (as countless concrete highways and sidewalks attest). Therefore, according to our definition of composites, concrete qualifies as a composite material even when it is not reinforced with steel bar or wire.

Fine aggregate. This material basically consists of sand, and according to ASTM Standard C125 (*Standard Definitions of Terms Relating to Concrete and Concrete Aggregates*), is defined as follows:

Aggregate passing a $\frac{3}{8}$-in. (9.5-mm) sieve, and almost entirely passing a No. 4 (4.75-mm) sieve, and predominately retained on the No. 200 (75-μm) sieve.

Essentially, this means particle sizes ranging from 75 μm to 9.5 mm, with most of this material between 75 μm and 4.75 mm in size. Such sampling should be conducted in accordance with ASTM Standard C136 (*Sieve or Screen Analysis of Fine and Coarse Aggregates*) using U.S.A. Standard series of sieves (ASTM E11).

A sand displaying proper gradation in size from coarse to fine is usually preferable to one that is either uniformly coarse or fine. The sand should be free of foreign matter, such as silt, clay, loam, and mica. If these contaminants coat the sand particles, good adherence of the cement is impeded. Also, the impurities, such as mica, may react with cement, degrading its properties or retarding the hardening reaction. Some impurities can be removed by washing. However, this process must be used judiciously, since too much washing tends to remove most of the fine particles, resulting in a relatively coarser gradation.

The relative fineness of the sand can be assessed by determining its *fineness modulus* (FM). This is an empirical factor obtained by adding the total percentages of a sample of aggregate retained on each of a specified series of sieves and dividing the sum by 100. The specific sieves used in this determination are given in Example 16-3. The value of fineness modulus is simply a general indicator of the fineness of sand (i.e., smaller values of FM indicate finer sand). Sand with FM values between 2.25 and 3.25 is usually appropriate for concrete purposes.

Example 16-3

Determine the fineness modulus (FM) of a 500-g sample of sand if the results of a sieve analysis are as follows:

Sieve size no.	4	8	16	30	50	100	Remainder (Bottom pan)
Opening (mm)	4.75	2.36	1.18	0.60	0.30	0.15	
Grams retained	6.8	56.4	104.6	86.7	132.9	92.0	20.6

Solution The cumulative percent retained on each sieve is obtained as follows:

Sieve size no.	Grams retained	Percent retained	Cumulative percent retained
4	6.8	1.4	1.4
8	56.4	11.3	12.7
16	104.6	20.9	33.6
30	86.7	17.3	50.9
50	132.9	26.6	77.5
100	92.0	18.4	95.9
			$\Sigma = 272.0$

The fineness modulus (FM) is the sum of the cumulative percent retained on these sieves divided by 100. Therefore,

$$FM = \frac{272.0}{100} = 2.72 \qquad Ans.$$

Coarse aggregate. According to ASTM Standard C125 coarse aggregate is defined as that material predominately retained on the No. 4 (4.75-mm) sieve. In ordinary concrete applications this aggregate may consist of crushed stone, gravel, or blast furnace slag (silicates and aluminosilicates of calcium). Normally, this type of aggregate may range up to about 3 in. (76.2 mm). But in very massive structures, larger sizes may be used for practical purposes, if special consideration is given to their distribution and location.

Cleanliness requirements for coarse aggregate are similar to those we discussed for fine aggregate. In general, the shape of the particles is less important than their size and hardness. However, thin, flat pieces should be avoided, and large amounts of shale are especially undesirable.

Water

The third major ingredient of concrete is water. A clean supply of water is necessary for concrete mixtures. Water contaminated with impurities such as acids, alkalis, or other organic matter generally should not be used, since they degrade the strength of concrete. The use of sea water should also be avoided, if possible. Salt water can affect the strength of concrete, and also may react corrosively with steel reinforcement in the concrete.

Concrete Mix Design

Indubitably, the final properties of concrete depend strongly on the quality of its individual components. Therefore, the cement, aggregates, and water must meet the appropriate standards or requirements for concrete applications and be capable of producing the desired properties.

Since cement is usually the most expensive constituent in concrete mixes, an efficiently designed concrete should contain just enough cement to properly bond the aggregate particles together. In theory this is fine, but it often takes considerable experience to consistently design successful concrete mixes, especially when the aggregates are subject to significant variability.

One method for "predetermining" the behavior of a concrete mix is to produce *trial batches* (in the laboratory) from the available materials. This technique simply consists of selecting the desired ratio of water to cement, and adding aggregate (fine and coarse) until the desired consistency is achieved in the mix.

Water–cement ratio. We have previously examined the reaction between water and the cement clinker. It is this hydration reaction that hardens the concrete and develops its strength. However, as we cautioned previously, when water in excess of the amount necessary to produce hydration is added to cement, it can evaporate, leaving behind interconnected porosity in the concrete. This condition is analogous to the polymeric foam we depicted in Figure 14-18(a). Although an excess amount of water tends to make the concrete mix more workable and easier to place, this condition also degrades the strength, and resistance to frost, of the resultant product. The effects of water–cement ratio and curing time on the compressive strength of concrete are illustrated in Figure 16-27. Notice that although compressive strength increases with curing time, increasing the water–cement ratio clearly lowers the strength of the resultant concrete. The astute student will also recognize that type III (high early

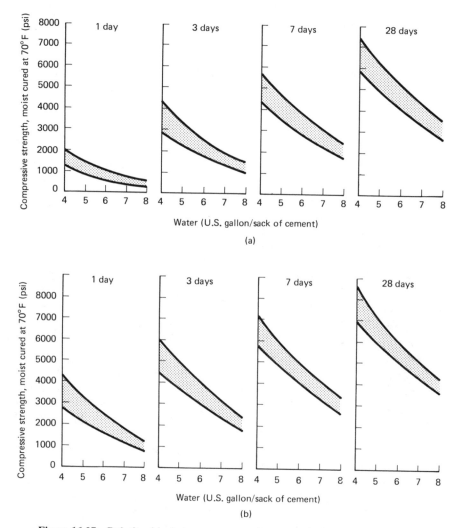

Figure 16-27 Relationship between compressive strength, water–cement ratio, and curing time: (a) type I general-purpose portland cement; (b) type III high early strength portland cement.

strength) cement produces a significantly stronger product than type I (general purpose) cement in 1 day's curing time. Yet in 28 days both cements (I and III) result in approximately the same strength level. In fact, concrete test specimens (cylinders) are generally "cured" for 28 days (after pouring) prior to compression testing.

Job curve. The job-curve method of designing a concrete mix consists of preparing trial batches from the intended constituents (as we previously described) with various water–cement ratios. Test cylinders 6 in. in diameter by 12 in. long are molded, cured appropriately, and compression tested to failure for each batch. A specific job curve is then plotted from these data. Once the job curve is established for the available materials, the corresponding "28-day strength" of this concrete can be reliably predicted for a specific water–cement ratio.

Slump test. To ensure that each batch of concrete has the same consistency, a quality control measure referred to as the *slump test* may be performed prior to placement of the mix. Essentially, this test consists of filling a truncated metal cone with the concrete mix in question, carefully removing the cone, then measuring the distance the concrete has slumped. This procedure (ASTM Standard C143) is illustrated in Figure 16-28. The consistencies recommended for various types of construction are given in Table 16-8.

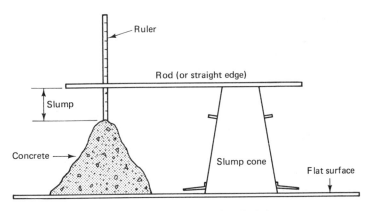

Figure 16-28 Schematic of typical slump test.

Proportioning by volumes. In the batching plant, concrete materials are proportioned by weight, because this method is accurate and practical. However, proportioning by volume is still a useful technique for small *on-site* batches. Based on experience with the materials involved, a volume ratio of cement, fine aggregate, and coarse aggregate is commonly specified. For example, a 1:2:3 mix consists of 1 part (by volume) cement, 2 parts sand, and 3 parts coarse aggregate. Water is added to this mixture until the desired consistency or workability is achieved.

Although such an arbitrary volume method is not precise, it is sufficient for many noncritical applications and jobs where plant-batched concrete is not feasible.

TABLE 16-8 RECOMMENDED SLUMP RANGES FOR VARIOUS CONCRETE CONSTRUCTION

Construction	Slump	
	in.	cm.
Reinforced foundation walls and footings	2–4	5–10
Unreinforced footings and substructure walls	1–3	2.5–7.6
Reinforced slabs, beams and walls	2–5	5–12.7
Building columns	3–5	7.6–12.7
Pavement	1–2	2.5–5
Sidewalks, driveways, slabs on grade	2–4	5–10

Typical arbitrary volumes have been recommended as guidelines and include the following applications:

Side walks	$1:2:4$
Concrete pavements	$1:2:3$
Mass concrete	$1:3:5$
Reinforced concrete	$1:2:4$

Example 16-4

A certain reinforced concrete structure calls for a $1:2:4$ mix with a water–cement ratio of $6\frac{1}{2}$ gal/sack. The weight of sand is assumed to be 105 lb/ft³, and the coarse aggregate 100 lb/ft³. If the specific gravity of the cement and aggregate are 3.15 and 2.65, respectively, how much of each constituent is needed per cubic yard of concrete?

Solution We must first determine the volume of concrete mix which is associated with one sack of cement (1 sack = 94 lb and is assumed to be 1 ft³ in volume).

Cement (based on 1 ft³): $\dfrac{(1\ ft^3)(94\ lb/ft^3)}{(3.15)(62.5\ lb/ft^3)} = 0.48\ ft^3$ abs. vol.

Sand (based on 2 ft³): $\dfrac{(2)(105)}{(2.65)(62.5)} = 1.27\ ft^3$

Coarse aggregate (based on 4 ft³): $\dfrac{(4)(100)}{(2.65)(62.5)} = 2.42\ ft^3$

Water (based on 6.5 gal): $\dfrac{6.5\ gal.}{7.5\ gal/ft^3} = 0.88\ ft^3$

Total volume $5.05\ ft^3$

Therefore, at this ratio one sack of cement will produce 5.05 ft³ of concrete. The materials needed per cubic yard are:

Cement: $\dfrac{27 \text{ ft}^3/\text{yd}^3}{5.05 \text{ ft}^3/\text{sack}} = 5.3$ sacks/yd^3 *Ans.*

Sand: $\dfrac{(2)(5.3) \text{ ft}^3}{27 \text{ ft}^3/\text{yd}^3} = 0.39$ yd^3 *Ans.*

Coarse aggregate: $\dfrac{(4)(5.3)}{27} = 0.78$ yd^3 *Ans.*

Water: $(6.5 \text{ gal})(5.3) = 34.4$ gal   *Ans.*

Comment: This example assumes that the aggregates are dry and that no adjustment is needed to the mixing water addition. In cases where the aggregates are wet, the mixing water is decreased by the amount of free water.

Properties of Concrete

In addition to the unique functional and economical aspects of concrete as a constructional material, certain mechanical and physical properties are important with respect to the application and performance of concrete. Let us briefly examine the major properties and characteristics of *unreinforced* concrete.

Compressive strength.   Concrete is used most effectively in compressive loading applications. Depending on the water–cement ratio and length of curing time, concrete typically attains compressive strengths between 2000 and 6000 psi (14 and 41 MPa). The strength of the composite concrete is limited by the strength of the cement and its bond with the aggregate. Compressive applications for concrete ordinarily include foundations, bases, footings, piers, walls, and columns.

Of course there are exceptions, such as *gunite*, which is a mixture of cement, sand, and water deposited (sprayed) under pressure through hoses. During such forceful deposition, excess water is literally driven off and a very dense mortar results. Gunite typically exhibits compressive strengths on the order of 9000 psi (62 MPa) and is utilized for irregularly shaped structures which are difficult to form and pour, such as a free-form swimming pool.

Tensile strength.   Generally, the tensile strength of concrete is roughly 10% of its compressive strength.[8] Since this property is relatively low, concrete is ordinarily not used in direct tensile applications. However, concrete is often utilized in applications involving bending or flexure stresses. In this respect, the flexure strength is virtually dependent on the tensile strength. Therefore, applications such as beams and slabs, which are subject to bending loads, are reinforced with steel wire or bars. In true composite fashion, steel reinforcing is embedded in concrete near the regions of highest expected tensile stresses. The incorporation of steel reinforcing bars in concrete is shown schematically in Figure 16-29(a). Thus, in essence, the steel carries

[8]The student is referred back to Figure 8-7.

Concrete   **617**

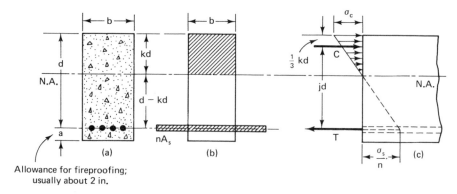

Allowance for fireproofing;
usually about 2 in.

Figure 16-29 Rectangular concrete beam reinforced with steel bar: (a) location of steel in beam; (b) equivalent section in terms of concrete; (c) resultant forces due to an applied load.

the tensile stress, allowing much greater bending stresses to be developed in the structure than if unreinforced concrete was used.

Analysis of reinforced concrete. The analysis of reinforced concrete components involves *transforming* the cross section from one consisting of multiple constituents to an equivalent component of one material. The transformation of a section is actually accomplished by changing the dimensions of a cross section parallel to the neutral axis in the ratio of elastic moduli of the materials involved. A transformed or equivalent section is shown in Figure 16-29(b). In this case, the ratio of moduli is expressed as follows:

$$n = \frac{E_s}{E_c} \qquad (16\text{-}9)$$

where E_s = modulus of steel
$\quad E_c$ = modulus of concrete

The value of E_c is generally taken as $1/15 E_s$, so that n is assumed to be 15. The shaded portions indicate the respective areas that are effective in resisting bending above and below the neutral axis (N.A.).

The neutral axis is located by applying the principle that the sum of the area moments (M) about the neutral axis is zero:

$$\Sigma M_{N.A.} = 0$$

Therefore,

concrete area moment = steel area moment

$$(b)(kd)\left(\frac{kd}{2}\right) = nA_s(d - kd) \qquad (16\text{-}10)$$

where b = width of beam
$\quad kd$ = distance from N.A. to top of beam
$\quad d$ = distance from top surface to centroid of steel bar

The resultant forces, compression (C) and tension (T), are shown in Figure 16-29(c).

The compressive force acts at a distance $\frac{1}{3}kd$ from the top surface, while the tensile force acts through the centroid of the steel bar. For equilibrium conditions, therefore, the resisting couple of opposed compressive and tensile forces have a moment arm (jd) as follows:

$$jd = d - \tfrac{1}{3}kd \tag{16-11}$$

The resisting moments in the concrete (M_c) and steel (M_s) are expressed as follows:

$$M_c = C(jd) \tag{16-12}$$

$$M_s = T(jd) \tag{16-13}$$

In terms of stresses, the average compressive stress in the concrete is $\frac{1}{2}\sigma_c$, where σ_c is the maximum compressive stress and occurs at the top surface. Thus the resisting compressive force (C) in the concrete is stress times area:

$$C = \tfrac{1}{2}\sigma_c(b)(kd) \tag{16-14}$$

Similarly, the resisting tensile force (T) in the steel is expressed as

$$T = \sigma_s A_s \tag{16-15}$$

where σ_s is the stress in the steel. In actual practice, the *conservative* bending moment will be the smaller of the two values M_c and M_s.

Example 16-5

A rectangular concrete beam with cross-sectional dimensions $b = 10$ in. and $d = 15$ in. is reinforced with four steel bars (0.75 in. diameter) as illustrated in Figure 16-29. If a bending moment of 750,000 in.-lb is applied to the beam, determine the maximum stresses produced in the concrete and the steel.

Solution The sum of the moment of area about the neutral axis is zero, so we can write from equation (16-10),

$$\text{concrete moment of area} = \text{steel moment of area}$$

$$10(kd)\left(\frac{kd}{2}\right) = nA_s(15 - kd)$$

n is taken as 15, and the area of steel is

$$A_s = 4\frac{\pi d_s^2}{4} = \pi(0.75 \text{ in.})^2 = 1.8 \text{ in.}^2$$

Substituting yields

$$10kd\left(\frac{kd}{2}\right) = (15)(1.8 \text{ in.}^2)(15 - kd)$$

$$5(kd)^2 = (27 \text{ in.}^2)(15 - kd)$$

which rearranges to

$$5(kd)^2 + 27kd = 405$$

$$(kd)^2 + 5.4kd - 81 = 0$$

This quadratic equation is in the form $ax^2 + bx + c = 0$, and may be solved by applying the quadratic formula:

$$x = \frac{-b \pm \sqrt{b^2 - 4ac}}{2a}$$

which yields $kd = 6.7$ in.

The moment arm of the resisting couple (jd) is then obtained from equation (16-11):

$$jd = 15 \text{ in.} - \tfrac{1}{3}(6.7 \text{ in.})$$
$$= 12.8 \text{ in.}$$

The resisting moment in the concrete is

$$M_c = C(jd)$$
$$= \tfrac{1}{2}\sigma_c(b)(kd)(jd)$$

Substituting values we have

$$750{,}000 \text{ in.-lb} = \tfrac{1}{2}\sigma_c(10 \text{ in.})(6.7 \text{ in.})(12.8 \text{ in.})$$

Solving for the stress in the concrete,

$$\sigma_c = \frac{(750{,}000)(2) \text{ in.-lb}}{(10)(6.7)(12.8) \text{ in.}^3} = 1749 \text{ psi or } (12 \text{ MPa}) \qquad Ans.$$

Correspondingly, the resisting moment in the steel is

$$M_s = T(jd)$$
$$= \sigma_s A_s(jd)$$

Substituting gives us

$$750{,}000 \text{ in.-lb} = \sigma_s(1.8 \text{ in.}^2)(12.8 \text{ in.})$$

Solving for stress in steel, we have

$$\sigma_s = \frac{750{,}000 \text{ in.-lb}}{(1.8)(12.8) \text{ in.}^3} = 32{,}552 \text{ psi } (225 \text{ MPa}) \qquad Ans.$$

Modulus of elasticity. The deformation of concrete is *not* directly proportional to stress for any range of loading. Thus the modulus of elasticity (E) is not a constant for any range of stress. However, for very small strains (< 0.001 in./in.) the modulus can be approximated by a tangent to the stress–strain curve (in compression) at its origin. Values of this *tangent modulus* range approximately from 1 million to 5 million psi (6900 to 34,500 MPa).

Even though concrete does not display perfectly elastic behavior for any range of loading, an approximate modulus of elasticity is useful in determining the relative stresses in reinforced concrete structures. For small deformations, the strain in the concrete equals the strain in the steel as long as the reinforcing bars do not "pull out." Under these conditions, the stresses in the respective components can be determined by applying Hooke's law [see equation (6-13)].

Durability. Ordinarily, concrete contains a small percentage (1 to 2 vol %) of void space in the form of pores. This porosity is the result of entrapped air and evaporation of excess water in the initial mix. As we have already indicated, when the porosity is interconnected, concrete will readily absorb water, and damage from subsequent freeze–thaw cycles can be severe. Freeze-thaw damage is typically manifested in cracking or spalling of the concrete surfaces. However, entrainment of air on the order of 2 to 6 vol % by the use of an air-entraining admixture can significantly increase the resistance of concrete to the disintegrating action of freeze–thaw cycles.

Air-entraining agents, which include rosin, stearates, and other foaming resins, produce a fine dispersion of *disconnected* voids throughout the concrete. Such a condition is analogous to the closed-cell polymeric foam pictured in Figure 14-18(b). These voids consequently provide a means to disperse the forces of expansion (hydrostatic pressure) when absorbed water freezes.

The effect of air entrainment on the relative durability of concrete is shown in Figure 16-30. Note that both air entrainment and low water–cement ratio are asso-

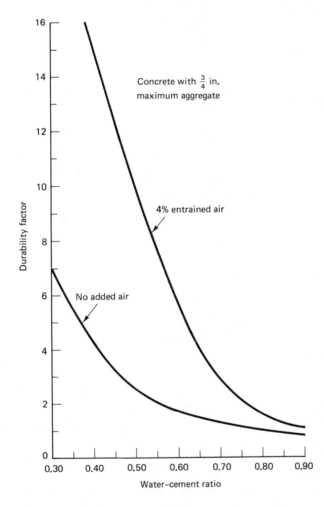

Figure 16-30 Effect of air entrainment and water–cement ratio on the relative durability of concrete. (From *Concrete Manual*, 7th ed., U.S. Dept. of the Interior, Bureau of Reclamation, Denver, Colo., 1966, p.37.)

ciated with high durability. We should also recognize that introduction of voids or pores in the concrete affects its strength, in much the same manner that gas porosity or shrinkage cavities affect the strength of metals and ceramics. Therefore, while durability benefits by air entrainment, the compressive strength of concrete can decrease by as much as 20%. Such attendant loss in strength must necessarily be considered in the overall design of a concrete structure.

Physical properties. Concrete exhibits certain physical characteristics which are important to its design and utilization. Thus it will be worthwhile to comment briefly on these properties.

For instance, the unit weight of concrete is significant in certain applications, such as dams and piers, which may rely to a great degree on their weight for stability. The Grand Coulee dam on the Columbia river for example, contains over 10,000,000 yd^3 of concrete. This property must also be considered in dead-weight design and analysis of concrete structural members. The unit weight of concrete exhibits substantial variations depending on several factors, including water–cement ratio, air content, and composition of the aggregates and their specific gravity. Typically, the unit weight of ordinary concrete (unreinforced) ranges from 137 to 160 lb/ft^3 (22 to 26 kN/m^3).

Lightweight versions of structural concrete are also commonly used in the construction industry, particularly in "high-rise" structures. This type of concrete employs lightweight aggregates such as cinders, shells, expanded slag, shale, and other natural lightweight aggregates. *Lightweight concrete* can weigh from 35 to 115 lb/ft^3 (6 to 18 kN/m^3) depending on the type of aggregate used and how it is produced. Principally applied to floors and roof slabs, lightweight concrete is usually considered whenever dead loads or insulating properties are an important design factor.

The thermal expansion of concrete is also an important property in the structural application of this material. Concrete's coefficient of thermal expansion (α) is approximately $5.5 \times 10^{-6}°F^{-1}$ ($9.9 \times 10^{-6}°C^{-1}$). Fortunately, this value is comparable to that of steel ($11 \times 10^{-6}°C^{-1}$). Therefore, concrete containing steel reinforcing bar does not experience harmful strains, due to differential distortions (between the steel and the concrete), when service temperatures fluctuate. Moreover, this compatability is an asset in the application of concrete as a fireproofing material, because the concrete sheathing tends to remain in contact with the structural steel, thereby providing a barrier between the structural steel and a hostile environment such as a fire. If unprotected, steel columns and beams could sustain permanent damage from "overheating," or could fail completely during a serious fire.

Prestressed concrete. Prestressing refers to the technique for developing stresses in a concrete structure which are opposite in sign to those resulting from service loads. These "beneficial" stresses are induced in the concrete before the material is put in service and allow the designer to utilize the composite of concrete and steel in a very efficient manner: concrete in compression, steel in tension. Applications for prestressed concrete include storage tanks, beams, and slabs.

Since concrete is notoriously weak in tension, only very small tensile stresses can ordinarily be sustained by concrete structures. But if the concrete is placed in compression before service loads that produce tension are actually encountered, the applied tensile stresses must first overcome the "built-in" compressive stress.[9] Only then does the concrete genuinely experience tension. Such prestressing also prevents

---

[9]The student may recall that this principle of intentionally producing *beneficial* residual stresses in an engineering material has been mentioned on several other occasions in this text [e.g., tempered glass (Chapter 15) and "shot-peening" in Chapter 9 (see Figure 9-31)].

cracking in the concrete members as long as the design loads are not exceeded. Therefore, this material is more resistant than ordinary reinforced concrete to weathering.

The production of prestressed concrete is based on one of two approaches: pretensioning or post-tensioning. This simply indicates whether the reinforcing steel is placed in tension prior to or after the concrete is poured. For example, let us consider the stages of prestressing a concrete beam, as illustrated in Figure 16-31. Initially, the steel reinforcing bar(s) are loaded (elastically deformed) in tension to the desired stress level (a); the concrete mix is then placed around these bars and allowed to harden. This process may be accelerated by the addition of suitable admixtures as long as such additions are not detrimental to the concrete or the steel. At this stage,

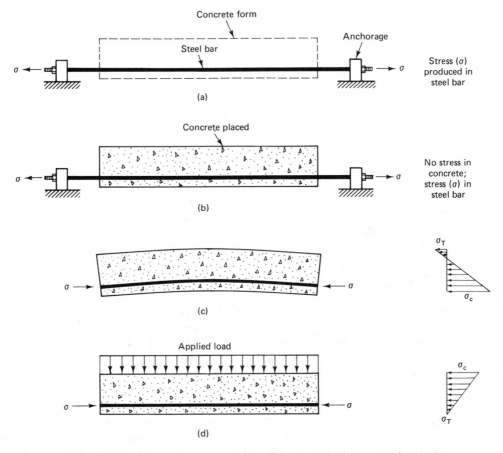

Figure 16-31 Schematic representation of a prestressed concrete beam: (a) steel bar stretched to produce tensile stress prior to concrete placement; (b) concrete placed and hardened, steel bar still anchored—no stress in concrete; (c) steel bar cut off—compressive stress produced in lower portion of beam, negligible tensile stress produced in upper portion; (a) beam shown in service with a hypothetical load applied—compressive stress results in upper portion of beam, negligible tensile stress results in lower portion.

(b), no intentional stress is produced in the concrete. After the concrete has hardened sufficiently, the bars are cut off from the tensioning device, allowing the steel to contract and induce a compressive stress in the beam, as shown in (c). When the beam is put in service (d), an applied load produces flexure (tensile stress) in the bottom portion. However, this applied stress is opposed by the induced compressive stress, and by design, no tensile stresses, or at the most a negligible tensile stress, results.

Post-tensioning results in the same effect as pretensioning, but the tensile stress is developed in the reinforcing bar *after* the concrete has been placed. This is actually accomplished by putting the steel rods through metal tubes embedded in the concrete. Next, the reinforcing bars are appropriately stressed and the tubes are grouted (filled with a cement mortar mix). Essentially, post-tensioning then follows the same stages we described for pretensioning [Figure 16-31(c) and (d)].

ASPHALT

Bituminous materials rank among the world's oldest and most common construction materials. These materials are mixtures of hydrocarbons in the form of liquid, semi-solid, or solid. Bitumens may occur naturally or they may be produced by the refinement of petroleum. *Asphalt* is a dark solid or semisolid, cementitious material composed chiefly of bitumens. Although asphalts are used extensively for roofing and waterproofing applications, we will limit our discussion of this construction material to asphalt concrete, which is used in roads and pavement.

The composition of asphalt can be divided into three general categories: asphaltenes, resins, and oils. The *asphaltenes* are large high-molecular-weight hydro-carbons which precipitate from asphalt, making up the "body" of the mixture. *Resins* are also hydrocarbon molecules that play a role in the adhesiveness and ductility of asphalt, while the *oils* influence its viscosity. Thus the asphalt is, in essence, a polymeric material that softens upon heating (as with a thermoplastic) in the approximate range 160 to 210°F (71 to 99°C).

Asphalt Concrete

When the appropriate amounts of fine and coarse aggregates are mixed with asphalt, a composite paving material referred to as *asphalt concrete* is produced. The process is conducted at elevated temperatures so that the mixture is a fluid (pliable) mass. Asphalt concrete is kept hot until placement so that the paving operation can be accomplished with the proper amount of compaction.

Asphalt concrete mixes are designed to produce certain properties depending on the specific application. For example, the requirements on a parking lot mixture would be less demanding than those imposed on a major thoroughfare which carries industrial traffic. Typical full-depth pavement courses for asphalt concrete on good subgrades are illustrated in Figure 16-32.

Asphalt concrete mixes can be modified slightly from the compositions used for highways and paving to produce a membrane impermeable to water and certain other liquids. These impermeable membranes make effective and efficient barriers for sani-

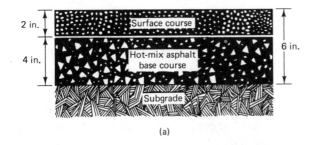

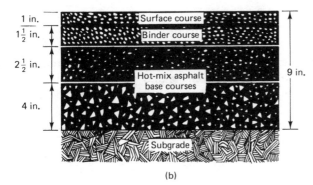

Figure 16-32 Typical cross section of asphalt concrete pavement on good subgrades: (a) residential street with light traffic; (b) industrial street subjected to heavy traffic. (Courtesy of the National Asphalt Pavement Association and the Asphalt Institute, Riverdale, Md.)

tary landfill liners and caps, toxic waste sites, pond liners, and storage pads for salt and other materials whose leachate may be harmful to the surrounding environment.

WOOD

Wood and wood products unquestionably make up a significant amount of the construction materials used throughout the world today. In addition, the main ingredient in wood (cellulose) is an important constituent of many other commercial and industrial products, such as paper, explosives, and certain synthetic materials (e.g., rayon).

Despite the large number of tree species, the ones of significant commercial value are very limited indeed. The bulk of wood utilized in structrual applications is derived primarily from the following varieties: pine, fir, cedar, cypress, spruce, hemlock, oak, hickory, maple, ash, and walnut.

Our examination of this time-honored engineering material will consist of exploring the basic structure of wood and the factors that affect its physical and mechanical behavior.

Types of Wood

All woods may be divided into two principal groups, *exogenous* and *endogenous*. The distinction between these groups is based on their manner of growth. Exogenous trees grow in diameter by the annual formation of new wood between the old wood and the bark. Virtually all commercially important trees are in this group. Endogenous trees grow in diameter but chiefly grow longitudinally by the addition of new wood fibers intermingled with the old. Only a few trees of this type are used for structural purposes, such as the *palm*, *yucca*, and *bamboo*. Therefore, we will concern ourselves with the former group, which can be further subdivided into two categories: *conifers* (softwoods) and *broad-leaved* (hardwoods).

Conifers (softwoods). The conifers are the needle-bearing trees, such as pines, firs, spruce, and redwood. This group is commonly referred to as *softwoods* because they are relatively soft and exhibit relatively low density. The conifers, which are chiefly evergreens, constitute the bulk of timber used for structural lumber.

Broad-leaved (hardwoods). This group of deciduous trees includes oak, maple, ash, and hickory. We commonly refer to these woods as *hardwoods* because they are relatively dense and hard. The hardwoods are used on a limited basis for structural purposes, but because of their properties and pleasing appearance (grain), they are more widely applied to furniture, cabinets, and other forms of interior finishing. Very often these applications utilize the hardwoods in the form of *veneer*, which is a very thin sheet of hardwood adhesively bonded to the exterior of a lesser-quality wood. A standard veneer thickness is $\frac{1}{28}$ in. (0.9 mm); however, some veneer may be sliced as thin as $\frac{1}{100}$ in. (0.25 mm).

Structure and Composition

The properties of wood that make it an extremely useful engineering material are strongly dependent on its chemical composition and internal structure. In fact, the *cellular* nature of wood imparts a fibrous structure to this material and is primarily responsible for its anisotropic behavior in mechanical properties, which is characteristic of all continuous fiber-reinforced composites.

Cellular structure. The cells that form wood structure are fairly tubular in shape and commonly appear rectangular with rounded corners. In most woods, the majority of these cells are oriented with their long axis parallel to the axis of the tree and are identified as *longitudinal tracheids*. The cellular structure of wood is illustrated in Figure 16-33. In addition, other cells may be oriented roughly perpendicular to the axis of the tree. These cellular groups form the *medullary rays*. Notice the similarity between this naturally tubular structure and the honeycomb core utilized in sandwich construction (see Figure 16-17). From our previous discussions of filamentary composite materials we should therefore expect wood properties, especially strength, to be highly anisotropic. As you will soon see, the compressive strength of wood is indeed much higher in a direction parallel to the long axis of the tree than perpendicular to this axis.

Furthermore, there are subtle but nevertheless important differences in the longitudinal groups of cells. For instance, if we consider cross-sectional area, the early season growth (spring wood) tends to produce larger cells, while the summer growth tends to result in more compact cells. Therefore, spring wood generally contains less cell wall per unit area than does summer wood. This difference in gross area of cell wall is reflected in the strength of the respective wood. As one might anticipate, the summer

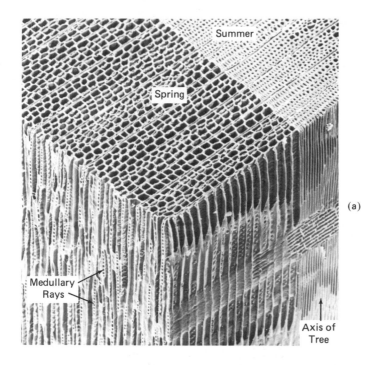

Summer

Spring

Medullary
Rays

Axis of
Tree

(a)

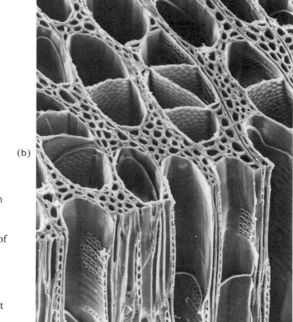

(b)

Figure 16-33 Cellular structure of wood as viewed in the SEM: (a) orthogonal section through western larch (softwood) showing difference in cell structure between summer wood and spring wood (80×); (b) close-up of section through the longitudinal tracheids in cottonwood (150×). (Courtesy of W. A. Coté, Jr., State University of New York, College of Environmental Science and Forestry at Syracuse.)

Wood

wood with more cell wall per unit area exhibits greater compressive strength than spring wood, in a direction parallel to the long axis of the tree.

Chemical composition. The cell walls of wood (both soft and hard) are composed primarily of two polymers, *cellulose* and *lignin*. The composition of wood is approximately 60% cellulose and 35% lignin, with the balance consisting of carbohydrates and minerals necessary for plant life.

The cellulose molecules $(C_6H_{10}O_5)_x$ actually form bundles of long helical chains with the following chemical structure:

These molecular chains are held together by the lignin matrix, which acts as a binder. It is this composite structure that produces the strength and elastic properties of wood.

Unfortunately, cellulose is not resistant to many strains of bacteria and forms of animal life, and can be decomposed by certain enzymes. Therefore, with the exception of a few varieties, wood generally must be protected from attack or degradation by the environment.

Physical Characteristics

We have previously examined the "micro" structure of wood and the role played by this structure. The "macro" structure of wood is also important from the standpoint of strength and appearance. These factors are affected by the orientation of the cutting plane with respect to the long axis of the tree (longitudinal tracheids).

Appearance. The types of saw cuts commonly made in timber to produce lumber are illustrated in Figure 16-34. Cross cuts are made perpendicular to the axis of the tree and contain the ends of longitudinal cells, which appear as the annular ring pattern. The *rift* cut is made parallel to the tree axis and in a radial plane. The *slash* cut is parallel to the axis of the tree, but tangential to the annular rings.

There are specific advantages to both types of cutting. For example, rift-cut lumber shrinks and swells less in width, does not check or split so badly, and wears more evenly than does slash-cut lumber. On the other hand, slash-cut lumber is more economical because it requires less time to cut and involves less waste, has a conspicuous grain resulting from the intersection of annular rings, and is less affected by by knots than is rift-cut lumber.

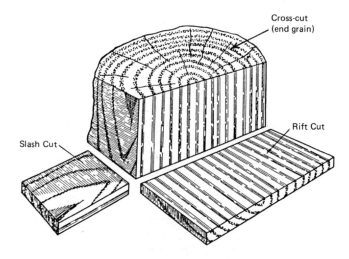

Cross-cut
(end grain)

Rift Cut

Slash Cut

Figure 16-34 Schematic example of cutting planes in timber. (From A. P. Mills, H. W. Hayward, and L. F. Rader, *Materials of Construction*, John Wiley & Sons, Inc., New York, 1939, p. 448.)

Defects. Lumber may contain certain structural imperfections which impair its strength and durability or detract from its appearance. For instance, *knots* are one of the most common defects in lumber, arising from encasement of a limb by subsequent annual growth. Depending on its location, this type of defect tends to reduce bending or flexure strength and can also impair tensile and compressive strength.

Wood may also contain cracks or separations. When these separations are located between the annular rings and parallel to the grain, they are termed *shakes*. A *check* is also a lengthwise separation, but is situated across the growth rings (see Figure 16-34). Other types of defects include distortion from nonuniform shrinkage, various holes, streaks, and stains.

Seasoning of wood. Structural wood is ordinarily *seasoned* or dried prior to use. This process is necessary for the wood to become dimensionally stabilized and achieve its optimum mechanical properties. Seasoning can occur naturally during long exposure periods in air, or the process can be artificially conducted by "kiln drying" at elevated temperatures. Kiln drying typically employs temperatures on the order of 70 to 82°C (158 to 180°F).

The shrinkage associated with seasoning of wood results from moisture loss in the cellular walls. As the moisture evaporates, the cell walls become thinner and the cross-sectional dimensions of the lumber decrease. When this shrinkage is nonuniform due to unequal drying or irregularities in the wood structure, *warping* results. Checks (separations) may also be produced during drying, as a consequence of internal strains that cannot be accommodated.

Furthermore, the moisture content of "green" wood is usually expressed as a percentage of the weight of the oven-dry wood. This value, generally well in excess of 100%, reflects the very high moisture content of unseasoned wood. In addition to the amount of material contained in the cellular walls and its specific gravity, the weight or specific gravity of wood is related to the amount of moisture it contains. Hence the specific gravity of wood decreases during the seasoning process. The weight of wood is an important factor in design consideration; therefore, control of the moisture content is essential for most applications. For example, at 70% moisture content, eastern white pine weighs 34.6 lb/ft³ (4.1 kN/m³). However, this wood weighs 25.4 lb/ft³ in the air-dry condition (contains 15% moisture), a weight difference of 36%.

Electrical properties. It should not be surprising that the important electrical characteristics of wood are its *resistivity* and *dielectric* properties. In general, the electrical resistance of wood varies inversely with moisture content and to a lesser extent, density. Since its electrical resistance therefore decreases with increasing moisture content, wood used in insulating applications (e.g., electrical power transmission usage) must be carefully controlled with regard to moisture content, and any other factors that would influence resistivity. At low moisture contents, wood is normally classified as an insulator or dielectric.

Thermal properties. The coefficient of linear expansion (α) in the longitudinal direction (parallel to the grain) generally varies from about 1.7×10^{-6} to 2.5×10^{-6} per °F (3.1 to 4.5 per °C). This range is substantially less than other constructional materials, such as concrete, window glass, and most metals. In addition, the thermal expansion (or contraction) of wood is much smaller than the swelling and shrinkage that occurs from ordinary exposure to weather and typical ambient conditions. As a consequence, the thermal expansion of wood is usually disregarded in most structural designs.

The thermal conductivity of wood is, however, another matter. When one considers heat energy conservation, the conductivity properties of wood used in many industrial and residential building applications become an important consideration indeed.

Thermal conductivity of wood is influenced by several factors, including specific gravity, moisture content, defects, and direction of the grain. Generally, the conductivity is approximately equal in the radial and tangential directions. Yet in the longitudinal direction (along the grain) thermal conductivity can be two to three times greater than the transverse directions. Also, the lighter woods tend to be better insulators, since the conductivity increases with specific gravity.

Preservation treatment. As a natural organic material, wood is susceptible to attack by various fungi and insects. Such attack is usually manifested as decay (rot) or degradation (holes, tunnels, etc.) of the wood, which left unchecked will eventually create an unserviceable condition or perhaps result in failure of the wood structure. Fortunately, wood can be protected from invasion by these harmful organisms by treatment with certain chemicals. The degree of protection depends mainly on the type of preservatives employed and the methods by which they are applied.

Preservatives used for protecting wood are generally classified as (1) oils, such as creosote[10] and petroleum solutions of pentachlorophenol; and (2) aqueous salt solutions (e.g., zinc chloride, ammoniacal copper arsenite, chromated copper arsenate). Historically, creosote solutions made from coal tar, or petroleum oil mixed with

---

[10]Coal-tar creosote is a black or brownish oil produced by the distillation of coal tar.

creosote, were applied at ambient pressures to the surface of wood products. Some of these "topical" preservative treatments have been very successful, as evidenced, for example, by the huge number of railroad ties that have endured long periods of exposure before experiencing any serious deterioration.

Pressure processes are also utilized to impregnate wood products with chemical preservatives, in order to obtain more uniform distribution of the chemicals throughout the wood. Typically, these processes are conducted in closed vessels (autoclaves) under pressure ranging from 100 to 200 psi, which forces the preservatives into the interior of the wood.

Wood products, such as structural lumber pressure treated with pentachlorophenol solutions, have demonstrated excellent resistance to attack under such conditions as atmospheric exposure and below ground. Furthermore, wood impregnated with aqueous solutions of phenol formaldehyde resin has displayed good resistance to chemical attack. The latter type of material is generally employed in chemcial processing applications such as tanks, vats, trays, and covers, where metals and plastics would be inappropriate from the standpoint of corrosivity or strength.

Mechanical Properties

Earlier in the section dealing with wood structure and composition, we indicated that the mechanical properties exhibit a high degree of anisotropy. In fact, we compared the cellular structure of wood with that of other filamentary-reinforced composite materials. The effective utilization of this engineering material, therefore, must necessarily take such "directionality" into account, when structures are designed and fabricated from wood.

Tensile strength.　Although wood is not frequently subjected to pure tensile stresses, its tensile strength is about three times greater than the compressive strength, along the grain. The actual tensile strength is difficult to measure because failure by compression across the grain, or shearing along the grain, usually occurs before the ultimate tensile strength is reached. The tensile loads that can be sustained across the grain are considerably lower (on the order of $\frac{1}{10}$ to $\frac{1}{20}$) than those which can be carried in tension parallel to the grain.

Compressive strength.　The compressive strength of wood is also very dependent on the direction of applied loading with respect to the grain. Wood loaded in compression along the grain can typically sustain about three times the load in compression across the grain. This condition is shown schematically in Figure 16-35. Applications such as posts and columns utilize the compressive strength of wood in its most favorable direction (parallel to the grain). Values of maximum compressive strength (*crush strength*) parallel to the grain are given for some commercially important woods in Table 16-9.

Flexure strength.　Wood is commonly used in applications that impose bending loads on it. Such uses include beams, joists, and trusses. However, when wood experiences flexure, the critical factors that determine how much load can be applied are as follows:

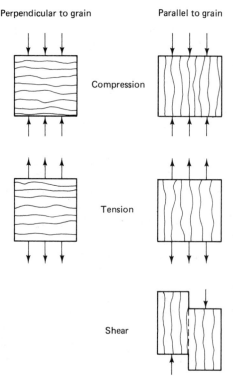

Perpendicular to grain   Parallel to grain

Compression

Tension

Shear

Figure 16-35 Examples of various loading situations in wood.

- Compressive strength parallel to the grain
- Shear strength parallel to the grain

Values of these properties are included in Table 16-9.

The *modulus of elasticity* of wood is also a measure of its inherent stiffness. This property is very important in bending applications, where deflections must be controlled within prescribed limits. Higher values of elastic modulus result in smaller deflections for the same bending or loading conditions. Values of elastic modulus for various types of wood are also given in Table 16-9.

Factors affecting mechanical properties.   In addition to the mechanical property anisotropy that results from orientation of the grain with respect to the loading direction, other factors can have a significant effect on wood's mechanical behavior.

For example, moisture content lowers the strength and elastic modulus of wood. Thus "green" wood has substantially lower properties than wood that has been properly dried. Drying wood to moisture values below the fiber saturation point[11] results in sizable gains in mechanical properties, as illustrated in Figure 16-36.

[11]This is the moisture content where all the *free* water is evaporated from the intercellular spaces and the cell cavities.

TABLE 16-9 MECHANICAL PROPERTIES OF VARIOUS WOODS (AT 12% MOISTURE CONTENT)

Species	Weight (Density)		Modulus of elasticity		Maximum compressive strength parallel to grain		Proportional limit compression perpendicular to grain		Maximum shear strength parallel to grain	
	lb/ft³	kN/m³	10³ psi	MPa	psi	MPa	psi	MPa	psi	MPa
Hardwoods										
Ash, white	42	6.6	1,770	12,213	7,410	51	1,410	10	1,950	13
Elm, White	35	5.5	1,340	9,246	5,520	38	850	6	1,510	10
Hickory, pignut	52	8.2	2,260	15,594	9,190	63	2,450	17	2,150	15
Maple, red	38	6.0	1,640	11,316	6,540	45	1,240	8	1,850	13
Maple, sugar	44	6.9	1,830	12,627	7,830	54	1,810	12	2,330	16
Oak, red	44	6.9	1,820	12,558	6,760	47	1,250	9	1,780	12
Oak, white	48	7.5	1,780	12,282	7,440	51	1,320	9	2,000	14
Walnut, black	38	6.0	1,680	11,592	7,580	52	1,250	9	1,370	9
Softwoods										
Cedar, western red	23	3.6	1,120	7,728	5,020	35	610	4	860	6
Cypress, southern	32	5.0	1,440	9,936	6,360	44	900	6	1,000	7
Douglas fir	34	5.3	1,920	13,248	7,420	51	910	6	1,140	8
Hemlock, eastern	28	4.4	1,200	8,280	5,410	37	800	6	1,060	7
Pine, longleaf	41	6.4	1,990	13,731	8,440	58	1,190	8	1,500	10
Pine, shortleaf	36	5.6	1,760	12,144	7,070	49	1,000	7	1,310	9
Pine, western white	27	4.2	1,510	10,419	5,620	39	540	4	850	6
Redwood	24	3.8	1,120	7,728	5,240	36	640	4	930	6
Spruce, sitka	28	4.4	1,570	10,833	5,610	39	710	5	1,150	8
Spruce, white	28	4.4	1,340	9,246	5,470	38	570	4	1,080	7
Tamarack	37	5.8	1,640	11,316	7,160	49	990	7	1,280	8

Source: Wood Handbook, USDA Handbook No. 72, U.S. Government Printing Office, Washington, D.C., 1955

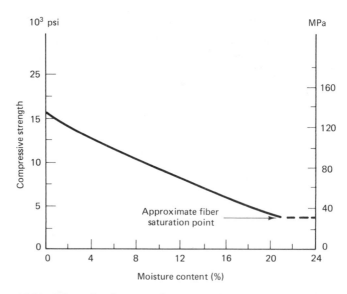

Figure 16-36 Effect of moisture on the compressive strength of longleaf yellow pine. (From A. P. Mills, H. W. Hayward, and L. F. Rader, *Materials of Construction*, John Wiley & Sons, Inc., New York, 1939, p. 448.)

The weight of wood also affects its strength. Generally, the greater the specific gravity, the stronger a wood behaves. This is simply based on a greater amount of cellular material per unit volume being available for load-carrying capacity. The weight factor is also related to the rate of growth of a tree. Most timber species used for structural purposes tend to exhibit an optimum number of annual growth rings per inch. This optimum rate of growth is associated with the greatest strength for a particular species.

Design considerations. Since wood used for structural purposes is subject to a number of strength-reducing factors and variability in mechanical properties, an empirical method for determining the *allowable working stress*, has been devised. The working stress is generally the criterion used in design calculations and analyses.

Essentially, the working stress is obtained for a type of wood by adjusting a *basic stress* for that species by a *strength ratio*. The basic stress is a working strength for clear (defect-free) wood that depends on both laboratory data and engineering experience in the field. The strength ratio of structural lumber represents the percentage of strength remaining after an allowance has been made for defects and other strength-reducing factors.

Therefore, the working stress that applies to a specific piece of lumber is obtained by adjusting the basic stress for that species. The basic stresses for commercially important woods are given in Table 16-10.

TABLE 16-10 BASIC STRESSES FOR CLEAR LUMBER

Species	Extreme fiber in bending (psi)	Modulus of elasticity (psi)	Compression parallel to grain, $L/d \leq 10$[a] (psi)[b]	Compression perpendicular to grain (psi)[b]	Maximum horizontal shear (psi)[b]
Hardwoods					
Ash, commercial white	1866	1,500,000	1466	500	167
Elm, white	1466	1,200,000	1066	250	133
Hickory, true and pecan	2533	1,800,000	2000	600	187
Maple, sugar and black	2000	1,600,000	1600	500	167
Oak, commercial red and white	1866	1,500,000	1333	500	167
Softwoods					
Cedar, western red	1200	1,000,000	933	200	106
Cypress, southern	1733	1,200,000	1466	300	133
Douglas fir, coast region	2000	1,600,000	1466	325	120
Fir, commercial white	1466	1,100,000	933	300	93
Hemlock, eastern	1466	1,100,000	933	300	93
Pine, western white, northern white, ponderosa, and sugar	1200	1,000,000	1000	250	113
Pine, Norway	1466	1,200,000	1066	300	113
Pine, southern yellow (Longleaf or shortleaf)	2000	1,600,000	1466	325	146
Redwood	1600	1,200,000	1333	250	93
Spruce, red, white, and Sitka	1466	1,200,000	1066	250	113
Tamarack	1600	1,300,000	1333	300	126

[a]L, unsupported length; d, smallest cross-sectional dimension.
[b]To obtain MPa, multiply psi values by 6.9×10^{-3}.

Example 16-6

The allowable working stresses for a certain grade of eastern hemlock structural lumber are 1100 psi for extreme fiber in bending and 60 psi horizontal shear. Determine the maximum length of rectangular beam that will meet these criteria if the loading conditions shown in the following sketch apply, and $b = 4$ in., $h = 4$ in.

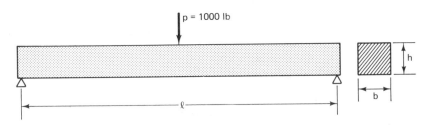

Wood

Solution The maximum flexure stress occurs at the midspan point and is determined as follows:

$$S_{max} = \frac{Mc}{I}$$

where M = maximum moment = $Pl/4$
 c = distance from neutral axis to outer fiber
 I = moment of inertia of cross section = $bh^3/12$

$$S_{max} = \frac{(1000\ lb)(2\ in.)l}{(4)(21.3\ in.^4)}$$

Solving for l, we have

$$l = \frac{(S_{max})(4)(21.3\ in.^4)}{(1000\ lb)(2\ in.)}$$

Substituting allowable working stress for bending results in

$$l = \frac{(1100\ lb)(4)(21.3\ in.^4)}{(1000\ lb)\ in.^2\ (2\ in.)}$$

$$= 46.9\ in.\ maximum\ length\ (span)\qquad Ans.$$

The maximum horizontal shear stress (H) produced in the beam acts at the neutral axis and may be calculated from the following expression:

$$H = \frac{3V}{2bh}$$

where V is the shear load at any location along the beam. In this situation, the maximum shear load occurs at the ends of the beam and equals $P/2$. Thus

$$H = \frac{3(500\ lb)}{(2)(4)(4)\ in.^2}$$

$$= 46.9\ psi\qquad Ans.$$

Both working stress criteria are met by this beam and the design condition imposed on it.

Example 16-7

A solid wood column with cross-sectional dimensions 6 in. × 6 in. (152 mm × 152 mm) is made from Douglas fir with an allowable working stress of 1200 psi (83 MPa) in compression parallel to the grain. The column is 16 ft long (4.9 m) and the material has a modulus of elasticity of 1.6 million psi (11,000 MPa). Determine the *safe* working load that can be supported by this member.

Solution The compressive load (P_c) that can be sustained by the column is given simply as

$$P_c = (\sigma_c)(A)$$

where σ_c = allowable working stress (compressive)
 A = area in compression

Thus

$$P = (1200\ psi)(36\ in.^2)$$

$$= 43,200\ lb$$

Ostensibly, we could safely load this column with 43,200 lb parallel to the grain. However, as an *unsupported* column gets more slender, the load-carrying capacity becomes increasingly dependent on the modulus of elasticity (E) and less dependent on the compressive strength of wood. *Buckling* becomes a prime consideration. Safe loads for long columns such as this are determined from a form of Euler's equation (adjusted for variability) as follows:

$$\frac{P}{A} = \frac{0.3E}{(L/d)^2}$$

where P = safe working load
    A = cross-sectional area
    L = unsupported length
    d = smallest cross-section dimension

Therefore, we obtain

$$P = \frac{(0.3)(1.6 \times 10^6 \text{ psi})(36 \text{ in.}^2)}{(192 \text{ in.}/6 \text{ in.})^2}$$

$$= 16,875 \text{ lb } (75.1 \text{ kN}) \qquad Ans.$$

Note that this value of P is considerably *less* than that determined from allowable compressive stress.

STUDY PROBLEMS

16.1. List some examples of composite materials (other than those cited at the beginning of this chapter) and specify the constituents contained in these materials. In each example, which constituents appear to be the stronger portion? Which appear to be the weaker?

16.2. A composite material containing continuous fibers is deformed in tension from 75.0 cm to 75.3 cm. What is the stress produced in the fibers and in the matrix if no slippage occurs between the fibers and the adjacent matrix? Given E_f = 72,400 MPa, E_m = 1000 MPa. Express your answer in psi. What have you assumed in arriving at your answer?
Answer: σ_f = 290 MPa, σ_m = 4 MPa

16.3. A stress of 56,000 psi is produced in the fibers of a 24-in.-long continuous fiber-reinforced composite shaft.
(a) If the elastic modulus of the fibers is 35 million psi, what is the resultant elongation in the fibers?
(b) What is the elongation in the matrix if no slippage occurs between the composite constituents?

16.4. A bar of composite material is produced by reinforcing epoxy resin with continuous alumina (Al_2O_3) fibers. The dimensions of this component are 4 cm $\times$ 2 cm $\times$ 50 cm and the composite contains 60 vol % fibers. Predict the elastic modulus of this material. How well does your prediction agree with the data given in Table 16-3? (See Tables 14-4 and 15-8 for the respective moduli of Al_2O_3 and epoxy.)

16.5. A tubular composite is made by embedding 0.025-mm-diameter continuous graphite fibers (high strength) in an aluminum matrix. The component measures 80 mm OD × 70 mm ID × 1000 mm long and contains 784 fibers per mm² of cross section. What is the stress produced in this composite if the tube elongates to 1003 mm during stage 1-type deformation? (See Table 16-1 for fiber and matrix properties.)
Answer: 404.3 MPa or 58,644 psi

16.6. Copper is reinforced with tungsten wire as shown in Figure 16-4.
   (a) For a fiber volume of 60%, predict the ultimate tensile strength of the composite. Assume that the UTS of the tungsten wire is 200,000 psi and copper is 25,000 psi, respectively.
   (b) What is the strength of the composite for $V_f = 0.05$?
   (c) How does your answer in part (a) compare with the strength measured in Figure 16-4?

16.7. Aluminum is reinforced with chopped steel wire ($\sigma_{UTS} = 350,000$ psi).
   (a) If the shear strength of the aluminum = 15,000 psi, what is the critical length of the steel fibers in terms of their diameter for this system?
   (b) If steel fibers 0.010 in. diameter by $\frac{3}{4}$ in. long are employed in this composite, what fiber concentration is necessary to obtain a tensile strength of 30,000 psi? (See Figure 16-6.)
   Answer: (a) $l_c = 11.7$ d (b) $V_f = 0.18$

16.8. The ultimate shear strength (τ_u) of a certain matrix material is 220 MPa. Compare the ultimate strength of the composite (σ_c) for a fiber orientation of 10°, with respect to σ, versus an orientation of 40° (see Figure 16-7).

16.9. What function do coupling agents perform in fiber-reinforced composite materials? Sketch a two-dimensional structural formula for a possible coupling agent. (*Hint:* See Figure 16-8.)

16.10. You are asked to produce a dispersion-strengthened composite from thoria (ThO_2) particles and a nickel matrix. This is an important material for elevated-temperature applications and is referred to as "TD Nickel." If the particles have an average diameter of 0.05 μm, what is the mean free path between particles for 10 vol % particle addition?

16.11. A composite material is made from Al_2O_3 particles with a diameter of 0.10 μm dispersed in a copper matrix. The mean free path between particles is determined to be 0.20 μm. What volume percent particles does this material contain?
Answer: 25 v/o

16.12. What is the shear stress (τ) that will cause dislocations to bow around the particles in problem 16.11 if the interparticle spacing equals the mean free path and the Burger's vector of a dislocation is approximately 3 Å? [*Hint:* See equation (6-17).]

16.13. Discuss the role of a matrix in a composite material. What are the advantages and disadvantages of polymeric matrices? Advantages and disadvantages of metallic matrix materials?

16.14. How are metal matrix composites typically produced? What is a particular type of problem that these materials can experience in elevated temperature service? What can be done to alleviate this problem?

16.15. What is a typical failure mode in laminated composite materials, and why does it occur?

16.16. How are directionally solidified eutectic composites produced? Why is this type of composite a very efficient system from the standpoint of load transfer and chemical reactivity? Name a serious limitation to the application of these type composites.

16.17. (a) A cement material is chemically analyzed in your laboratory with the following results: 50% Al_2O_3, 10% SiO_2, 40% CaO. Identify this type of cement.
  (b) A certain cement is formed from the $CaO-SiO_2-Al_2O_3$ system as denoted by the star in Figure 16-24. What is the composition of this material?

16.18. A mechanical sieve analysis is performed on a 500-g sample of sand to assess its applicability to concrete mixtures. The results of this analysis are:

Sieve no.	4	8	16	30	50	100
Grams retained	36.2	81.7	169.8	115.6	63.4	24.0

  (a) Calculate and report in table form (1) weight retained on each sieve, (2) the weight passing each sieve, (3) percent retained on each sieve, (4) percent passing each sieve, and (5) the cumulative percent retained on each sieve.
  (b) What is the fineness modulus of this sand?
  (c) What percentage of this sample is larger than 1.18 mm?
  (d) What percentage is finer than 0.30 mm?
  Answer: (b) 3.6 (c) 57.5% (d) 6.7%

16.19. Suppose that you wish to produce a concrete with a compressive strength of 4000 psi (27.6 MPa) using type I (general purpose) cement.
  (a) What is the range of U.S. gallons (H_2O) per sack of cement that will satisfy such a requirement in 7 days? In 28 days?
  (b) As a *safety factor*, water–cement ratios are generally selected to produce a concrete with 15% greater strength than the specified values. How does this affect the answers in part (a)?

16.20. A concrete pavement design calls for a 1:2:3 mix with a water/cement ratio of 6 gal/sack. Assuming that the aggregates are surface dry and exhibit properties corresponding to those in Example 16-4, how much of each material will be needed per cubic yard of concrete? In a concrete batching plant these materials are weighed prior to mixing. What is the weight of each material in a 10-yd^3 load? Total weight of concrete mix to be transported?

16.21. A concrete footing will be reinforced with $\frac{1}{2}$-in.-diameter steel bar. This footing will be 12 in. wide by 18 in. high and may be subjected to bending moments as large as 1×10^6 in.-lb. If the reinforcing bars are located 2 in. from the bottom, how many bars will be needed to accommodate such a moment if the stress in the concrete must not exceed 2500 psi?
  Answer: 4 bars

16.22. (a) What type of damage can occur in concrete that is exposed to the environment?
  (b) How is such damage alleviated?

16.23. A concrete column experiences a temperature change from 0°C to 50°C.
  (a) What percent linear expansion takes place in this column?
  (b) If the column is restrained from freely expanding, what stress may be produced by these conditions? (Assume that $E = 6900$ MPa.)

16.24. What is the principal advantage of prestressed concrete over conventionally poured and reinforced concrete?

16.25. What are the primary constituents that compose wood, and how do they function in composite fashion to give wood its anisotropic properties?

16.26. In which direction is the radial growth occurring for the wood displayed in Figure 16-33(a)?

16.27. Why must "green" wood be seasoned prior to being used as structural lumber? In addition to density, name some properties of wood that are affected by moisture contact. How are these properties affected?

16.28. A rectangular wood beam 6 in. wide by 10 in. high is simply supported over a span of 12 ft. If this member is made from commercial white fir as listed in Table 16-10, what is the maximum concentrated load that can be applied to the center of the span which will not exceed the basic stresses specified for extreme fiber bending or horizontal shear? (Neglect the weight of the beam itself for the purposes of this problem.) (*Hint:* See Example 16-5.)
Answer: 4,072 lb.

16.29. A simply supported wood beam made from southern pine is subjected to a uniformly distributed load of 120 lb/ft. If this beam extends 14 ft between supports, what is the minimum cross section that can safely support the applied load (neglecting the weight of the beam itself) if the maximum horizontal shear stress is not to exceed the value given in Table 16-10?

16.30. A rectangular wood post is made from spruce with the grain parallel to the long axis of the post.
  (a) If the applied load (*P*) is 222 kN, what is the minimum cross section that will withstand the compressive stress given in Table 16-10? Express your answer in SI units.
  (b) Based on your answer in part (a), would a beam with a square cross section be safe from buckling if its unsupported length was 6 m?

16.31. Wood is an energy-intensive resource which is utilized in many forms. List the ideal sequence of utilization for the following forms which minimizes degradation of this material.
  (a) Paper pulp
  (b) Heat energy
  (c) Synthetic cloth
  (d) Structural lumber
  (e) Chemical feedstock
  (f) Fiberboard

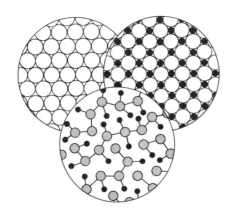

CHAPTER SEVENTEEN:
Corrosion of Engineering
Materials

From the standpoint of many producers, corrosion of a component after it has been manufactured frequently has little interest, yet from the standpoint of maximizing utility and function, corrosion is a subject that has immense importance. It has been stated that the economic losses resulting from corrosion run into the billions of dollars yearly, in the United States alone, to say nothing of the cost of preventative measures.

Corrosion may be broadly defined as the degradation of a material by its environment. This definition would properly include metals, ceramics, polymers, and composites; however, classical corrosion studies deal with the dissolution of metals. All metallic corrosion processes have as their basis, electrochemical reactions. Some corrosion mechanisms are primarily electrochemical, whereas others require the interaction of electrochemical and mechanical forces. This chapter describes the electrochemical basis for corrosion and discusses the various types of localized corrosion that can occur in metallic systems.

ELECTROCHEMICAL NATURE OF CORROSION

The electrochemical nature of corrosion can be demonstrated as follows. If a metal is placed in an active electrolyte, it will exhibit an electrochemical potential that can be measured relative to a standard reference electrode such as the hydrogen electrode, which has an electrode potential of 0.00 V. This reference electrode is constructed by placing a platinum tube or wire in a solution containing 1 molar (1 M) concentration of hydrogen ions and bubbling hydrogen gas through the solution. The platinum does not take part in the reaction, but merely serves as a substrate on which the reaction occurs.

641

In the case of iron, the potential difference can be demonstrated by immersing the iron in a divided cell containing a dilute solution of hydrochloric acid with Fe^+ at unit activity on one side and a hydrogen electrode on the other, as shown in Figure 17-1. The potential difference between the hydrogen electrode and iron is -0.44 V.

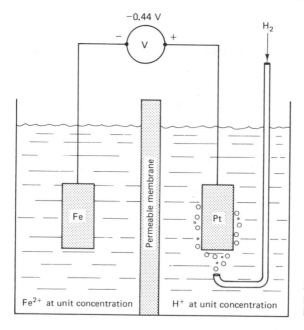

Figure 17-1 Electrochemical potential difference, iron versus hydrogen reference electrode.

It should be noted at this point that the cell potential is related to the free-energy change (ΔF) occurring in an electrochemical reaction, and although it will not indicate the *rate* of reaction, the sign of the free-energy change indicates whether the reaction will occur spontaneously.

Electrochemical Half-Cell Reactions

The reactions occurring at each side of the cell can be expressed as a half-cell reaction. The corrosion or dissolution of a metal may also be considered as an oxidation process, that is, the loss of electrons and the gain of positive valence.

$$M \longrightarrow M^{n+} + ne \qquad (17\text{-}1)$$

where n is the number of electrons. This reaction occurs where the metal is going into solution and is considered the *anode* reaction. The released electrons flow through the metal (or through conductors in the system) to a cathodic location, where they participate in a *cathode* reaction.

In the case of iron in an acid solution, the electrons react with hydrogen ions adjacent to the metal surface forming atomic hydrogen and ultimately hydrogen gas according to the reaction

$$2H^+ + 2e \longrightarrow H_2 \qquad (17\text{-}2)$$

Similarly, measurements can be made for other metals. Table 17-1 lists the anodic

TABLE 17-1 ELECTRODE HALF-CELL POTENTIALS FOR SEVERAL COMMON SYSTEMS AT UNIT CONCENTRATION AND AT 25°C

Anodic half-cell reaction[a]	Electrode potential (V)	
$Au \rightarrow Au^{3+} + 3e^-$	+1.50	
$2H_2O \rightarrow O_2 + 4H^+ + 4e^-$	+1.23	
$Pt \rightarrow Pt^{4+} + 4e^-$	+1.20	Cathodic (noble)
$Ag \rightarrow Ag^+ + e^-$	+0.80	
$Fe^{2+} \rightarrow Fe^{3+} + e^-$	+0.77	
$4(OH)^- \rightarrow O_2 + 2H_2O + 4e^-$	+0.40	
$Cu \rightarrow Cu^{2+} + 2e^-$	+0.34	
$H_2 \rightarrow 2H^+ + 2e^-$	0.000	Reference
$Pb \rightarrow Pb^{2+} + 2e^-$	−0.13	
$Sn \rightarrow Sn^{2+} + 2e^-$	−0.14	
$Ni \rightarrow Ni^{2+} + 2e^-$	−0.25	
$Fe \rightarrow Fe^{2+} + 2e^-$	−0.44	
$Cr \rightarrow Cr^{2+} + 2e^-$	−0.74	
$Zn \rightarrow Zn^{2+} + 2e^-$	−0.76	Anodic (active)
$Al \rightarrow Al^{3+} + 3e^-$	−1.66	
$Mg \rightarrow Mg^{2+} + 2e^-$	−2.36	
$Na \rightarrow Na^+ + e^-$	−2.71	
$K \rightarrow K^+ + e^-$	−2.92	
$Li \rightarrow Li^+ + e^-$	−2.96	

[a]For cathodic reaction, arrow direction is reversed.

half-cell reactions and their potentials for a number of metal systems. This table can be used to predict the spontaneous direction of the electrochemical reaction by the use of the following rule: The most negative (active) half-cell tends to be *oxidized* and the most positive (noble) tends to be *reduced*. A knowledge of these electrochemical reactions is essential in developing an understanding of the various forms of corrosion that will be discussed in later sections.

It should be noted that these half-cell potentials are for standard conditions, 25°C and unit molar concentrations. If the concentration of the metallic ions is less than 1 M, the tendency toward dissolution is greater because there are fewer ions available to facilitate the reverse reaction.

Under these conditions, the Nernst equation may be used to calculate the half-cell potential:

$$E = E_0 + 2.3 \frac{RT}{nF} \log \frac{[a]_{oxidized}}{[a]_{reduced}} \qquad (17\text{-}3)$$

where E_0 = standard half-cell potential (V)
R = gas constant (J/mol °K)
T = absolute temperature (°K)
n = number of electrons transferred
F = Faraday's constant (Coulomb)
a = activity of the oxidized and reduced species

This equation can be simplified by substituting the values of $R = 8.314\,\text{J/mol-}°\text{K}$, $F = 96,490$ Coulomb, and $T = 298°\text{F}$, so that

$$E = E_0 + \frac{0.0592}{n} \log \frac{[a]_{\text{oxidized}}}{[a]_{\text{reduced}}} \qquad (17\text{-}4)$$

In dilute solutions, the activities of the oxidized species can be taken as equal to their concentration and the activity of the solid metal (the reduced species) can be taken as equal to 1. Similarly, the activity of hydrogen gas is equal to its pressure, usually taken to be 1 atm., so that the activity of hydrogen is also 1, and

$$E = E_0 + \frac{0.0592}{n} \log [C_{\text{ion}}] \qquad (17\text{-}5)$$

where C is the concentration of the hydrogen ion.

Example 17-1

Calculate the corrected potential for the hydrogen electrode at pH $= 1, 2,$ and 3.

Solution   The reduction reaction to be considered is

$$2H^+ + 2e \longrightarrow H_2$$

Using equation (17-5), we obtain

$$E = E_0 + 0.0592 \log [C_{\text{ion}}]$$

Since pH $= 1/\log [H^+]$, then

> For pH $= 1$: $E = 0.0 - 0.0592(1) = -0.0592$ V   *Ans.*
>
> For pH $= 2$: $E = 0.0 - 0.0592(2) = -0.1184$ V   *Ans.*
>
> For pH $= 3$: $E = 0.0 - 0.0592(3) = -0.1776$ V   *Ans.*

Looking at the solution to example problem 17-1, it is apparent that as the hydrogen ion concentration decreases, the potential becomes more negative. Similarly, as the concentration of the oxidized species increases, the potential would become more positive. The change in potential corresponds to 0.0592 V for each tenfold increase in concentration where one electron is being transferred per ion.

ELECTRODE KINETICS

We have stated that half-cell potentials indicates the tendency of a particular reaction to occur; however, it is important to remember that although the spontaneous direction of the reaction may favor metallic corrosion, this does not mean that substantial corrosion will necessarily occur. Metals undergoing corrosion are not at equilibrium and therefore thermodynamic calculations do not apply. For example, the corrosion rate may be so infinitesimally small that it is negligible, and the metal will be essentially inert.

Since we cannot determine the corrosion rate from metals that are at equilibrium, let us examine what happens when we short-circuit the two electrodes shown in Figure 17-1 and permit current to flow. In this case, a reaction occurs; the iron electrode corrodes and is accompanied by a vigorous evolution of hydrogen at the platinum electrode. Electrons released from the iron oxidation reaction are moved through the wire connection to the platinum electrode, where they are utilized in the hydrogen reduction reaction, as shown in Figure 17-2. The movement of these electrons results

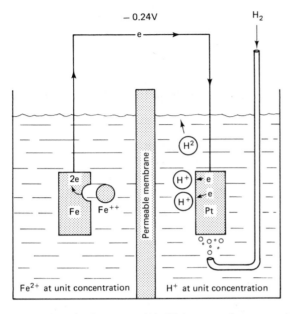

Figure 17-2   Short circuit of Fe and Pt electrode. Under these conditions, Fe corrodes and H_2 is generated at Pt electrode. Voltage drops from -0.44 V at equilibrium to -0.24 V.

in an electrical current (i_c). This corrosion current (i_c) may be related to the weight of metal transferred (m) by the use of Faraday's law:

$$m = Zi_c t \qquad (17\text{-}6)$$

where t = time (seconds)
  i_c = corrosion current (A)
  $Z = \dfrac{\text{(atomic weight)}}{\text{(number of electrons)}(96{,}500 \text{ A-sec})}$

In practice, the corrosion current is usually expressed as a current density (I/unit area) so that some knowledge of the surface area is required to determine the total amount of metal lost.

Example 17-2

How long will it take to corrode 6.00 g of silver from a silver electrode if the current is 15 A?

Solution   From equation (17-6):

$$t = \frac{m}{Zi_c}$$

$$= \frac{6 \text{ g}}{107.88(15 \text{ A})/(1)(96{,}500)}$$

$$= 358 \text{ sec} \qquad Ans.$$

Electrode Kinetics

Polarization

When a cell such as the Fe/H_2 cell just described is short-circuited and oxidation and reduction reactions are occurring on their respective surfaces, these electrodes will no longer be at their equilibrium potentials. This displacement from the equilibrium potential is termed *polarization* and is measured as the overvoltage.

Overvoltage (η) is the difference in potential between a metallic electrode at equilibrium and the metal in the displaced state, and can be positive or negative. To illustrate overvoltage, let us examine the iron and hydrogen electrodes at equilibrium as we see them in Figure 17-1. The potential of the system is -0.44 V. If we connect them and permit current to flow, the potential drops to -0.24 V; thus the overvoltage (η) is $+0.20$ V.

Activation polarization. Two types of polarization can occur, activation and concentration polarization. In activation polarization, the electrochemical reactions are controlled by an energy-requiring step in the reactions. The relationship for overvoltage as related to the reaction rate is

$$\eta = \pm \beta \log \frac{i}{i_0} \qquad (17\text{-}7)$$

where β = constant (Tafel constant)
$\quad i$ = rate of oxidation (or reduction)
$\quad i_0$ = exchange current density

This equation is illustrated graphically in Figure 17-3. You will note that if the current

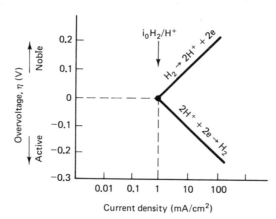

Figure 17-3 Activation polarization curve for a hydrogen electrode (on platinum) showing overvoltage as a function of current density.

is plotted logarithmically, the plot has a linear form, where β is the slope. The i_0 term (exchange current density) is the rate at which oxidation and reduction reactions are occurring at an electrode surface which is in equilibrium. It is expressed in terms of current density on the electrode surface, usually as A/cm^2. This value is determined experimentally for various electrode reactions and may be found in reference tables of electrochemical constants. It is important to note that the measurements of potential are being conducted with the use of an electrometer voltmeter, which is an instrument of such high internal resistance that essentially no current flow occurs as a result of the measurement.

Concentration polarization.　Under low rates of reactions, the ions participating in the reduction reaction exist in sufficient quantity in the region adjacent to the electrode so that they are generally available to take part in the reaction.

When the rate of reaction at the cathode is high, it is possible that the solution may be depleted of positive ions in the region adjacent to the cathode. This condition, termed concentration polarization, is depicted in Figure 17-4. A zone of depletion

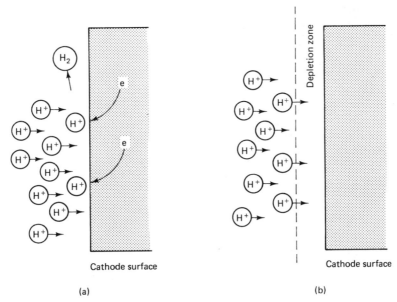

(a)　　　　　　　　　　　　　　　　　　　　　(b)

Figure 17-4 Schematic diagram illustrating (a) normal distribution of positive ions occurring with low rates of reduction and (b) depletion of positive ion with high rates of reduction.

occurs and the replenishment of positive ions to the electrode surface is limited by the diffusion rate of the ions. Factors that increase the rate of diffusion, such as increased temperature and increased flow rate, tend to decrease concentration polarization and thereby increase the reaction rate occurring at the cathode. As you will see in later sections, this has the effect of increasing the corrosion rate.

Mixed Potentials

In view of the foregoing, one might reasonably expect the corrosion rate (rate of oxidation) to have some relationship to the rate of reduction. In fact, this is true and the total rate of reduction must equal the total rate of oxidation.

This can be illustrated by the following example. If we consider zinc corroding in an acid solution, we see that there is a reduction reaction occurring at the zinc

Electrode Kinetics

surface in which hydrogen ions are being reduced according to the reaction

$$2H^+ + 2e \longrightarrow H^2$$

and the zinc is oxidizing according to the reaction

$$Zn \longrightarrow Zn^{2+} + 2e$$

with the electrons being furnished for the reduction reaction above.

These reactions can be described graphically as shown in Figure 17-5. The point

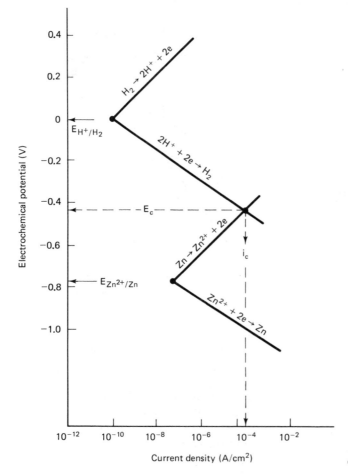

Figure 17-5 Kinetic behavior of zinc electrode in an acid solution.

in the system where the reduction rate equals the oxidation rate is at the corrosion potential (E_c). At this potential, the current density (i_c) represents the rate of corrosion of the anode. An examination of Figure 17-5 provides a review of the electrochemical characteristics of the zinc–acid system. As can be seen, the potential of the uncoupled hydrogen electrode would be at 0.0 V and that of the uncoupled zinc electrode would be at -0.76 V. When these are coupled together, as they would be if zinc were immersed in acid, the potential of the system becomes -0.44 V and represents the corrosion potential, E_c. The corrosion current at this point, i_c, is approximately 3×10^{-4} A/cm^2 and represents the corrosion rate of zinc.

PREDICTION OF CORROSION RATES

Concentration Polarization

The use of mixed-potential theory can be used to illustrate various unusual corrosion phenomena that occur under real conditions. For example, let us examine the system being affected by concentration polarization. In Figure 17-6, as the rate of stirring increases (e.g., from 1 to 2), the velocity increases and ion transport also increases.

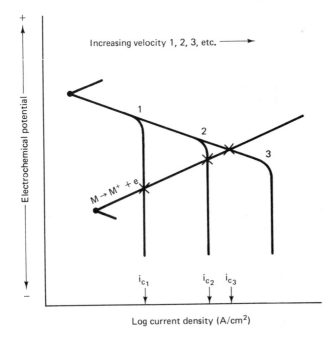

Figure 17-6 Concentration polarization resulting from diffusion-controlled cathodic process. Curve shows the effect of velocity on corrosion rate.

One can see from an examination of this figure that as the stirring rate increases, the corrosion current (i_c) also increases and it is possible to go from a situation controlled by concentration polarization to one controlled by activation polarization with a disproportionate increase in the corrosion rate. The phenomenon of concentration polarization is used to control corrosion in such systems as steam generation plants and high-performance pumps by limiting the amount of oxygen in the water so that it is not present to take part in the reduction reaction and by controlling velocities in certain critical components.

Passive Behavior

Another unusual type of corrosion phenomenon which occurs with some metals is that of *passivity*. Simply put, this is a displayed increase in corrosion resistance under certain conditions. Typically, a metal having an active–passive transition exhibits a

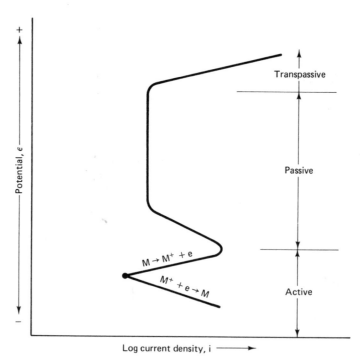

Figure 17-7 Polarization curve showing active–passive transition.

polarization curve resembling the one shown in Figure 17-7, which has an active region, a passive region, and a transpassive region. Among the metals displaying such behavior are iron, nickel, chromium, and titanium, as well as alloys in which they are the major elements.

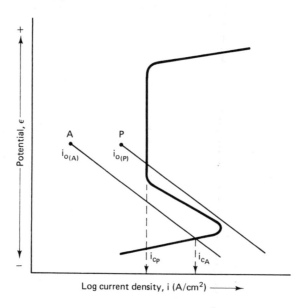

Figure 17-8 Corrosion of an active–passive metal.

If one has such a curve, it is easy to see how such a metal can display active corrosion characteristics under some conditions and passive behavior under some other conditions. Shown in Figure 17-8 is the polarization curve of an active–passive metal under two different corrosion conditions. The first, condition A, intersects the oxidation curve in the active region and results in a corrosive current of i_{c_A}.

In condition P, however, the reduction curve intersects the passive region and a corrosion current i_{c_P} results which is many times lower. Titanium is a metal which manifests such behavior. In acids containing oxidizers, it displays passive behavior, while in dilute acids with low oxygen or oxidizer concentration, it may exhibit active corrosion. This behavior is largely related to the development of corrosion-resistant oxide film at the metal surface.

FORMS OF CORROSION

Corrosion may occur in a number of forms which differ in appearance. In the following sections we shall discuss the various forms of corrosion that the materials scientist is likely to encounter.

General Corrosion

General corrosion is the most common form of corrosion that is likely to be encountered and the most significant in terms of economic losses. It is characterized by a more or less uniform attack over the entire exposed surface with only minimal variations in the depth of damage. A typical example of such attack is illustrated in Figure 17-9. In systems where general attack does occur, it does not usually result in sudden unexpected failure, since its occurrence and rate of attack can be determined and

Figure 17-9 General corrosion; remnant of a steel container immersed in fresh water for an extended time period.

predicted from laboratory tests. The results of these tests provide data on ordinary engineering materials in various environments. Such data expressed in the form of a depth of attack per surface area per unit time [e.g., milligrams/decimeter^2/day (mdd)] permit the thickness of the component to be considered relative to the expected rate of attack. In addition, coatings and electroplated finishes may be used to minimize corrosion attack.

Galvanic Corrosion

A serious type of corrosion damage may occur when two or more dissimilar metals are electrically coupled and placed in an electrolyte. This is known as galvanic corrosion and results from the existence of a potential difference between the coupled metals which causes a flow of current between them. The more active metal undergoes accelerated corrosion, while corrosion in the less active member of the couple is retarded or eliminated. Table 17-2 illustrates the change in corrosion rate which results by coupling iron first to one metal, then to another.

TABLE 17-2 COMPARISON OF THE CORROSION RATES WHEN IRON IS COUPLED TO A SECOND METAL IN 1% SODIUM CHLORIDE SOLUTION

Second metal	Weight loss (iron) (mg)	Weight loss (second metal) (mg)
Magnesium	0.0	3104.3
Zinc	0.4	688.0
Cadmium	0.4	307.9
Aluminum	9.8	105.9
Antimony	153.1	13.8
Tungsten	176.0	5.2
Lead	183.2	3.6
Tin	171.1	2.5
Nickel	181.1	0.2
Copper	183.1	0.0

In the design of a system involving dissimilar metals, it is essential to know which metal in the couple is likely to suffer accelerated corrosion. The basis for establishing the reactivity of various metals in the couple is the relative difference in electrochemical potential between the two members of the couple.

It is important when dealing with corrosion problems to distinguish between the electromotive series that was presented in Table 17-1 and the similar but different galvanic series. The former holds only for pure metals in specific concentrations of their own salts. One does not often encounter metals applied under these idealized conditions. In studying galvanic corrosion, we use a somewhat similar ranking based on the actual experience gained with a number of metals or alloys in a specific environment of interest. The relative position for a number of alloys in a seawater environment is shown in Table 17-3.

TABLE 17-3 GALVANIC SERIES OF METALS
AND ALLOYS IN SEAWATER

	Metal
Active	Magnesium
↑	Zinc
	Alclad 3S
	Aluminum 3S
	Aluminum 61S
	Aluminum 63S
	Aluminum 52
	Low carbon steel
	Alloy carbon steel
	Cast iron
	Type 410 (active)
	Type 430 (active)
	Type 304 (active)
	Type 316 (active)
	Muntz metal
	Yellow brass
	Admiralty brass
	Aluminum brass
	Red brass
	Copper
	Aluminum bronze
	Composition G bronze
	90/10 copper-nickel
	70/30 copper-nickel—low iron
	70/30 copper-nickel—high iron
	Nickel
	Inconel, nickel–chromium alloy 600
	Silver
	Type 410 (passive)
	Type 430 (passive)
	Type 304 (passive)
	Type 316 (passive)
	Monel, nickel–copper alloy 400
	Hastelloy, alloy C
	Titanium
	Graphite
↓	Gold
Noble	Platinum

In each couple, the metal nearest the active end of the series will be the anode and undergo accelerated corrosion, while the more noble member will receive some measure of protection. This table shows that a metal can be either protected or suffer greatly increased attack depending on its position in the table (i.e., what metal forms the other member of the couple).

In addition to the identity of the other metal, the relative surface areas is also of major importance in determining the rate of attack. If a cathodic metal having a large surface area is connected to an anodic metal having a relatively small surface area, the rate of attack increases markedly and in proportion to the relative differences in surface areas. This occurs because the currents that are generated at the cathode are focused on the smaller anode and result in higher currents/unit area.

Galvanic corrosion can occur when a structure made of one alloy has components of a different alloy even if the components are the same size. An example of such a component is shown in Figure 17-10. This is a line-tensioning device in which the body

Figure 17-10 Galvanic corrosion created when a stainless steel component is fastened to an aluminum body.

is made of aluminum and the loop was manufactured from 304 austenitic stainless steel. As the inset in the figure shows, the aluminum body is beginning to suffer severe corrosion.

The phenomenon of galvanic corrosion also has useful aspects. It is the basis for the use of sacrificial anodes in which replaceable anodes are used to protect structures, such as pipeline, drilling platforms, and steel tanks. These anodes corrode preferentially as shown in Figure 17-11 and thereby protect the more costly structure to which they are attached.

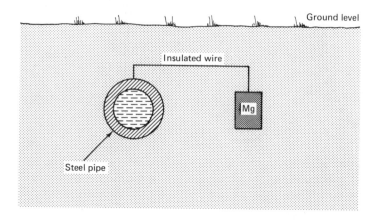

Figure 17-11 Use of sacrificial anode to protect underground piping. Steel pipe is cathodic, Mg rod is anodic.

Example 17-3

Steel rivets used to fasten an aluminum ladder together are exhibiting evidence of severe corrosive attack. According to Table 17-1, steel (iron) should be cathodic to aluminum. (a) How do you explain this phenomenon, and (b) how would you avoid this problem?

Solution (a) According to the series listed in Table 17-1, steel should exhibit a potential of -0.440 V versus a potential of -1.66 V for hydrogen and aluminum. However, the tendency of aluminum to form a strong adherent oxide means that the galvanic couple actually formed is aluminum *oxide*/steel, with steel being anodic to the aluminum oxide surface. This example also illustrates the dangerous practice of having a large cathode (aluminum) coupled with a small anode (steel), since failure of the fasteners would be very rapid. *Ans.*

(b) From the standpoint of both cost and corrosion resistance, the best choice would be high-strength aluminum alloy fasteners. *Ans.*

Crevice Corrosion

Crevice corrosion is a type of corrosive attack which occurs within confined spaces or crevices formed when components are in close contact.

Crevices can exist in any assembly, but there appear to be geometrical requirements. For crevice corrosion to occur, the crevice must be close fitting, having dimensions of less than a millimeter. Although the limits of the gap have not been defined, it is known that crevice corrosion does not occur in larger spaces.

It is not necessary for both approximating surfaces to be metal in order for crevice corrosion to occur. Crevice corrosion is also the major reason for the degradation of dental fillings. The crevice between the dental amalgam and the tooth is attacked by acids formed in the mouth and the filling weakened. An example of such attack is shown in Figure 17-12. Crevice corrosion has also been reported in crevices formed by a number of nonmetallic materials (e.g., polymers, glasses, rubber) in contact with metal surfaces. The fact that this can occur is of particular importance in the application and selection of gasketing materials.

A specific type of crevice corrosion can also occur beneath the shielded areas

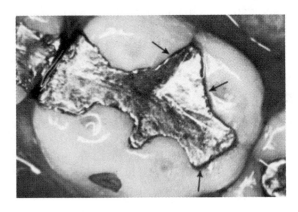

Figure 17-12 Crevice corrosion (arrows) in an ordinary silver-tin amalgam dental filling.

caused by the settling out of suspended solids upon a surface. This condition, called "poultice" corrosion, also occurs with automobiles. Road salts and debris collecting on ledges and pockets are kept moist by weather and severe body corrosion occurs in specific locations, as shown in Figure 17-13.

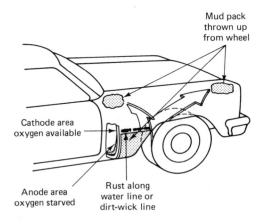

Figure 17-13 Poultice corrosion caused by mud and road debris packed against the underside of fenders or trapped in pockets in the body. (From J. C. Bittence, *Machine Design*, Vol. 48, No. 11, 1976, p. 149.)

The mechanisms of crevice corrosion may be divided into two stages. The initial increase in corrosion rate at the site is the result of a potential difference between the solution within the crevice and that outside the crevice. This potential difference may be the result of a difference in oxygen concentration in the two locations. This potential difference causes the metal in the interior of the crevice to undergo accelerated corrosion.

Although the initial action in the site is the result of oxygen concentration differences, the continued activity is dependent on an *autocatalytic* process. By an "autocatalytic process" we mean that the corrosion reaction in the crevice produces conditions that are favorable and necessary for continued corrosion.

Pitting

Pitting is a form of localized corrosion in which the attack is confined to numerous small cavities on the metal surface. The cavities created may vary in number, size, and form. However, it is commonly held that a true pit has a depth/width ratio equal to or greater than 1. Figure 17-14 shows the cross section and the profile of pits occurring in a martensitic (410 alloy) stainless steel plate. Pitting is a particularly insidious type of attack because although only a small amount of metal may be lost, failure due to perforation may occur. Pitting may contribute to general failure in another way. In highly stressed components, these pits can act as notches raising the localized stresses, resulting in the creation of fatigue cracks.

Pitting can occur with a number of metals and alloys, but stainless steels and aluminum alloys are particularly susceptible to this type of degradation. Pitting occurs most frequently in solutions of nearly neutral pH, containing halogen ions. In general, those factors favoring general corrosion (i.e., low pH, increased temperatures) do not

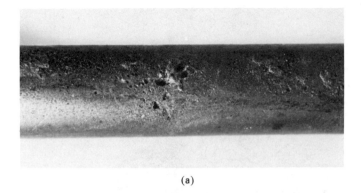

(a)

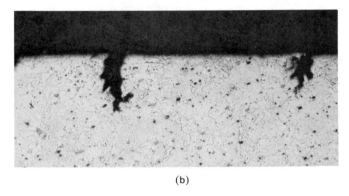

(b)

Figure 17-14 Pitting corrosion in a martensitic stainless steel: (a) top view ($1\frac{1}{2}\times$); (b) cross section showing the depth of pitting ($160\times$).

favor pitting attack. Pitting attack may be separated into the initiation stage and the propagative stage. The reasons for pit initiation are not clearly understood, but appear to be related to surface imperfection. The propagation of a pit is related to an auto-catalytic process occurring inside the pit. Metal dissolution occurs within the pit and is accompanied by O_2 reduction near the mouth of the pit or on the metal surface, as shown in Figure 17-15. This results in an excess of positive charge (M^+) at the pit base and migration of the chloride ion occurs in order to offset the imbalance. The net effect is an increase in metal chloride concentration in the pit. These metallic chlorides undergo hydrolysis according to the following reaction:

$$M^+Cl + H_2O \longrightarrow M(OH) + H^+Cl^-$$

This reaction increases the hydrogen ion concentration (increasing acidity) and favors increased rates of dissolution within the pit.

The best protection against pitting attack is the selection of a material having adequate pitting resistance. This requires that one must have ample data regarding the behavior of candidate materials in pitting environments. Such data are available from reference texts or may be obtained from laboratory evaluations of specific materials.

Forms of Corrosion

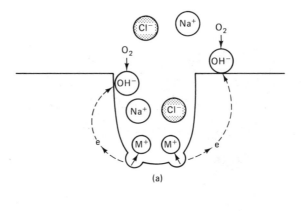

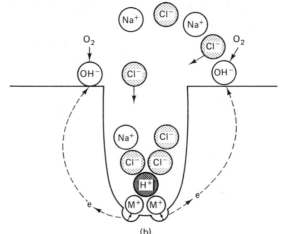

Figure 17-15 Schematic diagram illustrating pitting corrosion. (a) Initiation stage. (b) Growth stage.

Selective Dissolution

Selective dissolution refers to a type of corrosion in which one phase is preferentially attacked in an alloy or one element is preferentially dissolved from a solid solution. One example of this is the *dezincification* of brass, where brass, when exposed to sea-water, develops regions where the material is spongy in texture and primarily copper based. This occurs as a result of the dissolution of the brass in this region, with the zinc staying in solution and the copper atoms redepositing in the structure.

This type of corrosion occurs in several other systems as well: the dissolution of ferrite from austenitic and martensitic stainless steels, and the dissolution of ferrite from gray cast iron, called *graphitization*, which is shown in Figure 17-16.

Material science has created another group of materials in which selective dissolution can produce serious consequences: metallic composite structures. These are usually constructed so that the fibers and matrices possess widely differing chemical and mechanical characteristics. They may therefore be severely degraded and

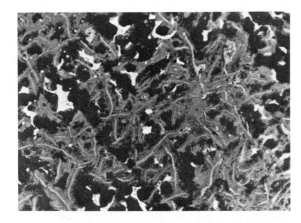

Figure 17-16 Photomicrograph illustrating graphitization; ferrite (white) in gray cast iron which has been corroded away leaving behind a graphite skeleton.

show sharp losses in strength if the reinforcing fibers or the matrix are selectively attacked.

The form of attack may vary; however, the severest degradation usually results when the attacked species or phase is present in a continuous network. There may be little change in the overall configuration or geometry of the component, but the mechanical properties are adversely affected in a very serious way.

Stress Corrosion Cracking

Stress corrosion cracking is a corrosive failure in which cracks form in a component under the combined action of mechanical stresses and an aggressive environment. Of all forms of localized attack, stress corrosion cracking (SCC) is by far the most serious and difficult to cope with. In SCC, the stresses and the corrosive environment conspire to cause rapid failure, when either condition alone would cause negligible damage.

The requisites for stress corrosion cracking are:

1. *A susceptible alloy:* This is an alloy that has a history of stress corrosion failure. High-strength steels, brasses, and austenitic stainless steels are examples of common alloys that exhibit stress corrosion under certain conditions.
2. *A specific environment:* Specific ions are usually necessary to cause cracking in each alloy. In the case of austenitic stainless steels, chloride ions are particularly effective in causing SCC. With brasses, it is the ammonium ion that causes stress corrosion problems.
3. *A source of tensile stress:* This may either be applied or residual. The applied stresses result from loads occurring while the component is in service, while residual stresses result from processing treatments such as cold working and uneven cooling during heat treatment. The presence of residual stress can result in totally unexpected failures. There are numerous cases where components have been placed in storage in sound condition and when withdrawn have been found to be cracked.

Stress corrosion failures generally exhibit little ductility and have the macroscopic appearance of a brittle fracture. There may be multiple cracks originating from the surface, but failure usually results from the progression of a single crack on a plane normal to the main tensile stress. Cracks may be either transgranular or intergranular. Figure 17-17 illustrates SCC in a cast manganese bronze alloy which exhibits transgranular fracture.

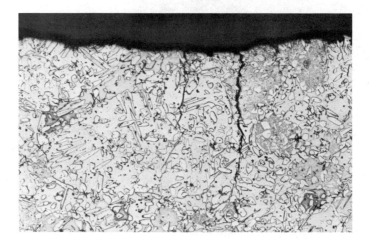

Figure 17-17 Transgranular stress corrosion cracks in a manganese bronze alloy (85×).

In austenitic stainless steels, cracks are usually transgranular, frequently but not always associated with a specific crystallographic plane. These alloys also exhibit intergranular cracking on certain media, notably caustic solutions and highly oxygenated chloride solutions.

Intergranular cracking is the predominant mode of failure for martensitic stainless steels. However, transgranular cracking has been observed in these alloys when tempered below 850°F (454°C).

In high-strength steels, the crack path is intergranular, as shown in Figure 17-18. It is widely believed that stress corrosion cracking in these materials is related to a *hydrogen embrittlement* mechanism as a result of hydrogen being generated at the crack tip as a corrosion product. The branched nature of the crack is also evident. This branching is also characteristic of stress corrosion cracking.

The latter aspect illustrates the problems that can occur with hydrogen embrittlement in high-strength steels and in other bcc systems. Hydrogen entering the system as the result of corrosion, electroplating, or welding can lead to low ductility and generally brittle behavior. A major source of the embrittling hydrogen in steels is from water vapor reacting with the molten steel. For extremely high-quality steels, this hydrogen is removed by vacuum degassing before solidification.

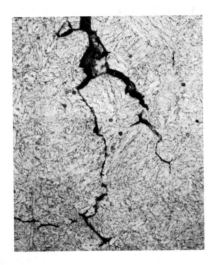

Figure 17-18 Intergranular stress corrosion crack in a high-strength steel (160×).

Intergranular Attack

Intergranular corrosion (IGA) is a condition in which the grain boundaries of a material are attacked preferentially and at a much greater rate than the bulk material. Since most structural metals and alloys are polycrystalline, this can be a serious problem.

The primary cause of grain boundary attack is an inhomogeneous chemical composition at the locus of attack. This inhomogeneity may be caused by segregation, as described in Chapter 6, or by intergranular precipitation, which may cause localized compositional disparities.

Since the actual volume of metal contained in the grain boundaries is quite small, penetration of the structure can be quite rapid, even with only small weight losses.

This form of attack is particularly prevalent with austenitic stainless steels. Intergranular attack in stainless steels is the result of the depletion of chromium at the grain boundary. This can occur if the stainless steel is subjected to improper thermal processing.

When unmodified austenitic stainless steels such as 302 or 304 are held in the temperature range 510 to 800°C (950 to 1500°F) or cooled slowly through it, chromium carbides precipitate at the grain boundaries, causing a depletion of chromium in solution adjacent to the carbide. This phenomenon, known as *sensitization*, was described in Chapter 12. The degree of sensitization is directly related to the temperature and duration of the thermal exposure. For example, at 510°C, several hours may be needed to create a sensitized condition. Sensitization can be caused by in-process heat treatments, exposure resulting from service, or as a result of welding procedures. In the latter case, zones adjacent to the weld can be subjected to the sensitization temperatures. The result of the depletion is that the matrix adjacent to the grain boundary has a lower chromium content than it originally possessed and is subject to rapid attack (see Figure 12-33).

Intergranular attack in austenitic stainless steel may be prevented by several techniques, among which are the following. The thermal processing can be controlled to eliminate extended exposures at the sensitizing temperatures. The carbon level may be reduced from the typical value of 0.08% maximum to levels of 0.04 or lower. This reduced carbon obviously results in lower carbide formation. The alloy may be modified with the addition of stabilizing elements such as titanium or niobium. These elements are strong carbide formers and tie up the available carbon so that it cannot react with chromium.

STUDY PROBLEMS

17.1. Copper and magnesium electrodes are placed in individual solutions containing their respective ions at unit activity. Calculate the potential of the resulting cell when these electrodes are coupled.
Answer: 2.70 V

17.2. In problem 17.1, which of the metals will corrode when connected? Write the resultant equations in their proper form.

17.3. Write the expected half-cell reactions for the following situations:
(a) Steel immersed in oxygenated water
(b) Galvanized nail in dilute solution of HCl
(c) Iron placed in a solution of copper sulfate
(d) An opened tin can containing orange juice

17.4. Copper, zinc, and aluminum are in ionic solutions of 0.1 M, 0.01 M, and 0.0001 M concentrations, respectively. Calculate the resultant electrode potential at 25°C.

17.5. Calculate the potential of the hydrogen electrode in two solutions, one having pH 4 and the other pH 7.

17.6. Calculate the electrochemical potential of a nickel electrode with respect to hydrogen immersed in an ionic solution containing 3.8 g of Ni^{+2} ions per liter.

17.7. A particular standard cell is composed of a magnesium electrode at 25°C and 1 M and a half-cell in which hydroxyl ions $(OH)^-$ are being produced via the reduction of water. What potential is being established between the two electrodes? Which is the anode?

17.8. A platinum electrode is placed in a solution of dilute sulfuric acid through which hydrogen gas is being bubbled. A cathodic current is imposed on the electrode and the following potentials were recorded with respect to a standard hydrogen electrode:

Current (A/m²)	Potential (V)
10^2	−0.1508
10^3	−0.1808

Calculate the Tafel constant for the polarization curve.
Answer: $\beta = -0.03$ (V-m^2)/A or (-0.03V/decade)

17.9. If the pH of the solution is 2 in problem 17.8, calculate the overvoltage at 10^{-2} A/cm^2 and the current density i_0 for hydrogen on the platinum electrode.

17.10. Describe several methods for protecting a susceptible structure such as a pipeline or offshore platform from corrosion. Give the advantages and disadvantages of each method.

17.11. A steel tank is corroding with a corrosion current (i_c) of 5.3×10^{-6} A/cm^2. Calculate the overvoltage and the current necessary to cathodically protect the tank. Assume

$$i_0 = 10^{-6} \text{ A/cm}^2$$

$$\beta_a = 0.050 \text{ V/decade}$$

$$E_{0(Fe)} = -0.510 \text{ V}$$

Answer: $\eta = 0.0362$ V, $i_c = 2.8 \times 10^{-5}$ A/cm^2

17.12. Plot the result obtained in problem 17.11. How do the results compare? What is the corrosion current?

17.13. A tin electrode having an area of 10 cm^2 is corroding at the rate of 4×10^{-5} A/cm^2. What is the weight loss of the electrode per hour?
Answer: 88.6×10^{-5} g/hr.

17.14. An aluminum water storage tank 1 m high and 40 cm in diameter exhibits a corrosion current of 4×10^{-6} A/cm^2. How much metal is being transferred to the contents of the tank per minute?

17.15. Assuming no localized corrosion, how many years will the tank described in problem 17.14 last if the wall thickness is 8 mm?

17.16. The wall thickness of a steel tank is measured monthly and the loss in thickness is approximately the same each month, 54 mdd. What is the most appropriate name for this type of corrosion? How long will the tank be expected to last if the initial wall thickness is 7.5 mm?
Answer: 29.9 yr.

17.17. A pump station may require the use of bronze, steel, copper, and aluminum in various parts of the system. Which of the metal combinations are permissible and which should be avoided from the standpoint of corrosion?

17.18. Some solid film lubricants contain graphite in their compositions. Explain why these lubricants may have an adverse effect when used with brass or aluminum in saline environments.

17.19. A special alloy exhibits a corrosion rate of 2×10^{-8} A/cm^2 when immersed in oxidizing acid such as concentrated HNO_3 and 3×10^{-5} A/cm^2 in dilute H_2SO_4. Explain the mechanism and draw a schematic of the anodic polarization diagram.

17.20. A process plant, in an attempt to alleviate a corrosion problem with the tank bottom, installed a fabricated tank with carbon steel sides and a stainless steel bottom. Their corrosion problems increased. They then painted the sides of the tank, whereupon the tank pitted so badly that it perforated. What is the explanation in each case? What should they have done with the fabricated tank in the first place?

17.21. What are the requisite conditions necessary in order for stress corrosion cracking to occur?

17.22. A manufacturer ships a packaged 304 stainless steel component across the country. Upon its arrival, the package is discovered to have been contaminated with seawater and the component exhibits numerous hairline fractures. What do you suspect is the cause?

17.23. From your own observations, list examples where aluminum is used in contact with steel and the steel corrodes preferentially. Why?

17.24. If the use of an undesirable couple cannot be avoided, what can be done to minimize the galvanic corrosion that may result?

17.25. A 304 stainless steel test coupon is immersed in a saline environment at pH 7. The weight loss is measured after 1000 hr and found to be negligible. Does this mean that the alloy may be used safely in this environment? Explain.

17.26. A certain application requires that a 304 stainless steel component be held at 1000°F for several hours. What is the potential problem, and how may it be eliminated?

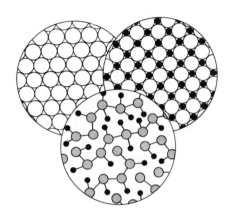

APPENDIX
Selected Mechanical and Physical Property Data for Engineering Materials[a]

Material	T_M (°C)	T_M (°F)	Density (g/cm^3)	Coefficient of linear thermal expansion $(10^{-6}/°C)$[b]	Thermal conductivity $\left(\frac{cal\text{-}cm}{cm^2\text{-}sec\text{-}°C}\right)$[c]	E (10^6 psi)	E (MPa)
Metals							
Aluminum	660	1220	2.70	24.0	0.53	10	69,000
Aluminum alloys (high strength)	548–660	1018–1220	2.80	22.8	0.46	10	69,000
Brass (70 Cu–30 Zn)	925–955	1697–1751	8.53	19.9	0.29	16	110,400
Cast iron (gray)	1148–1280	2098–2336	7.15	10.5	0.11	20	138,000
Copper	1083	1981	8.96	16.5	0.94	16	110,400
Gold	1063	1945	19.32	14.2	0.71	12	82,800
Iron	1539	2802	7.87	11.7	0.18	30	207,000
Magnesium	650	1202	1.74	26.0	0.38	6.5	44,850
Silver	960	1761	10.49	19.7	1.00	11	75,900
Steel, carbon			7.8	10–11.7	0.11	30	207,000
Steel, alloy			7.8	10–11	0.079	30	207,000
Steel, stainless, Type 302	1399–1421	2550–2590	7.9	17.3	0.037–0.050	28	193,200
Titanium	1668	3035	4.51	8.4	0.043	16.8	115,920
Tungsten	3410	6170	19.30	4.6	0.40	60	414,000
Ceramics							
Aluminum oxide	2015	3659	3.8	9.0	0.07	60	414,000
Brick, building	—		2.3		0.001		
Brick, fireclay	—		2.1	4.5	0.002		
Concrete	—		2.4	9.9	0.002	2	13,800
Glass							
Silica	1667[d]	3033	2.2	0.5	0.004	10	69,000
Soda–lime	730[d]	1346	2.5	8.5	0.002	10	69,000

Material	T_M °C	T_M °F	Density (g/cm³)	Coefficient of linear thermal expansion (10⁻⁶/°C)[b]	Thermal conductivity $\left(\frac{\text{cal-cm}}{\text{cm}^2\text{-sec-}°\text{C}}\right)$[c]	E 10⁶ psi	E MPa
Silicon carbide	2770	4892	3.2	4.5	0.003	68	469,200
Titanium carbide	3193	5779	4.5	7.0	0.07	50	345,000
Tungsten carbide (WC)	2871	5200	15.8		0.07	78	538,200
Polymers							
Nylon 66	267	513	1.2	100	0.00031	0.2	1380
Polyethylene (high density)	141	286	1.0	120	0.00064	0.1– 0.2	690– 1380
Polymethyl-methacrylate	160	320	1.2	90	0.00025	0.4– 0.5	2760– 3450
Polystyrene	240	464	1.1	63	0.00010	0.4– 0.6	2760– 4140

[a]Room-temperature data.

[b]To obtain °F⁻¹, divide by 1.8.

[c]To obtain (Btu-in./ft²-sec-°F), multiply by 0.79.

[d]Softens.

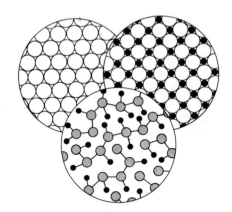

Index

L

Laminates, 600
  applications, 602
  bending, 602
  stresses, 600
Latent heat of fusion, 203
Lattice parameter, 48
LCD (liquid crystal display), 356
Lead, 488
Lead alloys:
  babbitt, 490
  calcium, 489
  low-melting ternary, 490
  tin, 489
  type metal, 489
LED (light emitting diode), 352
Lever arm rule, 181
Lignin, 628
Linear polymer, 498
Liquidus, definition, 180
Lodestones, 369
Luminescence, 351

M

Macromolecule, 496
Macrostructure, 208
Magnesium, 460
  alloys, 462
  properties, 463
Magnetic behavior, 369–78
  anisotropy, 378
  diamagnetic, 371
  ferromagnetic, 372
  field strength (H), 369
  flux density (B), 370
  paramagnetic, 372
  temperature effects, 378
Magnetic materials, 379
  amorphous metal alloys, 381
  ceramics, 382
  permanent magnets, 383
Magnetic permeability, 371
Magnetoresistive, 606

Magnetostriction, 377
Marage steel, 438
  properties, 439
Martempering (marquenching), 421
Martensite formation, 418
Martensitic transformation, 156
Materials, design, 12
  selection, 12
Medullary ray, 626
Meissner effect, 386
Mer, 497
Metal-matrix composites, 592
Metalloids, 16
Metals, 3
Methane, 38, 39
Microvoid coalescence, 251
Miller indices:
  cubic, 62
  hexagonal, 65
Modulus of elasticity, 143, 228
Modules of resilience, 231
Modules of rigidity, 145
Molecular arrangement, 40, 41
Molecular weight distribution, 511
Molecules, definition, 34
Monel, 483
Monomers, 497
Monotectic alloys, 200
Mott, Sir Neville, 95

N

NaC1 structure, 544
Neolithic Period, 1
Nernst-Einstein equation, 129, 339
Nernst equation, 643
Network modifiers, 564
Nickel, 482
Nickel alloys:
  composition, 486
  copper, 483
  iron, 485
  properties, 484, 487
Nickel-base superalloys, 485
Nickel-silver, 482

MOS, 356, 357
n-p-n, 355
p-n-p, 354
Transmission electron microscopy
(TEM), 80
Trimers, 497
True strain, 137
True stress, 142
Tunnel diode, 352
Twinning, 102

U

Ultimate tensile strength, definition, 228
Unidirectional solidification, 603
Unit cell, 45

V

Vacancies, 80
Valence electrons, 19
Van Der Waals forces, 32
Viscoelastic behavior, 512
Viscosity, glass, 562, 564, 566

W

Weight average molecular weight, 511

Whiskers, 560
Wood, 625–34
composition, 628
design considerations, 634
mechanical properties, 631, 633
physical characteristics, 628
physical properties, 630
structure, 626
types, 625
Wrought iron, 398
Wurtzite (ZnS structure), 547

X

X-ray diffraction, 70
methods, 70, 71

Y

Yield strength, 228
effect of K_{Ic}, 275
Young's modules, 143

Z

Zener diode, 352
Zirconia, 552, 553
stabilized, 556

PROPERTIES OF SELECTED ELEMENTS

Element	Sym-bol	Atomic number	Atomic mass (amu)	Density (g/cm³)	Crystal structure	Atomic radius (Å)	Ionic radius (Å)	Most common valence	Lattice parameter (Å) a	Lattice parameter (Å) c	Melting point (°C)	Average modulus of elasticity (10^6 psi)[a]
Aluminum	Al	13	26.98	2.70	fcc	1.431	0.51	+3	4.05		660	10
Beryllium	Be	4	9.01	1.85	hcp	1.143	0.35	+2	2.29	3.58	1277	44
Boron	B	5	10.82	2.34	Ortho.	0.970	0.23	+3			2030	—
Cadmium	Cd	48	112.41	8.65	hcp	1.489	0.97	+2	2.98	5.62	321	8
Calcium	Ca	20	40.08	1.55	fcc	1.973	0.99	+2	5.58		838	3.5
Carbon (graphite)	C	6	12.01	2.25	Hex	0.770	0.16	+4			>3500[b]	—
Cesium	Cs	55	132.91	1.90	bcc	2.620	1.67	+1	6.05		28	—
Chlorine	Cl	17	35.46	0.003	Tetra.	0.905	1.81	−1			−101	—
Chromium	Cr	24	52.01	7.19	bcc	1.249	0.63	+3	2.88		1875	36
Cobalt	Co	27	58.94	8.85	hcp	1.245	0.72	+2	2.51	4.07	1495	30
Copper	Cu	29	63.54	8.96	fcc	1.278	0.96	+1	3.61		1083	16
Fluorine	F	9	19.00	0.002	—	0.60	1.33	−1			−220	—
Gallium	Ga	31	69.72	5.91	Ortho.	1.224	0.62	+3			30	—
Germanium	Ge	32	72.60	5.32	Dia.	1.225	0.53	+4			937	—
Gold	Au	79	197.0	19.32	fcc	1.442	1.37	+1	4.08		1063	11.6
Hydrogen	H	1	1.008	—	—	0.460	1.54	+1			−259	—
Iron	Fe	26	55.85	7.87	bcc	1.241	0.74	+2	2.87		1537	29
					fcc	1.269	0.64	+3	3.66			
Lead	Pb	82	207.21	11.36	fcc	1.750	1.20	+2	4.95		327	—
Magnesium	Mg	12	24.32	1.74	hcp	1.604	0.66	+2	3.21	5.21	650	6
Manganese	Mn	25	54.94	7.43	Cubic	1.120	0.80	+2			1245	23
Molybdenum	Mo	42	95.95	10.22	bcc	1.362	0.70	+4	3.15		2610	47
Nickel	Ni	28	58.71	8.90	fcc	1.246	0.69	+2	3.52		1453	30
Nitrogen	N	7	14.01	0.001	Hex.	0.710	1.71	−3			−210	—
Oxygen	O	8	16.00	0.001	Cubic	0.600	1.32	−2			−218	—
Phosphorus	P	15	30.98	1.83	Cubic	1.090	0.35	+5			44	—
Platinum	Pt	78	195.09	21.45	fcc	1.388	0.80	+2	3.92		1769	27
Potassium	K	19	39.10	0.86	bcc	2.312	1.33	+1	5.25		64	—
Silicon	Si	14	28.09	2.33	Dia.	1.176	0.42	+4			1410	16
Silver	Ag	47	107.88	10.49	fcc	1.444	1.26	+1	4.09		961	11
Sodium	Na	11	22.99	0.97	bcc	1.857	0.97	+1	4.29		98	—
Sulfur	S	16	32.07	2.07	Ortho.	1.060	1.84	−2			119	—
Tantalum	Ta	73	180.95	16.6	bcc	1.430	0.68	+5	3.31		2996	27
Tin	Sn	50	118.70	7.30	Tetra.	1.509	0.71	+4			232	6
Titanium	Ti	22	47.90	4.51	hcp	1.475	0.68	+4	2.95	4.67	1668	16.8
Tungsten	W	74	183.86	19.3	bcc	1.367	0.70	+4	3.16		3410	50
Uranium	U	92	238.07	19.07	Ortho.	1.380	0.97	+4			1132	24
Vanadium	V	23	50.95	6.10	bcc	1.316	0.63	+4	3.02		1900	19
Zinc	Zn	30	65.38	7.13	hcp	1.332	0.74	+2	2.66	4.95	420	—
Zirconium	Zr	40	91.22	6.49	hcp	1.616	0.79	+4	3.23	5.15	1852	13.7

[a]To obtain MPa, multiply by 6.9 × 10³. [b]Sublimes.

CONSTANTS

Atomic mass unit (amu)	1.66×10^{-24} g
Avogadro's number (A_0)	6.02×10^{23} amu/g
Boltzmann's constant (k)	86.1×10^{-6} eV/°K
	13.8×10^{-24} J/°K
Bohr magneton	9.27×10^{-24} A-m^2
Density of water	1 g/cm^3
	62.4 lb/ft^3
Electron charge (q)	1.60×10^{-19} C
Electron mass (m_0)	9.11×10^{-28} g
Electron volt (eV)	1.60×10^{-19} J
Faraday	96.5×10^3 C
Gas constant (R)	1.987 cal/mol-°K
	8.314 J/mol-°K
Gravity (acceleration) g	9.80 m/sec^2
Permeability of vacuum (μ_0)	$4\pi \times 10^{-7}$ Wb/A-m
($4\pi\mu_0$)	1.23×10^{-5} N-m/C^2
Permittivity of vacuum (ϵ_0)	8.9×10^{-12} C^2/N-m^2
($1/4\pi\epsilon_0$)	9.0×10^9 N-m^2/C^2
Planck's constant (h)	6.62×10^{-34} J-sec
	4.14×10^{-15} eV-sec
Proton mass (mp)	1.672×10^{-24} g
Velocity of light (c)	3×10^{10} cm/sec
Wavelength of 1-eV photon (λ)	12,300 Å

SI UNIT PREFIXES

Multiple or submultiple	Prefix	Symbol
10^{18}	exa	E
10^{15}	peta	P
10^{12}	tera	T
10^9	giga	G
10^6	mega	M
10^3	kilo	k
10^2	hecto	h
10	deka	da
10^{-1}	deci	d
10^{-2}	centi	c
10^{-3}	milli	m
10^{-6}	micro	μ
10^{-9}	nano	n
10^{-12}	pico	p
10^{-15}	femto	f
10^{-18}	atto	a

SELECTED MATHEMATICAL RELATIONSHIPS

Natural logarithm base (e) = 2.718

$$\ln e^x = x$$

$$\ln x = 2.3 \log_{10} x$$

C^2